BEGINNING ALGEBRA HYBRID

Jerome E. Kaufmann
Karen L. Schwitters
SEMINOLE STATE COLLEGE OF FLORIDA
NICOLET COLLEGE

BROOKS/COLE
CENGAGE Learning™

Australia • Brazil • Japan • Korea • Mexico • Singapore • Spain • United Kingdom • United States

Beginning Algebra Hybrid
Jerome E. Kaufmann and
Karen L. Schwitters

Developmental Math Editor: Marc Bove
Assistant Editor: Shaun Williams
Editorial Assistant: Jennifer Cordoba
Media Editors: Heleny Wong and Guanglei Zhang
Marketing Manager: Gordon Lee
Marketing Assistant: Shannon Myers
Marketing Communications Manager: Darlene Macanan
Senior Content Project Manager: Tanya Nigh
Design Director: Rob Hugel
Art Director: Vernon Boes
Print Buyer: Judy Inouye
Rights Acquisitions Specialist: Tom McDonough
Production Service: Teresa Christie, MPS Limited, a Macmillan Company
Text Designer: Diane Beasley
Photo Researcher: PreMedia Global
Copy Editor: Martha Williams
Cover Designer: Denise Davidson
Cover Image: Lawrence Lawry/Getty Images
Compositor: MPS Limited, a Macmillan Company

© 2012, Brooks/Cole, Cengage Learning

ALL RIGHTS RESERVED. No part of this work covered by the copyright herein may be reproduced, transmitted, stored, or used in any form or by any means graphic, electronic, or mechanical, including but not limited to photocopying, recording, scanning, digitizing, taping, Web distribution, information networks, or information storage and retrieval systems, except as permitted under Section 107 or 108 of the 1976 United States Copyright Act, without the prior written permission of the publisher.

> For product information and technology assistance, contact us at
> **Cengage Learning Customer & Sales Support, 1-800-354-9706.**
>
> For permission to use material from this text or product,
> submit all requests online at www.cengage.com/permissions.
> Further permissions questions can be e-mailed to
> **permissionrequest@cengage.com.**

Library of Congress Control Number: 2010937954

ISBN-13: 978-0-8400-6589-6
ISBN-10: 0-8400-6589-2

Brooks/Cole
20 Davis Drive
Belmont, CA 94002-3098
USA

Cengage Learning is a leading provider of customized learning solutions with office locations around the globe, including Singapore, the United Kingdom, Australia, Mexico, Brazil, and Japan. Locate your local office at www.cengage.com/global.

Cengage Learning products are represented in Canada by Nelson Education, Ltd.

To learn more about Brooks/Cole, visit www.cengage.com/brooks/cole.
Purchase any of our products at your local college store or at our preferred online store www.cengagebrain.com.

Printed in the United States of America
1 2 3 4 5 6 7 14 13 12 11 10

Contents

1 Basic Concepts of Arithmetic and Algebra 1

1.1 Numerical and Algebraic Expressions 2
1.2 Addition, Subtraction, Multiplication, and Division of Integers 9
1.3 Use of Properties 20
1.4 Prime and Composite Numbers 27
1.5 Rational Numbers 33
1.6 Decimals and Real Numbers 44
Chapter 1 Review Problem Set 53
Chapter 1 Practice Test 56

2 Equations, Inequalities, and Problem Solving 58

2.1 Solving First-Degree Equations 59
2.2 Translating from English to Algebra 64
2.3 Equations and Problem Solving 69
2.4 More on Solving Equations and Problem Solving 75
2.5 Equations Involving Parentheses and Fractional Forms 80
2.6 Inequalities 87
2.7 Inequalities, Compound Inequalities, and Problem Solving 93
Chapter 2 Review Problem Set 99
Chapter 2 Practice Test 101
Chapters 1–2 Cumulative Review Problem Set 103

3 Ratio, Percent, Formulas, and Problem Solving 104

3.1 Ratio, Proportion, and Percent 105
3.2 More on Percents and Problem Solving 112
3.3 Formulas: Geometric and Others 117
3.4 Problem Solving 125
3.5 More about Problem Solving 131
Chapter 3 Review Problem Set 136
Chapter 3 Practice Test 139
Chapters 1–3 Cumulative Review Problem Set 141

4 Exponents and Polynomials 143

4.1 Addition and Subtraction of Polynomials 144
4.2 Multiplying Monomials 149
4.3 Multiplying Polynomials 155
4.4 Dividing by Monomials 161
4.5 Dividing by Binomials 163
4.6 Zero and Negative Integers as Exponents 168
Chapter 4 Review Problem Set 175
Chapter 4 Practice Test 177
Chapters 1–4 Cumulative Review Problem Set 178

5 Factoring, Solving Equations, and Problem Solving 179

5.1 Factoring by Using the Distributive Property 180
5.2 Factoring the Difference of Two Squares 186
5.3 Factoring Trinomials of the Form $x^2 + bx + c$ 190
5.4 Factoring Trinomials of the Form $ax^2 + bx + c$ 197
5.5 Factoring, Solving Equations, and Problem Solving 200
Chapter 5 Review Problem Set 207
Chapter 5 Practice Test 209

6 Rational Expressions and Rational Equations 210

6.1 Simplifying Rational Expressions 211
6.2 Multiplying and Dividing Rational Expressions 214
6.3 Adding and Subtracting Rational Expressions 219
6.4 More on Addition and Subtraction of Rational Expressions 224
6.5 Rational Equations and Problem Solving 231
6.6 More Rational Equations and Problem Solving 237
Chapter 6 Review Problem Set 245
Chapter 6 Practice Test 247
Chapters 1–6 Cumulative Review Problem Set 249

7 Coordinate Geometry 251

7.1 Cartesian Coordinate System 252
7.2 Graphing Linear Equations and Applications 259
7.3 Graphing Linear Inequalities 267

7.4 Slope of a Line   272
7.5 Writing Equations of Lines   280
Chapter 7 Review Problem Set   287
Chapter 7 Practice Test   289

8 Systems of Equations   291

8.1 Solving Linear Systems by Graphing   292
8.2 Elimination-by-Addition Method   297
8.3 Substitution Method   304
8.4 3×3 Systems of Equations   312
Chapter 8 Review Problem Set   321
Chapter 8 Practice Test   323

9 Roots and Radicals   325

9.1 Roots and Radicals   326
9.2 Simplifying Radicals   332
9.3 Simplifying Radicals of Quotients   338
9.4 Products and Quotients Involving Radicals   343
9.5 Solving Radical Equations   348
Chapter 9 Review Problem Set   353
Chapter 9 Practice Test   355
Chapters 1–9 Cumulative Review Problem Set   357

10 Quadratic Equations   359

10.1 Quadratic Equations   360
10.2 Completing the Square   367
10.3 Quadratic Formula   372
10.4 Solving Quadratic Equations—Which Method?   376
10.5 Solving Problems Using Quadratic Equations   380
Chapter 10 Review Problem Set   385
Chapter 10 Practice Test   386

Solutions to Warm-Up Problems   387

Solutions to Odd-Numbered Classroom Problem Sets   398

Answers to All Chapter Review, Chapter Practice Test, and Cumulative Review Problems   452

Index   I-1

Preface

Many traditional lecture-based Beginning Algebra courses are evolving into lecture-lab courses or into courses in which all homework and tests are delivered online. Distance learning delivery, where instruction and assessment are conducted primarily through the Internet, is growing rapidly. In this first edition, *Beginning Algebra Hybrid*, as in its companion traditional paperback series, our aim is to provide instructors and students with tools they can use to meet these new challenges.

What do we mean by **hybrid**? A hybrid text involves an integration of several products and can be used best in a course with a strong blend of lecture time and on-line course work.

We believe that this text will be more manageable and less expensive for the student who is spending his or her homework time on a computer. In preparing the text, we did not sacrifice the common thread that has made our series successful: **students learn the skill first, then use the skill to help solve equations, and, finally, use the equations to solve the application problems**. This thread influenced many of the decisions we made in preparing this text.

For the *Hybrid*, we kept the same philosophy while modifying the traditional paperback series to fit the model of online homework and assessment. In writing this hybrid text, one of our main goals is to encourage students to be active learners—to become engaged with learning the skills and drills of mastering mathematics. The major modifications are as follows:

- Each chapter opens with **Warm-Up Problems**, which will check and reinforce students' understanding of what they should know before beginning a new chapter.

- Recognizing that the lecture portion of a class may not be taught in a computer lab, we moved the Practice Your Skill in-text problems, collated these, and included a second problem of each type, into an end-of-section problem set called **Classroom Problem Sets**. This allows for more flexibility for the instructor to teach the example in lecture and allow students to try two similar problems in class. Each Classroom Problem Set problem refers students back to the chapter objective for reference to the concept and skill needed to solve the Classroom Problem Set problems.

- Homework exercises are provided through Enhanced WebAssign®—an easy-to-use online homework management system that allows for repeated homework, quizzing, study, and review. If you compare the paperback version of this text, you'll see that all the **odd-numbered end-of-section problem sets** appear in Enhanced WebAssign.

We have continued to include the Internet Project that follows every chapter opener and the Concept Quiz that ends every section as useful tools to engage students conceptually in the skills and to help them master the ideas and vocabulary presented in each section.

Special Features for *Beginning Algebra Hybrid*

All odd-numbered end-of-section problem sets as they appear in the traditional paperback series are now included in Enhanced WebAssign. In any mathematics course, the most important way to foster conceptual understanding is through the problems that the instructor assigns. To that end, we have provided a wide selection of exercises. Each exercise set is carefully graded, progressing from basic conceptual exercises and skill development problems to more challenging problems requiring synthesis of previously learned material with new concepts.

Each chapter of the text opens with *Warm-Up Problems* that allow students to check their understanding of previously covered material, particularly prerequisite material relevant to the upcoming chapter.

Key Concepts, which include highlighted chapter review material summarizing key concepts with matching examples for a quick study aid, appear at the back of the text on perforated pages.

Getting the Most from Your Enhanced WebAssign Experience, which includes directions and suggestions for best practices enhancing your engagement with our simple homework system, also appears at the back of the text on perforated pages.

For Instructors

Instructor's Guide
ISBN-13: 978-1-111-57324-9

The Instructor's Guide includes teaching tips and materials targeted toward the challenges of teaching an online course. A mathematics teacher for 26 years, author Tony Craig is currently Professor of Mathematics at Paradise Valley Community College in Phoenix, Arizona, where he teaches Developmental Mathematics, College Mathematics, Precalculus, and Calculus.

Complete Solutions Manual
ISBN-13: 978-0-495-38824-1

The Complete Solutions Manual was written for the traditional paperback series. This includes the odd-numbered problems found in Enhanced WebAssign and the even-numbered solutions from the traditional paperback series.

Text-Specific DVDs by Rena Petrello
ISBN-13: 978-0-495-38828-9

These 10- to 20-minute problem-solving lessons cover nearly every learning objective. Recipient of the Mark Dever Award for Excellence in Teaching, Rena Petrello presents each lesson using her experience teaching online mathematics courses. It was through this online teaching experience that Rena discovered the lack of suitable content for online instructors, which caused her to develop her own video lessons—and ultimately create this video project. These videos have won four awards: two Telly Awards, one Communicator Award, and one Aurora Award (an international honor). Students will love the additional guidance and support when they have missed a class or when they are preparing for an upcoming quiz or exam. The videos are available for purchase as a set of DVDs or online via CengageBrain.com.

PowerLecture CD-ROM with ExamView®
ISBN: 978-0-495-38822-7

This CD-ROM provides the instructor with dynamic media tools for teaching. Create, deliver, and customize tests (both print and online) in minutes with *ExamView® Computerized Testing Featuring Algorithmic Equations*. Easily build solution sets for homework or exams using *Solution Builder*'s online solutions manual. Microsoft® PowerPoint® lecture slides, figures from the book, and Test Bank, in electronic format, are also included on this CD-ROM.

Enhanced WebAssign
ISBN: 978-0-538-73810-1

Used by over 1 million students at more than 1100 institutions, Enhanced WebAssign enables you to assign, collect, grade, and record homework assignments via the Web. This proven and reliable homework system includes thousands of algorithmically generated homework problems, links to relevant textbook sections, video examples, problem-specific tutorials, and more. Diagnostic quizzing for each chapter identifies concepts that students still need to master and directs them to the appropriate review material. Students will appreciate the interactive eBook, which offers searching, highlighting, and note-taking functionality, as well as links to multimedia resources—all available to students when you choose Enhanced WebAssign.

For Students

Text-Specific DVDs by Rena Petrello
ISBN-13: 978-0-495-38828-9

These 10- to 20-minute problem-solving lessons cover nearly every learning objective. Recipient of the Mark Dever Award for Excellence in Teaching, Rena Petrello presents each lesson using her experience teaching online mathematics courses. It was through this online teaching experience that Rena discovered the lack of suitable content for online instructors, which caused her to develop her own video lessons—and ultimately create this video project. These videos have won four awards: two Telly Awards, one Communicator Award, and one Aurora Award (an international honor). Students will love the additional guidance and support when they have missed a class or when they are preparing for an upcoming quiz or exam. The videos are available for purchase as a set of DVDs or online via CengageBrain.com.

Student Solutions Manual
ISBN-13: 978-0-495-38823-4

The Student Solutions Manual provides worked-out solutions to the odd-numbered problems from the traditional paperback text, which have been added into the Enhanced WebAssign course.

Enhanced WebAssign
ISBN: 978-0-538-73810-1

Used by over 1 million students at more than 1100 institutions, Enhanced WebAssign enables you to assign, collect, grade, and record homework assignments via the Web. This proven and reliable homework system includes thousands of algorithmically generated homework problems, links to relevant textbook sections, video examples, problem-specific tutorials, and more. Diagnostic quizzing for each chapter identifies concepts that students still need to master and directs them to the appropriate review material. Students will appreciate the interactive eBook, which offers searching, highlighting, and note-taking functionality, as well as links to multimedia resources—all available to students when you choose Enhanced WebAssign.

Basic Concepts of Arithmetic and Algebra

Chapter 1 Warm-Up Problems

1. **(a)** Place the values 4, $2\frac{1}{2}$, and 1.75 on the number line.

   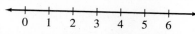

   (b) What is the value of the point marked x?

   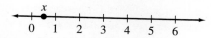

2. Identify the place value of the 3 in each number.
   (a) 130   **(b)** 4.132   **(c)** 2.013

3. Add.   **(a)** $12 + 7$ **(b)** $8 + 14 + 6$ **(c)** $23 + 0 + 18$

4. Subtract.   **(a)** $12 - 12$ **(b)** $37 - 11$ **(c)** $123 - 15$

5. Multiply.   **(a)** $(32)(4)$ **(b)** $(14)(5)$ **(c)** $(42)(0)$

6. Divide.   **(a)** $18 \div 3$ **(b)** $2 \div 3$ **(c)** $9 \div 0$

7. Multiply.   **(a)** $2 \cdot 2 \cdot 5 \cdot 7$ **(b)** $3 \cdot 5 \cdot 5 \cdot 11$ **(c)** $2 \cdot 2 \cdot 2 \cdot 3 \cdot 3 \cdot 3$

8. Divide.   **(a)** $6\overline{)96}$ **(b)** $14\overline{)238}$ **(c)** $4\overline{)194}$

9. Write the mixed number as an improper fraction.
   (a) $3\frac{2}{3}$ **(b)** $6\frac{5}{8}$ **(c)** $9\frac{1}{2}$

10. Reduce the lowest terms.   **(a)** $\frac{12}{20}$ **(b)** $\frac{6}{16}$ **(c)** $\frac{18}{36}$

- 1.1 Numerical and Algebraic Expressions
- 1.2 Addition, Subtraction, Multiplication, and Division of Integers
- 1.3 Use of Properties
- 1.4 Prime and Composite Numbers
- 1.5 Rational Numbers
- 1.6 Decimals and Real Numbers

In the first two chapters of this text, the concept of a **numerical expression** is used as a basis for reviewing addition, subtraction, multiplication, and division of various kinds of numbers. Then the concept of a **variable** enables us to move from numerical expressions to algebraic expressions—that is, to start the transition from arithmetic to algebra. Keep in mind that algebra is simply a generalized approach to arithmetic. Many algebraic concepts are extensions of arithmetic ideas. Therefore, we will build on your knowledge of arithmetic to help you with the study of algebra.

Video tutorials for all section learning objectives are available in a variety of delivery modes.

INTERNET PROJECT

Do an Internet search on fractions in the stock market. When the New York Stock Exchange began in 1792, the Buttonwood Agreement determined the basis for the stock values. What fraction was used for the basis or spread for stock prices when the NYSE began? Is this currently the basis for stock prices? How did the Common Cents Stock Pricing Act of 1997 change the pricing of stocks? Why do you think the basis for stock prices was changed?

1.1 Numerical and Algebraic Expressions

OBJECTIVES

1. Learn the Basic Vocabulary and Symbols for Sets
2. Simplify Numerical Expressions According to the Order of Operations
3. Evaluate Algebraic Expressions

Algebraic Expressions In arithmetic, we use symbols such as 4, 8, 17, and π to represent numbers. We indicate the basic operations of addition, subtraction, multiplication, and division by the symbols $+$, $-$, $\cdot$, and $\div$, respectively. For example, we can write the indicated sum of eight and four as $8 + 4$; $8 + 4$ is a **numerical expression**.

In algebra, the concept of a "variable" provides the basis for generalizing. By using x and y to represent *any* number, we can use the expression $x + y$ to represent the indicated sum of *any two* numbers. The x and y in this expression are called **variables**, and the phrase $x + y$ is called an **algebraic expression**. We commonly use letters of the alphabet such as x, y, z, and w as variables. The key idea is that they represent numbers; therefore, as we review various operations and properties pertaining to numbers, we are building the foundation for our study of algebra.

Many of the notational agreements made in arithmetic are extended to algebra with a few slight modifications. The following chart summarizes the notational agreements pertaining to the four basic operations. Note the variety of ways to write a product by using parentheses to indicate multiplication. Actually, the ab form is the simplest and the form probably used most often; expressions such as abc, $6x$, and $7xyz$ all indicate multiplication. Also note the various forms for indicating division; the fractional form, $\dfrac{c}{d}$, is generally used in algebra, although the other forms do serve a purpose at times.

Operation	Arithmetic	Algebra	Vocabulary
Addition	$4 + 6$	$x + y$	The *sum* of x and y
Subtraction	$7 - 2$	$w - z$	The *difference* of w and z
Multiplication	$9 \cdot 8$	$a \cdot b, a(b), (a)b,$ $(a)(b),$ or ab	The *product* of a and b
Division	$8 \div 2, \dfrac{8}{2}, 2\overline{)8}$	$c \div d, \dfrac{c}{d},$ or $d\overline{)c}$	The *quotient* of c and d

1 Learn the Basic Vocabulary and Symbols for Sets

As we review arithmetic ideas and introduce algebraic concepts, it is convenient to use some of the basic vocabulary and symbols associated with sets. A **set** is a collection of objects, and the objects are called **elements** or **members** of the set. In arithmetic and algebra, the elements of a set are often numbers. To communicate about sets, we use set braces, { }, to enclose the elements (or a description of the elements), and we use capital letters to name sets. For example, we can represent a set A, which consists of the vowels of the English alphabet, in these ways:

$A = \{\text{vowels of the English alphabet}\}$ Word description

$A = \{a, e, i, o, u\}$ List or roster description

We can modify the listing approach if the number of elements is quite large. For example, all the letters of the alphabet can be listed as

$\{a, b, c, \ldots, z\}$

We simply begin by writing enough elements to establish a pattern, and then the three dots indicate that the set continues in that pattern. The final entry indicates the last element of the pattern. If we write

$\{1, 2, 3, \ldots\}$

the set begins with the counting numbers, 1, 2, and 3. The three dots indicate that it continues in a like manner forever; there is no last element. A set that consists of no elements is called the **null set** (written $\emptyset$).

Two sets are said to be *equal* if they contain exactly the same elements. For example,

$\{1, 2, 3\} = \{2, 1, 3\}$

because both sets contain the same elements; the order in which the elements are written doesn't matter.

The slash mark through the equality symbol denotes *not equal to*. Thus if $A = \{1, 2, 3\}$ and $B = \{1, 2, 3, 4\}$, we can write $A \neq B$, which we read as "set A is not equal to set B."

2 Simplify Numerical Expressions According to the Order of Operations

Now let's simplify some numerical expressions that involve the set of **whole numbers** — that is, the set $\{0, 1, 2, 3, \ldots\}$.

EXAMPLE 1

Simplify $8 + 7 - 4 + 12 - 7 + 14$.

Solution

The additions and subtractions should be performed from left to right in the order in which they appear.

$$8 + 7 - 4 + 12 - 7 + 14 = 15 - 4 + 12 - 7 + 14$$
$$= 11 + 12 - 7 + 14$$
$$= 23 - 7 + 14$$
$$= 16 + 14$$
$$= 30$$

Thus $8 + 7 - 4 + 12 - 7 + 14$ simplifies to 30.

Chapter 1 Basic Concepts of Arithmetic and Algebra

EXAMPLE 2

Simplify $7(9 + 5)$.

Solution

The parentheses indicate the product of 7 and the quantity $9 + 5$. Perform the addition inside the parentheses first, and then multiply by 7.

$$7(9 + 5) = 7(14)$$
$$= 98$$

Thus $7(9 + 5)$ simplifies to 98. ∎

EXAMPLE 3

Simplify $(7 + 8) \div (4 - 1)$.

Solution

First, perform the operations inside the parentheses. Then divide the two quantities.

$$(7 + 8) \div (4 - 1) = 15 \div 3$$
$$= 5$$

Thus $(7 + 8) \div (4 - 1)$ simplifies to 5. ∎

We frequently express a problem like Example 3 in the form $\dfrac{7 + 8}{4 - 1}$. We don't need parentheses in this case, because the fraction bar indicates that the sum of 7 and 8 is to be divided by the difference $4 - 1$. A problem may, however, contain both parentheses and fraction bars, as the next example illustrates.

EXAMPLE 4

Simplify $\dfrac{(4 + 2)(7 - 1)}{9} + \dfrac{7}{10 - 3}$.

Solution

First simplify above and below the fraction bars, and then proceed to evaluate as follows:

$$\frac{(4 + 2)(7 - 1)}{9} + \frac{7}{10 - 3} = \frac{(6)(6)}{9} + \frac{7}{7}$$
$$= \frac{36}{9} + 1 = 4 + 1 = 5 \quad \blacksquare$$

EXAMPLE 5

Simplify $7 \cdot 9 + 5$.

Solution

If there are no parentheses to indicate otherwise, multiplication takes precedence over addition. First perform the multiplication, and then do the addition.

$$7 \cdot 9 + 5 = 63 + 5$$
$$= 68$$

Thus $7 \cdot 9 + 5$ simplifies to 68. ∎

(Compare Example 2 and Example 5, and note the difference in meaning.)

EXAMPLE 6

Simplify $8 + 4 \cdot 3 - 14 \div 2$.

Solution

The multiplication and division should be done first in the order in which they appear, from left to right. Then perform the addition and subtraction in the order in which they appear.

$$8 + 4 \cdot 3 - 14 \div 2 = 8 + 12 - 7$$
$$= 20 - 7$$
$$= 13$$

Therefore, $8 + 4 \cdot 3 - 14 \div 2$ simplifies to 13. ∎

EXAMPLE 7

Simplify $18 \div 3 \cdot 2 + 8 \cdot 10 \div 2$.

Solution

If we perform the multiplications and divisions first, in the order in which they appear, and then do the additions and subtractions, our work takes on the following format:

$$18 \div 3 \cdot 2 + 8 \cdot 10 \div 2 = 6 \cdot 2 + 80 \div 2$$
$$= 12 + 40$$
$$= 52$$

Thus $18 \div 3 \cdot 2 + 8 \cdot 10 \div 2$ simplifies to 52. ∎

EXAMPLE 8

Simplify $5 + 6[2(3 + 9)]$.

Solution

We use brackets for the same purpose as parentheses. In such a problem, we need to simplify *from the inside out*, performing the operations in the innermost parentheses first.

$$5 + 6[2(3 + 9)] = 5 + 6[2(12)]$$
$$= 5 + 6[24]$$
$$= 5 + 144 = 149$$

∎

Exponents are used in both arithmetic and algebra to indicate repeated multiplications. For example, we can write $5 \cdot 5 \cdot 5$ as 5^3, and the 3 indicates that 5 is to be used as a factor three times. For the expression 5^3, the 5 is referred to as the **base** and the 3 is called the **exponent**. The exponent is also sometimes referred to as a power. The expression 5^3 can be read "5 raised to the 3 power." Exponents are also used in algebraic expressions when the base is a variable. For example, x^4 means that x will be used as a factor four times, so $x^4 = x \cdot x \cdot x \cdot x$. An exponent of 1 is usually not written. For example, y^1 would just be written as y. When simplifying numerical expressions, exponential expressions should be evaluated before performing the operations of multiplication, division, addition, or subtraction. The following example presents an expression with exponents that needs simplifying.

EXAMPLE 9

Simplify $6^2 \cdot 5 + 2^4$.

Solution

First simplify the expressions with exponents by performing their repeated multiplications. Then do the indicated multiplication followed by the addition.

$$6^2 \cdot 5 + 2^4 = (6 \cdot 6) \cdot 5 + (2 \cdot 2 \cdot 2 \cdot 2)$$
$$= 36 \cdot 5 + 16$$
$$= 180 + 16$$
$$= 196$$

Let us now summarize the ideas presented in the preceding examples on simplifying numerical expressions. When we **simplify a numerical expression**, the operations should be performed in the following order:

Order of Operations

1. Perform the operations inside the symbols of inclusion (parentheses and brackets) and above and below each fraction bar. Start with the innermost inclusion symbol.
2. Evaluate all exponential expressions.
3. Perform all multiplications and divisions in the order in which they appear from left to right.
4. Perform all additions and subtractions in the order in which they appear from left to right.

---

3 Evaluate Algebraic Expressions

We can use the concept of a variable to generalize from numerical expressions to algebraic expressions. Each of the following is an example of an algebraic expression:

$$3x + 2y \qquad 5a - 2b + c \qquad 7(w + z) \qquad a^2 + b^2$$

$$\frac{5d + 3e}{2c - d} \qquad 2xy + 5yz \qquad (x + y)(x - y) \qquad 9x^2 - y^2$$

An algebraic expression takes on a numerical value whenever each variable in the expression is replaced by a specific number. For example, if x is replaced by 9 and z by 4, the algebraic expression $x - z$ becomes the numerical expression $9 - 4$, which simplifies to 5. We say that $x - z$ *has a value* of 5 when x equals 9 and z equals 4. The value of $x - z$, when x equals 25 and z equals 12, is 13. The general algebraic expression $x - z$ has a specific value each time x and z are replaced by numbers.

Consider the next examples, which illustrate the process of finding the value of an algebraic expression. This process is often referred to as **evaluating algebraic expressions**.

EXAMPLE 10

Find the value of $3x + 2y$ when x is replaced by 5 and y by 17.

Solution

The following format is convenient for such problems:

$$3x + 2y = 3(5) + 2(17) \quad \text{when } x = 5 \text{ and } y = 17$$
$$= 15 + 34$$
$$= 49$$

In Example 10, for the algebraic expression, $3x + 2y$, note that the multiplications "3 times x" and "2 times y" are implied without the use of parentheses. The algebraic expression switches to a numerical expression when numbers are substituted for variables; in that case, parentheses are used to indicate the multiplication. We could also use the raised dot to indicate multiplication; that is, $3(5) + 2(17)$ could be written as $3 \cdot 5 + 2 \cdot 17$. Furthermore, note that once we have substituted numbers for variables, we can start the process of simplifying numerical expressions.

EXAMPLE 11

Find the value of $12a - 3b$ when $a = 5$ and $b = 9$.

Solution

$$12a - 3b = 12(5) - 3(9) \quad \text{when } a = 5 \text{ and } b = 9$$
$$= 60 - 27$$
$$= 33$$

EXAMPLE 12

Evaluate $4xy + 2xz - 3yz$ when $x = 8$, $y = 6$, and $z = 2$.

Solution

$$4xy + 2xz - 3yz = 4(8)(6) + 2(8)(2) - 3(6)(2) \quad \text{when } x = 8, y = 6, \text{ and } z = 2$$
$$= 192 + 32 - 36$$
$$= 188$$

EXAMPLE 13

Evaluate $\dfrac{5c + d}{3c - d}$ for $c = 12$ and $d = 4$.

Solution

$$\frac{5c + d}{3c - d} = \frac{5(12) + 4}{3(12) - 4} \quad \text{for } c = 12 \text{ and } d = 4$$
$$= \frac{60 + 4}{36 - 4}$$
$$= \frac{64}{32}$$
$$= 2$$

EXAMPLE 14

Evaluate $(2x + 5y)(3x - 2y)$ when $x = 6$ and $y = 3$.

Solution

$$(2x + 5y)(3x - 2y) = (2 \cdot 6 + 5 \cdot 3)(3 \cdot 6 - 2 \cdot 3) \quad \text{when } x = 6 \text{ and } y = 3$$
$$= (12 + 15)(18 - 6)$$
$$= (27)(12)$$
$$= 324$$

EXAMPLE 15

Evaluate $5x^2 - 3x + 7$ when $x = 4$.

Solution

It is always a good practice to use parentheses when substituting a value for a variable. This is especially true when the variable is raised to an exponent.

$$5x^2 - 3x + 7 = 5(4)^2 - 3(4) + 7 \quad \text{when } x = 4$$
$$= 5(16) - 3(4) + 7$$
$$= 80 - 12 + 7$$
$$= 68 + 7$$
$$= 75$$

EXAMPLE 16 The formula for the area of a triangle is $A = \dfrac{bh}{2}$, where b is the length of the base of the triangle, and h represents the length of the height of the triangle. Use the formula $A = \dfrac{bh}{2}$ to determine the value of A when $b = 14$ and $h = 5$.

Solution

$$A = \frac{bh}{2} = \frac{14 \cdot 5}{2} \quad \text{when } b = 14 \text{ and } h = 5$$
$$= 35$$

CONCEPT QUIZ 1.1 For Problems 1–10, answer true or false.

1. The expression ab indicates the sum of a and b.
2. Any of the following notations, $(a)b, a \cdot b, a(b)$, can be used to indicate the product of a and b.
3. The phrase "$2x + y - 4z$" is called an algebraic expression.
4. A set is a collection of objects, and the objects are called "terms."
5. The sets $\{2, 4, 6, 8\}$ and $\{6, 4, 8, 2\}$ are equal.
6. The set $\{1, 3, 5, 7, \ldots\}$ has a last element of 99.
7. The null set has one element.
8. To evaluate $24 \div 6 \cdot 2$, the first operation that should be performed is to multiply 6 times 2.
9. To evaluate $6 + 8 \cdot 3$, the first operation that should be performed is to multiply 8 times 3.
10. The algebraic expression $2(x + y)$ simplifies to 24, if x is replaced by 10 and y is replaced by 0.

Section 1.1 Classroom Problem Set

Objective 2

1. Simplify $5 - 3 + 12 - 2 - 1$.
2. Simplify $12 - 5 - 1 + 15 - 4 - 3$.
3. Simplify $12(6 - 4)$.
4. Simplify $8(6 + 12)$.
5. Simplify $(9 - 1) \div (9 - 5)$.
6. Simplify $(11 - 3) \div (15 - 7)$.
7. Simplify $\dfrac{(8 - 5)(1 + 3)}{6} + \dfrac{12}{11 - 7}$.
8. Simplify $\dfrac{3(17 - 9)}{4} + \dfrac{9(16 - 7)}{3}$.
9. Simplify $16 \cdot 2 + 6$.
10. Simplify $9 \cdot 7 - 10$.
11. Simplify $3 + 5 \cdot 2 - 10 \div 2$.
12. Simplify $6 + 12 \div 3 - 3 \cdot 2$.
13. Simplify $24 \div 6 \cdot 2 + 4 \cdot 8 \div 2$.
14. Simplify $14 \div 7 \cdot 8 - 35 \div 7 \cdot 2$.
15. Simplify $12 + 4[5(2 + 8)]$.
16. Simplify $56 - [3(9 - 6)]$.
17. Simplify $3 \cdot 5^2 - 7 \cdot 2^3$.
18. Simplify $4^2 \cdot 3^2 - 9 \cdot 2^3$.

Objective 3

19. Evaluate $4x + 3y$ when $x = 5$ and $y = 2$.
20. Evaluate $7x + 4y$ when $x = 6$ and $y = 8$.
21. Evaluate $10a - 5b$ when $a = 8$ and $b = 3$.

22. Evaluate $16a - 9b$ when $a = 3$ and $b = 4$.

23. Evaluate $3xy - 2y + 4xyz$ when $x = 6, y = 2$ and $z = 4$.

24. Evaluate $9xy - 4xyz + 3yz$ when $x = 7, y = 3$, and $z = 2$.

25. Evaluate $\dfrac{7c + d}{2c - d}$ when $c = 4$ and $d = 2$.

26. Evaluate $\dfrac{9b - a}{a - b}$ when $a = 11$ and $b = 3$.

27. Evaluate $(x + 3y)(4x - 2y)$ when $x = 8$ and $y = 3$.

28. Evaluate $(5x - 2y)(3x + 4y)$ when $x = 3$ and $y = 6$.

29. Evaluate $a^2 + 2ab + b^2$ when $a = 7$ and $b = 4$.

30. Evaluate $y^2 + 2xy + x^2$ when $x = 3$ and $y = 4$.

31. Use the formula $A = \dfrac{bh}{2}$ to determine A when $b = 15$ and $h = 4$.

32. Use the formula $A = \dfrac{bh}{2}$ to determine A when $b = 18$ and $h = 13$.

THOUGHTS INTO WORDS

1. Explain the difference between a numerical expression and an algebraic expression.

2. Your friend keeps getting an answer of 45 when simplifying $3 + 2(9)$. What mistake is he making, and how would you help him?

Answers to the Concept Quiz
1. False **2.** True **3.** True **4.** False **5.** True **6.** False **7.** False **8.** False **9.** True **10.** False

1.2 Addition, Subtraction, Multiplication, and Division of Integers

OBJECTIVES

1. Know the Names for Certain Sets of Integers
2. Add Integers
3. Subtract Integers
4. Multiply Integers
5. Divide Integers
6. Simplify Numerical Expressions Involving Integers
7. Evaluate Algebraic Expressions for Integer Values
8. Use Integers to Solve Application Problems

1 Know the Names for Certain Sets of Integers

"A record temperature of 35° *below* zero was recorded on this date in 1904." "The IMDigital stock closed *down* 3 points yesterday." "On a first-down sweep around left end, Faulk *lost* 7 yards." "The West Coast Manufacturing Company reported *assets* of 50 million dollars and *liabilities* of 53 million dollars for 2007." These statements illustrate our need for negative numbers.

The number line is a helpful visual device for our work at this time. We can associate the set of whole numbers with evenly spaced points on a line as indicated in Figure 1.1. For each nonzero whole number, we can associate its *negative* to the left

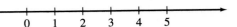

Figure 1.1

of zero; with 1 we associate -1, with 2 we associate -2, and so on, as indicated in Figure 1.2. The set of whole numbers, along with $-1, -2, -3$, and so on, is called the set of **integers**.

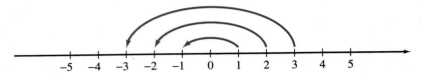

Figure 1.2

The following terminology is used with reference to the integers:

$\{\ldots, -3, -2, -1, 0, 1, 2, 3, \ldots\}$	Integers
$\{1, 2, 3, 4, \ldots\}$	Positive integers
$\{0, 1, 2, 3, 4, \ldots\}$	Nonnegative integers
$\{\ldots, -3, -2, -1\}$	Negative integers
$\{\ldots, -3, -2, -1, 0\}$	Nonpositive integers

The symbol -1 can be read as "negative one," "opposite of one," or "additive inverse of one." The opposite-of and additive-inverse-of terminology is very helpful when working with variables. For example, reading the symbol $-x$ as "opposite of x" or "additive inverse of x" emphasizes an important issue. Because x can be any integer, $-x$ (the opposite of x) can be zero, positive, or negative. If x is a positive integer, then $-x$ is negative. If x is a negative integer, then $-x$ is positive. If x is zero, then $-x$ is zero. These statements can be written and illustrated on the number lines as in Figure 1.3.

If $x = 3$,
then $-x = -(3) = -3$.

If $x = -3$,
then $-x = -(-3) = 3$.

If $x = 0$,
then $-x = -(0) = 0$.

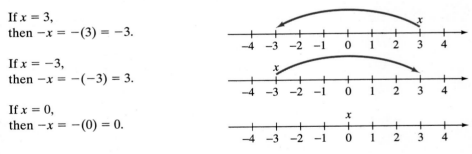

Figure 1.3

From this discussion we can recognize the following general property:

Property 1.1

If a is any integer, then

$$-(-a) = a$$

(The opposite of the opposite of any integer is the integer itself.)

---

2 Add Integers

The number line is also a convenient visual aid for interpreting the addition of integers. Consider the following examples and their number-line interpretations as shown in Figure 1.4.

1.2 Addition, Subtraction, Multiplication, and Division of Integers

Problem	Number line interpretation	Sum
3 + 2	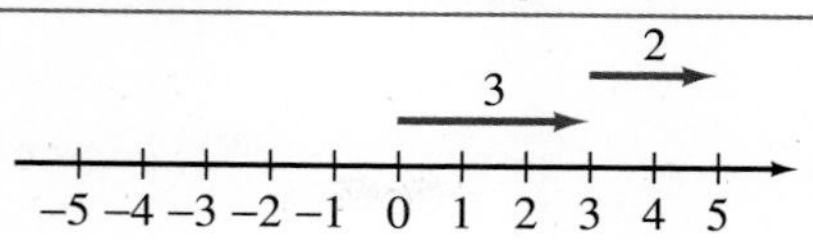	3 + 2 = 5
3 + (−2)	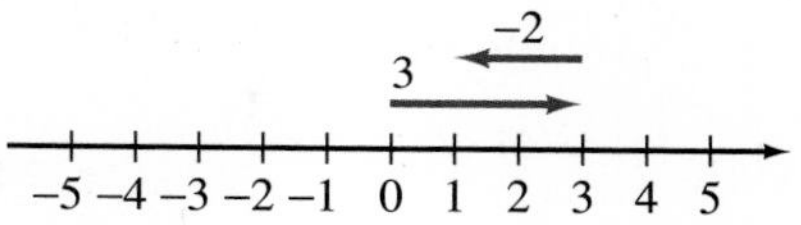	3 + (−2) = 1
−3 + 2	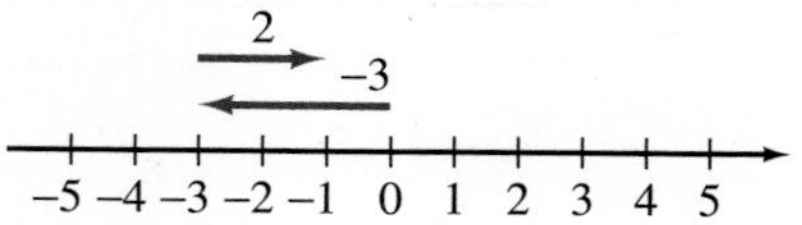	−3 + 2 = −1
−3 + (−2)	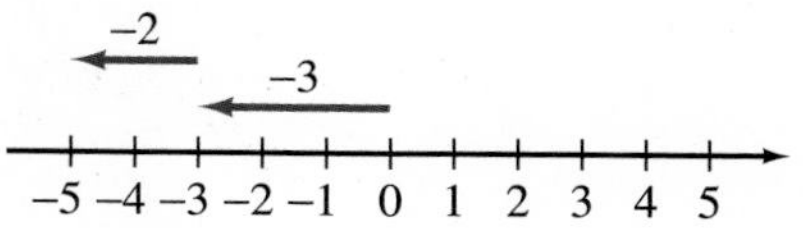	−3 + (−2) = −5

Figure 1.4

Once you get a feel for movement on the number line, simply forming a mental image of this movement is sufficient. Consider the next addition problems, and mentally picture the number-line interpretation. Be sure that you agree with all of our answers.

$5 + (-2) = 3 \qquad -6 + 4 = -2 \qquad -8 + 11 = 3$

$-7 + (-4) = -11 \qquad -5 + 9 = 4 \qquad 9 + (-2) = 7$

$14 + (-17) = -3 \qquad 0 + (-4) = -4 \qquad 6 + (-6) = 0$

The last example illustrates a general property that you should note: **Any integer plus its opposite equals zero**.

Remark: Profits and losses pertaining to investments also provide a good physical model for interpreting the addition of integers. A loss of $25 on one investment, along with a profit of $60 on a second investment, produces an overall profit of $35. We can express this as $-25 + 60 = 35$. You may want to check the preceding examples using a profit and loss interpretation.

Even though all problems that involve the addition of integers could be done by using the number-line interpretation, it is sometimes convenient to give a more precise description of the addition process. For this purpose, we need to consider briefly the concept of absolute value. The **absolute value** of a number is the distance between the number and zero on the number line. For example, the absolute value of 6 is 6. The absolute value of -6 is also 6. The absolute value of 0 is 0. Vertical bars on either side of a number denote absolute value. Thus we write

$|6| = 6 \qquad |-6| = 6 \qquad |0| = 0$

Note that the absolute value of a positive number is the number itself, but the absolute value of a negative number is its opposite. Thus the absolute value of any number except 0 is positive.

We can describe the process of **adding integers** by using the concept of absolute value.

Addition of Integers: Two Positive Integers

The sum of two positive integers is the sum of their absolute values. (The sum of two positive integers is a positive integer.)

$43 + 54 = |43| + |54| = 43 + 54 = 97$

Addition of Integers: Two Negative Integers

The sum of two negative integers is the opposite of the sum of their absolute values. (The sum of two negative integers is a negative integer.)

$$(-67) + (-93) = -(|-67| + |-93|)$$
$$= -(67 + 93)$$
$$= -160$$

Addition of Integers: One Positive and One Negative Integer

The sum of a positive integer and a negative integer can be found by subtracting the smaller absolute value from the larger absolute value and giving the result the sign of the original number that has the larger absolute value. If the integers have the same absolute value, then their sum is zero.

$$82 + (-40) = |82| - |-40|$$
$$= 82 - 40$$
$$= 42$$

$$74 + (-90) = -(|-90| - |74|)$$
$$= -(90 - 74)$$
$$= -16$$

$$(-17) + 17 = |-17| - |17|$$
$$= 17 - 17$$
$$= 0$$

Addition of Integers: Zero and Another Integer

The sum of zero and any integer is the integer itself.

$$0 + (-46) = -46$$
$$72 + 0 = 72$$

EXAMPLE 1   Add the integers to find their sum.

(a) $-18 + (-56)$ (b) $-71 + (-32)$ (c) $64 + (-49)$ (d) $-56 + 93$
(e) $-114 + 48$ (f) $45 + (-73)$ (g) $46 + (-46)$ (h) $0 + (-81)$

Solution

(a) $-18 + (-56) = -(|-18| + |-56|) = -(18 + 56) = -74$

(b) $-71 + (-32) = -(|-71| + |-32|) = -(71 + 32) = -103$

(c) $64 + (-49) = |64| - |-49| = 64 - 49 = 15$

(d) $-56 + 93 = |93| - |-56| = 93 - 56 = 37$

(e) $-114 + 48 = -(|-114| - |48|) = -(114 - 48) = -66$

(f) $45 + (-73) = -(|-73| - |45|) = -(73 - 45) = -28$

(g) $46 + (-46) = 0$

(h) $0 + (-81) = -81$ ∎

3 Subtract Integers

The following examples illustrate a relationship between addition and subtraction of *whole numbers*:

$$7 - 2 = 5 \quad \text{because } 2 + 5 = 7$$
$$9 - 6 = 3 \quad \text{because } 6 + 3 = 9$$
$$5 - 1 = 4 \quad \text{because } 1 + 4 = 5$$

This same relationship between addition and subtraction holds for *all integers*:

$$5 - 6 = -1 \quad \text{because } 6 + (-1) = 5$$
$$-4 - 9 = -13 \quad \text{because } 9 + (-13) = -4$$
$$-3 - (-7) = 4 \quad \text{because } -7 + 4 = -3$$
$$8 - (-3) = 11 \quad \text{because } -3 + 11 = 8$$

Now consider a further observation:

$$5 - 6 = -1 \quad \text{and} \quad 5 + (-6) = -1$$
$$-4 - 9 = -13 \quad \text{and} \quad -4 + (-9) = -13$$
$$-3 - (-7) = 4 \quad \text{and} \quad -3 + 7 = 4$$
$$8 - (-3) = 11 \quad \text{and} \quad 8 + 3 = 11$$

The previous examples help us realize that we can state the **subtraction of integers** in terms of the addition of integers. More precisely, a general description for the subtraction of integers follows:

Subtraction of Integers

If *a* and *b* are integers, then $a - b = a + (-b)$.

It may be helpful for you to read $a - b = a + (-b)$ as "*a* minus *b* is equal to *a* plus the opposite of *b*." Every subtraction problem can be changed into an equivalent addition problem, as illustrated by the next example.

EXAMPLE 2

Subtract the integers.

(a) $6 - 13$ **(b)** $9 - (-12)$ **(c)** $-8 - 13$ **(d)** $-7 - (-8)$

Solution

(a) $6 - 13 = 6 + (-13) = -7$

(b) $9 - (-12) = 9 + 12 = 21$

(c) $-8 - 13 = -8 + (-13) = -21$

(d) $-7 - (-8) = -7 + 8 = 1$

It should be apparent that addition of integers is a key operation. Being able to add integers effectively is indispensable for further work in algebra. ∎

4 Multiply Integers

Multiplication of whole numbers may be interpreted as repeated addition. For example, $3 \cdot 4$ means the sum of three 4s; thus $3 \cdot 4 = 4 + 4 + 4 = 12$. Consider the following examples, which use the idea of repeated addition to find the product of a positive integer and a negative integer.

$$3(-2) = -2 + (-2) + (-2) = -6$$
$$2(-4) = -4 + (-4) = -8$$
$$4(-1) = -1 + (-1) + (-1) + (-1) = -4$$

Note the use of parentheses to indicate multiplication. Sometimes both numbers are enclosed in parentheses; in this case we would have $(3)(-2)$.

When multiplying whole numbers, we realize that the order in which we multiply two factors does not change the product; in other words, $2(3) = 6$ and $3(2) = 6$. Using this idea, we can now handle a negative integer times a positive integer:

$$(-2)(3) = (3)(-2) = (-2) + (-2) + (-2) = -6$$
$$(-3)(2) = (2)(-3) = (-3) + (-3) = -6$$
$$(-4)(3) = (3)(-4) = (-4) + (-4) + (-4) = -12$$

Finally, let's consider the product of two negative integers. The following pattern helps us with the reasoning for this situation.

$$4(-3) = -12$$
$$3(-3) = -9$$
$$2(-3) = -6$$
$$1(-3) = -3$$
$$0(-3) = 0 \quad \text{The product of zero and any integer is zero}$$
$$(-1)(-3) = ?$$

Certainly, to continue this pattern, the product of -1 and -3 has to be 3. This type of reasoning helps us to realize that the product of any two negative integers is a positive integer.

Using the concept of absolute value, we can now precisely describe the **multiplication of integers**:

Multiplying Integers

1. The product of two positive integers or two negative integers is the product of their absolute values.
2. The product of a positive integer and a negative integer (either order) is the opposite of the product of their absolute values.
3. The product of zero and any integer is zero.

The next example illustrates this description of multiplication.

EXAMPLE 3   Find the product of the integers.

(a) $(-5)(-2)$ (b) $(7)(-6)$ (c) $(-8)(9)$ (d) $(-14)(0)$
(e) $(0)(-28)$

Solution

(a) $(-5)(-2) = |-5| \cdot |-2| = 5 \cdot 2 = 10$
(b) $(7)(-6) = -(|7| \cdot |-6|) = -(7 \cdot 6) = -42$

(c) $(-8)(9) = -(|-8| \cdot |9|) = -(8 \cdot 9) = -72$
(d) $(-14)(0) = 0$
(e) $(0)(-28) = 0$

■

This example showed a step-by-step process for multiplying integers. In reality, however, the key issue to remember is whether the product is positive or negative. In other words, we need to remember that *the product of two positive integers or two negative integers is a positive integer*, and that *the product of a positive integer and a negative integer (in either order) is a negative integer*. Then we can avoid the step-by-step analysis and simply write the results as follows:

$(7)(-9) = -63$

$(8)(7) = 56$

$(-5)(-6) = 30$

$(-4)(12) = -48$

5 Divide Integers

By looking back at our knowledge of whole numbers, we can get some guidance for our work with integers. We know, for example, that $\frac{8}{2} = 4$, because $2 \cdot 4 = 8$. In other words, we find the quotient of two whole numbers by looking at a related multiplication problem. In the following examples, we have used this same link between multiplication and division to determine the quotients:

$\dfrac{8}{-2} = -4$ because $(-2)(-4) = 8$

$\dfrac{-10}{5} = -2$ because $(5)(-2) = -10$

$\dfrac{-12}{-4} = 3$ because $(-4)(3) = -12$

$\dfrac{0}{-6} = 0$ because $(-6)(0) = 0$

$\dfrac{-9}{0}$ is undefined   because no number times 0 produces -9

$\dfrac{0}{0}$ is undefined   because any number times 0 equals 0.

Remember that division by zero is undefined!

Dividing Integers

1. The quotient of two positive integers or two negative integers is the quotient of their absolute values.
2. The quotient of a positive integer and a negative integer (or a negative and a positive) is the opposite of the quotient of their absolute values.
3. The quotient of zero and any nonzero integer (zero divided by any nonzero integer) is zero.

The next example illustrates this description of division.

16 Chapter 1 Basic Concepts of Arithmetic and Algebra

EXAMPLE 4

Find the quotient of the integers.

(a) $\dfrac{-8}{-4}$ (b) $\dfrac{-14}{2}$ (c) $\dfrac{15}{-3}$ (d) $\dfrac{0}{-4}$

Solution

(a) $\dfrac{-8}{-4} = \dfrac{|-8|}{|-4|} = \dfrac{8}{4} = 2$

(b) $\dfrac{-14}{2} = -\left(\dfrac{|-14|}{|2|}\right) = -\left(\dfrac{14}{2}\right) = -7$

(c) $\dfrac{15}{-3} = -\left(\dfrac{|15|}{|-3|}\right) = -\left(\dfrac{15}{3}\right) = -5$

(d) $\dfrac{0}{-4} = 0$ ∎

For practical purposes, it is important to determine whether the quotient is positive or negative. *The quotient of two positive integers or two negative integers is positive*, and *the quotient of a positive integer and a negative integer or of a negative integer and a positive integer is negative*. We can then simply write the quotients as follows without showing all of the steps:

$$\dfrac{-18}{-6} = 3 \qquad \dfrac{-24}{12} = -2 \qquad \dfrac{36}{-9} = -4$$

Remark: Occasionally people use the phrase "two negatives make a positive." We hope they realize that the reference is to multiplication and division only; in addition, the sum of two negative integers is still a negative integer. It is probably best to avoid such imprecise statements.

6 Simplify Numerical Expressions Involving Integers

Now we can simplify numerical expressions involving any or all of the four basic operations with integers. Keep in mind the agreements on the order of operations we stated in Section 1.1.

EXAMPLE 5

Simplify $-4(-3) - 7(-8) + 3(-9)$.

Solution

$$-4(-3) - 7(-8) + 3(-9) = 12 - (-56) + (-27)$$
$$= 12 + 56 + (-27)$$
$$= 41$$ ∎

EXAMPLE 6

Simplify $\dfrac{-8 - 4(5)}{-4}$.

Solution

$$\dfrac{-8 - 4(5)}{-4} = \dfrac{-8 - 20}{-4}$$
$$= \dfrac{-28}{-4}$$
$$= 7$$ ∎

7 Evaluate Algebraic Expressions for Integer Values

Evaluating algebraic expressions often involves using two or more operations with integers. The final examples of this section illustrate such situations. Be sure to use parentheses when substituting values for the variables.

EXAMPLE 7 Find the value of $3x + 2y$ when $x = 5$ and $y = -9$.

Solution

$$3x + 2y = 3(5) + 2(-9) \quad \text{when } x = 5 \text{ and } y = -9$$
$$= 15 + (-18)$$
$$= -3$$

EXAMPLE 8 Evaluate $-2a + 9b$ for $a = 4$ and $b = -3$.

Solution

$$-2a + 9b = -2(4) + 9(-3) \quad \text{when } a = 4 \text{ and } b = -3$$
$$= -8 + (-27)$$
$$= -35$$

EXAMPLE 9 The formula $C = \dfrac{5(F - 32)}{9}$ is used to convert temperatures from Fahrenheit to Celsius. Find the value of $\dfrac{5(F - 32)}{9}$ when $F = 5$.

Solution

Evaluate $\dfrac{5(F - 32)}{9}$ when $F = 5$.

$$\frac{5(F - 32)}{9} = \frac{5(5 - 32)}{9} \quad \text{when } F = 5$$
$$= \frac{5(-27)}{9} = \frac{-135}{9} = -15$$

8 Use Integers to Solve Application Problems

The addition, subtraction, multiplication, and division of integers can be used to solve problems that involve both positive and negative numbers. In business situations, profits are represented by positive numbers, and losses are represented by negative numbers. In sports such as football, a loss of 5 yards is represented as -5 yards, and in golf being 2 strokes under par would be represented as -2. In the problem set, you will apply your skill in adding, subtracting, multiplying, and dividing integers to solve application problems as shown in the next two examples.

EXAMPLE 10

Apply Your Skill

A nutrition and fitness instructor asked her clients to weigh in and record their weight loss or gain in pounds every Friday over a six-week period. Given the records for one of the clients (Dominic), determine his overall weight loss or gain.

	Aug. 12	Aug. 19	Aug. 26	Sept. 2	Sept. 9	Sept. 16
Dominic	loss of 3	gain of 1	loss of 5	loss of 2	gain of 3	loss of 4

Solution

Let's use negative numbers to represent the losses and positive numbers to represent the gains. Then Dominic's weekly records would be represented by $-3, +1, -5, -2, +3$, and -4. Adding the numbers will give the overall loss or gain.

$$-3 + (+1) + (-5) + (-2) + (+3) + (-4) = -10$$

So over the six weeks Dominic lost 10 pounds. ∎

EXAMPLE 11

Apply Your Skill

In a local amateur poker tournament, Dwayne lost $25 for three hands, he lost $14 for two hands, and he won $10 for four hands in the first round. Determine the average amount Dwayne won or lost for his hands in the first round of play.

Solution

Let's use negative numbers to represent the losses and positive numbers to represent the wins. Dwayne's first round would be represented by $3(-25) + 2(-14) + 4(+10)$. To find the average we would divide this expression by 9, the total number of hands played.

$$\frac{3(-25) + 2(-14) + 4(+10)}{9} = \frac{-75 - 28 + 40}{9}$$
$$= \frac{-63}{9}$$
$$= -7$$

So, on average, Dwayne lost $7 per hand in the first round. ∎

CONCEPT QUIZ 1.2

For Problems 1–10, answer true or false.

1. The number zero is considered a positive integer.
2. The absolute value of a number is the distance on the number line between the number and one.
3. The $|-4|$ is -4.
4. The opposite of -5 is 5.
5. a minus b is equivalent to a plus the opposite of b.
6. The product of a positive integer and a negative integer is a positive integer.
7. When multiplying three negative integers, the product is negative.
8. The rules for adding integers and the rules for multiplying integers are the same.
9. The quotient of two negative integers is negative.
10. The product of zero and any nonzero integer is zero.

Section 1.2 Classroom Problem Set

Objective 2

1. Find the sum.
   (a) $-24 + (-32)$
   (b) $-40 + 75$
   (c) $86 + (-31)$
   (d) $15 + (-15)$

2. Find the sum.
   (a) $-1 + (-100)$
   (b) $-40 + 92$
   (c) $11 + (-92)$
   (d) $74 + (-74)$

Objective 3

3. Subtract the integers.
   (a) $-20 - (-38)$
   (b) $35 - 75$
   (c) $81 - (-31)$
   (d) $25 - (-25)$

4. Subtract the integers.
   (a) $12 - 51$
   (b) $15 - (-37)$
   (c) $-19 - 22$
   (d) $-25 - (-11)$

Objective 4

5. Find the product.
   (a) $(6)(-2)$
   (b) $(-1)(-5)$
   (c) $(0)(-8)$
   (d) $(-4)(6)$

6. Find the product.
   (a) $(-4)(15)$
   (b) $(-3)(-12)$
   (c) $(9)(-9)$
   (d) $(0)(-7)$

Objective 5

7. Find the quotient.
   (a) $\dfrac{-18}{6}$
   (b) $\dfrac{0}{-8}$
   (c) $\dfrac{-50}{-2}$
   (d) $\dfrac{80}{-4}$

8. (a) $\dfrac{-25}{5}$
   (b) $\dfrac{-40}{40}$
   (c) $\dfrac{12}{-1}$
   (d) $\dfrac{0}{-100}$

Objective 6

9. Simplify $5(-3) - 4(-2) - 7(+1)$.

10. Simplify $-6(3) - 4(-6) - 7(-5)$.

11. Simplify $\dfrac{20 + 4(-2)}{-3}$.

12. Simplify $\dfrac{-12 - 6(-1)}{-6}$.

Objective 7

13. Find the value of $4x - 5y$ when $x = -2$ and $y = -6$.

14. Find the value of $-3x - 2y$ when $x = 4$ and $y = -3$.

15. Evaluate $-3a - 4b$ when $a = -1$ and $b = 5$.

16. Evaluate $7a - 8b$ when $a = -1$ and $b = -1$.

17. Find the value of $\dfrac{5(F - 32)}{9}$ when $F = -4$.

18. Find the value of $\dfrac{5(F - 32)}{9}$ when $F = -14$.

Objective 8

19. A boat manufacturer reported that the company incurred a loss of $835,000 in 2007, a gain of $320,000 in 2008, a gain of $410,000 in 2009, and a loss of $120,000 in 2010. Use the addition of integers to describe this situation, and determine the boat manufacturer's total loss or gain for the four-year period.

20. On five consecutive days in January 2010 in Madison, Wisconsin, the university recorded the following high temperatures: $-7°, -5°, 3°, 4°,$ and $10°$. Find the average high temperature for those five days in January.

21. In a golf tournament to raise funds for scholarships, Kay shot 2 strokes under par on four holes, 1 stroke under par on three holes, and 3 strokes over par on two holes. Use multiplication and addition of integers to describe this situation and determine Kay's score relative to par for the tournament.

22. Nadia holds shares in a company that invests heavily in gold. Last month the company recorded 5 days with a loss of $2, 3 days with a gain of $4, 6 days with a loss of $3, and 7 days with a gain of $1. Determine the loss or gain for the days indicated.

THOUGHTS INTO WORDS

1. The statement $-6 - (-2) = -6 + 2 = -4$ can be read as "negative six minus negative two equals negative six plus two, which equals negative four." Express each equation in words.

   (a) $8 + (-10) = -2$
   (b) $-7 - 4 = -7 + (-4) = 11$
   (c) $9 - (-12) = 9 + 12 = 21$
   (d) $-5 + (-6) = -11$

2. The algebraic expression $-x - y$ can be read "the opposite of x minus y." Give each expression in words.
   (a) $-x + y$
   (b) $x - y$
   (c) $-x - y + z$

3. Your friend keeps getting an answer of -7 when simplifying the expression $-6 + (-8) \div 2$. What mistake is she making and how would you help her?

4. Make up a problem that you can solve using $6(-4) = -24$.

Answers to the Concept Quiz
1. False 2. False 3. False 4. True 5. True 6. False 7. True 8. False 9. False 10. True

1.3 Use of Properties

OBJECTIVES

1. Recognize the Properties of Integers
2. Use the Properties of Integers to Simplify Numerical Expressions
3. Simplify Algebraic Expressions
4. Evaluate Algebraic Expressions

1 Recognize the Properties of Integers

We will begin this section by listing and briefly commenting on some of the basic properties of integers. We will then show how these properties facilitate manipulation with integers and also serve as a basis for some algebraic computation.

Commutative Property of Addition

If a and b are integers, then
$$a + b = b + a$$

Commutative Property of Multiplication

If a and b are integers, then
$$ab = ba$$

Addition and multiplication are said to be commutative operations. This means that the order in which you add or multiply two integers does not affect the result. For example, $3 + 5 = 5 + 3$ and $7(8) = 8(7)$. It is also important to realize that subtraction and division *are not* commutative operations; order does make a difference. For example, $8 - 7 \neq 7 - 8$ and $16 \div 4 \neq 4 \div 16$.

Associative Property of Addition

If a, b, and c are integers, then
$$(a + b) + c = a + (b + c)$$

Associative Property of Multiplication

If $a, b,$ and c are integers, then

$$(ab)c = a(bc)$$

Our arithmetic operations are binary operations. We only operate (add, subtract, multiply, or divide) on two numbers at a time. Therefore, when we need to operate on three or more numbers, the numbers must be grouped.

The associative properties can be thought of as grouping properties. For example, $(-8 + 3) + 9 = -8 + (3 + 9)$. Changing the grouping of the numbers for addition does not affect the result. This is also true for multiplication, as $[(-6)(5)](-4) = (-6)[(5)(-4)]$ illustrates. Addition and multiplication are associative operations. Subtraction and division *are not* associative operations. For example, $(8 - 4) - 7 = -3$, whereas $8 - (4 - 7) = 11$. An example showing that division is not associative is $(8 \div 4) \div 2 = 1$, whereas $8 \div (4 \div 2) = 4$.

Identity Property of Addition

If a is an integer, then

$$a + 0 = 0 + a = a$$

We refer to zero as the identity element for addition. This simply means that the sum of any integer and zero is exactly the same integer. For example, $-197 + 0 = 0 + (-197) = -197$.

Identity Property of Multiplication

If a is an integer, then

$$a(1) = 1(a) = a$$

We call one the identity element for multiplication. The product of any integer and one is exactly the same integer. For example, $(-573)(1) = (1)(-573) = -573$.

Additive Inverse Property

For every integer a, there exists an integer $-a$ such that

$$a + (-a) = (-a) + a = 0$$

The integer $-a$ is called the *additive inverse* of a or the *opposite* of a. Thus 6 and -6 are additive inverses, and their sum is 0. The additive inverse of 0 is 0.

Multiplication Property of Zero

If a is an integer, then

$$a(0) = (0)(a) = 0$$

In other words, the product of 0 and any integer is 0. For example, $(-873)(0) = (0)(-873) = 0$.

> **Multiplicative Property of Negative One**
>
> If a is an integer, then
>
> $$(a)(-1) = (-1)(a) = -a$$

The product of any integer and -1 is the opposite of the integer. For example, $(-1)(48) = (48)(-1) = -48$.

> **Distributive Property**
>
> If a, b, and c are integers, then
>
> $$a(b + c) = ab + ac$$

The distributive property involves both addition and multiplication. We say that *multiplication distributes over addition*. For example, $3(4 + 7) = 3(4) + 3(7)$. Because $b - c = b + (-c)$, it follows that *multiplication also distributes over subtraction*. This could be stated as $a(b - c) = ab - ac$. For example, $7(8 - 2) = 7(8) - 7(2)$.

Let's now consider some examples that use these properties to help with certain types of manipulations.

2 Use the Properties of Integers to Simplify Numerical Expressions

EXAMPLE 1   Find the sum $[43 + (-24)] + 24$.

Solution

In such a problem, it is much more advantageous to group -24 and 24. Thus

$[43 + (-24)] + 24 = 43 + [(-24) + 24]$ Associative property for addition
$= 43 + 0$
$= 43$ ∎

EXAMPLE 2   Find the product $[(-17)(25)](4)$.

Solution

In this problem, it is easier to group 25 and 4. Thus

$[(-17)(25)](4) = (-17)[(25)(4)]$ Associative property for multiplication
$= (-17)(100)$
$= -1700$ ∎

EXAMPLE 3   Find the sum $17 + (-24) + (-31) + 19 + (-14) + 29 + 43$.

Solution

Certainly we could add in the order in which the numbers appear. However, because addition is *commutative* and *associative*, we can change the order, and group in any convenient way. For example, we can add all the positive integers, add all the negative integers, and then add these two results. It might be convenient to use the vertical format:

$$\begin{array}{r} 17 \\ 19 \\ 29 \\ \underline{43} \\ 108 \end{array} \qquad \begin{array}{r} -24 \\ -31 \\ -14 \\ \underline{-69} \end{array} \qquad \begin{array}{r} 108 \\ \underline{-69} \\ 39 \end{array}$$

■

For a problem such as Example 3, it might be advisable first to work out the problem by adding in the order in which the numbers appear and then to use the rearranging and regrouping idea as a check. Don't forget the link between addition and subtraction: A problem such as $18 - 43 + 52 - 17 - 23$ can be changed to $18 + (-43) + 52 + (-17) + (-23)$.

EXAMPLE 4

Simplify $(-75)(-4 + 100)$.

Solution

For such a problem, it might be convenient to apply the *distributive property* and then to simplify. Thus

$$\begin{aligned} (-75)(-4 + 100) &= (-75)(-4) + (-75)(100) \\ &= 300 + (-7500) \\ &= -7200 \end{aligned}$$

■

EXAMPLE 5

Simplify $19(-26 + 25)$.

Solution

For this problem, we are better off *not* applying the distributive property but simply adding the numbers inside the parentheses and then finding the indicated product. Thus

$$19(-26 + 25) = 19(-1) = -19$$

■

EXAMPLE 6

Simplify $27(104) + 27(-4)$.

Solution

Keep in mind that the *distributive property* enables us to change from the form $a(b + c)$ to $ab + ac$, or from $ab + ac$ to $a(b + c)$. In this problem we want to use the latter change:

$$\begin{aligned} 27(104) + 27(-4) &= 27[104 + (-4)] \\ &= 27(100) \\ &= 2700 \end{aligned}$$

■

Examples 4, 5, and 6 demonstrate an important issue. Sometimes the form $a(b + c)$ is the most convenient, but at other times the form $ab + ac$ is better. A suggestion regarding this issue—a suggestion that also applies to the use of the other properties—is to *think first* and then decide whether or not you can use the properties to make the manipulations easier.

3 Simplify Algebraic Expressions

Algebraic expressions such as these:

$$3x \quad 5y \quad 7xy \quad -4abc \quad z$$

are called "terms." A **term** is an indicated product that may have any number of factors. We call the variables in a term **literal factors**, and we call the numerical factor the **numerical coefficient**. Thus in the term $7xy$, the x and y are literal factors, and 7 is the numerical coefficient. The numerical coefficient of the term $-4abc$ is -4. Because $z = 1(z)$, the numerical coefficient of the term z is 1. Terms that have the same literal factors are called **like terms** or **similar terms**. Some examples of similar terms are

$3x$ and $9x$ $\qquad$ $14abc$ and $29abc$

$7xy$ and $-15xy$ $\qquad$ $4z, 9z,$ and $-14z$

We can simplify algebraic expressions that contain similar terms by using a form of the distributive property. Consider these examples:

$$3x + 5x = (3 + 5)x$$
$$= 8x$$

$$-9xy + 7xy = (-9 + 7)xy$$
$$= -2xy$$

$$18abc - 27abc = (18 - 27)abc$$
$$= [18 + (-27)]abc$$
$$= -9abc$$

$$4x + x = (4 + 1)x \quad \text{Don't forget that } x = 1(x)$$
$$= 5x$$

More complicated expressions might first require that we rearrange terms by using the commutative property.

EXAMPLE 7

Simplify these algebraic expressions:

(a) $7x + 3y + 9x + 5y$ $\qquad$ **(b)** $9a - 4 - 13a + 6$

Solution

(a) $7x + 3y + 9x + 5y = 7x + 9x + 3y + 5y$ $\qquad$ Commutative property for addition
$$= (7 + 9)x + (3 + 5)y \qquad \text{Distributive property}$$
$$= 16x + 8y$$

(b) $9a - 4 - 13a + 6 = 9a + (-4) + (-13a) + 6$
$$= 9a + (-13a) + (-4) + 6 \qquad \text{Commutative property for addition}$$
$$= [9 + (-13)]a + 2 \qquad \text{Distributive property}$$
$$= -4a + 2 \qquad \blacksquare$$

As you become more adept at handling the various simplifying steps, you may want to do the steps mentally and thereby go directly from the given expression to the simplified form.

$$19x - 14y + 12x + 16y = 31x + 2y$$

$$17ab + 13c - 19ab - 30c = -2ab - 17c$$

$$9x + 5 - 11x + 4 + x - 6 = -x + 3$$

Simplifying some algebraic expressions requires repeated applications of the distributive property, as the next example demonstrates.

EXAMPLE 8

Simplify these algebraic expressions:

(a) $5(x - 2) + 3(x + 4)$ (b) $-7(y + 1) - 4(y - 3)$ (c) $5(x + 2) - (x + 3)$

Solution

(a) $5(x - 2) + 3(x + 4) = 5(x) - 5(2) + 3(x) + 3(4)$ Distributive property
$= 5x - 10 + 3x + 12$
$= 5x + 3x - 10 + 12$ Commutative property
$= 8x + 2$

(b) $-7(y + 1) - 4(y - 3) = -7(y) - 7(1) - 4(y) - 4(-3)$
$= -7y - 7 - 4y + 12$ Be careful with this sign.
$= -7y - 4y - 7 + 12$
$= -11y + 5$

(c) $5(x + 2) - (x + 3) = 5(x + 2) - 1(x + 3)$ Remember that $-a = -1a$.
$= 5(x) + 5(2) - 1(x) - 1(3)$
$= 5x + 10 - x - 3$
$= 5x - x + 10 - 3$
$= 4x + 7$

After you are sure of each step, you can use a more simplified format.

$5(a + 4) - 7(a - 2) = 5a + 20 - 7a + 14$
$= -2a + 34$

$9(z - 7) + 11(z + 6) = 9z - 63 + 11z + 66$
$= 20z + 3$

$-(x - 2) + (x + 6) = -x + 2 + x + 6$
$= 8$

4 Evaluate Algebraic Expressions

Simplifying by combining similar terms aids in the process of evaluating some algebraic expressions. The next examples of this section illustrate this idea.

EXAMPLE 9

Evaluate $8x - 2y + 3x + 5y$ for $x = 3$ and $y = -4$.

Solution

Let's first simplify the given expression.

$8x - 2y + 3x + 5y = 11x + 3y$

Now we can evaluate for $x = 3$ and $y = -4$.

$11x + 3y = 11(3) + 3(-4)$
$= 33 + (-12)$
$= 21$

EXAMPLE 10

Evaluate $2ab + 5c - 6ab + 12c$ for $a = 2, b = -3$, and $c = 7$.

Solution

$$2ab + 5c - 6ab + 12c = -4ab + 17c$$
$$= -4(2)(-3) + 17(7) \quad \text{When } a = 2, b = -3, \text{ and } c = 7$$
$$= 24 + 119$$
$$= 143$$

EXAMPLE 11

Evaluate $8(x - 4) + 7(x + 3)$ for $x = 6$.

Solution

$$8(x - 4) + 7(x + 3) = 8x - 32 + 7x + 21 \quad \text{Distributive property}$$
$$= 15x - 11$$
$$= 15(6) - 11 \quad \text{When } x = 6$$
$$= 79$$

CONCEPT QUIZ 1.3

For Problems 1–10, answer true or false.

1. Addition is a commutative operation.
2. Subtraction is a commutative operation.
3. $[(2)(-3)](7) = (2)[(-3)(7)]$ is an example of the associative property for multiplication.
4. $[(8)(5)](-2) = (-2)[(8)(5)]$ is an example of the associative property for multiplication.
5. Zero is the identity element for addition.
6. The integer $-a$ is the additive inverse of a.
7. The additive inverse of 0 is 0.
8. The numerical coefficient of the term $-8xy$ is 8.
9. The numerical coefficient of the term ab is 1.
10. $6xy$ and $-2xyz$ are similar terms.

Section 1.3 Classroom Problem Set

Objective 2

1. Find the sum $-37 + [37 + (-8)]$.
2. Find the sum $-14 + [-14 + 35]$.
3. Find the product $50[(-2)(-37)]$.
4. Find the product $(-5)[(8)(6)]$.
5. Find the sum $-18 + 24 + (-8) + (-12) + 32$.
6. Find the sum $-50 + 4 + (-7) + (-8) + 16 + (-28)$.
7. Simplify $(-15)(-2 + 10)$.
8. Simplify $(-125)(-4 + 10)$.
9. Simplify $-19(-34 + 34)$.
10. Simplify $-43(-3 + 2)$.

11. Simplify $27(12) + 27(-2)$.
12. Simplify $67(-5) + 67(-95)$.

Objective 3

13. Simplify these algebraic expressions:
    (a) $6a - 3a + 7 - 5a + 1$
    (b) $2y - 3x + 5x - 6y$
14. Simplify these algebraic expressions:
    (a) $10ab - 6c - 11ab + c$
    (b) $2y - 3x + 5x - 6y$
15. Simplify these algebraic expressions:
    (a) $2(a - 2) - 5(a + 6)$
    (b) $7(4 - 2b) - 4(5 - b)$

16. Simplify these algebraic expressions:
    (a) $-3(cd - 8) - (1 - 4cd)$
    (b) $-(x - y) - 2(y - x)$

Objective 4

17. Evaluate $5a + 3b - 2a - 7b$ for $a = 6$ and $b = -5$.
18. Evaluate $-4c + 5d - c - d$ for $c = -1$ and $d = -2$.
19. Evaluate $2xy + 3x - 2y - 4xy$ for $x = -3$ and $y = -1$.
20. Evaluate $2ac - b - 9ac + 5b$ for $a = -1, b = -2,$ and $c = -3$.
21. Evaluate $6(a - 3) - 2(a + 4)$ for $a = 7$.
22. Evaluate $-(c + 7) - 4(c - 3)$ for $c = -3$.

THOUGHTS INTO WORDS

1. State in your own words the associative property for addition of integers.
2. State in your own words the distributive property for multiplication over addition.

Answers to the Concept Quiz

1. True   2. False   3. True   4. False   5. True   6. True   7. True   8. False   9. True   10. False

1.4 Prime and Composite Numbers

OBJECTIVES

1. Understand Divisibility
2. Identify Prime or Composite Numbers
3. Factor a Whole Number into a Product of Prime Numbers
4. Find the Greatest Common Factor
5. Find the Least Common Multiple

1 Understand Divisibility

Occasionally terms in mathematics have a special meaning in the discussion of a particular topic. Such is the case with the term "divides" as it is used in this section.

We say that 6 *divides* 18 because 6 times the whole number 3 produces 18, but 6 *does not divide* 19 because there is no whole number such that 6 times the number produces 19. Likewise 5 *divides* 35 because 5 times the whole number 7 produces 35, but 5 *does not divide* 42 because there is no whole number such that 5 times the number produces 42. We can use this general definition:

Definition 1.1

Given that a and b are whole numbers, with a not equal to zero, a *divides* b if and only if there exists a whole number k such that $a \cdot k = b$.

Remark: Note the use of the variables a, b, and k in the statement of a general definition. Also note that the definition merely generalizes the concept of *divides*, which we introduced in the paragraph preceding the definition.

The following statements further clarify Definition 1.1. Pay special attention to the italicized words because they indicate some of the terminology used for this topic.

1. 8 *divides* 56, because $8 \cdot 7 = 56$.
2. 7 *does not divide* 38, because there is no whole number k such that $7 \cdot k = 38$.
3. 3 is a *factor* of 27, because $3 \cdot 9 = 27$.
4. 4 is *not a factor* of 38, because there is no whole number k such that $4 \cdot k = 38$.
5. 35 is a *multiple* of 5, because $5 \cdot 7 = 35$.
6. 29 is *not a multiple* of 7, because there is no whole number k such that $7 \cdot k = 29$.

The *factor* terminology is used extensively. We say that 7 and 8 are factors of 56 because $7 \cdot 8 = 56$; 4 and 14 are also factors of 56 because $4 \cdot 14 = 56$. The **factors** of a number are also the divisors of the number.

2 Identify Prime or Composite Numbers

Now consider two special kinds of whole numbers called "prime numbers" and "composite numbers" according to the following definition.

> **Definition 1.2   Prime Numbers vs. Composite Numbers**
>
> A **prime number** is a whole number, greater than 1, that has no factors (divisors) other than itself and 1. Whole numbers, greater than 1, that are not prime numbers are called **composite numbers**.

The prime numbers less than 50 are 2, 3, 5, 7, 11, 13, 17, 19, 23, 29, 31, 37, 41, 43, and 47. Note that each number has no factors other than itself and 1. An interesting point is that the set of prime numbers is an infinite set; that is, the prime numbers go on forever, and there is no *largest* prime number.

3 Factor a Whole Number into a Product of Prime Numbers

We can express every composite number as the indicated product of prime numbers. Let's consider the first several composite numbers, 4, 6, 8, 9, 10, and 12, and write them as an indicated product of prime numbers. Note the use of exponents to represent repeated multiplications.

$$4 = 2 \cdot 2 = 2^2 \qquad 6 = 2 \cdot 3 \qquad 8 = 2 \cdot 2 \cdot 2 = 2^3 \qquad 9 = 3 \cdot 3 = 3^2$$
$$10 = 2 \cdot 5 \qquad 12 = 2 \cdot 2 \cdot 3 = 2^2 \cdot 3$$

The indicated product of prime numbers is sometimes called the **prime factored form** of the number.

We can use various procedures to find the prime factors of a given composite number. For our purposes, the simplest technique is to factor the composite number into any two easily recognized factors and then to continue to factor each of these until we obtain only prime factors. The order in which we write the prime factors is not important, because the commutative property of multiplication can be used to reorder the factors. The order in which we write the prime factored form is smallest to largest. This will aid in comparing prime factored forms when we are looking for greatest common factors and least common multiples. Consider this example.

EXAMPLE 1

Factor these composite numbers into an indicated product of prime numbers:

(a) 18   (b) 27   (c) 24   (d) 150

Solution

(a) $18 = 2 \cdot 9 = 2 \cdot 3 \cdot 3 = 2 \cdot 3^2$
(b) $27 = 3 \cdot 9 = 3 \cdot 3 \cdot 3 = 3^3$
(c) $24 = 4 \cdot 6 = 2 \cdot 2 \cdot 2 \cdot 3 = 2^3 \cdot 3$
(d) $150 = 10 \cdot 15 = 2 \cdot 5 \cdot 3 \cdot 5 = 2 \cdot 3 \cdot 5^2$

Remark: It does not matter which two factors we choose first. For instance, we could have factored 18 by starting with the factors 3 and 6. Doing so would give us $18 = 3 \cdot 6 = 3 \cdot 2 \cdot 3 = 2 \cdot 3^2$. Either way, 18 contains two prime factors of 3 and one prime factor of 2. ∎

Familiarity with a few basic divisibility rules will be helpful for determining the prime factors of some numbers. For example, if you can quickly recognize that 51 is divisible by 3, then you can divide 51 by 3 to find another factor of 17. Because 3 and 17 are both prime numbers, we have $51 = 3 \cdot 17$. The divisibility rules for 2, 3, 5, and 9 are shown here.

Rule for 2

A whole number is divisible by 2 if and only if the units digit of its base-10 numeral is divisible by 2. (In other words, the units digit must be 0, 2, 4, 6, or 8.)

EXAMPLES 68 is divisible by 2, because 8 is divisible by 2.
57 is not divisible by 2, because 7 is not divisible by 2.

Rule for 3

A whole number is divisible by 3 if and only if the sum of the digits of its base-10 numeral is divisible by 3.

EXAMPLES 51 is divisible by 3, because $5 + 1 = 6$ and 6 is divisible by 3.
144 is divisible by 3, because $1 + 4 + 4 = 9$ and 9 is divisible by 3.
133 is not divisible by 3, because $1 + 3 + 3 = 7$ and 7 is not divisible by 3.

Rule for 5

A whole number is divisible by 5 if and only if the units digit of its base-10 numeral is divisible by 5. (In other words, the units digit must be 0 or 5.)

EXAMPLES 115 is divisible by 5, because 5 is divisible by 5.
172 is not divisible by 5, because 2 is not divisible by 5.

Rule for 9

A whole number is divisible by 9 if and only if the sum of the digits of its base-10 numeral is divisible by 9.

EXAMPLES 765 is divisible by 9, because $7 + 6 + 5 = 18$ and 18 is divisible by 9.
147 is not divisible by 9, because $1 + 4 + 7 = 12$ and 12 is not divisible by 9.

4 Find the Greatest Common Factor

We can use the prime factorization form of two composite numbers to conveniently find their **greatest common factor** (GCF). Consider this example:

$42 = 2 \cdot 3 \cdot 7$
$70 = 2 \cdot 5 \cdot 7$

Note that 2 is a factor of both, as is 7. Therefore, 14 (the product of 2 and 7) is the greatest common factor of 42 and 70. In other words, 14 is the largest whole number that divides both 42 and 70. The following examples show the process of finding the greatest common factor of two or more numbers.

EXAMPLE 2

Find the greatest common factor of 42 and 105.

Solution

$$42 = 6 \cdot 7 = 2 \cdot 3 \cdot 7$$
$$105 = 3 \cdot 35 = 3 \cdot 5 \cdot 7$$

Because 3 and 7 are common to both, the greatest common factor of 42 and 105 is $3 \cdot 7 = 21$. ∎

When exponents are used to write the prime factored form, the technique for finding the greatest common factor is slightly different. We actually make use of the exponents to find the number of times that factor is common. Consider this example of finding the greatest common factor of 36 and 120. First we write each number in prime factored form.

$$36 = 4 \cdot 9 = 2 \cdot 2 \cdot 3 \cdot 3 = 2^2 \cdot 3^2$$
$$120 = 4 \cdot 30 = 2 \cdot 2 \cdot 2 \cdot 15 = 2 \cdot 2 \cdot 2 \cdot 3 \cdot 5 = 2^3 \cdot 3 \cdot 5$$

We determine that 2 and 3 are common factors. To find the greatest common factor, take the common factors to their *lowest* exponent in the factorizations and find their product. Therefore, the product of 2^2 times 3 gives us the greatest common factor, which is 12. The following example should clarify the process of finding the greatest common factor when exponents are present in the prime factored form.

EXAMPLE 3

Find the greatest common factor of 80 and 120.

Solution

$$80 = 8 \cdot 10 = 2 \cdot 4 \cdot 2 \cdot 5 = 2 \cdot 2 \cdot 2 \cdot 2 \cdot 5 = 2^4 \cdot 5$$
$$120 = 10 \cdot 12 = 2 \cdot 5 \cdot 3 \cdot 4 = 2 \cdot 5 \cdot 3 \cdot 2 \cdot 2 = 2^3 \cdot 3 \cdot 5$$

The common factors are 2 and 5, so take 2 to its *lowest* exponent, which is 3, times 5. Therefore, the product of 2^3 times 5 gives us the greatest common factor, which is 40. ∎

EXAMPLE 4

Find the greatest common factor of 24 and 35.

Solution

$$24 = 2 \cdot 2 \cdot 2 \cdot 3 = 2^3 \cdot 3$$
$$35 = 5 \cdot 7$$

Because there are no common prime factors, the greatest common factor is 1. ∎

The concept of "greatest common factor" can be extended to more than two numbers, as the next example demonstrates.

EXAMPLE 5

Find the greatest common factor of 24, 28, and 120.

Solution

$$24 = 4 \cdot 6 = 2 \cdot 2 \cdot 2 \cdot 3 = 2^3 \cdot 3$$
$$28 = 4 \cdot 7 = 2 \cdot 2 \cdot 7 = 2^2 \cdot 7$$
$$120 = 10 \cdot 12 = 2 \cdot 5 \cdot 2 \cdot 6 = 2 \cdot 5 \cdot 2 \cdot 2 \cdot 3 = 2^3 \cdot 3 \cdot 5$$

The only factor that is common to all three numbers is 2. To find the greatest common factor, take 2 to its *lowest* exponent. Therefore $2^2 = 4$ is the greatest common factor. ∎

5 Find the Least Common Multiple

We stated that 35 is a *multiple* of 5 because $5 \cdot 7 = 35$. We can produce a set of all whole number multiples of 5 by multiplying 5 times each successive whole number. The following list shows the set of multiples of 5.

Whole number multiples of 5:   0, 5, 10, 15, 20, 25, 30, 35, 40, 45, ...

In a like manner, we can produce the set of all whole number multiples of 4. The following list shows the set of multiples of 4.

Whole number multiples of 4:   0, 4, 8, 12, 16, 20, 24, 28, 32, 36, 40, 44, 48, 52, ...

Inspection of the two lists shows us that the numbers 0, 20, and 40 are common to both lists; in this case, these numbers would be referred to as common multiples. The **least common multiple** (LCM) is the smallest common *nonzero* multiple of two or more numbers. Therefore the least common multiple of 5 and 4 is 20. Stated another way, 20 is the smallest *nonzero* whole number that is divisible by both 5 and 4. Let's find the least common multiple of 6 and 8. Start by listing the whole number multiples of 6 and 8.

Whole number multiples of 6:   0, 6, 12, 18, 24, 30, 36, 42, 48, ...
Whole number multiples of 8:   0, 8, 16, 24, 32, 40, 48, ...

By inspection we can determine that the least common multiple is 24.

Finding the least common multiple by inspection can be cumbersome. Another method to determine the least common multiple is to use the prime factored forms of numbers. The following steps show a systematic technique for finding the least common multiple. In the solutions to the following examples, these steps will be demonstrated.

Step 1   Express each number in prime factored form.

Step 2   The least common multiple contains each different prime factor as many times as the maximum number of times it appears in any one factorization. If the prime factored form is written using exponents, then we want each different prime factor raised to its *greatest* exponent.

EXAMPLE 6

Find the least common multiple of 24 and 36.

Solution

Let's first express each number in its prime factored form.

$24 = 4 \cdot 6 = 2 \cdot 2 \cdot 2 \cdot 3 = 2^3 \cdot 3$
$36 = 6 \cdot 6 = 2 \cdot 3 \cdot 2 \cdot 3 = 2^2 \cdot 3^2$

The different prime factors are 2 and 3. So for the least common multiple we need the product of 2 (with its greatest exponent) and 3 (with its greatest exponent). The least common multiple is $2^3 \cdot 3^2 = 8 \cdot 9 = 72$. In other words, 72 is the smallest nonzero number that is divisible by both 24 and 36. ■

EXAMPLE 7

Find the least common multiple of 48 and 84.

Solution

Let's first express each number in its prime factored form.

$48 = 4 \cdot 12 = 2 \cdot 2 \cdot 2 \cdot 6 = 2 \cdot 2 \cdot 2 \cdot 2 \cdot 3 = 2^4 \cdot 3$
$84 = 4 \cdot 21 = 2 \cdot 2 \cdot 3 \cdot 7 = 2^2 \cdot 3 \cdot 7$

The different prime factors are 2, 3, and 7. So for the least common multiple we need the product of 2 (with its greatest exponent), 3, and 7. The least common multiple is $2^4 \cdot 3 \cdot 7 = 16 \cdot 3 \cdot 7 = 336$. In other words, 336 is the smallest nonzero number that is divisible by both 48 and 84. ∎

EXAMPLE 8 Find the least common multiple of 12, 18, and 28.

Solution

Let's first express each number in its prime factored form.

$12 = 3 \cdot 4 = 3 \cdot 2 \cdot 2 = 2^2 \cdot 3$
$18 = 2 \cdot 9 = 2 \cdot 3 \cdot 3 = 2 \cdot 3^2$
$28 = 4 \cdot 7 = 2 \cdot 2 \cdot 7 = 2^2 \cdot 7$

The different prime factors are 2, 3, and 7. So for the least common multiple we need the product of 2 (with its greatest exponent), 3 (with its greatest exponent), and 7. The least common multiple is $2^2 \cdot 3^2 \cdot 7 = 4 \cdot 9 \cdot 7 = 252$. In other words, 252 is the smallest nonzero number that is divisible by 12, 18, and 28. ∎

EXAMPLE 9 Find the least common multiple of 8 and 9.

Solution

Let's first express each number in its prime factored form.

$8 = 2 \cdot 4 = 2 \cdot 2 \cdot 2 = 2^3$
$9 = 3 \cdot 3 = 3^2$

The different prime factors are 2 and 3. So for the least common multiple we need the product of 2 (with its greatest exponent) and 3 (with its greatest exponent). The least common multiple is $2^3 \cdot 3^2 = 8 \cdot 9 = 72$. In other words, 72 is the smallest nonzero number that is divisible by both 8 and 9. ∎

CONCEPT QUIZ 1.4 For Problems 1–5, answer true or false.

1. Every even whole number greater than 2 is a composite number.
2. Two is the only even prime number.
3. One is a prime number.
4. The prime factored form of 24 is $2 \cdot 2 \cdot 6$.
5. Some whole numbers are both prime and composite numbers.

Section 1.4 Classroom Problem Set

Objective 3

1. Factor these composite numbers into an indicated product of prime numbers:
   (a) 36   (b) 42

2. Factor these composite numbers into an indicated product of prime numbers:
   (a) 44   (b) 72

Objective 4

3. Find the greatest common factor of 42 and 60.
4. Find the greatest common factor of 42 and 70.
5. Find the greatest common factor of 63 and 54.
6. Find the greatest common factor of 24 and 84.
7. Find the greatest common factor of 49 and 80.
8. Find the greatest common factor of 35 and 66.

9. Find the greatest common factor of 36, 72, and 90.
10. Find the greatest common factor of 84, 90, and 120.

Objective 5

11. Find the least common multiple of 12 and 30.
12. Find the least common multiple of 48 and 54.
13. Find the least common multiple of 20 and 75.
14. Find the least common multiple of 75 and 105.
15. Find the least common multiple of 12, 18, and 30.
16. Find the least common multiple of 15, 25, and 40.
17. Find the least common multiple of 15 and 8.
18. Find the least common multiple of 12 and 25.

THOUGHTS INTO WORDS

1. How would you explain the concepts "greatest common factor" and "least common multiple" to a friend who missed class during that discussion?

2. Is it always true that the greatest common factor of two numbers is less than the least common multiple of those same two numbers? Explain your answer.

Answers to the Concept Quiz
1. True   2. True   3. False   4. False   5. False

1.5 Rational Numbers

OBJECTIVES

1. Reduce Fractions
2. Multiply Rational Numbers
3. Divide Rational Numbers
4. Add and Subtract Rational Numbers
5. Simplify Numerical Expressions
6. Solve Application Problems Involving Rational Numbers

1 Reduce Fractions

Fractions and rational numbers are encountered throughout the study of algebra. The form $\frac{a}{b}$ is called a fraction or sometimes a common fraction. The numbers we often call fractions are rational numbers. The precise definition of rational numbers follows.

Definition 1.3   Rational Numbers

Any number that can be written in the form $\frac{a}{b}$, where a and b are integers, and b is not zero, is a rational number.

Here are some examples of rational numbers:

$$\frac{1}{2} \quad \frac{7}{9} \quad \frac{15}{7} \quad \frac{-3}{4} \quad \frac{5}{-7} \quad \frac{-11}{-13}$$

All integers are rational numbers, because every integer can be expressed as the indicated quotient of two integers. For example,

$$6 = \frac{6}{1} = \frac{12}{2} = \frac{18}{3}, \text{etc.}$$

$$27 = \frac{27}{1} = \frac{54}{2} = \frac{81}{3}, \text{etc.}$$

$$0 = \frac{0}{1} = \frac{0}{2} = \frac{0}{3}, \text{etc.}$$

Our work in Section 1.2 with the division of negative integers helps with the next three examples:

$$-4 = \frac{-4}{1} = \frac{-8}{2} = \frac{-12}{3}, \text{etc.}$$

$$-6 = \frac{6}{-1} = \frac{12}{-2} = \frac{18}{-3}, \text{etc.}$$

$$10 = \frac{10}{1} = \frac{-10}{-1} = \frac{-20}{-2}, \text{etc.}$$

Observe the following general property:

Property 1.2

$$\frac{-a}{b} = \frac{a}{-b} = -\frac{a}{b} \quad \text{and} \quad \frac{-a}{-b} = \frac{a}{b}$$

Therefore, we can write the rational number $\frac{-2}{3}$, for example, as $\frac{2}{-3}$ or $-\frac{2}{3}$. (However, we seldom express rational numbers with negative denominators.)

We define multiplication of rational numbers in common fractional form as follows:

Definition 1.4 Multiplication of Rational Numbers

If a, b, c, and d are integers, and b and d do not equal to zero, then

$$\frac{a}{b} \cdot \frac{c}{d} = \frac{a \cdot c}{b \cdot d}$$

To multiply rational numbers in common fractional form, we simply multiply numerators and multiply denominators. Because the numerators and denominators are integers, our previous agreements pertaining to multiplication of integers hold for the rationals. That is, *the product of two positive rational numbers or of two negative rational numbers is a positive rational number. The product of a positive rational number and a negative rational number (in either order) is a negative rational number.* Furthermore, we see from the definition that the commutative and associative properties hold for the multiplication of rational numbers. We are free to rearrange and regroup factors as we do with integers.

The following examples illustrate the definition for multiplying rational numbers:

$$\frac{1}{3} \cdot \frac{2}{5} = \frac{1 \cdot 2}{3 \cdot 5} = \frac{2}{15}$$

$$\frac{3}{4} \cdot \frac{5}{7} = \frac{3 \cdot 5}{4 \cdot 7} = \frac{15}{28}$$

1.5 Rational Numbers

$$\frac{-2}{3} \cdot \frac{7}{9} = \frac{-2 \cdot 7}{3 \cdot 9} = \frac{-14}{27} \quad \text{or} \quad -\frac{14}{27}$$

$$\frac{1}{5} \cdot \frac{9}{-11} = \frac{1 \cdot 9}{5(-11)} = \frac{9}{-55} \quad \text{or} \quad -\frac{9}{55}$$

$$-\frac{3}{4} \cdot \frac{7}{13} = \frac{-3}{4} \cdot \frac{7}{13} = \frac{-3 \cdot 7}{4 \cdot 13} = \frac{-21}{52} \quad \text{or} \quad -\frac{21}{52}$$

$$\frac{3}{5} \cdot \frac{5}{3} = \frac{3 \cdot 5}{5 \cdot 3} = \frac{15}{15} = 1$$

The last example is a very special case. *If the product of two numbers is 1, the numbers are said to be **reciprocals** of each other.*

Using Definition 1.4 and applying the multiplication property of one, the fraction $\frac{a \cdot k}{b \cdot k}$, where b and k are nonzero integers, simplifies as shown:

$$\frac{a \cdot k}{b \cdot k} = \frac{a}{b} \cdot \frac{k}{k} = \frac{a}{b} \cdot 1 = \frac{a}{b}$$

This result is stated as Property 1.3: Fundamental Principle of Fractions.

Property 1.3   Fundamental Principle of Fractions

If b and k are nonzero integers, and a is any integer, then

$$\frac{a \cdot k}{b \cdot k} = \frac{a}{b}$$

When we work with fractions, we often use Property 1.3, which provides the basis for creating equivalent fractions. In the following examples, we will use this property to reduce fractions to lowest terms or express fractions in simplest or reduced form.

EXAMPLE 1   Reduce $\frac{12}{18}$ to lowest terms.

Solution

$$\frac{12}{18} = \frac{2 \cdot 6}{3 \cdot 6} = \frac{2}{3} \cdot \frac{6}{6} = \frac{2}{3} \cdot 1 = \frac{2}{3}$$

EXAMPLE 2   Change $\frac{14}{35}$ to simplest form.

Solution

$$\frac{14}{35} = \frac{2 \cdot 7}{5 \cdot 7} = \frac{2}{5} \qquad \text{Divide a common factor of 7 out of both the numerator and denominator}$$

EXAMPLE 3   Reduce $-\frac{72}{90}$.

Solution

$$-\frac{72}{90} = -\frac{2 \cdot 2 \cdot 2 \cdot 3 \cdot 3}{2 \cdot 3 \cdot 3 \cdot 5} = -\frac{4}{5} \qquad \text{Use the prime factored forms of the numerator and denominator to help recognize common factors}$$

36 Chapter 1 Basic Concepts of Arithmetic and Algebra

2 Multiply Rational Numbers

We are now ready to use Definition 1.4 to multiply fractions with the understanding that the final answer should be expressed in reduced form. Study the following examples carefully, because different methods are used to handle the problems.

EXAMPLE 4 Multiply $\dfrac{7}{9} \cdot \dfrac{5}{14}$.

Solution

$$\dfrac{7}{9} \cdot \dfrac{5}{14} = \dfrac{7 \cdot 5}{9 \cdot 14} = \dfrac{\cancel{7} \cdot 5}{3 \cdot 3 \cdot 2 \cdot \cancel{7}} = \dfrac{5}{18}$$

EXAMPLE 5 Find the product of $\dfrac{8}{9}$ and $\dfrac{18}{24}$.

Solution

$$\dfrac{\overset{1}{\cancel{8}}}{\underset{1}{\cancel{9}}} \cdot \dfrac{\overset{2}{\cancel{18}}}{\underset{3}{\cancel{24}}} = \dfrac{2}{3} \qquad \text{Divide a common factor of 8 out of 8 and 24 and a common factor of 9 out of 9 and 18}$$

EXAMPLE 6 Multiply $\left(-\dfrac{6}{8}\right)\left(\dfrac{14}{32}\right)$.

Solution

$$\left(-\dfrac{6}{8}\right)\left(\dfrac{14}{32}\right) = -\dfrac{\overset{3}{\cancel{6}} \cdot \overset{7}{\cancel{14}}}{\underset{4}{\cancel{8}} \cdot \underset{16}{\cancel{32}}} = -\dfrac{21}{64} \qquad \text{Divide a common factor of 2 out of 6 and 8 and a common factor of 2 out of 14 and 32}$$

EXAMPLE 7 Multiply $\left(-\dfrac{9}{4}\right)\left(-\dfrac{14}{15}\right)$.

Solution

$$\left(-\dfrac{9}{4}\right)\left(-\dfrac{14}{15}\right) = \dfrac{3 \cdot 3 \cdot 2 \cdot 7}{2 \cdot 2 \cdot 3 \cdot 5} = \dfrac{21}{10} \qquad \text{Immediately we recognize that a negative times a negative is positive}$$

---

3 Divide Rational Numbers

The following example motivates a definition for division of rational numbers in fractional form.

$$\dfrac{\frac{3}{4}}{\frac{2}{3}} = \left(\dfrac{\frac{3}{4}}{\frac{2}{3}}\right)\left(\dfrac{\frac{3}{2}}{\frac{3}{2}}\right) = \dfrac{\left(\frac{3}{4}\right)\left(\frac{3}{2}\right)}{\left(\frac{2}{3}\right)\left(\frac{3}{2}\right)} = \dfrac{\left(\frac{3}{4}\right)\left(\frac{3}{2}\right)}{1} = \left(\dfrac{3}{4}\right)\left(\dfrac{3}{2}\right) = \dfrac{9}{8}$$

↑

Notice that this is a form of 1 and $\dfrac{3}{2}$ is the reciprocal of $\dfrac{2}{3}$

In other words, $\frac{3}{4}$ divided by $\frac{2}{3}$ is equivalent to $\frac{3}{4}$ times $\frac{3}{2}$. The following definition for division should seem reasonable:

> **Definition 1.5 Division of Rational Numbers**
>
> If b, c, and d are nonzero integers, and a is any integer, then
>
> $$\frac{a}{b} \div \frac{c}{d} = \frac{a}{b} \cdot \frac{d}{c}$$

Note that to divide $\frac{a}{b}$ by $\frac{c}{d}$, we multiply $\frac{a}{b}$ times the reciprocal of $\frac{c}{d}$, which is $\frac{d}{c}$. The following example demonstrates the important steps of a division problem.

EXAMPLE 8   Find the quotient.

(a) $\dfrac{2}{3} \div \dfrac{1}{2}$ (b) $\dfrac{5}{6} \div \dfrac{3}{4}$ (c) $-\dfrac{9}{12} \div \dfrac{3}{6}$ (d) $\dfrac{6}{7} \div 2$

Solution

(a) $\dfrac{2}{3} \div \dfrac{1}{2} = \dfrac{2}{3} \cdot \dfrac{2}{1} = \dfrac{4}{3}$

(b) $\dfrac{5}{6} \div \dfrac{3}{4} = \dfrac{5}{6} \cdot \dfrac{4}{3} = \dfrac{5 \cdot 4}{6 \cdot 3} = \dfrac{5 \cdot 2 \cdot 2}{2 \cdot 3 \cdot 3} = \dfrac{10}{9}$

(c) $-\dfrac{9}{12} \div \dfrac{3}{6} = -\dfrac{\cancel{9}^{3}}{\cancel{12}_{2}} \cdot \dfrac{\cancel{6}^{1}}{\cancel{3}_{1}} = -\dfrac{3}{2}$

(d) $\dfrac{6}{7} \div 2 = \dfrac{6}{7} \cdot \dfrac{1}{2} = \dfrac{\cancel{6}^{3}}{7} \cdot \dfrac{1}{\cancel{2}_{1}} = \dfrac{3}{7}$ ∎

4 Add and Subtract Rational Numbers

Suppose that it is one-fifth of a mile between your dorm and the student center and two-fifths of a mile between the student center and the library, along a straight line as indicated in Figure 1.5. The total distance between your dorm and the library is three-fifths of a mile, and we write $\dfrac{1}{5} + \dfrac{2}{5} = \dfrac{3}{5}$.

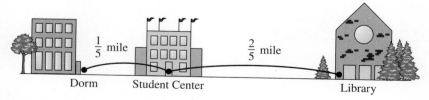

Figure 1.5

A pizza is cut into seven equal pieces and you eat two of the pieces (see Figure 1.6). How much of the pizza remains? We represent the whole pizza by $\dfrac{7}{7}$ and then conclude that $\dfrac{7}{7} - \dfrac{2}{7} = \dfrac{5}{7}$ of the pizza remains.

Figure 1.6

These examples motivate the following definition for addition and subtraction of rational numbers in $\dfrac{a}{b}$ form.

> **Definition 1.6 Addition and Subtraction of Rational Numbers**
>
> If a, b, and c are integers, and b is not zero, then
>
> $$\dfrac{a}{b} + \dfrac{c}{b} = \dfrac{a+c}{b} \quad \text{Addition}$$
>
> $$\dfrac{a}{b} - \dfrac{c}{b} = \dfrac{a-c}{b} \quad \text{Subtraction}$$

We say that rational numbers with common denominators can be added or subtracted by adding or subtracting the numerators and placing the results over the common denominator. Consider these examples:

$$\dfrac{3}{7} + \dfrac{2}{7} = \dfrac{3+2}{7} = \dfrac{5}{7}$$

$$\dfrac{7}{8} - \dfrac{2}{8} = \dfrac{7-2}{8} = \dfrac{5}{8}$$

$$\dfrac{2}{6} + \dfrac{1}{6} = \dfrac{2+1}{6} = \dfrac{3}{6} = \dfrac{1}{2} \quad \text{We agree to reduce the final answer}$$

$$\dfrac{3}{11} - \dfrac{5}{11} = \dfrac{3-5}{11} = \dfrac{-2}{11} = -\dfrac{2}{11}$$

How do we add or subtract if the fractions do not have a common denominator? We use the fundamental principle of fractions, $\dfrac{a}{b} = \dfrac{a \cdot k}{b \cdot k}$, and obtain equivalent fractions that have a common denominator. **Equivalent fractions** are fractions that name the same number. Consider the following example, which shows the details.

EXAMPLE 9 Add $\dfrac{1}{2} + \dfrac{1}{3}$.

Solution

$$\dfrac{1}{2} = \dfrac{1 \cdot 3}{2 \cdot 3} = \dfrac{3}{6} \quad \dfrac{1}{2} \text{ and } \dfrac{3}{6} \text{ are equivalent fractions that name the same number}$$

$$\dfrac{1}{3} = \dfrac{1 \cdot 2}{3 \cdot 2} = \dfrac{2}{6} \quad \dfrac{1}{3} \text{ and } \dfrac{2}{6} \text{ are equivalent fractions that name the same number}$$

$$\dfrac{1}{2} + \dfrac{1}{3} = \dfrac{3}{6} + \dfrac{2}{6} = \dfrac{3+2}{6} = \dfrac{5}{6}$$

∎

Note that in Example 9 we chose 6 as our common denominator, and 6 is the least common multiple of the original denominators 2 and 3. (Recall that the least common multiple is the smallest nonzero whole number divisible by the given numbers.) In general, we use the least common multiple of the denominators of the fractions to be added or subtracted as a **least common denominator** (LCD).

Recall from Section 1.4 that the least common multiple is found either by inspection or by using the prime factorization forms of the numbers. Let's consider some examples involving these procedures.

EXAMPLE 10 Add $\dfrac{1}{4} + \dfrac{2}{5}$.

Solution

By inspection we see that the LCD is 20. Thus both fractions can be changed to equivalent fractions that have a denominator of 20.

$$\frac{1}{4} + \frac{2}{5} = \frac{1 \cdot 5}{4 \cdot 5} + \frac{2 \cdot 4}{5 \cdot 4} = \frac{5}{20} + \frac{8}{20} = \frac{13}{20}$$

Use of fundamental principle of fractions ■

If the LCD is not obvious by inspection, then we can use the technique from Section 1.4 to find the least common multiple. We proceed as follows:

Step 1 Express each denominator as a product of prime factors.

Step 2 The LCD contains each different prime factor as many times as the *most* times it appears in any one of the factorizations from step 1.

EXAMPLE 11 Add $\dfrac{5}{18} + \dfrac{7}{24}$.

Solution

If we cannot find the LCD by inspection, then we can use the prime factorization forms.

$$\left. \begin{array}{l} 18 = 2 \cdot 3 \cdot 3 \\ 24 = 2 \cdot 2 \cdot 2 \cdot 3 \end{array} \right\} \rightarrow \text{LCD} = 2 \cdot 2 \cdot 2 \cdot 3 \cdot 3 = 72$$

$$\frac{5}{18} + \frac{7}{24} = \frac{5 \cdot 4}{18 \cdot 4} + \frac{7 \cdot 3}{24 \cdot 3} = \frac{20}{72} + \frac{21}{72} = \frac{41}{72}$$ ■

EXAMPLE 12 Subtract $\dfrac{3}{14} - \dfrac{8}{35}$.

Solution

$$\left. \begin{array}{l} 14 = 2 \cdot 7 \\ 35 = 5 \cdot 7 \end{array} \right\} \rightarrow \text{LCD} = 2 \cdot 5 \cdot 7 = 70$$

$$\frac{3}{14} - \frac{8}{35} = \frac{3 \cdot 5}{14 \cdot 5} - \frac{8 \cdot 2}{35 \cdot 2} = \frac{15}{70} - \frac{16}{70} = \frac{-1}{70} \quad \text{or} \quad -\frac{1}{70}$$ ■

EXAMPLE 13

Add $\dfrac{-5}{8} + \dfrac{3}{14}$.

Solution

$$\left.\begin{array}{l} 8 = 2 \cdot 2 \cdot 2 \\ 14 = 2 \cdot 7 \end{array}\right\} \rightarrow \quad \text{LCD} = 2 \cdot 2 \cdot 2 \cdot 7 = 56$$

$$\dfrac{-5}{8} + \dfrac{3}{14} = \dfrac{-5 \cdot 7}{8 \cdot 7} + \dfrac{3 \cdot 4}{14 \cdot 4} = \dfrac{-35}{56} + \dfrac{12}{56} = \dfrac{-23}{56} \quad \text{or} \quad -\dfrac{23}{56} \quad \blacksquare$$

EXAMPLE 14

Add $-3 + \dfrac{2}{5}$.

Solution

$$-3 + \dfrac{2}{5} = \dfrac{-3 \cdot 5}{1 \cdot 5} + \dfrac{2}{5} = \dfrac{-15}{5} + \dfrac{2}{5} = \dfrac{-15 + 2}{5} = \dfrac{-13}{5} \quad \text{or} \quad -\dfrac{13}{5} \quad \blacksquare$$

5 Simplify Numerical Expressions

Let's now consider simplifying numerical expressions that contain rational numbers. As with integers, we first do the multiplications and divisions and then perform the additions and subtractions. In these next examples only the major steps are shown, so be sure that you can fill in all of the details.

EXAMPLE 15

Simplify $\dfrac{3}{4} + \dfrac{2}{3} \cdot \dfrac{3}{5} - \dfrac{1}{2} \cdot \dfrac{1}{5}$.

Solution

$$\dfrac{3}{4} + \dfrac{2}{3} \cdot \dfrac{3}{5} - \dfrac{1}{2} \cdot \dfrac{1}{5} = \dfrac{3}{4} + \dfrac{2}{5} - \dfrac{1}{10} \quad \text{Perform the multiplications}$$

$$= \dfrac{15}{20} + \dfrac{8}{20} - \dfrac{2}{20} = \dfrac{15 + 8 - 2}{20} = \dfrac{21}{20} \quad \begin{array}{l}\text{Change to} \\ \text{equivalent} \\ \text{fractions and} \\ \text{combine the} \\ \text{numerators}\end{array} \quad \blacksquare$$

EXAMPLE 16

Simplify $\dfrac{3}{5} \div \dfrac{8}{5} + \left(-\dfrac{1}{2}\right)\left(\dfrac{1}{3}\right) + \dfrac{5}{12}$.

Solution

$$\dfrac{3}{5} \div \dfrac{8}{5} + \left(-\dfrac{1}{2}\right)\left(\dfrac{1}{3}\right) + \dfrac{5}{12} = \dfrac{3}{5} \cdot \dfrac{5}{8} + \left(-\dfrac{1}{2}\right)\left(\dfrac{1}{3}\right) + \dfrac{5}{12} \quad \begin{array}{l}\text{Change division} \\ \text{to multiply by} \\ \text{the reciprocal}\end{array}$$

$$= \dfrac{3}{8} + \dfrac{-1}{6} + \dfrac{5}{12}$$

$$= \dfrac{9}{24} + \dfrac{-4}{24} + \dfrac{10}{24} \quad \begin{array}{l}\text{Change to} \\ \text{equivalent} \\ \text{fractions}\end{array}$$

$$= \dfrac{9 + (-4) + 10}{24}$$

$$= \dfrac{15}{24} = \dfrac{5}{8} \quad \text{Reduce!} \quad \blacksquare$$

The distributive property, $a(b + c) = ab + ac$, holds true for rational numbers and (as with integers) can be used to facilitate manipulation.

EXAMPLE 17

Simplify $12\left(\dfrac{1}{3} + \dfrac{1}{4}\right)$.

Solution

For help in this situation, let's change the form by applying the distributive property.

$$12\left(\dfrac{1}{3} + \dfrac{1}{4}\right) = 12\left(\dfrac{1}{3}\right) + 12\left(\dfrac{1}{4}\right)$$
$$= 4 + 3$$
$$= 7$$

EXAMPLE 18

Simplify $\dfrac{5}{8}\left(\dfrac{1}{2} + \dfrac{1}{3}\right)$.

Solution

In this case it may be easier not to apply the distributive property but to work with the expression in its given form.

$$\dfrac{5}{8}\left(\dfrac{1}{2} + \dfrac{1}{3}\right) = \dfrac{5}{8}\left(\dfrac{3}{6} + \dfrac{2}{6}\right)$$
$$= \dfrac{5}{8}\left(\dfrac{5}{6}\right)$$
$$= \dfrac{25}{48}$$

Examples 13 and 14 emphasize a point made in Section 1.4. *Think first* and decide whether or not you can use the properties to make the manipulations easier.

6 Solve Application Problems Involving Rational Numbers

EXAMPLE 19

Apply Your Skill

Frank has purchased 50 candy bars to make s'mores for the Boy Scout troop. If he uses $\dfrac{2}{3}$ of a candy bar for each s'more, how many s'mores will he be able to make?

Solution

To find how many s'mores can be made, we need to divide 50 by $\dfrac{2}{3}$.

$$50 \div \dfrac{2}{3} = 50 \cdot \dfrac{3}{2} = \dfrac{\overset{25}{\cancel{50}}}{1} \cdot \dfrac{3}{\underset{1}{\cancel{2}}} = \dfrac{75}{1} = 75$$

So Frank can make 75 s'mores.

EXAMPLE 20

Apply Your Skill

Brian brought 5 cups of flour with him on a camping trip. He wants to make biscuits and cake for tonight's supper. It takes $\frac{3}{4}$ of a cup of flour for the biscuits and $2\frac{3}{4}$ cups of flour for the cake. How much flour will be left over for the rest of his camping trip?

Solution

Let's do this problem in two steps. First, add the amounts of flour needed for the biscuits and cake.

$$\frac{3}{4} + 2\frac{3}{4} = \frac{3}{4} + \frac{11}{4} = \frac{14}{4} = \frac{7}{2}$$

Then to find the amount of flour left over, we will subtract $\frac{7}{2}$ from 5.

$$5 - \frac{7}{2} = \frac{10}{2} - \frac{7}{2} = \frac{3}{2} = 1\frac{1}{2}$$

So $1\frac{1}{2}$ cups of flour remain. ∎

CONCEPT QUIZ 1.5

For Problems 1–10, answer true or false.

1. 6 is a rational number.
2. $\dfrac{-5}{3} = \dfrac{5}{-3}$
3. If the product of two rational numbers is 1, then the numbers are said to be reciprocals.
4. The reciprocal of $\dfrac{-3}{7}$ is $\dfrac{7}{3}$.
5. $\dfrac{10}{12}$ is reduced to lowest terms.
6. To add rational numbers with common denominators, add the numerators and place the result over the common denominator.
7. When adding or subtracting fractions, the least common multiple of the denominators can always be used as a common denominator.
8. To subtract $\dfrac{1}{5}$ from $\dfrac{3}{8}$, we need to find equivalent fractions with a common denominator.
9. To multiply $\dfrac{5}{7}$ and $\dfrac{2}{3}$, we need to find equivalent fractions with a common denominator.
10. When adding $\dfrac{1}{4}$ and $\dfrac{3}{5}$, either 20, 40, or 60 can be used as a common denominator, but 20 is the least common denominator.

Section 1.5 Classroom Problem Set

Objective 1

1. Reduce $\dfrac{28}{32}$ to lowest terms.
2. Reduce $\dfrac{15}{25}$ to lowest terms.
3. Reduce $\dfrac{25}{35}$ to lowest terms.
4. Reduce $\dfrac{27}{36}$ to lowest terms.

5. Reduce $-\dfrac{60}{144}$ to lowest terms.

6. Reduce $-\dfrac{45}{150}$ to lowest terms.

Objective 2

7. Multiply $\dfrac{5}{7} \cdot \dfrac{3}{10}$.

8. Multiply $\dfrac{3}{8} \cdot \dfrac{12}{15}$.

9. Find the product of $\dfrac{5}{6}$ and $\dfrac{12}{25}$.

10. Find the product of $\dfrac{3}{8}$ and $\dfrac{16}{27}$.

11. Multiply $\left(\dfrac{6}{18}\right)\left(-\dfrac{10}{14}\right)$.

12. Multiply $\left(-\dfrac{7}{20}\right)\left(-\dfrac{15}{28}\right)$.

13. Multiply $\left(-\dfrac{10}{3}\right)\left(-\dfrac{42}{18}\right)$.

14. Multiply $\left(-\dfrac{6}{7}\right)\left(-\dfrac{21}{24}\right)$.

Objective 3

15. Find the quotient.
    (a) $\dfrac{1}{5} \div \dfrac{2}{15}$ (b) $-\dfrac{3}{8} \div 4$

16. Find the quotient.
    (a) $\dfrac{2}{3} \div \dfrac{4}{15}$ (b) $-10 \div \dfrac{1}{4}$

Objective 4

17. Add $\dfrac{1}{7} + \dfrac{2}{3}$.

18. Add $\dfrac{3}{5} + \dfrac{1}{4}$.

19. Add $\dfrac{2}{5} + \dfrac{1}{3}$.

20. Add $\dfrac{3}{7} + \dfrac{1}{6}$.

21. Add $\dfrac{13}{60} + \dfrac{5}{18}$.

22. Add $\dfrac{5}{18} + \dfrac{8}{27}$.

23. Subtract $\dfrac{8}{15} - \dfrac{3}{10}$.

24. Subtract $\dfrac{7}{36} - \dfrac{1}{24}$.

25. Add $\dfrac{3}{10} + \left(\dfrac{-7}{12}\right)$.

26. Add $\dfrac{1}{21} + \left(\dfrac{-3}{14}\right)$.

27. Add $\dfrac{5}{8} + (-2)$.

28. Add $\dfrac{3}{4} - 6$.

Objective 5

29. Simplify $\dfrac{3}{14} + \dfrac{1}{4} \cdot \dfrac{4}{7} - \dfrac{1}{2} \cdot \dfrac{3}{7}$.

30. Simplify $\dfrac{2}{3} + \dfrac{1}{2} \cdot \dfrac{2}{5} - \dfrac{1}{3} \cdot \dfrac{1}{5}$.

31. Simplify $\dfrac{3}{11} \div \dfrac{2}{11} + \left(\dfrac{-1}{3}\right)\left(\dfrac{1}{4}\right) + \dfrac{5}{6}$.

32. Simplify $\dfrac{2}{7} \div \dfrac{3}{7} + \left(\dfrac{-1}{2}\right)\left(\dfrac{1}{3}\right) + \dfrac{6}{7}$.

33. Simplify $20\left(\dfrac{1}{2} + \dfrac{1}{5}\right)$.

34. Simplify $24\left(\dfrac{3}{4} - \dfrac{1}{6}\right)$.

35. Simplify $\dfrac{3}{8}\left(\dfrac{1}{3} + \dfrac{1}{5}\right)$.

36. Simplify $\dfrac{3}{13}\left(\dfrac{2}{3} - \dfrac{1}{6}\right)$.

Objective 6

37. A spool of coaxial cable contains 120 feet of cable. If $\dfrac{3}{4}$ of a foot of cable is needed for each installation, how many installations can be done with the spool of coaxial cable?

38. Sandra's recipe calls for $3\dfrac{1}{2}$ cups of milk. If she wants to make one-half of the recipe, how much milk should she use?

39. Victor is constructing an outdoor deck. From a board 12 feet in length he needs to cut off a piece $1\dfrac{3}{8}$ feet in length for an upright post and another piece $6\dfrac{1}{4}$ feet in length for the header. What is the length of the piece remaining?

40. For exercise Scott jogs $2\dfrac{3}{4}$ miles, walks $\dfrac{1}{2}$ mile, and then finishes the exercise by jogging $3\dfrac{3}{4}$ miles. Find the total distance that Scott covers.

THOUGHTS INTO WORDS

1. State in your own words the property
$$-\dfrac{a}{b} = \dfrac{-a}{b} = \dfrac{a}{-b}$$

2. Find the mistake in this simplification process:
$$\dfrac{1}{2} \div \left(\dfrac{2}{3}\right)\left(\dfrac{3}{4}\right) \div 3 = \dfrac{1}{2} \div \dfrac{1}{2} \div 3 = \dfrac{1}{2} \cdot 2 \cdot \dfrac{1}{3} = \dfrac{1}{3}$$
How would you correct the error?

3. Give a step-by-step description of how to add the rational numbers $\frac{3}{8}$ and $\frac{5}{18}$.

4. The will of a deceased collector of antique automobiles specified that his cars be left to his three children. Half were to go to his elder son, $\frac{1}{3}$ to his daughter, and $\frac{1}{9}$ to his younger son. At the time of his death, 17 cars were in the collection. The administrator of his estate borrowed a car to make 18. Then he distributed the cars as follows:

Elder son: $\frac{1}{2}(18) = 9$

Daughter: $\frac{1}{3}(18) = 6$

Younger son: $\frac{1}{9}(18) = 2$

This totaled 17 cars, so he then returned the borrowed car. Where is the error in this problem?

Answers to the Concept Quiz
1. True   2. True   3. True   4. False   5. False   6. True   7. True   8. True   9. False   10. True

1.6 Decimals and Real Numbers

OBJECTIVES

1. Classify Real Numbers
2. Expand the Properties of Numbers to Include Real Numbers
3. Add, Subtract, Multiply, and Divide Decimals
4. Combine Similar Terms with Decimal Coefficients
5. Evaluate Algebraic Expressions
6. Solve Application Problems That Involve Rational Numbers in Decimal Form

1 Classify Real Numbers

We classify decimals—also called decimal fractions—as **terminating**, **repeating**, or **nonrepeating**. Here are examples of these classifications:

Terminating decimals	Repeating decimals	Nonrepeating decimals
0.3	0.333333...	0.5918654279...
0.26	0.5466666...	0.26224222722229...
0.347	0.14141414...	0.1451172111193111148...
0.9865	0.237237237...	0.645751311...

A repeating decimal has a block of digits that repeats indefinitely. This repeating block of digits may contain any number of digits, and it may or may not begin repeating immediately after the decimal point. Technically, a terminating decimal can be thought of as repeating zeros after the last digit. For example, 0.3 = 0.30 = 0.300 = 0.3000, and so on.

In Section 1.5 we defined a rational number to be any number that can be written in the form $\frac{a}{b}$, where a and b are integers and b is not zero. *A rational number can also be defined as any number that has a terminating or repeating decimal representation.* Thus we can express rational numbers in either common-fraction form or decimal-fraction form, as the next examples illustrate. A repeating decimal can be written by using a bar over the digits that repeat, for example, $0.\overline{14}$.

1.6 Decimals and Real Numbers

Terminating decimals

$\dfrac{3}{4} = 0.75$

$\dfrac{1}{8} = 0.125$

$\dfrac{5}{16} = 0.3125$

$\dfrac{7}{25} = 0.28$

$\dfrac{2}{5} = 0.4$

Repeating decimals

$\dfrac{1}{3} = 0.3333\ldots$

$\dfrac{2}{3} = 0.66666\ldots$

$\dfrac{1}{6} = 0.166666\ldots$

$\dfrac{1}{12} = 0.08333\ldots$

$\dfrac{14}{99} = 0.14141414\ldots$

The nonrepeating decimals are called **irrational numbers**, and they do appear in forms other than decimal form. For example, $\sqrt{2}$, $\sqrt{3}$, and π are irrational numbers. An approximate decimal representation for each of these follows.

$\left.\begin{array}{l}\sqrt{2} = 1.414213562373\ldots \\ \sqrt{3} = 1.73205080756887\ldots \\ \pi = 3.14159265358979\ldots\end{array}\right\}$ Nonrepeating decimals

We will do more work with irrational numbers in Chapter 7.

The rational numbers together with the irrationals form the set of **real numbers**. This tree diagram of the real number system is helpful for summarizing some basic ideas.

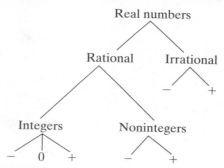

Any real number can be traced down through the diagram as follows:

5 is real, rational, an integer, and positive

-4 is real, rational, an integer, and negative

$\dfrac{3}{4}$ is real, rational, a noninteger, and positive

0.23 is real, rational, a noninteger, and positive

$-0.161616\ldots$ is real, rational, a noninteger, and negative

$\sqrt{7}$ is real, irrational, and positive

$-\sqrt{2}$ is real, irrational, and negative

In Section 1.2, we associated the set of integers with evenly spaced points on a line as indicated in Figure 1.7. This idea of associating numbers with points on a line can be extended so that there is a one-to-one correspondence between points on a line and the entire set of real numbers (as shown in Figure 1.8). That is to say, to each real number there corresponds one and only one point on the line, and to each point on the

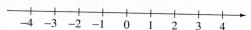

Figure 1.7

line there corresponds one and only one real number. The line is often referred to as the **real number line**, and the number associated with each point on the line is called the **coordinate** of the point.

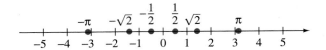

Figure 1.8

2 Expand the Properties of Numbers to Include Real Numbers

The properties of integers that we discussed in Section 1.4 are true for all real numbers. We restate them here for your convenience. The multiplicative inverse property has been added to the list. A discussion follows.

Commutative Property of Addition

If a and b are real numbers, then $a + b = b + a$.

Commutative Property of Multiplication

If a and b are real numbers, then $ab = ba$.

Associative Property of Addition

If a, b, and c are real numbers, then $(a + b) + c = a + (b + c)$.

Associative Property of Multiplication

If a, b, and c are real numbers then $(ab)c = a(bc)$.

Identity Property of Addition

If a is any real number, then $a + 0 = 0 + a = a$.

Identity Property of Multiplication

If a is any real number, then $a(1) = 1(a) = a$.

Additive Inverse Property

For every real number a, there exists an integer $-a$ such that $a + (-a) = (-a) + a = 0$.

> **Multiplication Property of Zero**
>
> If a is any real number, then $a(0) = (0)(a) = 0$.

> **Multiplicative Property of Negative One**
>
> If a is any real number, then $(a)(-1) = (-1)(a) = -a$.

> **Multiplicative Inverse Property**
>
> For every nonzero real number a, there exists a real number $\frac{1}{a}$, such that $a\left(\frac{1}{a}\right) = \frac{1}{a}(a) = 1$.

> **Distributive Property**
>
> If a, b, and c are real numbers, then $a(b + c) = ab + ac$.

The number $\frac{1}{a}$ is called the **multiplicative inverse of a** or the **reciprocal of a**. For example, the reciprocal of 2 is $\frac{1}{2}$ and $2\left(\frac{1}{2}\right) = \frac{1}{2}(2) = 1$. Likewise, the reciprocal of $\frac{1}{2}$ is $\frac{1}{\frac{1}{2}}$. Therefore, 2 and $\frac{1}{2}$ are said to be reciprocals (or multiplicative inverses) of each other. Also, $\frac{2}{5}$ and $\frac{5}{2}$ are multiplicative inverses and $\left(\frac{2}{5}\right)\left(\frac{5}{2}\right) = 1$. Because division by zero is undefined, zero does not have a reciprocal.

3 Add, Subtract, Multiply, and Divide Decimals

The basic operations with decimals may be related to the corresponding operations with common fractions. For example, $0.3 + 0.4 = 0.7$ because $\frac{3}{10} + \frac{4}{10} = \frac{7}{10}$, and $0.37 - 0.24 = 0.13$ because $\frac{37}{100} - \frac{24}{100} = \frac{13}{100}$. In general, to add or subtract decimals, we add or subtract the hundredths, the tenths, the ones, the tens, and so on. To keep place values aligned, we line up the decimal points.

EXAMPLE 1

Find the sum.

(a) $3.42 + 18.2 + 0.08$ (b) $6.123 + 0.005 + 42.1 + 8.43$

Solution

Line up the numbers in a vertical format, keeping the decimal points aligned.

(a) 3.42
    18.2
   +0.08

   21.70

(b) 6.123
     0.005
    42.1
    +8.43

    56.658

EXAMPLE 2

Find the difference for each of the following.

(a) $6.84 - 1.3$ (b) $72.85 - 2.412$

Solution

Line up the numbers in a vertical format, keeping the decimal points aligned. Add a zero so that each number has the same number of places after the decimal point.

(a) 6.84
    −1.30
    ─────
     5.54

(b) 72.850
    − 2.412
    ──────
     70.438

We can use examples such as the following to help formulate a general rule for multiplying decimals.

Because $\dfrac{7}{10} \cdot \dfrac{3}{10} = \dfrac{21}{100}$, then $(0.7)(0.3) = 0.21$.

Because $\dfrac{9}{10} \cdot \dfrac{23}{100} = \dfrac{207}{1000}$, then $(0.9)(0.23) = 0.207$.

Because $\dfrac{11}{100} \cdot \dfrac{13}{100} = \dfrac{143}{10{,}000}$, then $(0.11)(0.13) = 0.0143$.

In general, to multiply decimals, we (1) multiply the numbers and ignore the decimal points, and then (2) insert the decimal point in the product so that the number of digits to the right of the decimal point in the product is equal to the sum of the numbers of digits to the right of the decimal point in each factor.

(0.7)	×	(0.3)	=	0.21
↑		↑		↑
One digit to right	+	One digit to right	=	Two digits to right
(0.9)	×	(0.23)	=	0.207
↑		↑		↑
One digit to right	+	Two digits to right	=	Three digits to right
(0.11)	×	(0.13)	=	0.0143
↑		↑		↑
Two digits to right	+	Two digits to right	=	Four digits to right

EXAMPLE 3

Multiply using a vertical format.

(a) $41.2(0.13)$ (b) $0.021(0.03)$

Solution

Stack the numbers, align them on the right side, and multiply as usual. Add together the number of digits to the right of the decimal point in each number; that sum determines the placement of the decimal point in the product. In other words, if there is a total of three digits, insert the decimal point so that there are three digits to the right of it.

(a) 41.2   One digit to right
     0.13   Two digits to right
    ─────
     1236
     412
    ─────
     5.356   Three digits to right

(b) 0.021   Three digits to right
     0.03   Two digits to right
    ───────
     0.00063   Five digits to right

Note that in part b, we actually multiplied 3 · 21 and then inserted three 0s to the left so that there would be five digits to the right of the decimal point. ∎

Once again let's look at some links between common fractions and decimals.

Because $\dfrac{6}{10} \div 2 = \dfrac{\overset{3}{\cancel{6}}}{10} \cdot \dfrac{1}{2} = \dfrac{3}{10}$, we have $2\overline{)0.6}$ with quotient 0.3.

Because $\dfrac{39}{100} \div 13 = \dfrac{\overset{3}{\cancel{39}}}{100} \cdot \dfrac{1}{13} = \dfrac{3}{100}$, we have $13\overline{)0.39}$ with quotient 0.03.

Because $\dfrac{85}{100} \div 5 = \dfrac{\overset{17}{\cancel{85}}}{100} \cdot \dfrac{1}{\cancel{5}} = \dfrac{17}{100}$, we have $5\overline{)0.85}$ with quotient 0.17.

In general, to divide a decimal by a nonzero whole number, we (1) place the decimal point in the quotient directly above the decimal point in the dividend

$$\text{Divisor}\overline{)\text{Dividend}}^{\text{Quotient}} \quad \text{Format for long division}$$

and then (2) divide as with whole numbers, except that in the division process, 0s are placed in the quotient immediately to the right of the decimal point in order to show the correct place value.

$$\begin{array}{r} 0.121 \\ 4\overline{)0.484} \end{array} \qquad \begin{array}{r} 0.24 \\ 32\overline{)7.68} \\ \underline{6\;4} \\ 1\;28 \\ \underline{1\;28} \end{array} \qquad \begin{array}{r} 0.019 \\ 12\overline{)0.228} \\ \underline{12} \\ 108 \\ \underline{108} \end{array} \quad \text{Zero needed to show the correct place value}$$

Don't forget that *you can check division by multiplication*. For example, because $(12)(0.019) = 0.228$, we know that our last division example is correct.

We can easily handle problems involving division by a decimal by changing to an equivalent problem that has a whole-number divisor. Consider the following examples, in which we have changed the original division problem to fractional form to show the reasoning involved in the procedure.

$$0.6\overline{)0.24} \rightarrow \dfrac{0.24}{0.6} = \left(\dfrac{0.24}{0.6}\right)\left(\dfrac{10}{10}\right) = \dfrac{2.4}{6} \rightarrow 6\overline{)2.4}^{\,0.4}$$

$$0.12\overline{)0.156} \rightarrow \dfrac{0.156}{0.12} = \left(\dfrac{0.156}{0.12}\right)\left(\dfrac{100}{100}\right) = \dfrac{15.6}{12} \rightarrow \begin{array}{r} 1.3 \\ 12\overline{)15.6} \\ \underline{12} \\ 3\;6 \\ \underline{3\;6} \end{array}$$

$$1.3\overline{)0.026} \rightarrow \dfrac{0.026}{1.3} = \left(\dfrac{0.026}{1.3}\right)\left(\dfrac{10}{10}\right) = \dfrac{0.26}{13} \rightarrow \begin{array}{r} 0.02 \\ 13\overline{)0.26} \\ \underline{26} \end{array}$$

EXAMPLE 4

Divide using a long division format.

(a) $1.176 \div 0.21$ (b) $0.148 \div 3.7$

Solution

Move the decimal point in the divisor to the end of the number. Then move the decimal point in the dividend the same number of places. The decimal point for the quotient will then line up with the decimal point in the dividend as shown below.

(a) $0.21 \overline{)1.17.6}$ gives 5.6

The arrows indicate that both the divisor and the dividend were multiplied by 100, which changes the divisor to a whole number

(b) $3.7 \overline{)0.1.48}$ gives 0.04

The divisor and dividend were multiplied by 10

Our agreements for operating with positive and negative integers extend to all real numbers. For example, the product of two negative real numbers is a positive real number. Make sure that you agree with the following results. (You may need to do some work on scratch paper, because the steps are not shown.)

$0.24 + (-0.18) = 0.06$ $\qquad$ $(-0.4)(0.8) = -0.32$

$-7.2 + 5.1 = -2.1$ $\qquad$ $(-0.5)(-0.13) = 0.065$

$-0.6 + (-0.8) = -1.4$ $\qquad$ $(1.4) \div (-0.2) = -7$

$2.4 - 6.1 = -3.7$ $\qquad$ $(-0.18) \div (0.3) = -0.6$

$0.31 - (-0.52) = 0.83$ $\qquad$ $(-0.24) \div (-4) = 0.06$

$(0.2)(-0.3) = -0.06$

4 Combine Similar Terms with Decimal Coefficients

Numerical and algebraic expressions may contain the decimal form as well as the fractional form of rational numbers. For numerical expressions, we continue to follow the rule of doing the multiplications and divisions *first* and then the additions and subtractions, unless parentheses indicate otherwise.

You have already combined similar terms when the coefficients were in fractional form. The next example shows how to combine similar terms for algebraic expressions when the coefficients are in decimal form.

EXAMPLE 5

Simplify $-1.4x + 0.03y - 0.6x + 21.5y$ by combining similar terms.

Solution

Use the commutative property to rearrange the terms. Then combine the like terms by applying the distributive property.

$$-1.4x + 0.03y - 0.6x + 21.5y = -1.4x + 0.03y - 0.6x + 21.5y$$
$$= -1.4x - 0.6x + 0.03y + 21.5y$$
$$= (-1.4 - 0.6)x + (0.03 + 21.5)y$$
$$= -2x + 21.53y$$

5 Evaluate Algebraic Expressions

The following examples illustrate a variety of situations that involve both the decimal and the fractional forms of rational numbers.

1.6 Decimals and Real Numbers

EXAMPLE 6 Evaluate $\frac{3}{5}a - \frac{1}{7}b$ for $a = \frac{5}{2}$ and $b = -1$.

Solution

$$\frac{3}{5}a - \frac{1}{7}b = \frac{3}{5}\left(\frac{5}{2}\right) - \frac{1}{7}(-1) \quad \text{for } a = \frac{5}{2} \text{ and } b = -1$$

$$= \frac{3}{2} + \frac{1}{7}$$

$$= \frac{21}{14} + \frac{2}{14}$$

$$= \frac{23}{14}$$

EXAMPLE 7 Evaluate $\frac{1}{2}x + \frac{2}{3}x - \frac{1}{5}x$ for $x = -\frac{3}{4}$.

Solution

First, let's *combine similar terms* by using the distributive property.

$$\frac{1}{2}x + \frac{2}{3}x - \frac{1}{5}x = \left(\frac{1}{2} + \frac{2}{3} - \frac{1}{5}\right)x$$

$$= \left(\frac{15}{30} + \frac{20}{30} - \frac{6}{30}\right)x$$

$$= \frac{29}{30}x$$

Now we can evaluate.

$$\frac{29}{30}x = \frac{29}{30}\left(-\frac{3}{4}\right) \quad \text{when } x = -\frac{3}{4}$$

$$= \frac{29}{\underset{10}{\cancel{30}}}\left(-\frac{\overset{1}{\cancel{3}}}{4}\right) = -\frac{29}{40}$$

EXAMPLE 8 Evaluate $2x + 3y$ for $x = 1.6$ and $y = 2.7$.

Solution

$$2x + 3y = 2(1.6) + 3(2.7) \quad \text{when } x = 1.6 \text{ and } y = 2.7$$
$$= 3.2 + 8.1$$
$$= 11.3$$

EXAMPLE 9 Evaluate $0.9x + 0.7x - 0.4x + 1.3x$ for $x = 0.2$.

Solution

First, let's *combine similar terms* by using the distributive property.

$$0.9x + 0.7x - 0.4x + 1.3x = (0.9 + 0.7 - 0.4 + 1.3)x = 2.5x$$

Now we can evaluate.

$$2.5x = (2.5)(0.2) \quad \text{for } x = 0.2$$
$$= 0.5$$

6 Solve Application Problems That Involve Rational Numbers in Decimal Form

EXAMPLE 10 Apply Your Skill

A layout artist is putting together a group of images. She has four images with widths of 1.35 centimeters, 2.6 centimeters, 5.45 centimeters, and 3.2 centimeters, respectively. If the images are set side by side, what will be their combined width?

Solution

To find the combined width, we need to add the four widths.

$$\begin{array}{r} 1.35 \\ 2.6 \\ 5.45 \\ + \ 3.2 \\ \hline 12.60 \end{array}$$

So the combined width would be 12.6 centimeters. ∎

CONCEPT QUIZ 1.6

For Problems 1–10, answer true or false.

1. A rational number can be defined as any number that has a terminating or repeating decimal representation.
2. A repeating decimal has a block of digits that repeat only once.
3. Every irrational number is also classified as a real number.
4. The rational numbers along with the irrational numbers form the set of natural numbers.
5. 0.141414 . . . is a rational number.
6. $-\sqrt{5}$ is real, irrational, and negative.
7. 0.35 is real, rational, integer, and positive.
8. The reciprocal of c, where $c \neq 0$, is also the multiplicative inverse of c.
9. Any number multiplied by its multiplicative inverse gives a result of 0.
10. Zero does not have a multiplicative inverse.

Section 1.6 Classroom Problem Set

Objective 3

1. Find the sum for each of the following.
   (a) $17.3 + 21.05 + 0.4$
   (b) $12.45 + 1.8 + 0.04 + 1.261$

2. Find the sum for each of the following.
   (a) $0.37 + 56.01 + 0.17$
   (b) $19.35 + 2.98$

3. Find the difference for each of the following.
   (a) $13.6 - 10.35$
   (b) $8.427 - 0.32$

4. Find the difference for each of the following.
   (a) $25.03 - 14.96$
   (b) $5.327 - 0.86$

5. Multiply each of the following.
   (a) $28.4(1.32)$
   (b) $0.017(0.02)$

6. Multiply each of the following.
   (a) $0.76(4.92)$
   (b) $0.147(0.019)$

7. Divide using a long division format.
   (a) $29.16 \div 1.2$
   (b) $20.528 \div 0.04$

8. Divide using a long division format.
   (a) $-2.94 \div 0.6$
   (b) $-0.126 \div -0.9$

Objective 4

9. Simplify $5.3a - 3.06b - 0.08b - 0.01a$.
10. Simplify $5.7x + 9.4y - 6.2x - 4.4y$.

Objective 5

11. Evaluate $\frac{2}{3}x - \frac{1}{4}y$ for $x = \frac{3}{7}$ and $y = -2$.
12. Evaluate $\frac{2}{7}a + \frac{1}{3}b$ for $a = \frac{1}{2}$ and $b = -6$.
13. Evaluate $\frac{1}{4}x + \frac{3}{5}x - \frac{1}{3}x$ for $x = -\frac{6}{7}$.
14. Evaluate $\frac{3}{5}y - \frac{2}{3}y - \frac{7}{15}y$ for $y = -\frac{5}{2}$.
15. Evaluate $-2x + 4y$ for $x = 2.1$ and $y = 0.07$.

16. Evaluate $5x - 6y$ for $x = -0.01$ and $y = 0.34$.
17. Evaluate $5.6x + 0.8x - 1.8x - 0.4$ for $x = -0.3$.
18. Evaluate $1.2x + 2.3x - 1.4x - 7.6x$ for $x = -2.5$.

Objective 6

19. A pharmacist is mixing a liquid preparation to fill a prescription. The mixture requires 4.6 ounces of magnesium citrate solution, 1.25 ounces of sodium chloride solution, 8 ounces of water, and 0.02 ounces of vanilla flavoring. How many ounces is the mixture that the pharmacist prepared?

20. Garrett bought 2 pounds of Red Delicious apples for $1.79 per pound and 5 pounds of Jonathan apples for $1.59 per pound. How much did he spend?

THOUGHTS INTO WORDS

1. At this time how would you describe the difference between arithmetic and algebra?
2. How have the properties of the real numbers been used thus far in your study of arithmetic and algebra?
3. Do you think that $2\sqrt{2}$ is a rational or an irrational number? Defend your answer.

Answers to the Concept Quiz
1. True 2. False 3. True 4. False 5. True 6. True 7. False 8. True 9. False 10. True

Chapter 1 Review Problem Set

For Problems 1–4, given the word description of the set, write its list description as a set.

1. A = {whole numbers less than 5}
2. B = {odd whole numbers less than 11}
3. C = {prime numbers less than 11}
4. D = {even whole numbers}

For Problems 5–8, insert the symbol for "equal" or "not equal" to make each of the following a true statement.

5. $\{-1, -2, -3, -4, -5\}$ ____ $\{\ldots, -5, -4, -3, -2, -1\}$
6. $\{1, 3, 5, \ldots\}$ ____ $\{5, 3, 1\}$
7. $\{8, 6, 4, 2\}$ ____ $\{2, 4, 6, 8\}$
8. $\{1, 3, 5\}$ ____ $\{1, 1, 3, 3, 5\}$

For Problems 9–14, classify the real numbers by tracing down the chart on page 45.

9. -8 10. 0 11. -0.45
12. $\frac{7}{5}$ 13. $\sqrt{2}$ 14. $\sqrt[3]{-16}$

For Problems 15–24, add the integers.

15. $7 + (-10)$ 16. $(-12) + (-13)$
17. $(-18) + 5$ 18. $17 + (-29)$

19. $(-12) + (-10)$ **20.** $(-4) + 10$

21. $25 + (-50)$ **22.** $(-21) + (-18)$

23. $(-12) + 36$ **24.** $(-18) + 8$

For Problems 25–34, subtract the integers.

25. $8 - (+13)$ **26.** $-6 - (+9)$

27. $-12 - (-11)$ **28.** $-17 - (-19)$

29. $35 - (-10)$ **30.** $-18 - (-7)$

31. $-32 - (+24)$ **32.** $16 - (+4)$

33. $-24 - (-58)$ **34.** $10 - (+55)$

For Problems 35–40, perform the indicated multiplication or division.

35. $(13)(-12)$ **36.** $(-14)(-18)$

37. $(-72) \div (-12)$ **38.** $117 \div (-9)$

39. $(-12)(-4)$ **40.** $(56) \div (-8)$

41. A record high temperature of 125°F occurred in Laughlin, Nevada, on June 29, 1994. A record low temperature of −50°F occurred in San Jacinto, Nevada, on January 8, 1937. Find the difference between the record high and low temperatures.

42. In North America the highest elevation, which is the top of Mt. McKinley, Alaska, is 20,320 feet above sea level. The lowest elevation in North America, which is the base of Death Valley, California, is 282 feet below sea level. Find the absolute value of the difference in elevation between Mt. McKinley and Death Valley.

43. As a running back in a football game, Marquette carried the ball 7 times. On two plays he gained 6 yards each play; on another play he lost 4 yards; on the next three plays he gained 8 yards per play; and on the last play he lost 1 yard. Write a numerical expression that gives Marquette's overall yardage for the game, and simplify the expression to give the total yardage.

44. Shelley started the month with $3278 in her checking account. During the month she deposited $175 each week for 4 weeks but had debit charges of $50, $189, $160, $20, and $115. What is the balance in her checking account after these deposits and debits?

45. State the property of integers demonstrated by $(4 + 6) + 8 = (6 + 4) + 8$.

46. State the property of integers demonstrated by $-4(3 + 7) = -4(3) + (-4)(7)$.

For Problems 47–52, classify each number as prime or composite.

47. 73 **48.** 87 **49.** 63

50. 81 **51.** 49 **52.** 91

For Problems 53–58, express each number as the product of prime factors.

53. 24 **54.** 63

55. 120 **56.** 135

57. 84 **58.** 64

59. Find the greatest common factor of 36 and 54.

60. Find the greatest common factor of 48, 60, and 84.

61. Find the least common multiple of 18 and 20.

62. Find the least common multiple of 12, 20, and 30.

For Problems 63–72, perform the indicated operations and express answers in reduced form.

63. $\left(\dfrac{2}{5}\right)\left(-\dfrac{3}{4}\right)$ **64.** $\left(-\dfrac{1}{8}\right)\left(-\dfrac{5}{2}\right)$

65. $\left(\dfrac{3}{4}\right) \div \left(\dfrac{7}{8}\right)$ **66.** $\left(-\dfrac{3}{8}\right) \div \left(\dfrac{1}{2}\right)$

67. $\left(\dfrac{12}{5}\right) \div \left(\dfrac{2}{15}\right)$ **68.** $\left(-\dfrac{18}{5}\right) \div \left(\dfrac{3}{2}\right)$

69. $\left(-\dfrac{12}{15}\right)\left(-\dfrac{4}{3}\right)$ **70.** $\left(-\dfrac{12}{7}\right)\left(\dfrac{14}{36}\right)$

71. $\left(-\dfrac{10}{3}\right) \div \left(\dfrac{3}{2}\right)$ **72.** $\left(\dfrac{4}{5}\right) \div \left(-\dfrac{4}{7}\right)$

For Problems 73–82, perform the indicated operations and express answers in reduced form.

73. $\dfrac{3}{8} - \dfrac{5}{8}$ **74.** $-\dfrac{11}{7} - \dfrac{1}{7}$

75. $\dfrac{13}{32} + \dfrac{15}{32}$ **76.** $\dfrac{9}{24} + \dfrac{7}{24}$

77. $\dfrac{11}{12} - \dfrac{1}{3}$ **78.** $\dfrac{1}{2} - \dfrac{7}{10}$

79. $-\dfrac{5}{12} + \dfrac{7}{9}$ **80.** $-\dfrac{17}{18} + \dfrac{5}{8}$

81. $\dfrac{-3}{2} + \dfrac{5}{9}$ **82.** $\dfrac{-2}{3} + \dfrac{7}{8}$

For Problems 83–88, perform the indicated operations.

83. $13.7 + 63.48 + 0.34$ **84.** $3.56 + 2.8 + 25.04$

85. $16.4 - 12.32$ **86.** $34.5 - 8.45$

87. $0.016 + 0.8 - 0.04$ **88.** $4.08 - 1.8 + 0.007$

For Problems 89–94, perform the indicated operations.

89. $(24.6)(1.2)$ **90.** $(6.42)(1.8)$

91. $(-5.07)(-0.03)$ **92.** $(0.045)(0.2)$

93. $(0.084)(-1.2)$ **94.** $(-0.2)(0.2)$

For Problems 95–100, perform the indicated operations.

95. $32.43 \div 2.3$

96. $0.9492 \div 0.6$

97. $37.92 \div 12$

98. $111.72 \div 21$

99. $240 \div 2.4$

100. $864 \div 1.2$

For Problems 101–104, evaluate each numerical expression.

101. 2^6

102. $(-3)^4$

103. $(3+2)^2$

104. $(-0.4)^2$

For Problems 105–114, simplify each numerical expression by following the order of operations.

105. $-6 + 4 \cdot 2 - 12 \div 2$

106. $8 + 4(-6) + 4 \div 2$

107. $36 - 24 \div 12 - 3 \cdot 2$

108. $-16 + 24 \div (-8) - 3 \cdot 2$

109. $\dfrac{4}{5} \div \dfrac{1}{5} \cdot \dfrac{2}{3} - \dfrac{1}{4}$

110. $\dfrac{2}{3} \cdot \dfrac{1}{4} \div \dfrac{1}{2} + \dfrac{2}{3} \cdot \dfrac{1}{4}$

111. $21 - [-2(14 \div 7)]$

112. $15 - [(-42 \div 7) \cdot 3]$

113. $\dfrac{3 \cdot 6 + 9 \cdot 2}{6 + 3}$

114. $\dfrac{4 \cdot 12 + 6 \cdot 8}{2 + 4}$

For Problems 115–124, simplify each algebraic expression by combining like terms.

115. $8x + 5y - 13x - y$

116. $5xy - 9xy + xy - y$

117. $5(x-4) - 3(x-9)$

118. $-3(x-2) - 4(x+6)$

119. $-2x - 3(x-4) + 2x$

120. $-(a-1) + 3(a-2) - 4a + 1$

121. $\dfrac{3}{8}x - \dfrac{2}{5}y - \dfrac{1}{4}x - \dfrac{3}{10}y$

122. $\dfrac{2}{3}\left(x + \dfrac{1}{4}y\right) - \dfrac{1}{6}x$

123. $1.4(a-b) + 0.8(a+b)$

124. $0.24ab + 0.73bc - 0.82ab - 0.37bc$

For Problems 125–132, evaluate each algebraic expression for the given values of the variables.

125. $5a + 6b - 7a - 2b$ for $a = -1$ and $b = 5$

126. $2xy + 6 + 5xy - 8$ for $x = -1$ and $y = 2$

127. $7(x+6) - 9(x+1)$ for $x = -2$

128. $-3(x-4) - 2(x+8)$ for $x = 7$

129. $\dfrac{-5x - 2y}{-2x - 7}$ for $x = 6$ and $y = 4$

130. $\dfrac{-3x + 4y}{3x}$ for $x = -4$ and $y = -6$

131. $a^3 + b^2$ for $a = -\dfrac{1}{2}$ and $b = \dfrac{1}{3}$

132. $2x^2 - 3y^2$ for $x = 5$ and $y = 2$

133. A caterer received $840 for catering a birthday party. If her chef is paid one-fifth of that amount, how much will the chef receive?

134. Julio has $5\dfrac{7}{8}$ yards of leather to reupholster automobile seats. If a seat requires $1\dfrac{3}{4}$ yards of leather, how much leather will be left after Julio reupholsters 3 seats?

135. Kyle took his race car engine to a machine shop to have the cylinder head bored out. The machine shop measured and found that the diameter of the cylinder is 3.257 inches. If the diameter of the cylinder is increased by 0.12 inches, how large will the diameter of the cylinder be?

136. When Marci was planning her trip to Europe, each U.S. dollar was worth 0.66 euros. At this exchange rate, how many euros will Marci have if she exchanges $450?

Practice Test

For Problems 1–5, simplify each of the numerical expressions.

1. $5(-7) - (-3)(8)$
2. $4 + 20 \div 2 \cdot 5$
3. $6(2 - 5) - 8$
4. $(8 - 3)^2$
5. $(-14)(4) \div 4 + (-6)$

6. It was reported on the 5 o'clock weather show that the current temperature was 7°F. The temperature was forecast to drop 13 degrees by 6:00 A.M. If the forecast is correct, what will be the temperature at 6:00 A.M.?

7. State the property of integers demonstrated by $(5 + 7) + (-8) = (7 + 5) + (-8)$

For Problems 8–11, simplify by combining similar terms.

8. $-7x + 9y - y + 2x - 2y - 7x$
9. $-2(x - 4) - 5(x + 7) - 6(x - 1)$
10. $3a^2 - 7b^2 - 4b^2 - a^2$
11. $5x^2 - 3x + 4 + 10x - 9$

12. Express $\dfrac{42}{54}$ in reduced form.

13. Classify 79 as prime or composite.

14. Express 360 as a product of prime numbers.

15. Find the greatest common factor of 36, 60, and 84.

16. Find the least common multiple of 9 and 24.

For Problems 17–20, perform the indicated operations, and express your answer in reduced form.

17. $\dfrac{5}{12} \div \dfrac{15}{8}$

18. $\dfrac{4}{15} + \dfrac{3}{10} - \dfrac{1}{6}$

19. $-\dfrac{2}{3} - \dfrac{1}{2}\left(\dfrac{3}{4}\right) + \dfrac{5}{6}$

20. $1.4(-2.3) + (0.8)^2$

For Problems 21–23, evaluate each algebraic expression for the given values of the variables.

21. $3x - 2y + xy$ for $x = 0.5$ and $y = -0.9$
22. $a^2 - ab + b^2$ for $a = 4$ and $b = -3$
23. $3(x - 2) - 5(x - 4) + 6(x - 1)$ for $x = -3$

24. Joni, a pediatric nurse, is mixing a solution of medicine for a patient. Each dose is a mixture of $\frac{1}{8}$ teaspoon of cough syrup, $\frac{1}{2}$ teaspoon of antihistamine, and $\frac{1}{4}$ teaspoon of decongestant. How many teaspoons of medicine will there be for 16 doses?

24. _____

25. On a recent fishing trip, Hector purchased gasoline daily from the marina. His purchases for the four days were 13.2 gallons, 18.4 gallons, 12.8 gallons, and 16.3 gallons. Find the cost of Hector's gasoline purchases if the marina charged $3.30 for a gallon of gasoline.

25. _____

2 Equations, Inequalities, and Problem Solving

- 2.1 Solving First-Degree Equations
- 2.2 Translating from English to Algebra
- 2.3 Equations and Problem Solving
- 2.4 More on Solving Equations and Problem Solving
- 2.5 Equations Involving Parentheses and Fractional Forms
- 2.6 Inequalities
- 2.7 Inequalities, Compound Inequalities, and Problem Solving

Chapter 2 Warm-Up Problems

1. **(a)** Write each number in words.

   i. 13,456 **ii.** 4.091 **iii.** $\dfrac{5}{4}$

   (b) Write each of the following with digits instead of words.

   i. one million, twenty thousand, one hundred sixty-four

   ii. two-thirds

   iii. eighty-three and two hundred five thousandths

2. Add or subtract.

   (a) $0.57 - 0.29$ **(b)** $1.73 + 0.097$ **(c)** $0.62 - 0.62$

3. Add or subtract.

   (a) $-5 - 5$ **(b)** $-15 + 7$ **(c)** $-6 - (-6)$

4. Find the least common multiple (LCM).

   (a) $6, 3, 8$ **(b)** $20, 12$ **(c)** $9, 15, 21$

5. Add the fractions. **(a)** $\dfrac{1}{4} + \dfrac{2}{3}$ **(b)** $\dfrac{6}{11} + \dfrac{3}{11}$ **(c)** $\dfrac{3}{8} + \dfrac{5}{14}$

6. Multiply or divide. **(a)** $\left(\dfrac{6}{7}\right)\left(\dfrac{7}{6}\right)$ **(b)** $(13)\left(\dfrac{1}{13}\right)$ **(c)** $\left(-\dfrac{3}{10}\right)\left(\dfrac{10}{3}\right)$

7. Simplify. **(a)** $18\left(\dfrac{2x}{3}\right)$ **(b)** $\dfrac{3}{2}(24y)$ **(c)** $9\left(\dfrac{5n}{6} - \dfrac{2n}{3}\right)$

8. Apply the distributive property.

   (a) $5(2x + 7)$

   (b) $-4(7x + 10)$

   (c) $-1(3n - 2)$

9. Simplify. **(a)** $3x + 2y - x$ **(b)** $16m - 9m$ **(c)** $2(x + 7) + 3x$

10. Find the average. **(a)** $8, 21, 16$ **(b)** $92, 83, 79, 94$ **(c)** $138, 174, x$

Throughout this book we follow a common theme: Develop some new skills, use the skills to help solve equations and inequalities, and finally, use the equations and inequalities to solve applied problems. In this chapter we want to use the skills we developed in the first chapter to solve equations and inequalities and to expand our work with applied problems.

Video tutorials for all section learning objectives are available in a variety of delivery modes.

INTERNET PROJECT

Some of the application problems in this chapter concern supplementary and complementary angles. Do an Internet search to find a site that has an interactive demonstration of supplementary or complementary angles. Such a site might allow you to change the size of one of the supplementary angles and see how the other supplementary angle changes. Also search for the definitions of acute angles and obtuse angles. Can both (of two) supplementary angles be obtuse angles? Can both (of two) supplementary angles be acute angles?

2.1 Solving First-Degree Equations

OBJECTIVES

1. Solve Equations Using the Addition-Subtraction Property of Equality
2. Solve Equations Using the Multiplication-Division Property of Equality

1 Solve Equations Using the Addition-Subtraction Property of Equality

These are examples of **numerical statements**:

$$3 + 4 = 7 \qquad 5 - 2 = 3 \qquad 7 + 1 = 12$$

The first two are true statements, and the third is a false statement.

When you use x as a variable, statements like

$$x + 3 = 4 \qquad 2x - 1 = 7 \qquad x^2 = 4$$

are called **algebraic equations** in x. We call a number a a **solution** or **root** of an equation if a true numerical statement is formed when we substitute a for x. (We also say that a satisfies the equation.) For example, 1 is a solution of $x + 3 = 4$ because substituting 1 for x produces the true numerical statement $1 + 3 = 4$. We call the set of all solutions of an equation its **solution set**. Thus the solution set of $x + 3 = 4$ is $\{1\}$. Likewise, the solution set of $2x - 1 = 7$ is $\{4\}$, and the solution set of $x^2 = 4$ is $\{-2, 2\}$. **Solving an equation** refers to the process of determining the solution set. Remember that a set that consists of no elements is called the **empty** or **null set** and is denoted by $\emptyset$. Thus we say that the solution set of $x = x + 1$ is $\emptyset$; that is, there are no real numbers that satisfy $x = x + 1$.

In this chapter we will consider techniques for solving **first-degree equations of one variable**. This means that the equations contain only one variable, and this variable has an exponent of 1. Here are some examples of first-degree equations of one variable:

$$3x + 4 = 7 \qquad 0.8w + 7.1 = 5.2w - 4.8$$

$$\frac{1}{2}y + 2 = 9 \qquad 7x + 2x - 1 = 4x - 1$$

Equivalent equations are equations that have the same solution set. For example,

$$5x - 4 = 3x + 8$$
$$2x = 12$$
$$x = 6$$

are all equivalent equations; this can be verified by showing that 6 is the solution for all three equations.

As we work with equations, we can use the following properties of equality:

> **Property 2.1   Properties of Equality**
>
> For all real numbers a, b, and c,
>
> 1. $a = a$ Reflexive property
> 2. If $a = b$, then $b = a$ Symmetric property
> 3. If $a = b$ and $b = c$, then $a = c$ Transitive property
> 4. If $a = b$, then a may be replaced by b, or b may be replaced by a, in any statement, without changing the meaning of the statement.   Substitution property

The general procedure for solving an equation is to continue replacing the given equation with equivalent but simpler equations until we obtain an equation of the form **variable = constant** or **constant = variable**. Thus in the preceding example, $5x - 4 = 3x + 8$ was simplified to $2x = 12$, which was further simplified to $x = 6$, from which the solution of 6 is obvious. The exact procedure for simplifying equations is our next concern.

Two properties of equality play an important role in the process of solving equations. The first of these is the **addition-subtraction property of equality**, which we state as follows:

> **Property 2.2   Addition-Subtraction Property of Equality**
>
> For all real numbers a, b, and c,
>
> 1. $a = b$ if and only if $a + c = b + c$
> 2. $a = b$ if and only if $a - c = b - c$

Property 2.2 states that *any number can be added to or subtracted from both sides of an equation, and the result is an equivalent equation.* Consider the use of this property in the next four examples.

EXAMPLE 1

Solve $x - 8 = 3$.

Solution

$$x - 8 = 3$$
$$x - 8 + 8 = 3 + 8 \quad \text{Add 8 to both sides}$$
$$x = 11$$

The solution set is $\{11\}$.

Remark: It is true that a simple equation like Example 1 can be solved by *inspection*. That is to say, we could think, "some number minus 8 produces 3," and obviously, the number is 11. However, as the equations become more complex, the technique of solving by inspection becomes ineffective. This is why it is necessary to develop more formal techniques for solving equations. Therefore we will begin developing such techniques even with very simple equations.

EXAMPLE 2

Solve $x + 14 = -8$.

Solution

$$x + 14 = -8$$
$$x + 14 - 14 = -8 - 14 \quad \text{Subtract 14 from both sides}$$
$$x = -22$$

The solution set is $\{-22\}$. ∎

EXAMPLE 3

Solve $n - \dfrac{1}{3} = \dfrac{1}{4}$.

Solution

$$n - \frac{1}{3} = \frac{1}{4}$$
$$n - \frac{1}{3} + \frac{1}{3} = \frac{1}{4} + \frac{1}{3} \quad \text{Add } \tfrac{1}{3} \text{ to both sides}$$
$$n = \frac{3}{12} + \frac{4}{12}$$
$$n = \frac{7}{12}$$

The solution set is $\left\{\dfrac{7}{12}\right\}$. ∎

EXAMPLE 4

Solve $0.72 = y + 0.35$.

Solution

$$0.72 = y + 0.35$$
$$0.72 - 0.35 = y + 0.35 - 0.35 \quad \text{Subtract 0.35 from both sides}$$
$$0.37 = y$$

The solution set is $\{0.37\}$. ∎

Note in Example 4 that the final equation is $0.37 = y$ instead of $y = 0.37$. Technically, the **symmetric property of equality** (if $a = b$, then $b = a$) would permit us to change from $0.37 = y$ to $y = 0.37$, but such a change is not necessary to determine that the solution is 0.37. You should also realize that you could apply the symmetric property to the original equation. Thus $0.72 = y + 0.35$ becomes $y + 0.35 = 0.72$, and subtracting 0.35 from both sides would produce $y = 0.37$.

At this time we should make one other comment that pertains to Property 2.2. Because subtracting a number is equivalent to adding its opposite, Property 2.2 could be stated only in terms of addition. Thus, to solve an equation such as Example 4 we could add -0.35 to both sides rather than subtracting 0.35 from both sides.

2 Solve Equations Using the Multiplication-Division Property of Equality

The other important property for solving equations is the **multiplication-division property of equality**.

> **Property 2.3 Multiplication-Division Property of Equality**
>
> For all real numbers a, b, and c, where $c \neq 0$,
>
> 1. $a = b$ if and only if $ac = bc$
> 2. $a = b$ if and only if $\dfrac{a}{c} = \dfrac{b}{c}$

Property 2.3 states that *we get an equivalent equation whenever both sides of a given equation are multiplied or divided by the same nonzero real number.* The following examples illustrate the use of this property.

EXAMPLE 5 Solve $\dfrac{3}{4}x = 6$.

Solution

$$\dfrac{3}{4}x = 6$$

$$\dfrac{4}{3}\left(\dfrac{3}{4}x\right) = \dfrac{4}{3}(6) \quad \text{Multiply both sides by } \dfrac{4}{3} \text{ because } \left(\dfrac{4}{3}\right)\left(\dfrac{3}{4}\right) = 1$$

$$x = 8$$

The solution set is {8}.

EXAMPLE 6 Solve $5x = 27$.

Solution

$$5x = 27$$

$$\dfrac{5x}{5} = \dfrac{27}{5} \quad \text{Divide both sides by 5}$$

$$x = \dfrac{27}{5} \quad \dfrac{27}{5} \text{ could be expressed as } 5\dfrac{2}{5} \text{ or } 5.4$$

The solution set is $\left\{\dfrac{27}{5}\right\}$.

EXAMPLE 7 Solve $-\dfrac{2}{3}p = \dfrac{1}{2}$.

Solution

$$-\dfrac{2}{3}p = \dfrac{1}{2}$$

$$\left(-\dfrac{3}{2}\right)\left(-\dfrac{2}{3}p\right) = \left(-\dfrac{3}{2}\right)\left(\dfrac{1}{2}\right) \quad \begin{array}{l}\text{Multiply both sides by } -\dfrac{3}{2} \\ \text{because } \left(-\dfrac{3}{2}\right)\left(-\dfrac{2}{3}\right) = 1\end{array}$$

$$p = -\dfrac{3}{4}$$

The solution set is $\left\{-\dfrac{3}{4}\right\}$.

EXAMPLE 8

Solve $-26 = \dfrac{x}{2}$.

Solution

$$-26 = \dfrac{x}{2}$$

$$2(-26) = 2\left(\dfrac{x}{2}\right) \quad \text{Multiply both sides by 2}$$

$$-52 = x$$

The solution set is $\{-52\}$.

Look back at Examples 5–8, and you will notice that whenever the coefficient was an integer we divided both sides of the equation by the coefficient of the variable—otherwise, we used the multiplication part of Property 2.3. Technically, because dividing by a number is equivalent to multiplying by its reciprocal, Property 2.3 could be stated only in terms of multiplication. Thus, to solve an equation such as $5x = 27$, we could multiply both sides by $\dfrac{1}{5}$ instead of dividing both sides by 5.

EXAMPLE 9

Solve $0.2n = 15$.

Solution

$$0.2n = 15$$

$$\dfrac{0.2n}{0.2} = \dfrac{15}{0.2} \quad \text{Divide both sides by 0.2}$$

$$n = 75$$

The solution set is $\{75\}$.

CONCEPT QUIZ 2.1

For Problems 1–10, answer true or false.

1. Equivalent equations have the same solution set.
2. $x^2 = 9$ is a first-degree equation.
3. The set of all solutions is called a solution set.
4. If the solution set is the null set, then the equation has at least one solution.
5. Solving an equation refers to obtaining any other equivalent equation.
6. If 5 is a solution, then a true numerical statement is formed when 5 is substituted for the variable in the equation.
7. Any number can be subtracted from both sides of an equation, and the result is an equivalent equation.
8. Any number can divide both sides of an equation to obtain an equivalent equation.
9. By the reflexive property, if $y = 2$ then $2 = y$.
10. By the transitive property, if $x = y$ and $y = 4$, then $x = 4$.

Section 2.1   Classroom Problem Set

Objective 1

1. Solve $x - 11 = -4$.
2. Solve $x + 13 = -4$.
3. Solve $y + 6 = 2$.
4. Solve $6 = n + 19$.
5. Solve $x - \dfrac{1}{5} = \dfrac{1}{2}$.
6. Solve $x + \dfrac{3}{5} = \dfrac{1}{3}$.
7. Solve $0.08 = x + 0.45$.
8. Solve $n - 3.6 = -7.3$.

Objective 2

9. Solve $\frac{2}{3}x = 12$.

10. Solve $-\frac{3}{8}n = 33$.

11. Solve $3x = 49$.

12. Solve $54y = 7$.

13. Solve $\frac{3}{5}y = -\frac{1}{4}$.

14. Solve $-\frac{6}{5}x = -\frac{10}{14}$.

15. Solve $\frac{x}{4} = -\frac{3}{2}$.

16. Solve $\frac{-y}{8} = \frac{2}{3}$.

17. Solve $0.3y = 18.6$.

18. Solve $0.9x = 22.5$.

THOUGHTS INTO WORDS

1. Describe the difference between a numerical statement and an algebraic equation.

2. Are the equations $6 = 3x + 1$ and $1 + 3x = 6$ equivalent equations? Defend your answer.

Answers to the Concept Quiz

1. True 2. False 3. True 4. False 5. False 6. True 7. True 8. False 9. False 10. True

2.2 Translating from English to Algebra

OBJECTIVES

1. Translate Algebraic Expressions into English Phrases
2. Translate English Phrases into Algebraic Expressions
3. Write Algebraic Expressions That Change the Units of Measure

1 Translate Algebraic Expressions into English Phrases

In order to use the tools of algebra for solving problems, we must be able to translate back and forth between the English language and the language of algebra. In this section we will translate algebraic expressions into English phrases (word phrases) and English phrases into algebraic expressions.

Let's begin by considering the following translations from algebraic expressions to word phrases.

Algebraic expression	Word phrase
$x + y$	The sum of x and y
$x - y$	The difference of x and y
$y - x$	The difference of y and x
xy	The product of x and y
$\frac{x}{y}$	The quotient of x and y
$3x$	The product of 3 and x
$x^2 + y^2$	The sum of x squared and y squared
$2xy$	The product of 2, x, and y
$2(x + y)$	Two times the quantity x plus y
$x - 3$	Three less than x

2 Translate English Phrases into Algebraic Expressions

Now let's consider the reverse process: translating from word phrases to algebraic expressions. Part of the difficulty in translating from English to algebra is that different word phrases translate into the same algebraic expression. Thus we need to become familiar with *different ways of saying the same thing*, especially when referring to the four fundamental operations. The next examples should acquaint you with some of the phrases used in the basic operations.

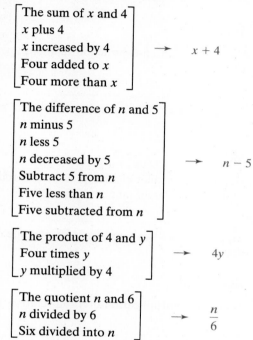

$$\left.\begin{array}{l}\text{The sum of } x \text{ and } 4\\ x \text{ plus } 4\\ x \text{ increased by } 4\\ \text{Four added to } x\\ \text{Four more than } x\end{array}\right\} \rightarrow x + 4$$

$$\left.\begin{array}{l}\text{The difference of } n \text{ and } 5\\ n \text{ minus } 5\\ n \text{ less } 5\\ n \text{ decreased by } 5\\ \text{Subtract } 5 \text{ from } n\\ \text{Five less than } n\\ \text{Five subtracted from } n\end{array}\right\} \rightarrow n - 5$$

$$\left.\begin{array}{l}\text{The product of } 4 \text{ and } y\\ \text{Four times } y\\ y \text{ multiplied by } 4\end{array}\right\} \rightarrow 4y$$

$$\left.\begin{array}{l}\text{The quotient } n \text{ and } 6\\ n \text{ divided by } 6\\ \text{Six divided into } n\end{array}\right\} \rightarrow \frac{n}{6}$$

Often a word phrase indicates more than one operation. Furthermore, the standard vocabulary of *sum*, *difference*, *product*, and *quotient* may be replaced by other terminology. Study the following translations very carefully. Also remember that the commutative property holds for addition and multiplication but not for subtraction and division. Therefore the phrase "x plus y" can be written as $x + y$ or $y + x$. However the phrase "x minus y" means that y must be subtracted from x, and the phrase is written as $x - y$. So be very careful of phrases that involve subtraction or division.

Word phrase	Algebraic expression
The sum of two times x and three times y	$2x + 3y$
The sum of the squares of a and b	$a^2 + b^2$
Five times x divided by y	$\dfrac{5x}{y}$
Two more than the square of x	$x^2 + 2$
Three less than the cube of b	$b^3 - 3$
Five less than the product of x and y	$xy - 5$
Nine minus the product of x and y	$9 - xy$
Four times the sum of x and 2	$4(x + 2)$
Six times the quantity w minus 4	$6(w - 4)$

Suppose you are told that the sum of two numbers is 12, and one of the numbers is 8. What is the other number? The other number is $12 - 8$, which equals 4. Now suppose you are told that the product of two numbers is 56, and one of the numbers is 7. What is the other number? The other number is $56 \div 7$, which equals 8. The following examples illustrate the use of these addition-subtraction and multiplication-division relationships.

EXAMPLE 1   The sum of two numbers is 83, and one of the numbers is x. What is the other number?

Solution

Using the addition-subtraction relationship, we can represent the other number by $83 - x$. ∎

EXAMPLE 2   The difference of two numbers is 14. The smaller number is n. What is the larger number?

Solution

Because the smaller number plus the difference must equal the larger number, we can represent the larger number by $n + 14$. ∎

EXAMPLE 3   The product of two numbers is 39, and one of the numbers is y. Represent the other number.

Solution

Using the multiplication-division relationship, we can represent the other number by $\dfrac{39}{y}$. ∎

The English statement may not contain key words such as *sum*, *difference*, *product*, or *quotient*. Instead the statement may describe a physical situation; from this description you need to deduce the operations involved. We now make some suggestions for handling such situations.

EXAMPLE 4   Arlene can type 70 words per minute. How many words can she type in m minutes?

Solution

In 10 minutes she would type $70(10) = 700$ words. In 50 minutes she would type $70(50) = 3500$ words. Thus in m minutes she would type $70m$ words. ∎

Note the use of some specific examples [$70(10) = 700$ and $70(50) = 3500$] to help formulate the general expression. This technique of first formulating some specific examples and then generalizing can be very effective.

EXAMPLE 5   Lynn has n nickels and d dimes. Express, in cents, this amount of money.

Solution

Three nickels and 8 dimes would be $5(3) + 10(8) = 95$ cents. Thus n nickels and d dimes would be $5n + 10d$ cents. ∎

EXAMPLE 6

A cruise ship travels at the rate of r miles per hour. How far will it travel in 8 hours?

Solution

Suppose that a cruise ship travels at 50 miles per hour. Using the formula *distance equals rate times time*, we find that it would travel $50 \cdot 8 = 400$ miles. Therefore, at r miles per hour, it would travel $r \cdot 8$ miles. We usually write the expression $r \cdot 8$ as $8r$. ∎

EXAMPLE 7

At a local restaurant the cost for a dinner to serve 8 people is d dollars. What is the cost per person for the dinner?

Solution

The price per person is figured by dividing the total cost by the number of people. Therefore, we represent the price per person by $\frac{d}{8}$. ∎

3 Write Algebraic Expressions That Change the Units of Measure

The English statement to be translated into algebra may contain some geometric ideas. For example, suppose that we want to express in inches the length of a line segment that is f feet long. Because 1 foot = 12 inches, we can represent f feet by 12 times f, written as $12f$ inches.

Table 2.1 lists some of the basic relationships pertaining to linear measurements in the English and metric systems. (Additional listings of both systems are located inside the back cover of this book.)

Table 2.1

English system	Metric system
12 inches = 1 foot	1 kilometer = 1000 meters
3 feet = 36 inches = 1 yard	1 hectometer = 100 meters
5280 feet = 1760 yards = 1 mile	1 dekameter = 10 meters
	1 decimeter = 0.1 meter
	1 centimeter = 0.01 meter
	1 millimeter = 0.001 meter

EXAMPLE 8

The distance between two cities is k kilometers. Express this distance in meters.

Solution

Because 1 kilometer equals 1000 meters, we need to multiply k by 1000. Therefore, $1000k$ represents the distance in meters. ∎

EXAMPLE 9

The length of a sofa is i inches. Express that length in yards.

Solution

To change from inches to yards, we must divide by 36. Therefore $\frac{i}{36}$ represents in yards the length of the sofa. ∎

EXAMPLE 10 The width of a rectangle is w centimeters, and the length is 5 centimeters less than twice the width. What is the length of the rectangle? What is the perimeter of the rectangle?

Solution

We can represent the length of the rectangle by $2w - 5$. Now we can sketch a rectangle as in Figure 2.1 and record the given information. The perimeter of a rectangle is the sum of the lengths of the four sides. Therefore, the perimeter, in centimeters, is given by $2w + 2(2w - 5)$, which can be written as $2w + 4w - 10$ and then simplified to $6w - 10$.

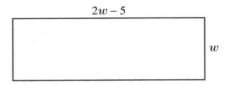

Figure 2.1

EXAMPLE 11 The length of a side of a square is x feet. Express the length of a side in inches. What is the area of the square in square inches?

Solution

Because 1 foot equals 12 inches, we need to multiply x by 12. Therefore, $12x$ represents the length of a side in inches. The area of a square is the length of a side squared. So the area in square inches is given by $(12x)^2 = (12x)(12x) = 12 \cdot 12 \cdot x \cdot x = 144x^2$.

CONCEPT QUIZ 2.2 For Problems 1–10, match the English phrase with its algebraic expression.

1. The product of x and y
2. Two less than x
3. x subtracted from 2
4. The difference of x and y
5. The quotient of x and y
6. The sum of x and y
7. Two times the sum of x and y
8. Two times x plus y
9. x squared minus y
10. Two more than x

A. $x - y$
B. $x + y$
C. $\dfrac{x}{y}$
D. $x - 2$
E. xy
F. $x^2 - y$
G. $2(x + y)$
H. $2 - x$
I. $x + 2$
J. $2x + y$

Section 2.2 Classroom Problem Set

Objective 2

1. The sum of two numbers is 56, and one of the numbers is y. Write an expression for the other number.

2. The sum of two numbers is 97, and one of the numbers is n. Write an expression for the other number.

3. The difference of two numbers is 20. The smaller number is x. Write an expression for the larger number.

4. The difference of two numbers is 5. The smaller number is y. Write an expression for the larger number.

5. The product of two numbers is 45, and one of the numbers is x. Write the other number as an algebraic expression.

6. The product of two numbers is 22, and one of the numbers is n. Write the other number as an algebraic expression.

7. A fax machine can transmit 12 pages in one minute. Represent the number of pages that can be sent in y minutes.

8. A printer can print 20 pages in one minute. Represent the number of pages that can be printed in c minutes.

9. Ramon has q quarters and d dimes. Express, in cents, the amount of money that Ramon has.

10. Cesar has d dimes and n nickels. Express, in cents, the amount of money that Cesar has.

11. An electric automobile travels at p miles per hour. How far will it travel in 4 hours?

12. An agility dog can run at x yards per second. How far can the dog run in 35 seconds?

13. The cost of an 8-day European cruise is e euros. Express the cost per day in euros.

14. A bicycle can travel b miles in t hours. Find the rate of the bicycle.

Objective 3

15. Two bolts on a machine are x centimeters apart. Express this distance in millimeters.

16. The side of a square has length of y meters. Write the length of the side of the square in centimeters.

17. The height of a plasma television is p inches. Express the height in feet.

18. The length of a bookcase is h inches. Write that length in yards.

19. The width of a rectangle is x inches, and the length is 3 inches more than twice the width. Write an expression to represent the length of the rectangle. Find the perimeter of the rectangle.

20. The width of a rectangle is w feet, and its length is four times the width. What is the perimeter of the rectangle in feet?

21. The length of a side of a square is s yards. Express the length of a side in feet. What is the area of the square in square feet?

22. The length of the side of a square is d feet. Find the area of the square in square inches.

THOUGHTS INTO WORDS

1. What does the phrase "translating from English to algebra" mean to you?

2. Your friend is having trouble with the following problem: A rectangular plot of ground is f feet long. What is its length in yards? She doesn't know if the answer should be $3f$ or $\frac{f}{3}$. What can you do to help her?

Answers to the Concept Quiz

1. E 2. D 3. H 4. A 5. C 6. B 7. G 8. J 9. F 10. I

2.3 Equations and Problem Solving

OBJECTIVES

1. Solve Equations Using Both the Addition-Subtraction and the Multiplication-Division Properties
2. Solve Word Problems

1 Solve Equations Using Both the Addition-Subtraction and the Multiplication-Division Properties

We often need to use more than one property of equality to help find the solution of an equation. Consider the next examples.

EXAMPLE 1 Solve $3x + 1 = 7$.

Solution

$$3x + 1 = 7$$
$$3x + 1 - 1 = 7 - 1 \quad \text{Subtract 1 from both sides}$$
$$3x = 6$$
$$\frac{3x}{3} = \frac{6}{3} \quad \text{Divide both sides by 3}$$
$$x = 2$$

We can check the potential solution by substituting it into the original equation to see whether we get a true numerical statement.

✔ **Check**

$$3x + 1 = 7$$
$$3(2) + 1 \stackrel{?}{=} 7$$
$$6 + 1 \stackrel{?}{=} 7$$
$$7 = 7$$

Now we know that the solution set is $\{2\}$. ∎

EXAMPLE 2 Solve $5x - 6 = 14$.

Solution

$$5x - 6 = 14$$
$$5x - 6 + 6 = 14 + 6 \quad \text{Add 6 to both sides}$$
$$5x = 20$$
$$\frac{5x}{5} = \frac{20}{5} \quad \text{Divide both sides by 5}$$
$$x = 4$$

✔ **Check**

$$5x - 6 = 14$$
$$5(4) - 6 \stackrel{?}{=} 14$$
$$20 - 6 \stackrel{?}{=} 14$$
$$14 = 14$$

The solution set is $\{4\}$. ∎

EXAMPLE 3 Solve $4 - 3a = 22$.

Solution

$$4 - 3a = 22$$
$$4 - 3a - 4 = 22 - 4 \quad \text{Subtract 4 from both sides}$$
$$-3a = 18$$

$$\frac{-3a}{-3} = \frac{18}{-3} \quad \text{Divide both sides by } -3$$

$$a = -6$$

✔ **Check**

$$4 - 3a = 22$$
$$4 - 3(-6) \stackrel{?}{=} 22$$
$$4 + 18 \stackrel{?}{=} 22$$
$$22 = 22$$

The solution set is $\{-6\}$. ■

Note that in Examples 1, 2, and 3, we first used the addition-subtraction property and then used the multiplication-division property. In general, this sequence of steps provides the easiest format for solving such equations. Perhaps you should convince yourself of that fact by doing Example 1 again, this time using the multiplication-division property first and then the addition-subtraction property.

EXAMPLE 4

Solve $19 = 2n + 4$.

Solution

$$19 = 2n + 4$$
$$19 - 4 = 2n + 4 - 4 \quad \text{Subtract 4 from both sides}$$
$$15 = 2n$$
$$\frac{15}{2} = \frac{2n}{2} \quad \text{Divide both sides by 2}$$
$$\frac{15}{2} = n$$

✔ **Check**

$$19 = 2n + 4$$
$$19 \stackrel{?}{=} 2\left(\frac{15}{2}\right) + 4$$
$$19 \stackrel{?}{=} 15 + 4$$
$$19 = 19$$

The solution set is $\left\{\frac{15}{2}\right\}$. ■

2 Solve Word Problems

In the previous section we translated English phrases into algebraic expressions. We are now ready to extend that idea to the translation of English *sentences* into algebraic *equations*. Such translations enable us to use the concepts of algebra to solve word problems. Let's consider some examples.

EXAMPLE 5　Apply Your Skill

A certain number added to 17 yields a sum of 29. What is the number?

Solution

Let n represent the number to be found. The sentence "a certain number added to 17 yields a sum of 29" translates to the algebraic equation $17 + n = 29$. To solve this equation, we use these steps:

$$17 + n = 29$$
$$17 + n - 17 = 29 - 17$$
$$n = 12$$

The solution is 12, which is the number asked for in the problem. ∎

We often refer to the statement "let n represent the number to be found" as **declaring the variable**. We need to choose a letter to use as a variable and indicate what it represents for a specific problem—this may seem like an insignificant idea, but as the problems become more complex, the process of declaring the variable becomes even more important. We could solve a problem such as Example 5 without setting up an algebraic equation; however, as problems increase in difficulty, the translation from English to algebra becomes critical. Therefore, even with these relatively simple problems, we need to concentrate on the translation process.

EXAMPLE 6　Apply Your Skill

Six years ago Bill was 13 years old. How old is he now?

Solution

Let y represent Bill's age now; therefore, $y - 6$ represents his age six years ago. Thus

$$y - 6 = 13$$
$$y - 6 + 6 = 13 + 6$$
$$y = 19$$

Bill is presently 19 years old. ∎

EXAMPLE 7　Apply Your Skill

Betty worked 8 hours Saturday and earned $66. How much did she earn per hour?

Solution A

Let x represent the amount Betty earned per hour. The number of hours worked times the wage per hour yields the total earnings. Thus

$$8x = 66$$
$$\frac{8x}{8} = \frac{66}{8}$$
$$x = 8.25$$

Betty earned $8.25 per hour.

Solution B

Let y represent the amount Betty earned per hour. The wage per hour equals the total wage divided by the number of hours. Thus

$$y = \frac{66}{8}$$
$$y = 8.25$$

Betty earned $8.25 per hour.

Sometimes we can use more than one equation to solve a problem. In Solution A we set up the equation in terms of multiplication, whereas in Solution B we were thinking in terms of division. ∎

EXAMPLE 8   Apply Your Skill

If 2 is subtracted from five times a certain number, the result is 28. Find the number.

Solution

Let n represent the number to be found. Translating the first sentence in the problem into an algebraic equation, we obtain

$$5n - 2 = 28$$

To solve this equation we proceed as follows:

$$5n - 2 + 2 = 28 + 2$$
$$5n = 30$$
$$\frac{5n}{5} = \frac{30}{5}$$
$$n = 6$$

The number to be found is 6. ∎

EXAMPLE 9   Apply Your Skill

The cost of a five-day vacation cruise package was $534. This cost included $339 for the cruise and an amount for two nights of lodging on shore. Find the cost per night of the on-shore lodging.

Solution

Let n represent the cost for one night of lodging; then $2n$ represents the total cost of lodging. Thus the total cost for the cruise and lodging is $534. We can proceed as follows:

Cost of cruise + Cost of lodging = $534
$$339 \quad + \quad 2n \quad = 534$$

To solve this equation we proceed as follows:

$$339 + 2n = 534$$
$$2n = 195$$
$$\frac{2n}{2} = \frac{195}{2}$$
$$n = 97.50$$

The cost of lodging per night is $97.50. ∎

CONCEPT QUIZ 2.3

For Problems 1–5, answer true or false.

1. Only one property of equality is necessary to solve any equation.
2. Substituting the solution into the original equation to obtain a true numerical statement can be used to check potential solutions.
3. The statement "let x represent the number" is referred to as checking the variable.
4. Sometimes there can be two approaches to solving a word problem.
5. To solve the equation, $\frac{1}{3}x - 2 = 7$, you could begin by either adding 2 to both sides of the equation or by multiplying both sides of the equation by 3.

For Problems 6–10, match the English sentence with its algebraic equation.

6. Three added to a number is 24.
7. The product of three and a number is 24.
8. Three less than a number is 24.
9. The quotient of a number and three is 24.
10. A number subtracted from three is 24.

A. $3x = 24$
B. $3 - x = 24$
C. $x + 3 = 24$
D. $x - 3 = 24$
E. $\frac{x}{3} = 24$

Section 2.3 Classroom Problem Set

Objective 1

1. Solve $4x + 3 = -17$.
2. Solve $17 = 2t + 5$.
3. Solve $3x - 2 = 13$.
4. Solve $5n - 6 = 19$.
5. Solve $1 - 5y = 36$.
6. Solve $17 - 2x = -19$.
7. Solve $24 = 4a - 3$.
8. Solve $17x - 41 = -37$.

Objective 2

9. A certain number added to 21 gives a sum of 78. What is the number?
10. Fifteen added to a certain number is 49. What is the number?
11. Five years ago Dahlia was 19 years old. How old is she now?
12. Twelve years ago Rachel was 5 years old. What is her present age?
13. Damien talked for 34 minutes on his cell phone and was charged $2.72. How much did he pay per minute for his phone call?
14. Aaron earns $1550 per week. How much does he earn per day?
15. If 6 is subtracted from four times a certain number, the result is 30. Find the number.
16. If 2 is subtracted from five times a certain number, the result is 48. Find the number.
17. An outlet store is selling a package of a high-definition DVD player and five DVD movies for $439.00. Find the cost for each DVD movie if the DVD player cost $349.00.
18. The Computer Store is selling a laptop computer and seven programs for a total of $1655.00. Find the cost of each program if the price of the laptop is $1200.

THOUGHTS INTO WORDS

1. Give a step-by-step description of how you would solve the equation $17 = -3x + 2$.
2. What does the phrase "declare a variable" mean when it refers to solving a word problem?
3. Suppose that you are helping a friend with his homework and he solves the equation $19 = 14 - x$ like this:

$$19 = 14 - x$$
$$19 + x = 14 - x + x$$
$$19 + x = 14$$
$$19 + x - 19 = 14 - 19$$
$$x = -5$$

The solution set is $\{-5\}$.

Does he have a correct solution set? What would you tell him about his method of solving the equation?

Answers to the Concept Quiz

1. False **2.** True **3.** False **4.** True **5.** True **6.** C **7.** A **8.** D **9.** E **10.** B

2.4 More on Solving Equations and Problem Solving

OBJECTIVES

1. Solve Equations by First Simplifying One Side of the Equation
2. Solve Equations of the Form $ax + b = cx + d$
3. Solve Word Problems That Involve Consecutive Numbers and Geometry

1 Solve Equations by First Simplifying One Side of the Equation

As equations become more complex, we will need additional steps to solve them, so it is important that we organize our work carefully to minimize the chances for error. Let's begin this section with some suggestions for solving equations, and then we will illustrate a *solution format* that is effective.

We can summarize the process of solving first-degree equations of one variable as follows:

Step 1 Simplify both sides of the equation as much as possible.

Step 2 Use the addition-subtraction property of equality to isolate a term that contains the variable on one side of the equation and a constant on the other.

Step 3 Use the multiplication-division property of equality to make the coefficient of the variable 1.

The following examples illustrate this step-by-step process for solving equations. Study them carefully and be sure that you understand each step.

EXAMPLE 1   Solve $5y - 4 + 3y = 12$.

Solution

$$5y - 4 + 3y = 12$$
$$8y - 4 = 12 \qquad \text{Combine similar terms on the left side}$$
$$8y - 4 + 4 = 12 + 4 \qquad \text{Add 4 to both sides}$$
$$8y = 16$$
$$\frac{8y}{8} = \frac{16}{8} \qquad \text{Divide both sides by 8}$$
$$y = 2$$

The solution set is $\{2\}$. You can do the check alone now! ∎

2 Solve Equations of the Form $ax + b = cx + d$

EXAMPLE 2   Solve $7x - 2 = 3x + 9$.

Solution

Note that both sides of the equation are in simplified form; thus we can begin by using the subtraction property of equality.

$$7x - 2 = 3x + 9$$
$$7x - 2 - 3x = 3x + 9 - 3x \quad \text{Subtract } 3x \text{ from both sides}$$
$$4x - 2 = 9$$
$$4x - 2 + 2 = 9 + 2 \quad \text{Add 2 to both sides}$$
$$4x = 11$$
$$\frac{4x}{4} = \frac{11}{4} \quad \text{Divide both sides by 4}$$
$$x = \frac{11}{4}$$

The solution set is $\left\{\dfrac{11}{4}\right\}$. ∎

EXAMPLE 3   Solve $5n + 12 = 9n - 16$.

Solution

$$5n + 12 = 9n - 16$$
$$5n + 12 - 9n = 9n - 16 - 9n \quad \text{Subtract } 9n \text{ from both sides}$$
$$-4n + 12 = -16$$
$$-4n + 12 - 12 = -16 - 12 \quad \text{Subtract 12 from both sides}$$
$$-4n = -28$$
$$\frac{-4n}{-4} = \frac{-28}{-4} \quad \text{Divide both sides by } -4$$
$$n = 7$$

The solution set is $\{7\}$. ∎

3 Solve Word Problems That Involve Consecutive Numbers and Geometry

As we expand our skill in solving equations, we also expand our ability to solve word problems. No one definite procedure will ensure success at solving word problems, but the following suggestions can be helpful.

> **Suggestions for Solving Word Problems**
>
> 1. Read the problem carefully and make sure that you understand the meanings of all the words. Be especially alert for any technical terms in the statement of the problem.
> 2. Read the problem a second time (perhaps even a third time) to get an overview of the situation described and to determine the known facts as well as what is to be found.

2.4 More on Solving Equations and Problem Solving

> 3. Sketch any figure, diagram, or chart that might be helpful in analyzing the problem.
> 4. Choose a meaningful variable to represent an unknown quantity in the problem (perhaps t if time is an unknown quantity); represent any other unknowns in terms of that variable.
> 5. Look for a **guideline** that you can use to set up an equation. A guideline might be a formula such as *distance equals rate times time*, or a statement of a relationship, such as *the sum of the two numbers is 28*. A guideline may also be indicated by a figure or diagram that you sketch for a particular problem.
> 6. Form an equation that contains the variable and that translates the conditions of the guideline from English to algebra.
> 7. Solve the equation and use the solution to determine all facts requested in the problem.
> 8. **Check all answers against the original statement of the problem.**

If you decide not to check an answer, at least use the *reasonableness of answer* idea as a partial check. That is, ask yourself, "Is this answer reasonable?" For example, if the problem involves two investments that total $10,000, then an answer of $12,000 for one investment is certainly *not reasonable*.

Now let's consider some problems and use these suggestions.

EXAMPLE 4   Apply Your Skill

Find two consecutive even numbers whose sum is 74.

Solution

To solve this problem, we must know the meaning of the technical phrase "two consecutive even numbers." Two consecutive even numbers are two even numbers that have one and only one whole number between them. For example, 2 and 4 are consecutive even numbers. Now we can proceed as follows: Let n represent the first even number; then $n + 2$ represents the next even number. Because their sum is 74, we can set up and solve the following equation:

$$n + (n + 2) = 74$$
$$2n + 2 = 74$$
$$2n + 2 - 2 = 74 - 2$$
$$2n = 72$$
$$\frac{2n}{2} = \frac{72}{2}$$
$$n = 36$$

If $n = 36$, then $n + 2 = 38$; thus the numbers are 36 and 38.

✔ **Check**

To check your answers for Example 4, determine whether they satisfy the conditions stated in the original problem. Because 36 and 38 are two consecutive even numbers, and $36 + 38 = 74$ (their sum is 74), we know that the answers are correct. ■

Suggestion 5 in our list of problem-solving suggestions was to "look for a *guideline* that can be used to set up an equation." The guideline may not be explicitly stated in the problem but may instead be implied by the nature of the problem. Consider the next example.

EXAMPLE 5   Apply Your Skill

Barry sells bicycles on a salary-plus-commission basis. He receives a monthly salary of $300 and a commission of $15 for each bicycle that he sells. How many bicycles must he sell in a month to have a total monthly salary of $750?

Solution

Let b represent the number of bicycles to be sold in a month. Then $15b$ represents Barry's commission for those bicycles. The *guideline* "fixed salary plus commission equals total monthly salary" generates the following equation:

Fixed salary + Commission = Total monthly salary
$$\$300 + 15b = \$750$$

Let's solve this equation.

$$300 + 15b - 300 = 750 - 300$$
$$15b = 450$$
$$\frac{15b}{15} = \frac{450}{15}$$
$$b = 30$$

He must sell 30 bicycles per month. (Does this number check?) ∎

Sometimes the guideline for setting up an equation to solve a problem is based on a geometric relationship. Several basic geometric relationships pertain to angle measure. Let's state three of these relationships and then consider some problems.

1. Two angles that together measure 90° (the symbol ° indicates degrees) are called **complementary angles**.
2. Two angles that together measure 180° are called **supplementary angles**.
3. The sum of the measures of the three angles of a triangle is 180°.

EXAMPLE 6   Apply Your Skill

One of two complementary angles is 14° larger than the other. Find the measure of each of the angles.

Solution

If we let a represent the measure of the smaller angle, then $a + 14$ represents the measure of the larger angle. Because they are complementary angles, their sum is 90° and we can proceed as follows:

$$a + a + 14 = 90$$
$$2a + 14 = 90$$
$$2a + 14 - 14 = 90 - 14$$
$$2a = 76$$
$$\frac{2a}{2} = \frac{76}{2}$$
$$a = 38$$

If $a = 38$, then $a + 14 = 52$, and the angles have measures of 38° and 52°. ∎

2.4 More on Solving Equations and Problem Solving

EXAMPLE 7   Apply Your Skill

Find the measures of the three angles of a triangle if the second is three times the first and the third is twice the second.

Solution

If we let a represent the measure of the smallest angle, then $3a$ and $2(3a)$ represent the measures of the other two angles. Therefore we can set up and solve the following equation:

$$a + 3a + 2(3a) = 180$$
$$a + 3a + 6a = 180$$
$$10a = 180$$
$$\frac{10a}{10} = \frac{180}{10}$$
$$a = 18$$

If $a = 18$, then $3a = 54$ and $2(3a) = 108$, so the angles have measures of $18°$, $54°$, and $108°$. ∎

CONCEPT QUIZ 2.4

For Problems 1–8, answer true or false.

1. If n represents a whole number, then $n + 1$ would represent the next consecutive whole number.
2. If n represents an odd whole number, then $n + 1$ would represent the next consecutive odd whole number.
3. If n represents an even whole number, then $n + 2$ would represent the next consecutive even whole number.
4. The sum of the measures of two complementary angles is $90°$.
5. The sum of the measures of two supplementary angles is $360°$.
6. The sum of the measures of the three angles in a triangle is $120°$.
7. When checking word problems, it is sufficient to check the solution in the equation.
8. For a word problem, the reasonableness of an answer is appropriate as a partial check.

Section 2.4   Classroom Problem Set

Objective 1

1. Solve $2x - 7 - 5x = 14$.
2. Solve $5n - 2 - 8n = 31$.

Objective 2

3. Solve $x - 5 = 6x - 45$.
4. Solve $-3y + 5 = -5y - 8$.
5. Solve $2a + 24 = 6a + 44$.
6. Solve $4x - 3 + 2x = 8x - 3 - x$.

Objective 3

7. Find two consecutive even numbers whose sum is 126.
8. Find three consecutive odd numbers whose sum is 159.
9. A pizza delivery driver gets paid $50.00 a night plus $0.75 for each delivery. How many deliveries must the driver make to earn $77.00 for the night?
10. The sum of a number and five times the number equals 18 less than three times the number. Find the number.
11. One of two supplementary angles is $56°$ smaller than the other angle. Find the measure of each angle.
12. If two angles are complementary, and the difference of their measures is $62°$, find the measure of each angle.
13. The measure of the third angle of a triangle is three times as large as the first angle. The second angle is $5°$ more than the third angle. Find the measures of all three angles.
14. One of the angles of a triangle has a measure of $40°$. Find the measures of the other two angles if the difference of their measures is $10°$.

THOUGHTS INTO WORDS

1. Give a step-by-step description of how you would solve the equation $3x + 4 = 5x - 2$.

2. Suppose your friend solved the problem "find two consecutive odd integers whose sum is 28" like this:

   $x + x + 1 = 28$
   $2x = 27$
   $x = \dfrac{27}{2} = 13\dfrac{1}{2}$

   She claims that $13\dfrac{1}{2}$ will check in the equation. Where has she gone wrong, and how would you help her?

Answers to the Concept Quiz
1. True   2. False   3. True   4. True   5. False   6. False   7. False   8. True

---

2.5 Equations Involving Parentheses and Fractional Forms

OBJECTIVES

1. Solve Equations That Involve the Use of the Distributive Property
2. Solve Equations That Involve Fractional Forms
3. Solve Equations That Are Contradictions or Identities
4. Solve Word Problems

1 Solve Equations That Involve the Use of the Distributive Property

We will use the distributive property frequently in this section as we expand our techniques for solving equations. Recall that in symbolic form, the distributive property states that $a(b + c) = ab + ac$. The following examples illustrate the use of this property to *remove parentheses*. Pay special attention to the last two examples, which involve a negative number in front of the parentheses.

$$3(x + 2) = \boxed{3 \cdot x + 3 \cdot 2} = 3x + 6$$
$$5(y - 3) = \boxed{5 \cdot y - 5 \cdot 3} = 5y - 15 \quad [a(b - c) = ab - ac]$$
$$2(4x + 7) = \boxed{2(4x) + 2(7)} = 8x + 14$$
$$-1(n + 4) = \boxed{(-1)(n) + (-1)(4)} = -n - 4$$
$$-6(x - 2) = \boxed{(-6)(x) - (-6)(2)} = -6x + 12$$

↓

Do this step mentally!

It is often necessary to solve equations in which the variable is part of an expression enclosed in parentheses. The distributive property is used to remove the parentheses, and then we proceed in the usual way. Consider the following examples. (Note that when solving an equation, we are beginning to show only the major steps.)

2.5 Equations Involving Parentheses and Fractional Forms

EXAMPLE 1   Solve $4(x + 3) = 2(x - 6)$.

Solution

$$4(x + 3) = 2(x - 6)$$
$4x + 12 = 2x - 12$ Applied distributive property on each side
$2x + 12 = -12$ Subtracted $2x$ from both sides
$2x = -24$ Subtracted 12 from both sides
$x = -12$ Divided both sides by 2

The solution set is $\{-12\}$.

It may be necessary to use the distributive property to remove more than one set of parentheses and then to combine similar terms. Consider the next two examples.

EXAMPLE 2   Solve $6(x - 7) - 2(x - 4) = 13$.

Solution

$6(x - 7) - 2(x - 4) = 13$ Be careful with this sign!
$6x - 42 - 2x + 8 = 13$ Distributive property
$4x - 34 = 13$ Combined similar terms
$4x = 47$ Added 34 to both sides
$x = \dfrac{47}{4}$ Divided both sides by 4

The solution set is $\left\{\dfrac{47}{4}\right\}$.

2 Solve Equations That Involve Fractional Forms

In a previous section we solved equations such as $x - \dfrac{2}{3} = \dfrac{3}{4}$ by adding $\dfrac{2}{3}$ to both sides. If an equation contains several fractions, then it is usually easier to *clear the equation of all fractions* by multiplying both sides by the least common denominator of all the denominators. Perhaps several examples will clarify this idea.

EXAMPLE 3   Solve $\dfrac{1}{2}x + \dfrac{2}{3} = \dfrac{5}{6}$.

Solution

$$\dfrac{1}{2}x + \dfrac{2}{3} = \dfrac{5}{6}$$

$6\left(\dfrac{1}{2}x + \dfrac{2}{3}\right) = 6\left(\dfrac{5}{6}\right)$ 6 is the LCD of 2, 3, and 6

$6\left(\dfrac{1}{2}x\right) + 6\left(\dfrac{2}{3}\right) = 6\left(\dfrac{5}{6}\right)$ Distributive property
                    Note how the equation has been *cleared*
$3x + 4 = 5$ *of all fractions*
$3x = 1$
$x = \dfrac{1}{3}$

The solution set is $\left\{\dfrac{1}{3}\right\}$.

EXAMPLE 4

Solve $\dfrac{5n}{6} - \dfrac{1}{4} = \dfrac{3}{8}$.

Solution

$$\dfrac{5n}{6} - \dfrac{1}{4} = \dfrac{3}{8}$$ Remember $\dfrac{5n}{6} = \dfrac{5}{6}n$

$$24\left(\dfrac{5n}{6} - \dfrac{1}{4}\right) = 24\left(\dfrac{3}{8}\right)$$ 24 is the LCD of 6, 4, and 8

$$24\left(\dfrac{5n}{6}\right) - 24\left(\dfrac{1}{4}\right) = 24\left(\dfrac{3}{8}\right)$$ Distributive property

$$20n - 6 = 9$$
$$20n = 15$$
$$n = \dfrac{15}{20} = \dfrac{3}{4}$$

The solution set is $\left\{\dfrac{3}{4}\right\}$. ∎

We use many of the ideas presented in this section to help solve the equations in the next two examples. Study the solutions carefully and be sure that you can supply reasons for each step.

EXAMPLE 5

Solve $\dfrac{x+3}{2} + \dfrac{x+4}{5} = \dfrac{3}{10}$.

Solution

$$\dfrac{x+3}{2} + \dfrac{x+4}{5} = \dfrac{3}{10}$$

$$10\left(\dfrac{x+3}{2} + \dfrac{x+4}{5}\right) = 10\left(\dfrac{3}{10}\right)$$ 10 is the LCD of 2, 5, and 10

$$10\left(\dfrac{x+3}{2}\right) + 10\left(\dfrac{x+4}{5}\right) = 10\left(\dfrac{3}{10}\right)$$ Distributive property

$$5(x+3) + 2(x+4) = 3$$
$$5x + 15 + 2x + 8 = 3$$
$$7x + 23 = 3$$
$$7x = -20$$
$$x = -\dfrac{20}{7}$$

The solution set is $\left\{-\dfrac{20}{7}\right\}$. ∎

EXAMPLE 6

Solve $\dfrac{x-1}{4} - \dfrac{x-2}{6} = \dfrac{2}{3}$.

Solution

$$\dfrac{x-1}{4} - \dfrac{x-2}{6} = \dfrac{2}{3}$$

$$12\left(\dfrac{x-1}{4} - \dfrac{x-2}{6}\right) = 12\left(\dfrac{2}{3}\right)$$ 12 is the LCD of 4, 6, and 3

2.5 Equations Involving Parentheses and Fractional Forms

$$12\left(\frac{x-1}{4}\right) - 12\left(\frac{x-2}{6}\right) = 12\left(\frac{2}{3}\right) \quad \text{Distributive property}$$
$$3(x-1) - 2(x-2) = 8$$
$$3x - 3 - 2x + 4 = 8 \quad \text{Be careful with this sign!}$$
$$x + 1 = 8$$
$$x = 7$$

The solution set is {7}.

3 Solve Equations That Are Contradictions or Identities

All the equations we have solved thus far are conditional equations. For instance, the equation $3x = 12$ is a true statement under the condition that $x = 4$. Now we will consider two other types of equations—contradictions and identities. When the equation is not true under any condition, then the equation is called a **contradiction**. The solution set for a contradiction is the empty or null set and is denoted by $\varnothing$. When an equation is true for any permissible value of the variable for which the equation is defined, the equation is called an **identity**, and the solution set for an identity is the set of all real numbers for which the equation is defined. We will denote the set of all real numbers as {All reals}. The following examples show the solutions for these types of equations.

EXAMPLE 7

Solve $4x + 5 = 2(2x - 8)$.

Solution

$$4x + 5 = 2(2x - 8)$$
$$4x + 5 = 4x - 16 \quad \text{Distributive property}$$
$$5 = -16 \quad \text{Subtracted } 4x \text{ from both sides}$$

The result is a false statement. Therefore the equation is a contradiction. There is no value of x that will make the equation a true statement, and hence the solution set is the empty set, $\varnothing$.

EXAMPLE 8

Solve $5(x + 3) + 2x - 4 = 7x + 11$.

Solution

$$5(x + 3) + 2x - 4 = 7x + 11 \quad \text{Distributive property}$$
$$5x + 15 + 2x - 4 = 7x + 11$$
$$7x + 11 = 7x + 11 \quad \text{Combined similar terms}$$
$$11 = 11 \quad \text{Subtracted } 7x \text{ from both sides}$$

The last step gives an equation with no variable terms, but the equation is a true statement. This equation is an identity, and any real number is a solution. The solution set would be written as {All reals}.

4 Solve Word Problems

We are now ready to solve some word problems using equations of the different types presented in this section.

EXAMPLE 9   Apply Your Skill

Loretta has 19 coins (quarters and nickels) that amount to $2.35. How many coins of each kind does she have?

Solution

Let q represent the number of quarters. Then $19 - q$ represents the number of nickels. We can use the following guideline to help set up an equation:

Value of quarters in cents + Value of nickels in cents = Total value in cents

$$25q + 5(19 - q) = 235$$

We can solve the equation in this way:

$$25q + 95 - 5q = 235$$
$$20q + 95 = 235$$
$$20q = 140$$
$$q = 7$$

If $q = 7$, then $19 - q = 12$, so she has 7 quarters and 12 nickels. ∎

EXAMPLE 10   Apply Your Skill

Find a number such that 4 less than two-thirds the number is equal to one-sixth the number.

Solution

Let n represent the number. Then $\frac{2}{3}n - 4$ represents 4 less than two-thirds the number, and $\frac{1}{6}n$ represents one-sixth the number.

$$\frac{2}{3}n - 4 = \frac{1}{6}n$$
$$6\left(\frac{2}{3}n - 4\right) = 6\left(\frac{1}{6}n\right)$$
$$4n - 24 = n$$
$$3n - 24 = 0$$
$$3n = 24$$
$$n = 8$$

The number is 8. ∎

EXAMPLE 11   Apply Your Skill

Lance is paid $1\frac{1}{2}$ times his normal hourly rate for each hour he works in excess of 40 hours in a week. Last week he worked 50 hours and earned $462. What is his normal hourly rate?

Solution

Let x represent his normal hourly rate. Then $\frac{3}{2}x$ represents $1\frac{1}{2}$ times his normal hourly rate. We can use the following guideline to help set up the equation:

Regular wages for first 40 hours + Wages for 10 hours of overtime = Total wages

$$40x + 10\left(\frac{3}{2}x\right) = 462$$

We get

$$40x + 15x = 462$$
$$55x = 462$$
$$x = 8.40$$

His normal hourly rate is $8.40.

EXAMPLE 12 Apply Your Skill

Find three consecutive whole numbers such that the sum of the first plus twice the second plus three times the third is 134.

Solution

Let n represent the first whole number. Then $n + 1$ represents the second whole number, and $n + 2$ represents the third whole number. We have

$$n + 2(n + 1) + 3(n + 2) = 134$$
$$n + 2n + 2 + 3n + 6 = 134$$
$$6n + 8 = 134$$
$$6n = 126$$
$$n = 21$$

The numbers are 21, 22, and 23.

Keep in mind that the problem-solving suggestions we offered in Section 2.4 simply outline a general algebraic approach to solving problems. You will add to this list throughout this course and in any subsequent mathematics courses that you take. Furthermore, you will be able to pick up additional problem-solving ideas from your instructor and from fellow classmates as you discuss problems in class. Always be on the alert for any ideas that might help you become a better problem solver.

CONCEPT QUIZ 2.5

For Problems 1–10, answer true or false.

1. To solve an equation of the form $a(x + b) = 14$, the associative property would be applied to remove the parentheses.
2. Multiplying both sides of an equation by the common denominator of all fractions in the equation clears the equation of all fractions.
3. If Jack has 15 coins (dimes and quarters), and x represents the number of dimes, then $x - 15$ represents the number of quarters.
4. The equation $3(x + 1) = 3x + 3$ has an infinite number of solutions.
5. The equation $2x = 0$ has no solution.
6. The equation $4x + 5 = 4x + 3$ has no solution.
7. The solution set for an equation that is a contradiction is the null set.
8. For a conditional equation, the solution set is the set of all real numbers.
9. When an equation is true for any permissible value of the variable, then the equation is called an identity.
10. When an equation is true for only certain values of the variable, then the equation is called a contradiction.

Section 2.5 Classroom Problem Set

Objective 1

1. Solve $3(x - 4) = -5(x + 2)$.
2. Solve $-(x + 7) = -2(x + 10)$.
3. Solve $2(x - 4) - 7(x + 2) = 3$.
4. Solve $3(n - 10) - 5(x + 12) = -86$.

Objective 2

5. Solve $\dfrac{4}{5}x + \dfrac{2}{3} = \dfrac{7}{15}$.
6. Solve $\dfrac{1}{2}x - \dfrac{3}{5} = \dfrac{3}{4}$.
7. Solve $\dfrac{7y}{12} - \dfrac{5}{6} = \dfrac{3}{8}$.
8. Solve $\dfrac{n}{6} + \dfrac{3n}{8} = \dfrac{5}{12}$.
9. Solve $\dfrac{x - 4}{3} + \dfrac{x + 2}{5} = \dfrac{7}{10}$.
10. Solve $\dfrac{x + 2}{3} + \dfrac{x + 3}{4} = \dfrac{13}{3}$.
11. Solve $\dfrac{x - 1}{8} - \dfrac{x - 4}{3} = \dfrac{5}{4}$.
12. Solve $\dfrac{x + 8}{2} - \dfrac{x + 10}{7} = \dfrac{3}{4}$.

Objective 3

13. Solve $6x + 1 = 2(3x + 2)$.
14. Solve $4x = 6x - 2(x - 8)$.
15. Solve $3(x - 1) + 2x + 7 = 5x + 4$.
16. Solve $3x + 2(x + 6) = 5x + 12$.

Objective 4

17. Michaela has 20 coins (quarters and dimes) that amount to $3.95. How many coins of each kind does she have?

18. Ike has some nickels and dimes, amounting to $2.90. The number of dimes is 1 less than twice the number of nickels. How many coins of each kind does he have?

19. Find a number such that 12 less than three-fourths of the number is equal to one-half of the number.

20. The sum of three-eighths of a number and five-sixths of the same number is 29. Find the number.

21. Ramon is paid $1\dfrac{1}{2}$ times his normal hourly rate for each hour he works in excess of 40 hours a week. Last week he worked 52 hours and earned $696. Find his normal hourly rate.

22. Ellen is paid "time and a half" for each hour over 40 hours worked in a week. Last week she worked 44 hours and earned $391. What is Ellen's normal hourly rate?

23. Find three consecutive whole numbers such that the sum of three times the first number plus twice the second number plus the third number is 76.

24. Find four consecutive whole numbers such that the sum of the first three equals the fourth number.

THOUGHTS INTO WORDS

1. Discuss how you would solve the equation
$$3(x - 2) - 5(x + 3) = -4(x + 9)$$

2. Why must potential answers to word problems be checked back in the original statement of the problem?

3. Consider these two solutions:

$$3(x + 2) = 9 \qquad 3(x - 4) = 7$$
$$\dfrac{3(x + 2)}{3} = \dfrac{9}{3} \qquad \dfrac{3(x - 4)}{3} = \dfrac{7}{3}$$
$$x + 2 = 3 \qquad x - 4 = \dfrac{7}{3}$$
$$x = 1 \qquad x = \dfrac{19}{3}$$

Are both of these solutions correct? How effective is the approach?

4. Make up an equation whose solution set is the null set. Explain why the solution set is null.

5. Make up an equation whose solution set is the set of all real numbers. Explain why the solution set is all real numbers.

Answers to the Concept Quiz
1. False   **2.** True   **3.** False   **4.** True   **5.** False   **6.** True   **7.** True   **8.** False   **9.** True   **10.** False

2.6 Inequalities

OBJECTIVES

1. Show the Solution Set of an Inequality by Using Set-Builder Notation and by Graphing
2. Solve Inequalities

Just as we use the symbol = to represent *is equal to*, we also use the symbols < and > to represent *is less than* and *is greater than*, respectively. The following are examples of **statements of inequality**. Note that the first four are true statements, and the last two are false.

$6 + 4 > 7$	True Statement
$8 - 2 < 14$	True Statement
$4 \cdot 8 > 4 \cdot 6$	True Statement
$5 \cdot 2 < 5 \cdot 7$	True Statement
$5 + 8 > 19$	False Statement
$9 - 2 < 3$	False Statement

Algebraic inequalities contain one or more variables. Here are some examples of algebraic inequalities:

$x + 3 > 4$

$2x - 1 < 6$

$x^2 + 2x - 1 > 0$

$2x + 3y < 7$

$7ab < 9$

An algebraic inequality such as $x + 1 > 2$ is neither true nor false as it stands; it is called an **open sentence**. Each time a number is substituted for x, the algebraic inequality $x + 1 > 2$ becomes a numerical statement that is either true or false. For example, if $x = 0$, then $x + 1 > 2$ becomes $0 + 1 > 2$, which is false. If $x = 2$, then $x + 1 > 2$ becomes $2 + 1 > 2$, which is true. **Solving an inequality** refers to the process of finding the numbers that make an algebraic inequality a true numerical statement. We say that such numbers, which are called the solutions of the inequality, *satisfy* the inequality. The set of all solutions of an inequality is called its **solution set**.

1 Show the Solution Set of an Inequality by Using Set-Builder Notation and by Graphing

We often state solution sets for inequalities with set-builder notation. For example, the solution set for $x + 1 > 2$ is the set of real numbers greater than 1, expressed as $\{x|x > 1\}$. The set builder notation $\{x|x > 1\}$ is read "the set of all x such that x is greater than 1." We sometimes graph solution sets for inequalities on a number line; the solution set for $\{x|x > 1\}$ is pictured in Figure 2.2.

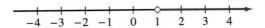

Figure 2.2

The open circle at 1 indicates that 1 is *not* a solution, and the red part of the line to the right of 1 indicates that all real numbers greater than 1 are solutions. We refer to the red portion of the number line as the *graph* of the solution set $\{x|x > 1\}$.

The solution set for $x + 2 \geq 5$ is the set of real numbers greater than or equal to 3; it is written as $\{x|x \geq 3\}$ in set-builder notation. The graph of the solution set $\{x|x \geq 3\}$ is shown in Figure 2.3. The solid dot at 3 indicates that 3 is a solution, and the red part of the line to the right of 3 indicates that all real numbers greater than 3 are solutions.

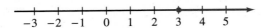

Figure 2.3

The examples in the table (Figure 2.4) below contain some simple algebraic inequalities, their solution sets, and graphs of the solution sets. Look them over very carefully to be sure you understand the symbols.

Algebraic inequality	Solution set	Graph of solution set
$x < 2$	$\{x\|x < 2\}$	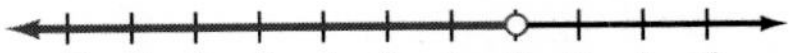
$x > -1$	$\{x\|x > -1\}$	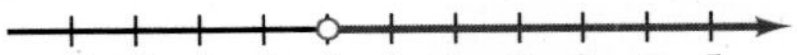
$3 < x$	$\{x\|x > 3\}$	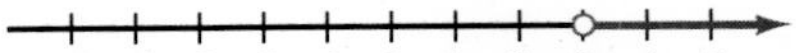
$x \geq 1$ ($\geq$ is read "greater than or equal to")	$\{x\|x \geq 1\}$	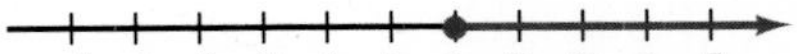
$x \leq 2$ ($\leq$ is read "less than or equal to")	$\{x\|x \leq 2\}$	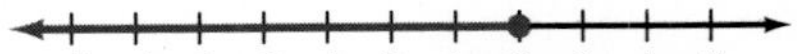
$1 \geq x$	$\{x\|x \leq 1\}$	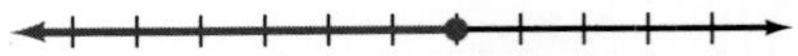

Figure 2.4

2 Solve Inequalities

The general process for solving inequalities closely parallels that for solving equations. We continue to replace the given inequality with equivalent but simpler inequalities. For example,

$$2x + 1 > 9 \tag{1}$$
$$2x > 8 \tag{2}$$
$$x > 4 \tag{3}$$

are all equivalent inequalities; that is, they have the same solutions. Thus to solve (1) we can find the solutions of (3), which are obviously all numbers greater than 4. The exact procedure for simplifying inequalities is based primarily on two properties. The first of these is the **addition-subtraction property of inequality**.

Property 2.4 Addition-Subtraction Property of Inequality

For all real numbers a, b, and c,

1. $a > b$ if and only if $a + c > b + c$.
2. $a > b$ if and only if $a - c > b - c$.

Property 2.4 states that any number can be added to or subtracted from both sides of an inequality, and the result is an equivalent inequality. The property is stated in terms of $>$, but analogous properties exist for $<$, $\geq$, and $\leq$. Consider the use of this property in the next three examples.

EXAMPLE 1

Solve $x - 3 > -1$ and graph the solutions.

Solution

$$x - 3 > -1$$
$$x - 3 + 3 > -1 + 3 \quad \text{Add 3 to both sides}$$
$$x > 2$$

The solution set is $\{x | x > 2\}$, and it can be graphed as shown in Figure 2.5.

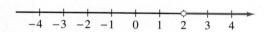

Figure 2.5

EXAMPLE 2

Solve $x + 4 \leq 5$ and graph the solutions.

Solution

$$x + 4 \leq 5$$
$$x + 4 - 4 \leq 5 - 4 \quad \text{Subtract 4 from both sides}$$
$$x \leq 1$$

The solution set is $\{x | x \leq 1\}$, and it can be graphed as shown in Figure 2.6.

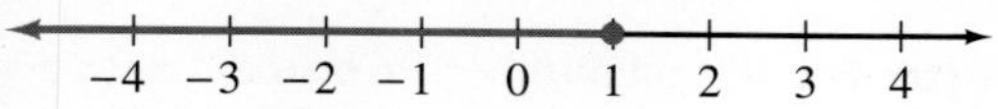

Figure 2.6

EXAMPLE 3

Solve $5 > 6 + x$ and graph the solutions.

Solution

$$5 > 6 + x$$
$$5 - 6 > 6 + x - 6 \quad \text{Subtract 6 from both sides}$$
$$-1 > x$$

Because $-1 > x$ is equivalent to $x < -1$, the solution set is $\{x|x < -1\}$. It can be graphed as shown in Figure 2.7.

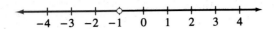

Figure 2.7

Now let's look at some numerical examples to see what happens when both sides of an inequality are multiplied or divided by some number.

$$4 > 3 \quad \rightarrow \quad 5(4) > 5(3) \quad \rightarrow \quad 20 > 15$$

$$-2 > -3 \quad \rightarrow \quad 4(-2) > 4(-3) \quad \rightarrow \quad -8 > -12$$

$$6 > 4 \quad \rightarrow \quad \frac{6}{2} > \frac{4}{2} \quad \rightarrow \quad 3 > 2$$

$$8 > -2 \quad \rightarrow \quad \frac{8}{4} > \frac{-2}{4} \quad \rightarrow \quad 2 > -\frac{1}{2}$$

Note that multiplying or dividing both sides of an inequality by a positive number produces an inequality of the same sense. This means that if the original inequality is *greater than*, then the new inequality is *greater than*, and if the original is *less than*, then the resulting inequality is *less than*.

Now note what happens when we multiply or divide both sides by a negative number:

$$3 < 5 \quad \rightarrow \quad -2(3) > -2(5) \quad \rightarrow \quad -6 > -10$$

$$-4 < 1 \quad \rightarrow \quad -5(-4) > -5(1) \quad \rightarrow \quad 20 > -5$$

$$14 > 2 \quad \rightarrow \quad \frac{14}{-2} < \frac{2}{-2} \quad \rightarrow \quad -7 < -1$$

$$-3 > -6 \quad \rightarrow \quad \frac{-3}{-3} < \frac{-6}{-3} \quad \rightarrow \quad 1 < 2$$

Multiplying or dividing both sides of an inequality *by a negative number reverses the sense of the inequality*. Property 2.5 summarizes these ideas.

Property 2.5 Multiplication-Division Property of Inequality

(a) For all real numbers a, b, and c, with $c > 0$,

  1. $a > b$ if and only if $ac > bc$
  2. $a > b$ if and only if $\dfrac{a}{c} > \dfrac{b}{c}$

(b) For all real numbers a, b, and c, with $c < 0$,

  1. $a > b$ if and only if $ac < bc$
  2. $a > b$ if and only if $\dfrac{a}{c} < \dfrac{b}{c}$

Similar properties hold when each inequality is reversed or when $>$ is replaced with $\geq$, and when $<$ is replaced with $\leq$. For example, if $a \leq b$ and $c < 0$, then $ac \geq bc$ and $\dfrac{a}{c} \geq \dfrac{b}{c}$.

Observe the use of Property 2.5 in the next three examples.

EXAMPLE 4

Solve $2x > 4$.

Solution

$$2x > 4$$
$$\frac{2x}{2} > \frac{4}{2} \quad \text{Divide both sides by 2}$$
$$x > 2$$

The solution set is $\{x | x > 2\}$.

EXAMPLE 5

Solve $\frac{3}{4}x \leq \frac{1}{5}$.

Solution

$$\frac{3}{4}x \leq \frac{1}{5}$$
$$\frac{4}{3}\left(\frac{3}{4}x\right) \leq \frac{4}{3}\left(\frac{1}{5}\right) \quad \text{Multiply both sides by } \frac{4}{3}$$
$$x \leq \frac{4}{15}$$

The solution set is $\left\{x | x \leq \frac{4}{15}\right\}$.

EXAMPLE 6

Solve $-3x > 9$.

Solution

$$-3x > 9$$
$$\frac{-3x}{-3} < \frac{9}{-3} \quad \text{Divide both sides by } -3, \text{ which reverses the inequality}$$
$$x < -3$$

The solution set is $\{x | x < -3\}$.

As we mentioned earlier, many of the same techniques used to solve equations may be used to solve inequalities. However, we must be extremely careful when we apply Property 2.5. Study the following examples and note the similarities between solving equations and solving inequalities.

EXAMPLE 7

Solve $4x - 3 > 9$.

Solution

$$4x - 3 > 9$$
$$4x - 3 + 3 > 9 + 3 \quad \text{Add 3 to both sides}$$
$$4x > 12$$
$$\frac{4x}{4} > \frac{12}{4} \quad \text{Divide both sides by 4}$$
$$x > 3$$

The solution set is $\{x | x > 3\}$.

EXAMPLE 8

Solve $-3n + 5 < 11$.

Solution

$$-3n + 5 < 11$$
$$-3n + 5 - 5 < 11 - 5 \quad \text{Subtract 5 from both sides}$$
$$-3n < 6$$
$$\frac{-3n}{-3} > \frac{6}{-3} \quad \text{Divide both sides by } -3, \text{ which reverses the inequality}$$
$$n > -2$$

The solution set is $\{n | n > -2\}$. ∎

Checking the solutions for an inequality presents a problem. Obviously we cannot check all the infinitely many solutions for a particular inequality. However, by checking at least one solution, especially when the multiplication-division property is used, we might catch the common mistake of forgetting to reverse the sense of the inequality. In Example 8 we are claiming that all numbers greater than -2 will satisfy the original inequality. Let's check one such number in the original inequality, say, -1.

$$-3n + 5 < 11$$
$$-3(-1) + 5 \stackrel{?}{<} 11$$
$$3 + 5 \stackrel{?}{<} 11$$
$$8 < 11$$

Thus -1 satisfies the original inequality. If we had forgotten to reverse the sense of the inequality when we divided both sides by -3, our answer would have been $n < -2$, and the check would have detected the error.

CONCEPT QUIZ 2.6

For Problems 1–10, answer true or false.

1. Numerical statements of inequality are always true.
2. The algebraic statement $x + 4 > 6$ is called an open sentence.
3. The algebraic inequality $2x > 10$ has one solution.
4. The algebraic inequality $x < 3$ has an infinite number of solutions.
5. The set-builder notation $\{x | x < -5\}$ is read "the set of variables that are particular to $x < -5$."
6. When graphing the solution set of an inequality, a solid dot is used to include the endpoint.
7. The properties for solving inequalities are the same as the properties for solving equations.
8. The solution set of the inequality $x \leq 6$ includes the number 6.
9. When multiplying both sides of an inequality by a negative number, the sense of the inequality stays the same.
10. When adding a negative number to both sides of an inequality, the sense of the inequality stays the same.

Section 2.6   Classroom Problem Set

Objective 2

1. Solve $x + 5 > 2$ and graph the solutions.
2. Solve $x + 2 \geq 2$ and graph the solutions.
3. Solve $x - 3 \leq -1$ and graph the solutions.
4. Solve $x + 1 \leq 0$ and graph the solutions.
5. Solve $8 > 5 + x$ and graph the solutions.
6. Solve $19 > 27 + y$ and graph the solutions.
7. Solve $7x > 21$.
8. Solve $8x > 28$.

9. Solve $\frac{2}{3}x \leq \frac{4}{7}$.
10. Solve $\frac{4}{5}x \geq \frac{8}{9}$.
13. Solve $2x - 1 > 7$.
14. Solve $8x + 3 > 25$.
11. Solve $-5x > 10$.
12. Solve $-9n \geq -63$.
15. Solve $-4y + 5 < 17$.
16. Solve $-x - 4 - 3x > 5$.

THOUGHTS INTO WORDS

1. Do the *greater-than* and *less-than* relations possess the symmetric property? Explain your answer.

2. Is the solution set for $x < 3$ the same as that for $3 > x$? Explain your answer.

3. How would you convince someone that it is necessary to reverse the sense of the inequality when multiplying both sides of an inequality by a negative number?

Answers to the Concept Quiz
1. False 2. True 3. False 4. True 5. False 6. True 7. False 8. True 9. False 10. True

2.7 Inequalities, Compound Inequalities, and Problem Solving

OBJECTIVES

1. Solve Inequalities
2. Solve Inequalities That Involve the Use of the Distributive Property
3. Solve Inequalities That Involve Fractional Forms
4. Solve Compound Inequalities
5. Solve Word Problems That Translate into Inequality Statements

1 Solve Inequalities

Let's begin this section by solving four inequalities with the same basic steps we used with equations. Again, be careful when applying the multiplication-division property of inequality.

EXAMPLE 1 Solve $5x + 8 \leq 3x - 10$.

Solution

$$5x + 8 \leq 3x - 10$$
$$5x + 8 - 3x \leq 3x - 10 - 3x \qquad \text{Subtract } 3x \text{ from both sides}$$
$$2x + 8 \leq -10$$
$$2x + 8 - 8 \leq -10 - 8 \qquad \text{Subtract 8 from both sides}$$
$$2x \leq -18$$
$$\frac{2x}{2} \leq \frac{-18}{2} \qquad \text{Divide both sides by 2}$$
$$x \leq -9$$

The solution set is $\{x | x \leq -9\}$. ∎

2 Solve Inequalities That Involve the Use of the Distributive Property

EXAMPLE 2 Solve $4(x + 3) + 3(x - 4) \geq 2(x - 1)$.

Solution

$$4(x + 3) + 3(x - 4) \geq 2(x - 1)$$
$$4x + 12 + 3x - 12 \geq 2x - 2 \quad \text{Distributive property}$$
$$7x \geq 2x - 2 \quad \text{Combine similar terms}$$
$$7x - 2x \geq 2x - 2 - 2x \quad \text{Subtract } 2x \text{ from both sides}$$
$$5x \geq -2$$
$$\frac{5x}{5} \geq \frac{-2}{5} \quad \text{Divide both sides by 5}$$
$$x \geq -\frac{2}{5}$$

The solution set is $\left\{ x \mid x \geq -\frac{2}{5} \right\}$. ∎

3 Solve Inequalities That Involve Fractional Forms

EXAMPLE 3 Solve $-\frac{3}{2}n + \frac{1}{6}n < \frac{3}{4}$.

Solution

$$-\frac{3}{2}n + \frac{1}{6}n < \frac{3}{4}$$
$$12\left(-\frac{3}{2}n + \frac{1}{6}n\right) < 12\left(\frac{3}{4}\right) \quad \text{Multiply both sides by 12, the LCD of all denominators}$$
$$12\left(-\frac{3}{2}n\right) + 12\left(\frac{1}{6}n\right) < 12\left(\frac{3}{4}\right) \quad \text{Distributive property}$$
$$-18n + 2n < 9$$
$$-16n < 9$$
$$\frac{-16n}{-16} > \frac{9}{-16} \quad \text{Divide both sides by } -16, \text{ which reverses the inequality}$$
$$n > -\frac{9}{16}$$

The solution set is $\left\{ n \mid n > -\frac{9}{16} \right\}$. ∎

In Example 3 we are claiming that all numbers greater than $-\frac{9}{16}$ will satisfy the original inequality. Let's check one number, say, 0.

$$-\frac{3}{2}n + \frac{1}{6}n < \frac{3}{4}$$
$$-\frac{3}{2}(0) + \frac{1}{6}(0) \overset{?}{<} \frac{3}{4}$$
$$0 < \frac{3}{4}$$

Therefore, 0 satisfies the original inequality. If we had forgotten to reverse the inequality sign when we divided both sides by -16, then our answer would have been $n < -\dfrac{9}{16}$, and the check would have detected the error.

EXAMPLE 4

Solve $\dfrac{x+2}{5} - \dfrac{x+1}{10} \leq \dfrac{3}{10}$.

Solution

$$\dfrac{x+2}{5} - \dfrac{x+1}{10} \leq \dfrac{3}{10}$$

$$10\left(\dfrac{x+2}{5} - \dfrac{x+1}{10}\right) \leq 10\left(\dfrac{3}{10}\right) \quad \text{Multiply both sides by 10}$$

$$10\left(\dfrac{x+2}{5}\right) - 10\left(\dfrac{x+1}{10}\right) \leq 10\left(\dfrac{3}{10}\right) \quad \text{Distributive property}$$

$$2(x+2) - 1(x+1) \leq 3 \quad \text{Don't forget to distribute the } -1$$

$$2x + 4 - x - 1 \leq 3$$

$$x + 3 \leq 3 \quad \text{Subtract 3 from both sides}$$

$$x \leq 0$$

The solution set is $\{x | x \leq 0\}$. ∎

4 Solve Compound Inequalities

The words "and" and "or" are used in mathematics to form compound statements. We use "and" and "or" to join two inequalities to form a compound inequality.

Consider the compound inequality

$$x > 2 \quad \text{and} \quad x < 5$$

For the solution set, we must find values of x that make both inequalities true statements. The solution set of a compound inequality formed by the word "and" is the **intersection** of the solution sets of the two inequalities. The intersection of two sets, denoted by ∩, contains the elements that are common to both sets. For example, if $A = \{1, 2, 3, 4, 5, 6\}$ and $B = \{0, 2, 4, 6, 8, 10\}$, then $A \cap B = \{2, 4, 6\}$. So to find the solution set of the compound inequality $x > 2$ and $x < 5$, we find the solution set for each inequality and then determine the solutions that are common to both solution sets.

EXAMPLE 5

Graph the solution set for the compound inequality $x > 2$ and $x < 5$, and write the solution set in set-builder notation.

Solution

$x > 2$ (a)

$x < 5$ (b)

$x > 2$ and $x < 5$ (c)

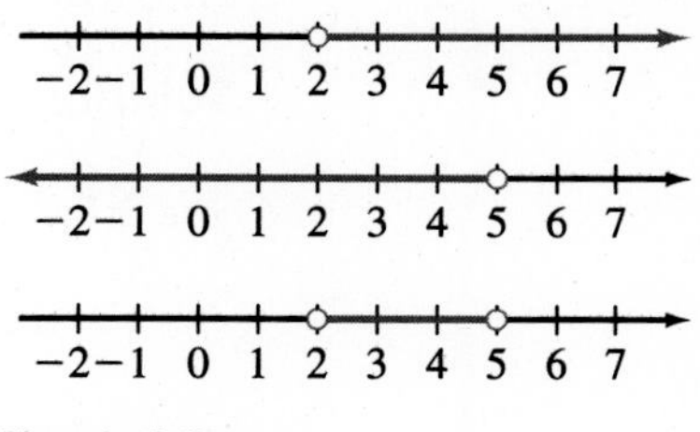

Figure 2.8

Thus all numbers greater than 2 and less than 5 are included in the solution set $\{x | 2 < x < 5\}$, and the graph is shown in Figure 2.8(c). ∎

EXAMPLE 6

Graph the solution set for the compound inequality $x \leq 1$ and $x \leq 4$ and write the solution in set-builder notation.

Solution

$x \leq 1$ (a)

$x \leq 4$ (b)

$x \leq 1$ and $x \leq 4$ (c)

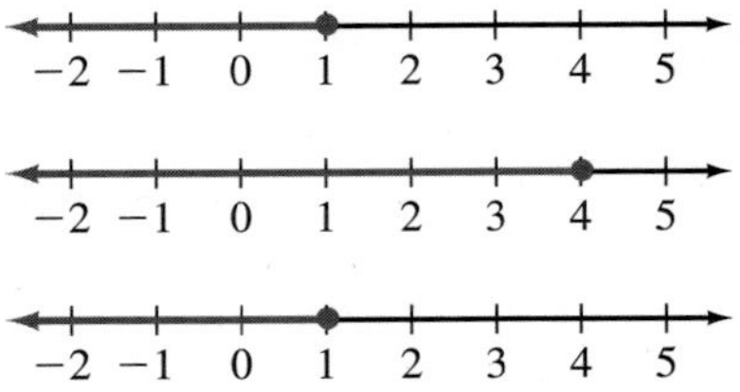

Figure 2.9

The intersection of the two solution sets is $x \leq 1$. The solution set $\{x | x \leq 1\}$ contains all the numbers that are less than or equal to 1, and the graph is shown in Figure 2.9(c). ∎

The solution set of a compound inequality formed by the word "or" is the **union** of the solution sets of the two inequalities. The union of two sets, denoted by ∪, contains all the elements in both sets. For example, if $A = \{0, 1, 2\}$ and $B = \{1, 2, 3, 4\}$, then $A \cup B = \{0, 1, 2, 3, 4\}$. Note that even though 1 and 2 are in both set A and set B, there is no need to write them twice in $A \cup B$.

To find the solution set of the compound inequality

$x > 1$ or $x > 3$

we find the solution set for each inequality and then take all the values that satisfy either inequality or both.

EXAMPLE 7

Graph the solution set for $x > 1$ or $x > 3$ and write the solution in set-builder notation.

Solution

$x > 1$ (a)

$x > 3$ (b)

$x > 1$ or $x > 3$ (c)

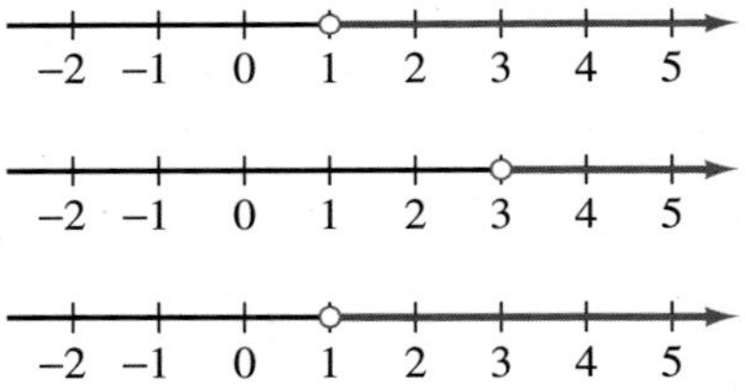

Figure 2.10

Thus all numbers greater than 1 are included in the solution set $\{x | x > 1\}$, and the graph is shown in Figure 2.10(c). ∎

EXAMPLE 8

Graph the solution set for $x \leq 0$ or $x \geq 2$, and write the solution in set-builder notation.

Solution

$x \leq 0$ (a)

$x \geq 2$ (b)

$x \leq 0$ or $x \geq 2$ (c)

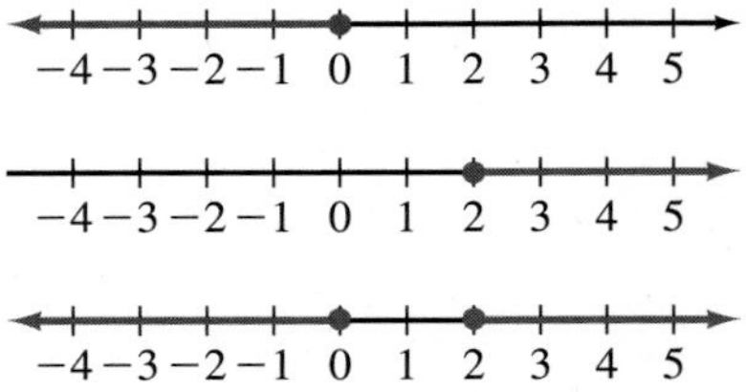

Figure 2.11

Thus all numbers less than or equal to 0 and all numbers greater than or equal to 2 are included in the solution set $\{x | x \leq 0 \text{ or } x \geq 2\}$, and the graph is shown in Figure 2.11(c).

5 Solve Word Problems That Translate into Inequality Statements

Let's consider some word problems that translate into inequality statements. The suggestions for solving word problems in Section 2.4 apply here, except that the situations described in these problems will translate into inequalities instead of equations.

EXAMPLE 9 Apply Your Skill

Ashley had scores of 95, 82, 93, and 84 on her first four exams of the semester. What score must she get on the fifth exam to have an average of 90 or higher for the five exams?

Solution

Let s represent the score needed on the fifth exam. Because we find the average by adding all five scores and dividing by 5 (the number of exams), we can solve this inequality:

$$\frac{95 + 82 + 93 + 84 + s}{5} \geq 90$$

We use the following steps:

$$\frac{354 + s}{5} \geq 90 \quad \text{Simplify numerator of left side}$$

$$5\left(\frac{354 + s}{5}\right) \geq 5(90) \quad \text{Multiply both sides by 5}$$

$$354 + s \geq 450$$

$$354 + s - 354 \geq 450 - 354 \quad \text{Subtract 354 from both sides}$$

$$s \geq 96$$

She must receive a score of 96 or higher on the fifth exam.

EXAMPLE 10 Apply Your Skill

The Cubs have won 40 baseball games and have lost 62 games. They have 60 more games to play. To win more than 50% of all their games, how many of the remaining 60 games must they win?

Solution

Let w represent the number of games the Cubs must win out of the 60 games remaining. Because they are playing a total of $40 + 62 + 60 = 162$ games, to win more than 50% of their games, they will have to win more than 81 games. Thus we have the inequality

$$w + 40 > 81$$

Solving this yields

$$w > 41$$

The Cubs need to win at least 42 of the remaining 60 games.

CONCEPT QUIZ 2.7

For Problems 1–5, answer true or false.

1. The solution set of a compound inequality formed by the word "and" is an intersection of the solution sets of the two inequalities.
2. The solution set of a compound inequality formed by the words "and" or "or" is a union of the solution sets of the two inequalities.
3. The intersection of two sets contains the elements that are common to both sets.
4. The union of two sets contains all the elements in both sets.
5. The intersection of set A and set B is denoted by $A \cap B$.

For Problems 6–10, match the compound statement with the graph of its solution set (Figure 2.12).

6. $x > 4$ or $x < -1$
7. $x > 4$ and $x > -1$
8. $x > 4$ or $x > -1$
9. $x \leq 4$ and $x \geq -1$
10. $x > 4$ or $x \geq -1$

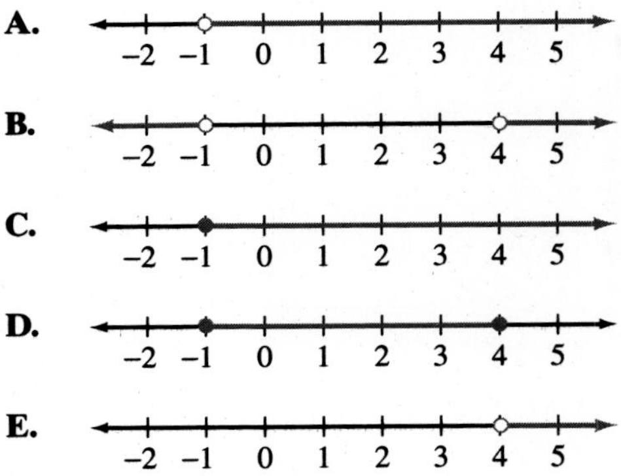

A. number line from −2 to 5, open circle at −1, arrow left
B. number line from −2 to 5, open circles at −1 and 4
C. number line from −2 to 5, closed circle at −1, arrow right
D. number line from −2 to 5, closed circles at −1 and 4
E. number line from −2 to 5, open circle at 4, arrow right

Figure 2.12

Section 2.7   Classroom Problem Set

Objective 1

1. Solve $7x + 8 \leq 3x + 12$.
2. Solve $6t + 14 \leq 8t - 16$.

Objective 2

3. Solve $2(x + 1) + 4(x - 2) \geq 3(x + 6)$.
4. Solve $4(n - 5) - 2(n - 1) < 13$.

Objective 3

5. Solve $-\dfrac{4}{3}y + \dfrac{5}{6}y < \dfrac{7}{6}$.
6. Solve $\dfrac{3}{4}n - \dfrac{5}{6}n < \dfrac{3}{8}$.
7. Solve $\dfrac{x + 4}{3} - \dfrac{x + 2}{4} \leq \dfrac{5}{12}$.
8. Solve $\dfrac{x - 1}{5} - \dfrac{x + 2}{6} \geq \dfrac{7}{15}$.

Objective 4

9. Graph the solution set for the compound inequality $x > 0$ and $x < 5$, and write the solution set in set-builder notation.
10. Graph the solution set for the compound inequality $x > -2$ and $x \leq 2$, and write the solution set in set-builder notation.
11. Graph the solution set for the compound inequality $x \leq 3$ and $x \leq 5$, and write the solution set in set-builder notation.
12. Graph the solution set for the compound inequality $x < 2$ and $x < 3$, and write the solution set in set-builder notation.
13. Graph the solution set for the compound inequality $x > 2$ or $x > 4$, and write the solution set in set-builder notation.
14. Graph the solution set for the compound inequality $x < 2$ or $x < 4$, and write the solution set in set-builder notation.

15. Graph the solution set for the compound inequality $x \leq -1$ or $x \geq 1$, and write the solution set in set-builder notation.

16. Graph the solution set for the compound inequality $x < 0$ or $x > 3$, and write the solution set in set-builder notation.

Objective 5

17. Felix had scores of 88, 95, and 90 on the first three exams in biology. What score must he get on the fourth exam to have an average of 92 or higher for the four exams?

18. This semester Sheila has scores of 96, 90, and 94 on her first three geometry exams. What must she average on the last two exams to have an average greater than 92 for all five exams?

19. The Red Sox want to win 70% or more of the 160 games they will play this season. So far this season, they have won 32 games. How many of the remaining games must they win to achieve their goal?

20. A computer business has costs of $4000 plus $32 per sale. The business receives revenue of $48 per sale. How many sales would ensure that the revenues exceed the costs?

THOUGHTS INTO WORDS

1. Give a step-by-step description of how you would solve the inequality $3x - 2 > 4(x + 6)$.

2. Find the solution set for each of the following compound statements, and in each case explain your reasoning.

   a. $x > 2$ and $5 > 4$
   b. $x > 2$ or $5 > 4$
   c. $x > 2$ and $4 > 10$
   d. $x > 2$ or $4 > 10$

Answers to the Concept Quiz
1. True 2. False 3. True 4. True 5. True 6. B 7. E 8. A 9. D 10. C

Chapter 2 Review Problem Set

For Problems 1–26, solve each of the equations.

1. $9x - 2 = -29$
2. $-3 = -4y + 1$
3. $7 - 4x = 10$
4. $6y - 5 = 4y + 13$
5. $4n - 3 = 7n + 9$
6. $3x - 4x - 2 = 7x - 14 - 9x$
7. $7(y - 4) = 4(y + 3)$
8. $2(x + 1) + 5(x - 3) = 11(x - 2)$
9. $-3(x + 6) = 5x - 3$
10. $-2(x - 4) = -3(x + 8)$
11. $5(n - 1) - 4(n + 2) = -3(n - 1) + 3n + 5$
12. $-(t - 3) - (2t + 1) = 3(t + 5) - 2(t + 1)$
13. $3(2t - 4) + 2(3t + 1) = -2(4t + 3) - (t - 1)$
14. $\dfrac{2}{5}n - \dfrac{1}{2}n = \dfrac{7}{10}$
15. $\dfrac{3n}{4} + \dfrac{5n}{7} = \dfrac{1}{14}$
16. $\dfrac{x - 3}{6} + \dfrac{x + 5}{8} = \dfrac{11}{12}$
17. $\dfrac{n}{2} - \dfrac{n - 1}{4} = \dfrac{3}{8}$
18. $\dfrac{x - 3}{9} = \dfrac{x + 4}{8}$
19. $\dfrac{x - 1}{-3} = \dfrac{x + 2}{-4}$
20. $\dfrac{2x - 1}{3} = \dfrac{3x + 2}{2}$
21. $4x - 3 = 6x + 7 - 2x$
22. $5x + 8 - 2x = 3x + 8$

23. $-2(x - 5) = 10 - 2x$

24. $-(x - 3) + 6x = 5x + 7$

25. $x + 2 = 3x - 2(x - 1)$

26. $3x - (2x + 5) = x$

For Problems 27–30, write the solution set in set-builder notation and graph the solution.

27. $x > 1$
28. $x \leq -2$
29. $x \leq 0$
30. $x > -3$

For Problems 31–46, solve each inequality.

31. $3x - 2 > 10$
32. $-2x - 5 < 3$
33. $2x - 9 \geq x + 4$
34. $3x + 1 \leq 5x - 10$
35. $-16 < 8 + 2y - 3y$
36. $-24 > 5x - 4 - 7x$
37. $6(x - 3) > 4(x + 13)$
38. $2(x + 3) + 3(x - 6) < 14$
39. $-3(n - 4) > 5(n + 2) + 3n$
40. $-4(n - 2) - (n - 1) < -4(n + 6)$
41. $-12 > -4(x - 1) + 2$
42. $36 < -3(x + 2) - 1$
43. $\dfrac{2n}{5} - \dfrac{n}{4} < \dfrac{3}{10}$
44. $\dfrac{n + 4}{5} + \dfrac{n - 3}{6} > \dfrac{7}{15}$
45. $\dfrac{3}{4}n - 6 \leq \dfrac{2}{3}n + 4$
46. $\dfrac{1}{2}n - \dfrac{1}{3}n - 4 \geq \dfrac{3}{5}n + 2$

For Problems 47–54, graph the solution set for each of the compound inequalities.

47. $x > -3$ and $x < 2$
48. $x \geq -2$ and $x \leq 5$
49. $x \leq 6$ and $x \leq 3$
50. $x > 1$ and $x > 0$
51. $x < -1$ or $x > 4$
52. $x < 2$ or $x > 0$
53. $x > -3$ or $x > 4$
54. $x \leq 1$ or $x \leq 4$

For Problems 55–64, translate each word phrase into an algebraic expression.

55. Five less than n
56. Five less n
57. Ten times the quantity, x minus 2
58. Ten times x minus 2
59. x minus 3
60. d divided by r
61. x squared plus 9
62. x plus 9, the quantity squared
63. The sum of the cubes of x and y
64. Four less than the product of x and y

For Problems 65–76, set up an equation or an inequality and solve the problem.

65. Three-fourths of a number equals 18. Find the number.

66. Nineteen is 2 less than three times a certain number. Find the number.

67. The difference of two numbers is 21. If 12 is the smaller number, find the other number.

68. One subtracted from nine times a certain number is the same as 15 added to seven times the number. Find the number.

69. The sum of two numbers is 40. Six times the smaller number equals four times the larger. Find the numbers.

70. Find a number such that 2 less than two-thirds of the number is 1 more than one-half of the number.

71. Miriam has 30 coins consisting of nickels and dimes amounting to $2.60. How many coins of each kind does she have?

72. Suppose that Russ has a bunch of nickels, dimes, and quarters amounting to $15.40. The number of dimes is 1 more than three times the number of nickels, and the number of quarters is twice the number of dimes. How many coins of each kind does he have?

73. The supplement of an angle is 14° more than three times the complement of the angle. Find the measure of the angle.

74. Pam rented a car from a rental agency that charges $25 a day and $0.20 per mile. She kept the car for 3 days and her bill was $215. How many miles did she drive during that 3-day period?

75. Monica had scores of 83, 89, 78, and 86 on her first four exams. What score must she get on the fifth exam so that her average for all five exams is 85 or higher?

76. Ameya's average score for her first three psychology exams was 84. What must she get on the fourth exam so that her average for the four exams is 85 or higher?

Chapter 2 Practice Test

For Problems 1–12, solve each of the equations.

1. $7x - 3 = 11$
2. $-7 = -3x + 2$
3. $4n + 3 = 2n - 15$
4. $3n - 5 = 8n + 20$
5. $4(x - 2) = 5(x + 9)$
6. $9(x + 4) = 6(x - 3)$
7. $5(y - 2) + 2(y + 1) = 3(y - 6)$
8. $\dfrac{3}{5}x - \dfrac{2}{3} = \dfrac{1}{2}$
9. $\dfrac{x - 2}{4} = \dfrac{x + 3}{6}$
10. $\dfrac{x + 2}{3} + \dfrac{x - 1}{2} = 2$
11. $\dfrac{x - 3}{6} - \dfrac{x - 1}{8} = \dfrac{13}{24}$
12. $-5(n - 2) = -3(n + 7)$

For Problems 13–18, solve each of the inequalities.

13. $3x - 2 < 13$
14. $-2x + 5 \geq 3$
15. $3(x - 1) \leq 5(x + 3)$
16. $-4 > 7(x - 1) + 3$
17. $-2(x - 1) + 5(x - 2) < 5(x + 3)$
18. $\dfrac{1}{2}n + 2 \leq \dfrac{3}{4}n - 1$

For Problems 19 and 20, graph the solution set for each compound inequality.

19. $x \geq -2$ and $x \leq 4$
20. $x < 1$ or $x > 3$

For Problems 21–25, set up an equation or an inequality and solve each problem.

21. A car repair bill without the tax was $441. This included $153 for parts and 4 hours of labor. Find the hourly rate that was charged for labor.

22. Suppose that a triangular plot of ground is enclosed by 70 meters of fencing. The longest side of the lot is two times the length of the shortest side, and the third side is 10 meters longer than the shortest side. Find the length of each side of the plot.

23. Tina had scores of 86, 88, 89, and 91 on her first four history exams. What score must she get on the fifth exam to have an average of 90 or higher for the five exams?

1. _____
2. _____
3. _____
4. _____
5. _____
6. _____
7. _____
8. _____
9. _____
10. _____
11. _____
12. _____
13. _____
14. _____
15. _____
16. _____
17. _____
18. _____
19. _____
20. _____
21. _____
22. _____
23. _____

24. _____

24. Sean has 103 coins consisting of nickels, dimes, and quarters. The number of dimes is 1 less than twice the number of nickels, and the number of quarters is 2 more than three times the number of nickels. How many coins of each kind does he have?

25. _____

25. In triangle ABC, the measure of angle C is one-half the measure of angle A, and the measure of angle B is 30° more than the measure of angle A. Find the measure of each angle of the triangle.

Chapters 1–2 Cumulative Review Problem Set

For Problems 1–4, simplify each numerical expression.

1. $3(-4) - 2 + (-3)(-6) - 1$
2. $-(2)^7$
3. $6.2 - 7.1 - 3.4 + 1.9$
4. $-\dfrac{2}{3} + \dfrac{1}{2} - \dfrac{1}{4}$

For Problems 5–7, evaluate each algebraic expression for the given values of the variables.

5. $-4x + 2y - xy$ for $x = -2$ and $y = 3$
6. $\dfrac{1}{5}x - \dfrac{2}{3}y$ for $x = -\dfrac{1}{2}$ and $y = \dfrac{1}{6}$
7. $0.2(x - y) - 0.3(x + y)$ for $x = 0.1$ and $y = -0.2$
8. Find the greatest common factor of 48, 60, and 96.
9. Find the least common multiple of 9 and 12.
10. Simplify $\dfrac{3}{8}x + \dfrac{3}{7}y - \dfrac{5}{12}x - \dfrac{3}{4}y$ by combining similar terms.

For Problems 11–14, simplify each expression by applying the distributive property and combining like terms.

11. $-(x - 2) + 6(x + 4) - 2(x - 7)$
12. $5(2x + 1) + 3(4x - 7)$
13. $\dfrac{1}{2}(2x - 8) + \dfrac{3}{4}(20x + 4)$
14. $-3(x + 5) - (2x + 7)$

For Problems 15–18, solve each equation.

15. $-3(x + 4) = -4(x - 1)$
16. $\dfrac{x + 1}{4} - \dfrac{x - 2}{3} = -2$
17. $2(2x - 1) + 3(x - 3) = -4(x + 7)$
18. $3(2x + 8) - (x + 4) = 5x + 12$

For Problems 19 and 20, solve the inequality.

19. $2(x - 1) \geq 3(x - 6)$
20. $-2 < -(x - 1) - 4$
21. Express 300 as the product of prime factors.
22. Graph the solutions for the compound inequality $x \geq -1$ and $x < 3$.

For Problems 23–25, use an equation or inequality to help solve each problem.

23. On Friday and Saturday nights, the police made a total of 42 arrests at a DUI checkpoint. On Saturday night they made 6 more than three times the arrests of Friday night. Find the number of arrests for each night.
24. For a wedding reception, the caterer charges a $125 fee plus $35 per person for dinner. If Peter and Rynette must keep the cost of the caterer to less than $2500, how many people can attend the reception?
25. Find two consecutive odd numbers whereby the smaller plus five times the larger equals 76.

3 Ratio, Percent, Formulas, and Problem Solving

3.1 Ratio, Proportion, and Percent

3.2 More on Percents and Problem Solving

3.3 Formulas: Geometric and Others

3.4 Problem Solving

3.5 More about Problem Solving

Chapter 3   Warm-Up Problems

1. Simplify.   (a) $\dfrac{5}{40}$ (b) $\dfrac{400}{0.25}$ (c) $(0.36)(50)$

2. Simplify.   (a) $0.3(1.2)$ (b) $0.27\left(\dfrac{1}{9}\right)$ (c) $\dfrac{4.4}{-0.2}$

3. Simplify.
   (a) $15 + 0.4(120)$ (b) $256 + 0.25(256)$ (c) $0.15(2.3) - 0.78$

4. Simplify.
   (a) $10(0.1n + 2)$ (b) $100(0.75x - 0.5y)$ (c) $0.04(2000) + 0.025(1500)$

5. Simplify.   (a) $\dfrac{1}{2}(12)(9)$ (b) $2(7)^2 + 2(7)(5)$ (c) $\dfrac{1}{2}(15)(8 + 12)$

6. Simplify.
   (a) $-3(4x - 9)$ (b) $6\left(\dfrac{1}{2}n - 5\right)$ (c) $\dfrac{2}{3}(63 - p)$

7. Simplify.
   (a) $x + \dfrac{3}{4}x - 2$ (b) $\dfrac{3}{8}n + \dfrac{5}{8}(n + 2)$ (c) $x + (x + 2) + (x + 4)$

8. Simplify.   (a) $x - 0.7x$ (b) $n + 0.6n$ (c) $0.03(400 + p)$

9. Write the mixed number as an improper fraction.
   (a) $3\dfrac{2}{3}$ (b) $6\dfrac{5}{8}$ (c) $9\dfrac{1}{2}$

10. Reduce to lowest terms.
    (a) $\dfrac{12}{20}$ (b) $\dfrac{6}{16}$ (c) $\dfrac{18}{36}$

We used the formula *distance equals rate times time*, which is usually expressed as $d = rt$, to set up the equation $7t = 5\left(t + \dfrac{1}{2}\right)$. Throughout this chapter we will use a variety of formulas in a problem-solving setting to connect algebraic and geometric concepts.

Video tutorials for all section learning objectives are available in a variety of delivery modes.

INTERNET PROJECT

In this chapter there are suggestions for solving problems. Do an Internet search on Polya problem solving. What are Polya's four principles of problem solving? How do Polya's four principles align with the suggestions given in this chapter for solving word problems?

3.1 Ratio, Proportion, and Percent

OBJECTIVES

1. Solve Proportions
2. Solve Word Problems Using Proportions
3. Use a Proportion to Convert a Fraction to a Percent
4. Solve Basic Percent Problems

1 Solve Proportions

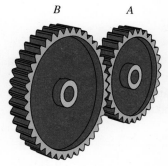

Figure 3.1

Ratio In Figure 3.1, as gear A revolves four times, gear B will revolve three times. We say that the gear ratio of A to B is 4 to 3, or the gear ratio of B to A is 3 to 4. Mathematically, a **ratio** is the comparison of two numbers by division. We can write the gear ratio of A to B in these equivalent expressions:

$$4 \text{ to } 3 \qquad 4{:}3 \qquad \frac{4}{3}$$

We express ratios as fractions in reduced form. For example, if there are 7500 women and 5000 men at a certain university, then the ratio of women to men is $\frac{7500}{5000} = \frac{3}{2}$.

Proportion A statement of equality between two ratios is called a **proportion**. For example,

$$\frac{2}{3} = \frac{8}{12}$$

is a proportion that states that the ratios $\frac{2}{3}$ and $\frac{8}{12}$ are equal. In general, if we have a proportion

$$\frac{a}{b} = \frac{c}{d} \qquad b \neq 0 \text{ and } d \neq 0$$

and we multiply both sides of the equation by the common denominator, bd, we obtain

$$(bd)\left(\frac{a}{b}\right) = (bd)\left(\frac{c}{d}\right)$$
$$ad = bc$$

Let's state this as a property of proportions.

$$\frac{a}{b} = \frac{c}{d} \quad \text{if and only if } ad = bc, \text{ where } b \neq 0 \text{ and } d \neq 0$$

The products ad and bc are commonly called **cross products**. Thus according to the property, cross products in a proportion are equal.

EXAMPLE 1

Solve $\dfrac{x}{20} = \dfrac{3}{4}$.

Solution

$$\dfrac{x}{20} = \dfrac{3}{4}$$
$$4x = 60 \qquad \text{Cross products are equal}$$
$$x = 15$$

The solution set is $\{15\}$. ∎

EXAMPLE 2

Solve $\dfrac{x-3}{5} = \dfrac{x+2}{4}$.

Solution

$$\dfrac{x-3}{5} = \dfrac{x+2}{4}$$
$$4(x-3) = 5(x+2) \qquad \text{Cross products are equal}$$
$$4x - 12 = 5x + 10 \qquad \text{Distributive property}$$
$$-12 = x + 10 \qquad \text{Subtracted } 4x \text{ from both sides}$$
$$-22 = x \qquad \text{Subtracted 10 from both sides}$$

The solution set is $\{-22\}$. ∎

If a variable appears in one or both of the denominators, then certain restrictions must be imposed to avoid division by zero, as the next example illustrates.

EXAMPLE 3

Solve $\dfrac{7}{a-2} = \dfrac{4}{a+3}$.

Solution

$$\dfrac{7}{a-2} = \dfrac{4}{a+3} \qquad a \neq 2 \text{ and } a \neq -3$$
$$7(a+3) = 4(a-2) \qquad \text{Cross products are equal}$$
$$7a + 21 = 4a - 8 \qquad \text{Distributive property}$$
$$3a + 21 = -8 \qquad \text{Subtracted } 4a \text{ from both sides}$$
$$3a = -29 \qquad \text{Subtracted 21 from both sides}$$
$$a = -\dfrac{29}{3} \qquad \text{Divided both sides by 3}$$

The solution set is $\left\{-\dfrac{29}{3}\right\}$. ∎

EXAMPLE 4

Solve $\dfrac{x}{4} + 3 = \dfrac{x}{5}$.

Solution

This is *not* a proportion. This example demonstrates the importance of *thinking first before pushing the pencil*. Because the equation is not in the form of a proportion, we need to revert to a previous technique for solving it. Let's multiply both sides by 20, the least common denominator, to clear the equation of all fractions.

$$\frac{x}{4} + 3 = \frac{x}{5}$$

$$20\left(\frac{x}{4} + 3\right) = 20\left(\frac{x}{5}\right) \quad \text{Multiply both sides by 20}$$

$$20\left(\frac{x}{4}\right) + 20(3) = 20\left(\frac{x}{5}\right) \quad \text{Distributive property}$$

$$5x + 60 = 4x$$

$$x + 60 = 0 \quad \text{Subtracted } 4x \text{ from both sides}$$

$$x = -60 \quad \text{Subtracted 60 from both sides}$$

The solution set is $\{-60\}$.

2 Solve Word Problems Using Proportions

Some word problems can be conveniently set up and solved using the concepts of ratio and proportion. Consider the next examples.

EXAMPLE 5   Apply Your Skill

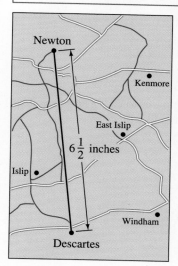

Figure 3.2

On the map in Figure 3.2, 1 inch represents 20 miles. If Newton and Descartes are $6\frac{1}{2}$ inches apart on the map, find the number of miles between the cities.

Solution

Let m represent the number of miles between the two cities. Now let's set up a proportion where one ratio compares distances in inches on the map, and the other ratio compares *corresponding* distances in miles on land.

$$\frac{1}{6\frac{1}{2}} = \frac{20}{m}$$

To solve this equation, we equate the cross products.

$$m(1) = \left(6\frac{1}{2}\right)(20)$$

$$m = \left(\frac{13}{2}\right)(20) = 130$$

The distance between the two cities is 130 miles.

EXAMPLE 6   Apply Your Skill

A sum of $1750 is to be divided between two people in the ratio of 3 to 4. How much does each person receive?

Solution

Let d represent the amount of money to be received by one person. Then $1750 - d$ represents the amount for the other person. We set up this proportion:

$$\frac{d}{1750 - d} = \frac{3}{4}$$

$$4d = 3(1750 - d)$$

$$4d = 5250 - 3d$$

$$7d = 5250$$

$$d = 750$$

If $d = 750$, then $1750 - d = 1000$; therefore, one person receives \$750, and the other person receives \$1000. ■

3 Use a Proportion to Convert a Fraction to a Percent

The word **percent** means *per one hundred*, and we use the symbol % to express it. For example, we write 7 percent as 7%, which means $\frac{7}{100}$, or 0.07. In other words, percent is a special kind of ratio—a ratio in which the denominator is always 100. Proportions provide a convenient basis for changing common fractions to percents. Consider the following examples.

EXAMPLE 7

Express $\frac{7}{20}$ as a percent.

Solution

We are asking, "What number compares to 100 as 7 compares to 20?" Therefore, if we let n represent that number, we can set up the proportion like this:

$$\frac{n}{100} = \frac{7}{20}$$

$$20n = 700 \quad \text{Cross products are equal}$$

$$n = 35$$

Thus $\frac{7}{20} = \frac{35}{100} = 35\%$. ■

EXAMPLE 8

Express $\frac{5}{6}$ as a percent.

Solution

$$\frac{n}{100} = \frac{5}{6}$$

$$6n = 500 \quad \text{Cross products are equal}$$

$$n = \frac{500}{6} = \frac{250}{3} = 83\frac{1}{3}$$

Therefore $\frac{5}{6} = 83\frac{1}{3}\%$. ■

3.1 Ratio, Proportion, and Percent

4 Solve Basic Percent Problems

What is 8% of 35? Fifteen percent of what number is 24? Twenty-one is what percent of 70? These are the three basic types of percent problems. Each of these problems can be solved easily by translating it into, and solving, a simple algebraic equation.

EXAMPLE 9

Apply Your Skill

What is 8% of 35?

Solution

Let n represent the number to be found. The word "is" refers to equality, and the word "of" implies multiplication. Thus the question translates into

$$n = (8\%)(35)$$

which can be solved as follows:

$$n = (0.08)(35)$$
$$= 2.8$$

Therefore 2.8 is 8% of 35.

EXAMPLE 10

Apply Your Skill

Fifteen percent of what number is 24?

Solution

Let n represent the number to be found.

$$(15\%)(n) = (24)$$
$$0.15n = 24$$
$$15n = 2400 \quad \text{Multiplied both sides by 100}$$
$$n = 160$$

Therefore 15% of 160 is 24.

EXAMPLE 11

Apply Your Skill

Twenty-one is what percent of 70?

Solution

Let r represent the percent to be found.

$$21 = r(70)$$
$$\frac{21}{70} = r$$
$$\frac{3}{10} = r \quad \text{Reduce!}$$
$$\frac{30}{100} = r \quad \text{Changed } \frac{3}{10} \text{ to } \frac{30}{100}$$
$$30\% = r$$

Therefore 21 is 30% of 70.

EXAMPLE 12

Apply Your Skill

Seventy-two is what percent of 60?

Solution

Let r represent the percent to be found.

$$72 = r(60)$$

$$\frac{72}{60} = r$$

$$\frac{6}{5} = r$$

$$\frac{120}{100} = r \qquad \text{Changed } \frac{6}{5} \text{ to } \frac{120}{100}$$

$$120\% = r$$

Therefore, 72 is 120% of 60. ∎

Get into the habit of checking answers for *reasonableness*. We also suggest that you alert yourself to a potential computational error by *estimating* the answer before you actually do the problem. For example, prior to doing Example 12, you may have estimated: "Because 72 is larger than 60, the answer has to be greater than 100%. Furthermore 1.5 (or 150%) times 60 equals 90." Therefore, you can estimate the answer to be somewhere between 100% and 150%. That may seem rather broad, but such an estimate will often detect a computational error.

CONCEPT QUIZ 3.1

For Problems 1–8, answer true or false.

1. A ratio is the comparison of two numbers by division.
2. The ratio of 7 to 3 can be written 3:7.
3. A proportion is a statement of equality between two ratios.
4. For the proportion $\frac{x}{3} = \frac{y}{5}$, the cross product would be $5x = 3y$.
5. The algebraic equation $\frac{w}{2} = \frac{w}{5} + 1$ is a proportion.
6. The word "percent" means parts per one thousand.
7. One hundred twenty percent of 30 is 24.
8. If the cross product of a proportion is $wx = yz$, then $\frac{x}{z} = \frac{y}{w}$.

Section 3.1 Classroom Problem Set

Objective 1

1. Solve $\frac{x}{18} = \frac{5}{6}$.

2. Solve $\frac{7}{8} = \frac{n}{16}$.

3. Solve $\frac{x-2}{3} = \frac{x-4}{5}$.

4. Solve $\frac{x-6}{7} = \frac{x+9}{8}$.

5. Solve $\frac{8}{a-1} = \frac{3}{a-2}$.

6. Solve $\frac{3}{2x-1} = \frac{2}{3x+2}$.

7. Solve $\frac{x}{4} - 1 = \frac{x}{2}$.

8. Solve $-3 - \frac{x+4}{5} = \frac{3}{2}$.

Objective 2

9. The plot plan for a new subdivision is scaled so that 1 inch represents 12 feet. Two driveways on the plot plan are $3\frac{1}{4}$ inches apart. What is the distance in feet between the two driveways?

10. A house plan has a scale in which 1 inch represents 6 feet. Find the dimensions of a rectangular room that measures $2\frac{1}{2}$ inches by $3\frac{1}{4}$ inches on the house plan.

11. A banker is told to deposit a $1500.00 paycheck into two accounts in the ratio of 2:3. How much does each account receive?

12. It was reported that a flu epidemic is affecting 6 out of every 10 college students at a certain university. At this rate, how many people will be affected if there are 15,000 students at the university?

Objective 3

13. Express $\frac{13}{25}$ as a percent.

14. Express $\frac{7}{20}$ as a percent.

15. Express $\frac{7}{12}$ as a percent.

16. Express $\frac{5}{7}$ as a percent.

Objective 4

17. What is 12% of 58?
18. What is 120% of 50?
19. Eight percent of what number is 56?
20. Fifty-five percent of what number is 38.5?
21. Thirty-two is what percent of 80?
22. Seventy-two is what percent of 120?
23. Sixty is what percent of 40?
24. Twenty-six is what percent of 20?

THOUGHTS INTO WORDS

1. Explain the difference between a ratio and a proportion.
2. What is wrong with this solution?

$$\frac{x}{2} + 4 = \frac{x}{6}$$
$$6\left(\frac{x}{2} + 4\right) = 2(x)$$
$$3x + 24 = 2x$$
$$x = -24$$

Explain how it should be solved.

3. Estimate an answer for each of the following problems, and explain how you arrived at your estimate. Then work out the problem to see how well you estimated.

   a. The ratio of female students to male students at a small private college is 5 to 3. If there is a total of 1096 students, find the number of male students.

   b. If 15 pounds of fertilizer will cover 1200 square feet of lawn, how many pounds are needed for 3000 square feet?

   c. An investment of $5000 earns $300 interest in a year. At the same rate, how much money must be invested to earn $450?

   d. If the ratio of the length of a rectangle to its width is 5 to 3, and the length is 70 centimeters, find its width.

Answers to the Concept Quiz

1. True   2. False   3. True   4. True   5. False   6. False   7. False   8. True

3.2 More on Percents and Problem Solving

OBJECTIVES

1. Solve Equations Involving Decimals
2. Solve Word Problems Involving Discount, Selling Price, Cost, or Profit
3. Solve Simple Interest Problems

1 Solve Equations Involving Decimals

We can solve the equation $x + 0.35 = 0.72$ by subtracting 0.35 from both sides of the equation. Another technique for solving equations that contain decimals is to *clear the equation of all decimals* by multiplying both sides by an appropriate power of 10. The following examples demonstrate the use of that strategy in a variety of situations.

EXAMPLE 1 Solve $0.5x = 14$.

Solution

$0.5x = 14$
$5x = 140$ Multiplied both sides by 10
$x = 28$ Divided both sides by 5

The solution set is $\{28\}$.

EXAMPLE 2 Solve $x + 0.07x = 0.13$.

Solution

$x + 0.07x = 0.13$
$100(x + 0.07x) = 100(0.13)$ Multiply both sides by 100
$100(x) + 100(0.07x) = 100(0.13)$ Distributive property
$100x + 7x = 13$
$107x = 13$
$x = \dfrac{13}{107}$

The solution set is $\left\{\dfrac{13}{107}\right\}$.

EXAMPLE 3 Solve $0.08y + 0.09y = 3.4$.

Solution

$0.08y + 0.09y = 3.4$
$8y + 9y = 340$ Multiplied both sides by 100
$17y = 340$
$y = 20$

The solution set is $\{20\}$.

EXAMPLE 4

Solve $0.10t = 560 - 0.12(t + 1000)$.

Solution

$$0.10t = 560 - 0.12(t + 1000)$$
$$10t = 56{,}000 - 12(t + 1000) \quad \text{Multiplied both sides by 100}$$
$$10t = 56{,}000 - 12t - 12{,}000 \quad \text{Distributive property}$$
$$22t = 44{,}000$$
$$t = 2000$$

The solution set is $\{2000\}$. ∎

2 Solve Word Problems Involving Discount, Selling Price, Cost, or Profit

We can solve many consumer problems with an equation approach. For example, here is a general guideline regarding discount sales:

> Original selling price − Discount = Discount sale price

Let's work some examples using our algebraic techniques along with this basic guideline.

EXAMPLE 5 Apply Your Skill

Amy bought a dress at a 30% discount sale for $35. What was the original price of the dress?

Solution

Let p represent the original price of the dress. We can use the basic discount guideline to set up an algebraic equation:

Original selling price − Discount = Discount sale price
$$(100\%)(p) \quad - (30\%)(p) = \quad \$35$$

Solving this equation, we get

$$(100\%)(p) - (30\%)(p) = 35$$
$$(70\%)(p) = 35$$
$$0.7p = 35$$
$$7p = 350$$
$$p = 50$$

The original price of the dress was $50. ∎

Don't forget that if an item is on sale for 30% off, then you are going to pay $100\% - 30\% = 70\%$ of the original price. Thus at a 30% off sales event, you can buy a $50 dress for $(70\%)(\$50) = \$35$. (Note that we just checked our answer for Example 5.)

EXAMPLE 6

Apply Your Skill

Find the cost of a $60 pair of jogging shoes on sale for 20% off.

Solution

Let x represent the discount sale price. Because the shoes are on sale for 20% off, we must pay 80% of the original price.

$$x = (80\%)(60)$$
$$= (0.8)(60) = 48$$

The sale price is $48.

Here is another equation that is useful for solving consumer problems:

$$\text{Selling price} = \text{Cost} + \text{Profit}$$

Profit (also called *markup*, *markon*, *margin*, and *margin of profit*) may be stated in different ways: as a percent of the selling price, as a percent of the cost, or simply in terms of dollars and cents. Let's consider some problems where the profit is either a percent of the selling price or a percent of the cost.

EXAMPLE 7

Apply Your Skill

A retailer has some shirts that cost him $20 each. He wants to sell them at a profit of 60% of the cost. What should the selling price be on the shirts?

Solution

Let s represent the selling price. The basic relationship *selling price equals cost plus profit* can be used as a guideline:

Selling price = Cost + Profit
s = $20 + (60\%)(20)$

Solving this equation, we obtain

$$s = 20 + (60\%)(20)$$
$$= 20 + (0.6)(20)$$
$$= 20 + 12$$
$$= 32$$

The selling price should be $32.

EXAMPLE 8

Apply Your Skill

Kathrin bought a painting for $120 and later decided to resell it. She made a profit of 40% of the selling price. How much money did she receive for the painting?

Solution

We can use the same basic relationship as a guideline, except this time the profit is a percent of the selling price. Let s represent the selling price.

Selling price = Cost + Profit
s = $120 + (40\%)(s)$

We can solve this equation:

$$s = 120 + (40\%)(s)$$
$$s = 120 + 0.4s$$
$$0.6s = 120 \quad \text{Subtracted } 0.4s \text{ from both sides}$$
$$s = \frac{120}{0.6} = 200$$

She received $200 for the painting.

3 Solve Simple Interest Problems

We can also translate certain types of investment problems into algebraic equations. In some of these problems, we use the simple interest formula $i = Prt$, where i represents the amount of interest earned by investing P dollars at a yearly rate of r percent for t years.

EXAMPLE 9 — Apply Your Skill

John invested $9300 for 2 years and received $1395 in interest. Find the annual interest rate John received on his investment.

Solution

$$i = Prt$$
$$1395 = 9300r(2)$$
$$1395 = 18600r$$
$$\frac{1395}{18600} = r$$
$$0.075 = r$$

The annual interest rate is 7.5%.

EXAMPLE 10 — Apply Your Skill

How much principal must be invested to receive $1500 in interest when the investment is made for 3 years at an annual interest rate of 6.25%?

Solution

$$i = Prt$$
$$1500 = P(0.0625)(3)$$
$$1500 = P(0.1875)$$
$$\frac{1500}{0.1875} = P$$
$$8000 = P$$

The principal must be $8000.

EXAMPLE 11 — Apply Your Skill

How much monthly interest will be charged on a credit card bill with a balance of $754 when the credit card company charges an 18% annual interest rate?

Solution

$$i = Prt$$
$$i = 754(0.18)\left(\frac{1}{12}\right) \quad \text{Remember, 1 month is } \frac{1}{12} \text{ of a year}$$
$$i = 11.31$$

The interest charge would be $11.31.

CONCEPT QUIZ 3.2

For Problems 1–5, answer true or false.

1. To clear the decimals from the equation, $0.5x + 1.24 = 0.07x + 1.8$, you would multiply both sides of the equation by 10.
2. If an item is on sale for 35% off, then you are going to pay 65% of the original price.
3. Profit is always a percent of the selling price.
4. In the formula $i = Prt$, the r represents the interest return.
5. The basic relationship *selling price equals cost plus profit* can be used whether the profit is based on selling price or cost.

Section 3.2   Classroom Problem Set

Objective 1

1. Solve $0.4x = 36$.
2. Solve $9 = 0.3y$.
3. Solve $3x + 0.2x = 128$.
4. Solve $20 = 0.1a - 0.2a$.
5. Solve $0.04a + 0.07a = 16.5$.
6. Solve $0.09x + 0.1(2x) = 130.5$.
7. Solve $0.08y = -0.15(y + 200) + 99$.
8. Solve $0.08x = 580 - 0.1(6000 - x)$.

Objective 2

9. Shawna bought a notebook computer at a 20% off sales event for $640. What was the original price of the notebook computer?
10. Maryanne bought a dress for $140, which represents a 20% discount off the original price. What was the original price of the dress?
11. Find the cost of a $140 pair of sunglasses on sale for 30% off.
12. Greg bought a $32 putter on sale for 35% off. How much did he pay for the putter?
13. A tire dealer has some alloy wheels that cost him $450. The dealer wants to sell the wheels at a profit of 80% of the cost. What selling price should be marked on the wheels?
14. A video store has some video games that cost $25 each. The store wants to sell them at a profit of 80% of the cost. What should the video store charge for the video games?
15. Carlos bought an autographed NASCAR photo for $80. He will sell the photo only if he makes a profit of 60% of the selling price. What will the selling price have to be for Carlos to sell the photo?
16. If a box of candy costs a retailer $2.50, and he wants to make a profit of 50% based on the selling price, what price should he charge for the candy?

Objective 3

17. After three years, Maribel received $1620 in interest on her $10,000 investment. Find the annual interest rate Maribel received on her investment.
18. Find the annual interest rate if $560 in interest is earned when $3500 is invested for 2 years.
19. How much principal must be invested for 2 years to receive $715 in interest when the annual interest rate is 6.5%?
20. For how many years must $2000 be invested at an annual interest rate of 5.4% to earn $162?
21. How much monthly interest will be charged on a student loan with a balance of $14,000 when the loan company charges 6% annual interest?
22. What will be the interest earned on a $5000 certificate of deposit invested at 6.8% annual interest for 10 years?

THOUGHTS INTO WORDS

1. What is wrong with this solution?

    $$1.2x + 2 = 3.8$$
    $$10(1.2x) + 2 = 10(3.8)$$
    $$12x + 2 = 38$$
    $$12x = 36$$
    $$x = 3$$

    How should it be solved?

2. From a consumer's standpoint, would you prefer that a retailer figure his profit on the basis of cost or the selling price of an item? Explain your answer.

Answers to the Concept Quiz

1. False   2. True   3. False   4. False   5. True

3.3 Formulas: Geometric and Others

OBJECTIVES

1. Evaluate Formulas
2. Apply Geometric Formulas
3. Solve Formulas for a Specific Variable
4. Solve an Equation for a Specific Variable

1 Evaluate Formulas

To find the distance traveled in 3 hours at a rate of 50 miles per hour, we multiply the rate by the time. Thus the distance is $50(3) = 150$ miles. We usually state the rule *distance equals rate times time* as the formula $d = rt$. **Formulas** are simply rules that we state in symbolic language and express as equations. Thus the formula $d = rt$ is an equation that involves three variables: d, r, and t.

As we work with formulas, it is often necessary to solve for a specific variable when we know numerical values for the remaining variables. Consider the next examples.

EXAMPLE 1

Solve $d = rt$ for r if $d = 330$ and $t = 6$.

Solution

Substitute 330 for d and 6 for t in the given formula:

$$330 = r(6)$$

Solving this equation yields

$$330 = 6r$$
$$55 = r$$

EXAMPLE 2 Solve $C = \frac{5}{9}(F - 32)$ for F if $C = 10$. (This formula expresses the relationship between the Fahrenheit and Celsius temperature scales.)

Solution

Substitute 10 for C to obtain

$$10 = \frac{5}{9}(F - 32)$$

We can solve this equation:

$$\frac{9}{5}(10) = \frac{9}{5}\left(\frac{5}{9}\right)(F - 32) \quad \text{Multiply both sides by } \frac{9}{5}$$

$$18 = F - 32$$

$$50 = F$$

2 Apply Geometric Formulas

There are several formulas in geometry that we use quite often. We review them below; we will use them periodically throughout the remainder of the text. These formulas (along with some others) and Figures 3.3 through 3.13 are also shown in the inside front cover of this text.

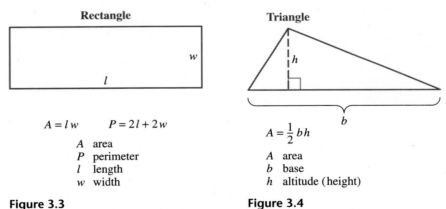

Rectangle

$A = lw \quad P = 2l + 2w$

- A area
- P perimeter
- l length
- w width

Figure 3.3

Triangle

$A = \frac{1}{2}bh$

- A area
- b base
- h altitude (height)

Figure 3.4

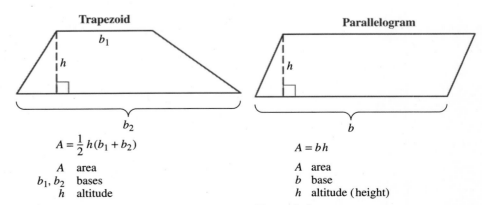

Trapezoid

$A = \frac{1}{2}h(b_1 + b_2)$

- A area
- b_1, b_2 bases
- h altitude

Figure 3.5

Parallelogram

$A = bh$

- A area
- b base
- h altitude (height)

Figure 3.6

3.3 Formulas: Geometric and Others   119

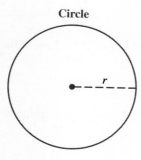

$A = \pi r^2 \qquad C = 2\pi r$

A area
C circumference
r radius

Figure 3.7

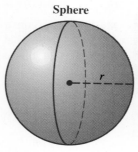

$V = \frac{4}{3}\pi r^3 \qquad S = 4\pi r^2$

S surface area
V volume
r radius

Figure 3.8

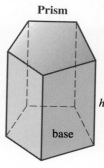

$V = Bh$

V volume
B area of base
h altitude (height)

Figure 3.9

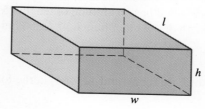

$V = lwh \qquad S = 2hw + 2hl + 2lw$

V   volume
S   total surface area
w   width
l   length
h   altitude (height)

Figure 3.10

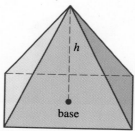

$V = \frac{1}{3}Bh$

V volume
B area of base
h altitude (height)

Figure 3.11

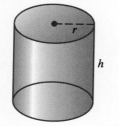

$V = \pi r^2 h \qquad S = 2\pi r^2 + 2\pi rh$

V volume
S total surface area
r radius
h altitude (height)

Figure 3.12

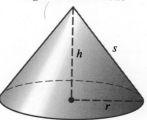

$V = \frac{1}{3}\pi r^2 h \qquad S = \pi r^2 + \pi rs$

V volume
S total surface area
r radius
h altitude (height)
s slant height

Figure 3.13

EXAMPLE 3

Apply Your Skill

Mark wants to construct a garden plot in the shape of a triangle. If the length of the altitude is 18 feet, what must the length of the base of the triangle be if he wants the garden to be 720 square feet in area?

Solution

We substitute 18 for h and 720 for A in the formula to find the area of a triangle.

$$A = \frac{1}{2}bh$$

$$720 = \frac{1}{2}b(18)$$

$$720 = 9b$$

$$80 = b$$

The length of the base should be 80 feet. ∎

EXAMPLE 4

Apply Your Skill

A contractor needs to determine the area of a foyer in a church that is in the shape of a trapezoid. The trapezoid has an altitude of length 25 feet, the length of one base is 30 feet, and the length of the other base is 42 feet. Find the area of the foyer.

Solution

We substitute 25 for h, 30 for b_1, and 42 for b_2 in the formula for finding the area of a trapezoid.

$$A = \frac{1}{2}h(b_1 + b_2)$$

$$A = \frac{1}{2}(25)(30 + 42)$$

$$A = \frac{1}{2}(25)(72)$$

$$A = 900$$

The area of the foyer is 900 square feet. ∎

EXAMPLE 5

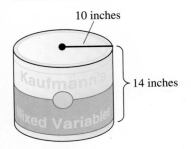

Figure 3.14

Find the total surface area of a right circular cylinder that has a radius of 10 inches and a height of 14 inches.

Solution

Let's sketch a right circular cylinder and record the given information as in Figure 3.14. We substitute 10 for r and 14 for h in the formula for finding the total surface area of a right circular cylinder.

$$S = 2\pi r^2 + 2\pi rh$$

$$= 2\pi(10)^2 + 2\pi(10)(14)$$

$$= 200\pi + 280\pi$$

$$= 480\pi$$

The total surface area is 480π square inches. ∎

In Example 5 we used Figure 3.14 to record the given information; you can also consult Figure 3.12 for the geometric formula. Now let's consider an example where a figure is very useful in the analysis of the problem.

EXAMPLE 6

A sidewalk 3 feet wide surrounds a rectangular plot of ground that measures 75 feet by 100 feet. Find the area of the sidewalk.

Solution

Let's make a sketch and record the given information as in Figure 3.15.

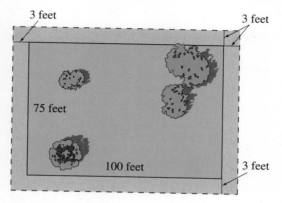

Figure 3.15

We can find the area of the sidewalk by subtracting the area of the rectangular plot from the area of the plot plus the sidewalk (the large dashed rectangle). The width of the large rectangle is $75 + 3 + 3 = 81$ feet, and its length is $100 + 3 + 3 = 106$ feet, so

$$A = (81)(106) - (75)(100)$$
$$= 8586 - 7500$$
$$= 1086$$

The area of the sidewalk is 1086 square feet. ∎

3 Solve Formulas for a Specific Variable

Sometimes it is convenient to change a formula's form by using the properties of equality. For example, we can change the formula $d = rt$ as follows:

$$d = rt$$
$$\frac{d}{r} = \frac{rt}{r} \quad \text{Divide both sides by } r$$
$$\frac{d}{r} = t$$

We say that the formula $d = rt$ "has been solved for the variable t." The formula can also be *solved for r*.

$$d = rt$$
$$\frac{d}{t} = \frac{rt}{t} \quad \text{Divide both sides by } t$$
$$\frac{d}{t} = r$$

∎

EXAMPLE 7
Solve $C = 2\pi r$ for r.

Solution

$$C = 2\pi r$$

$$\frac{C}{2\pi} = \frac{2\pi r}{2\pi} \quad \text{Divide both sides by } 2\pi$$

$$\frac{C}{2\pi} = r$$

EXAMPLE 8
Solve $V = \frac{1}{3}Bh$ for h.

Solution

$$V = \frac{1}{3}Bh$$

$$3(V) = 3\left(\frac{1}{3}Bh\right) \quad \text{Multiply both sides by 3}$$

$$3V = Bh$$

$$\frac{3V}{B} = \frac{Bh}{B} \quad \text{Divide both sides by } B$$

$$\frac{3V}{B} = h$$

EXAMPLE 9
Solve $P = 2l + 2w$ for w.

Solution

$$P = 2l + 2w$$

$$P - 2l = 2l + 2w - 2l \quad \text{Subtract } 2l \text{ from both sides}$$

$$P - 2l = 2w$$

$$\frac{P - 2l}{2} = \frac{2w}{2} \quad \text{Divide both sides by 2}$$

$$\frac{P - 2l}{2} = w$$

4 Solve an Equation for a Specific Variable

In Chapter 4 you will be working with equations that contain two variables. At times you will need to solve for one variable in terms of the other variable—that is, to change the form of the equation as we have been doing with formulas. The next examples illustrate how we can use the properties of equality for such situations.

EXAMPLE 10
Solve $3x + y = 4$ for x.

Solution

$$3x + y = 4$$

$$3x + y - y = 4 - y \quad \text{Subtract } y \text{ from both sides}$$

3.3 Formulas: Geometric and Others

$$3x = 4 - y$$

$$\frac{3x}{3} = \frac{4-y}{3} \quad \text{Divide both sides by 3}$$

$$x = \frac{4-y}{3}$$

EXAMPLE 11

Solve $4x - 5y = 7$ for y.

Solution

$$4x - 5y = 7$$

$$4x - 5y - 4x = 7 - 4x \quad \text{Subtract } 4x \text{ from both sides}$$

$$-5y = 7 - 4x$$

$$\frac{-5y}{-5} = \frac{7-4x}{-5} \quad \text{Divide both sides by } -5$$

$$y = \frac{7-4x}{-5}\left(\frac{-1}{-1}\right) \quad \text{Multiply numerator and denominator of fraction on the right by } -1$$

$$y = \frac{4x-7}{5} \quad \text{We commonly do this so that the denominator is positive}$$

EXAMPLE 12

Solve $y = mx + b$ for m.

Solution

$$y = mx + b$$

$$y - b = mx + b - b \quad \text{Subtract } b \text{ from both sides}$$

$$y - b = mx$$

$$\frac{y-b}{x} = \frac{mx}{x} \quad \text{Divide both sides by } x$$

$$\frac{y-b}{x} = m$$

CONCEPT QUIZ 3.3

For Problems 1–10, match the correct formula for each.

1. Area of a rectangle
2. Circumference of a circle
3. Volume of a rectangular prism
4. Area of a triangle
5. Area of a circle
6. Volume of a right circular cylinder
7. Perimeter of a rectangle
8. Volume of a sphere
9. Area of a parallelogram
10. Area of a trapezoid

A. $A = \pi r^2$
B. $V = lwh$
C. $P = 2l + 2w$
D. $V = \frac{4}{3}\pi r^3$
E. $A = lw$
F. $A = bh$
G. $A = \frac{1}{2}h(b_1 + b_2)$
H. $A = \frac{1}{2}bh$
I. $C = 2\pi r$
J. $V = \pi r^2 h$

Section 3.3 Classroom Problem Set

Objective 1

1. Solve $d = rt$ for t if $d = 150$ and $r = 60$.
2. Solve $d = rt$ for r if $d = 486$ and $t = 9$.
3. Solve $A = \frac{1}{2}h(B_1 + 18)$ for B_1 if $A = 150$ and $h = 10$.
4. Solve $A = P + Prt$ for t if $A = 5080$, $P = 4000$, and $r = 0.03$.

Objective 2

5. Find the height of a triangle whose area is 52 square feet and whose base is 16 feet in length.
6. A lawn is in the shape of a triangle with one side 130 feet long and the altitude to that side 60 feet long. Find the number of square feet the lawn covers.
7. John is building a deck that is in the shape of a trapezoid. Find the area of the deck if the trapezoid has an altitude whose length is 10 feet and bases with lengths of 12 feet and 16 feet.
8. A flower garden is in the shape of a trapezoid with bases of 6 yards and 10 yards. The distance between the bases is 4 yards. Find the area of the garden.
9. Find the total surface area of a right circular cylinder that has a radius of 15 feet and a height of 24 feet.
10. Find the total surface area and volume of a tin can if the radius of the base is 3 centimeters, and the height of the can is 10 centimeters. Write your answers in terms of π.
11. A deck 4 feet wide surrounds a rectangular swimming pool that measures 20 feet by 30 feet. Find the area of the deck.
12. A circular pool is 34 feet in diameter and has a flagstone walk around it that is 3 feet wide. Find the area of the walk. Express the answer in terms of π.

Objective 3

13. Solve $i = Prt$ for P.
14. Solve $V = \pi r^2 h$ for h.
15. Solve $A = \frac{1}{2}bh$ for b.
16. Solve $V = \frac{1}{3}\pi r^2 h$ for h.
17. Solve $S = 2\pi r^2 + 2\pi rh$ for h.
18. Solve $F = \frac{9}{5}C + 32$ for C.

Objective 4

19. Solve $2x + y = -3$ for x.
20. Solve $5x + 2y = 12$ for x.
21. Solve $3x - 8y = 5$ for y.
22. Solve $3x - 5y = 19$ for y.
23. Solve $y = ax + b$ for a.
24. Solve $ax - by - c = 0$ for y.

THOUGHTS INTO WORDS

1. Suppose that both the length and width of a rectangle are doubled. How does this affect the perimeter of the rectangle? Defend your answer.
2. Suppose that the length of the radius of a circle is doubled. How does this affect the area of the circle? Defend your answer.
3. To estimate the change from a Fahrenheit temperature reading to a Celsius reading, some people *subtract 32 and then divide by 2*. How good is this estimate?

Answers to the Concept Quiz

1. E 2. I 3. B 4. H 5. A 6. J 7. C 8. D 9. F 10. G

3.4 Problem Solving

OBJECTIVES

1. Solve Equations
2. Solve Word Problems Involving Simple Interest
3. Solve Word Problems Involving Geometry
4. Solve Word Problems Involving Motion

1 Solve Equations

Before we delve into word problems, let's take a look at solving equations. The next example shows specific types of equations that will often show up in word problems.

EXAMPLE 1

Solve:

(a) $y + \dfrac{3}{4}y - 2 = 48$

(b) $n + (n - 1) + (2n - 3) = 124$

(c) $\dfrac{3}{7}x + \dfrac{2}{7}(x + 3) = 6$

Solution

(a)
$$y + \dfrac{3}{4}y - 2 = 48$$

$$\dfrac{4}{4}y + \dfrac{3}{4}y - 2 = 48 \qquad \text{Change 1 to } \dfrac{4}{4}$$

$$\dfrac{7}{4}y - 2 = 48$$

$$\dfrac{7}{4}y = 50 \qquad \text{Added 2 to each side}$$

$$\dfrac{4}{7}\left(\dfrac{7}{4}y\right) = \dfrac{4}{7}(50)$$

$$y = \dfrac{200}{7}$$

The solution set is $\left\{\dfrac{200}{7}\right\}$.

(b)
$$n + (n - 1) + (2n - 3) = 124$$
$$n + n - 1 + 2n - 3 = 124$$
$$4n - 4 = 124$$
$$4n = 128 \qquad \text{Added 4 to each side}$$
$$n = 32 \qquad \text{Divided both sides by 4}$$

The solution set is {32}.

(c) $$\frac{3}{7}x + \frac{2}{7}(x + 3) = 6$$

$$7\left[\frac{3}{7}x + \frac{2}{7}(x + 3)\right] = 7(6) \quad \text{Multiply both sides by 7}$$

$$7\left(\frac{3}{7}x\right) + 7\left[\frac{2}{7}(x + 3)\right] = 7(6) \quad \text{Distributive property}$$

$$3x + 7\left(\frac{2}{7}\right)(x + 3) = 42$$

$$3x + 2(x + 3) = 42 \quad \text{Multiply 7 times } \frac{2}{7}$$

$$3x + 2x + 6 = 42 \quad \text{Distributive property}$$

$$5x + 6 = 42$$

$$5x = 36$$

$$x = \frac{36}{5}$$

The solution set is $\left\{\dfrac{36}{5}\right\}$. ■

Let's restate the suggestions for solving word problems that we offered in Section 2.4.

Suggestions for Solving Word Problems

1. Read the problem carefully, and make sure that you understand the meanings of all the words. Be especially alert for any technical terms used in the statement of the problem.
2. Read the problem a second time (perhaps even a third time) to get an overview of the situation being described and to determine the known facts as well as what is to be found.
3. Sketch any figure, diagram, or chart that might be helpful in analyzing the problem.
4. Choose a meaningful variable to represent an unknown quantity in the problem (perhaps t if time is the unknown quantity); represent any other unknowns in terms of that variable.
5. Look for a guideline that can be used to set up an equation. A guideline might be a formula, such as *selling price equals cost plus profit*, or a relationship, such as *interest earned from a 9% investment plus interest earned from a 10% investment equals total amount of interest earned*. A guideline may also be illustrated by a figure or diagram that you sketch for a particular problem.
6. Form an equation that contains the variable that translates the conditions of the guideline from English into algebra.
7. Solve the equation, and use the solution to determine all the facts requested in the problem.
8. **Check all answers back in the original statement of the problem.**

We emphasize the importance of suggestion 5. Determining the guideline to follow when setting up the equation is key to analyzing the problem. Sometimes the guideline is a formula—such as one of the formulas we presented in the previous section and accompanying problem set. Let's consider a problem of that type.

2 Solve Word Problems Involving Simple Interest

EXAMPLE 2

Apply Your Skill

How long will it take $500 to double itself if it is invested at 8% simple interest?

Solution

Let's use the basic simple interest formula, $i = Prt$, where i represents interest, P is the principal (money invested), r is the rate (percent), and t is the time in years. For $500 to "double itself" means that we want the original $500 to earn another $500 in interest. Thus, using $i = Prt$ as a guideline, we can proceed as follows:

$$i = Prt$$
$$500 = 500(8\%)(t)$$

Now let's solve this equation:

$$500 = 500(0.08)(t)$$
$$1 = 0.08t$$
$$100 = 8t$$
$$\frac{100}{8} = t$$
$$12\frac{1}{2} = t$$

It will take $12\frac{1}{2}$ years. ∎

3 Solve Word Problems Involving Geometry

If the problem involves a geometric formula, then a sketch of the figure is helpful for recording the given information and analyzing the problem. Try it in the next example.

EXAMPLE 3

Apply Your Skill

The length of a football field is 40 feet more than twice its width, and the perimeter of the field is 1040 feet. Find the length and width of the field.

Solution

Because the length is stated in terms of the width, we can let w represent the width, and then $2w + 40$ represents the length (see Figure 3.16 to see how we labeled the diagram). A guideline for this problem is the perimeter formula $P = 2l + 2w$. Thus we use the following equation to set up and solve the problem:

$$P = 2l + 2w$$
$$1040 = 2(2w + 40) + 2w$$
$$1040 = 4w + 80 + 2w$$
$$1040 = 6w + 80$$
$$960 = 6w$$
$$160 = w$$

If $w = 160$, then $2w + 40 = 2(160) + 40 = 360$. Thus the football field is 360 feet long and 160 feet wide.

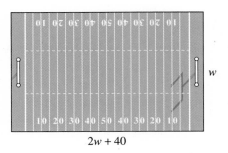

Figure 3.16

4 Solve Word Problems Involving Motion

Sometimes the formulas we use when we are analyzing a problem are different from those we use as a guideline for setting up the equation. For example, uniform-motion problems involve the formula $d = rt$, but the main guideline for setting up an equation for such problems is usually a statement about either *times*, *rates*, or *distances*. Let's consider an example.

EXAMPLE 4

Apply Your Skill

Pablo leaves city A on a moped and travels toward city B at 18 miles per hour. At the same time, Cindy leaves city B on a moped and travels toward city A at 23 miles per hour. The distance between the two cities is 123 miles. How long will it take before Pablo and Cindy meet on their mopeds?

Solution

First, sketch a diagram as in Figure 3.17. Then let t represent the time that Pablo travels and also the time that Cindy travels.

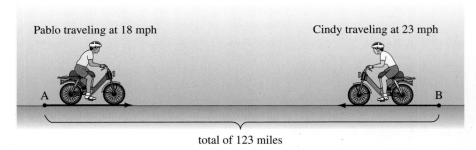

Figure 3.17

$$\text{Distance Pablo travels} + \text{Distance Cindy travels} = \text{Total distance}$$
$$18t + 23t = 123$$

Solving this equation yields

$$18t + 23t = 123$$
$$41t = 123$$
$$t = 3$$

They both travel for 3 hours.

Some people find it helpful to use a chart to organize the known and unknown facts in a uniform-motion problem. We will illustrate with the next example.

EXAMPLE 5   Apply Your Skill

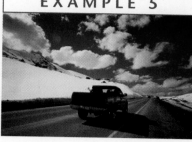

A car leaves a town traveling at 60 kilometers per hour. How long will it take a second car traveling at 75 kilometers per hour to catch the first car if the second car leaves 1 hour later and travels the same route?

Solution

Let t represent the time of the second car. Then $t + 1$ represents the time of the first car, because it travels 1 hour longer. We can now record the information of the problem in a chart.

	Rate	Time	Distance ($d = rt$)
First car	60	$t + 1$	$60(t + 1)$
Second car	75	t	$75t$

Because the second car is to overtake the first car, the distances must be equal.

Distance of second car = Distance of first car
$$75t = 60(t + 1)$$

We can solve this equation:

$$75t = 60(t + 1)$$
$$75t = 60t + 60$$
$$15t = 60$$
$$t = 4$$

The second car should overtake the first car in 4 hours. (Check the answer!) ∎

We offer one bit of advice at this time. Don't become discouraged if solving word problems is giving you trouble. Problem solving is not a skill that you can develop overnight. It takes time, patience, hard work, and an open mind. Keep giving it your best shot, and gradually you'll become more confident in your approach to such problems. Furthermore, we realize that some (perhaps many) of these problems may not seem "practical" to you; however, keep in mind that the real goal here is to develop your skill in applying problem-solving techniques. Finding and using a guideline, sketching a figure to record information and help in the analysis, estimating an answer before attempting to solve the problem, and using a chart to record information are important skills to develop.

CONCEPT QUIZ 3.4

Arrange the following steps in the correct order to solve word problems.

A. Declare a variable and represent any other unknown quantities in terms of that variable.
B. Check the answer back into the original statement of the problem.
C. Write an equation for the problem, and remember to look for a formula or guideline that could be used to write the equation.
D. Read the problem carefully, and be sure that you understand all the terms in the stated problem.

E. Sketch a diagram or figure that helps you analyze the problem.
F. Solve the equation, and figure out the answer being asked for in the problem.

Section 3.4 Classroom Problem Set

Objective 1

1. Solve.
   (a) $\dfrac{2}{5}b + b + 4 = 39$
   (b) $n + (n + 1) + (n + 2) = 153$
   (c) $\dfrac{1}{3}c + \dfrac{2}{3}(c + 4) = 15$

2. Solve.
   (a) $x + \dfrac{2}{3}x + 1 = 41$
   (b) $t + (3t - 2) + (4t - 4) = 42$
   (c) $16a + 8\left(\dfrac{9}{2} - a\right) = 60$

Objective 2

3. How long will it take $3000 to earn $525 interest if it is invested at 10% simple interest?

4. How many years will it take $500 to earn $750 in interest if it is invested at 6% simple interest?

Objective 3

5. The length of a rectangular computer display screen is twice its width, and the perimeter is 48 inches. Find the length and width of the screen.

6. The width of a rectangle is 1 foot more than one-third of its length. If the perimeter of the rectangle is 74 feet, find the area of the rectangle.

Objective 4

7. Justin leaves the park and walks toward home at 5 miles per hour. At the same time, Hope leaves home and heads to the park walking at 2 miles per hour. The distance between the park and home is 3.5 miles. How long will it take for Justin and Hope to meet?

8. Two cars start from the same place traveling in opposite directions. One car travels 4 miles per hour faster than the other car. Find their speeds if after 5 hours they are 520 miles apart.

9. A student left home to return to the university traveling at 60 miles per hour. Fifteen minutes later his mother discovered that he had forgotten his laptop computer and she left home traveling at 70 miles a hour to catch up to him. How long will it take her to catch up to him?

10. Stan starts jogging at 5 miles per hour. Half an hour later Felicia starts jogging on the same route at 7 miles per hour. How long will it take Felicia to catch Stan?

THOUGHTS INTO WORDS

1. Summarize the new ideas about problem solving that you have acquired thus far in this course.

Answers to the Concept Quiz
D E A C F B

3.5 More about Problem Solving

OBJECTIVES

1. Solve Word Problems Involving Mixture
2. Solve Word Problems Involving Interest
3. Solve Word Problems Involving Age

Let's begin this section with an important but often overlooked facet of problem solving: the importance of *looking back* over your solution and considering some of the following questions:

1. Is your answer to the problem a reasonable answer? Does it agree with the answer you estimated before doing the problem?
2. Have you checked your answer by substituting it back into the conditions stated in the problem?
3. Do you see another plan that you could use to solve the problem? Is there another guideline that could be used?
4. Is this problem closely related to another problem that you have previously solved?
5. Have you tucked away for future reference the technique you used to solve this problem?

Looking back over the solution of a newly solved problem can often lay important groundwork for solving similar problems in the future.

1 Solve Word Problems Involving Mixture

Now let's consider three problems that we often refer to as mixture problems. No basic formula applies for all of these problems, but the suggestion that *you think in terms of a pure substance* is often helpful in setting up a guideline. For example, a phrase such as "a 30% solution of acid" means that 30% of the solution is acid and the remaining 70% is water.

EXAMPLE 1

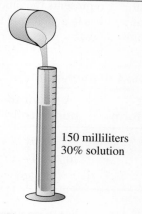

150 milliliters
30% solution

Figure 3.18

Apply Your Skill

How many milliliters of pure acid must be added to 150 milliliters of a 30% solution of acid to obtain a 40% solution of acid? (See Figure 3.18.)

Solution

To solve this problem, we can use the following guideline:

| Amount of pure acid in original solution | + | Amount of pure acid to be added | = | Amount of pure acid in final solution |

To apply this guideline, it is necessary to multiply 150 ml by 30% to determine the amount of pure acid in the solution. We also need to recognize that adding "pure acid" is equivalent to a 100% acid solution.

Let p represent the amount of pure acid to be added. Then the amount of the final solution is 150 milliliters plus the amount of pure acid added. Let's organize the information that we have in a table and use the formula, *amount of solution × percent = amount of pure acid*.

	Amount of solution × Percent = Amount of pure acid		
	Amount of solution	Percent (in decimal form)	Amount of pure acid
Original solution	150	0.30	0.30(150)
Pure acid	p	1.00	1.00p
Final solution	$150 + p$	0.40	$0.40(150 + p)$

Note that the last column of the table yields the expressions for the guideline. Using the guideline and the last column of the table we can write the equation:

$$0.30(150) + 1.00p = 0.40(150 + p)$$

Now let's solve the equation to determine the amount of pure acid to be added.

$$0.30(150) + 1.00p = 0.40(150 + p)$$
$$45 + p = 60 + 0.40p$$
$$0.60p = 15$$
$$p = \frac{15}{0.60} = 25$$

We must add 25 milliliters of pure acid. (You should check this answer.) ∎

EXAMPLE 2 Apply Your Skill

Suppose that you have a supply of a 30% solution of alcohol and a 70% solution of alcohol. How many quarts of each should be mixed to produce a 20-quart solution that is 40% alcohol?

Solution

We can use a guideline similar to the one in Example 1:

$$\boxed{\text{Pure alcohol in 30\% solution}} + \boxed{\text{Pure alcohol in 70\% solution}} = \boxed{\text{Pure alcohol in 40\% solution}}$$

Let x represent the amount of 30% solution. Then $20 - x$ represents the amount of 70% solution. Now, let's organize the information that we have in a table and use the formula, *amount of solution × percent = amount of pure alcohol*.

	Amount of solution × Percent = Amount of pure alcohol		
	Amount of solution	Percent (in decimal form)	Amount of pure alcohol
30% Solution	x	0.30	$0.30x$
70% Solution	$20 - x$	0.70	$0.70(20 - x)$
40% Solution	20	0.40	$0.40(20)$

The last column of the table yields the expressions for the guideline. Using the guideline and the last column of the table we can write the equation:

$$0.30x + 0.70(20 - x) = 0.40(20)$$

Solving this equation to determine the amount of the solutions to be mixed, we obtain

$$0.30x + 0.70(20 - x) = 0.40(20)$$
$$0.30x + 0.70(20 - x) = 8$$
$$30x + 70(20 - x) = 800 \quad \text{Multiplied both sides of the equation by 100}$$
$$30x + 1400 - 70x = 800$$
$$-40x = -600$$
$$x = 15$$

Therefore, $20 - x = 5$. We should mix 15 quarts of the 30% solution with 5 quarts of the 70% solution. ∎

2 Solve Word Problems Involving Interest

EXAMPLE 3 **Apply Your Skill**

A woman has a total of $5000 to invest. She invests part of it at 4%, and the remainder at 6%. Her total yearly interest from the two investments is $260. How much did she invest at each rate?

Solution

Let x represent the amount invested at 6%. Then $5000 - x$ represents the amount invested at 4%. Use the following guideline:

$$\underbrace{(6\%)(x)}_{\text{Interest earned from 6\% investment}} + \underbrace{(4\%)(\$5000 - x)}_{\text{Interest earned from 4\% investment}} = \underbrace{\$260}_{\text{Total interest earned}}$$

Solving this equation yields

$$(6\%)(x) + (4\%)(5000 - x) = 260$$
$$0.06x + 0.04(5000 - x) = 260$$
$$6x + 4(5000 - x) = 26{,}000$$
$$6x + 20{,}000 - 4x = 26{,}000$$
$$2x + 20{,}000 = 26{,}000$$
$$2x = 6000$$
$$x = 3000$$

Therefore, $5000 - x = 2000$.

She invested $3000 at 6% and $2000 at 4%. ∎

EXAMPLE 4 **Apply Your Skill**

An investor invests a certain amount of money at 3%. Then he finds a better deal and invests $5000 more than that amount at 5%. His yearly income from the two investments is $650. How much did he invest at each rate?

Solution

Let x represent the amount invested at 3%. Then $x + 5000$ represents the amount invested at 5%.

$$(3\%)(x) + (5\%)(x + 5000) = 650$$
$$0.03x + 0.05(x + 5000) = 650$$
$$3x + 5(x + 5000) = 65{,}000$$
$$3x + 5x + 25{,}000 = 65{,}000$$
$$8x + 25{,}000 = 65{,}000$$
$$8x = 40{,}000$$
$$x = 5000$$

Therefore, $x + 5000 = 10{,}000$.

He invested $5000 at 3% and $10,000 at 5%. ∎

3 Solve Word Problems Involving Age

The key to solving the problem in the next example is to represent the various unknown quantities in terms of one variable.

EXAMPLE 5   Apply Your Skill

Jody is 6 years younger than her sister Cathy, and in 7 years Jody will be three-fourths as old as Cathy is at that time. Find their present ages.

Solution

By letting c represent Cathy's present age, we can represent all of the unknown quantities like this:

c: Cathy's present age
$c - 6$: Jody's present age
$c + 7$: Cathy's age in 7 years
$c - 6 + 7$ or $c + 1$: Jody's age in 7 years

The statement "in 7 years Jody will be three-fourths as old as Cathy is at that time" serves as the guideline, so we can set up and solve the following equation:

$$c + 1 = \frac{3}{4}(c + 7)$$
$$4c + 4 = 3(c + 7)$$
$$4c + 4 = 3c + 21$$
$$c = 17$$

Therefore, Cathy's present age is 17, and Jody's present age is $17 - 6 = 11$. ∎

CONCEPT QUIZ 3.5

For Problems 1–7, answer true or false.

1. The phrase "a 40% solution of alcohol" means that 40% of the amount of the solution is alcohol.
2. The amount of pure acid in 300 ml of a 30% solution is 100 ml.

3. If we want to produce 10 quarts by mixing solution A and solution B, the amount of solution A needed could be represented by x, and the amount of solution B would then be represented by 10 − x.
4. The formula $d = rt$ is equivalent to $r = \dfrac{d}{t}$.
5. The formula $d = rt$ is equivalent to $t = \dfrac{r}{d}$.
6. If y represents John's current age, then his age 4 years ago would be represented by y − 4.
7. If Shane's current age is represented by x, then his age in 10 years would be represented by 10x.

Section 3.5 Classroom Problem Set

Objective 1

1. A new preschool regulation requires that juice mixtures be at least 20% pure juice. How many quarts of pure orange juice must be added to 36 quarts of a 10% juice mixture to obtain a 20% juice mixture?

2. How many gallons of 15% salt solution must be mixed with 8 gallons of a 20% salt solution to obtain a 17% salt solution?

3. An oil company obtains fuel from two different refineries. One refinery supplies fuel that is 10% ethanol and the other refinery supplies fuel that is 20% ethanol. How many gallons of each should be mixed to produce 6000 gallons of fuel that is 16% ethanol?

4. Thirty ounces of a punch containing 10% grapefruit juice is added to 50 ounces of a punch containing 20% grapefruit juice. Find the percent of grapefruit juice in the resulting mixture.

Objective 2

5. A sum of $8000 is invested in two different accounts. The total yearly interest from the two investments is $524. How much was invested in each account if the first account paid 7% interest and the second account paid 5% interest?

6. Sharron received an inheritance of $12,000 from her grandfather. She invested part of it at 6% interest and the remainder at 8%. If the total yearly interest from both investments was $860, how much did she invest at each rate?

7. Mrs. Stewart invested a certain amount of money at 6%. Then she found an investment that paid 9%. Because that investment carried more risk, she invested $2000 less than the amount she invested at 6%. Her yearly income from the two investments is $870. Find the amount invested in each investment.

8. If $2000 is invested at 4% interest, how much money must be invested at 7% interest so that the total return for both investments averages 6%?

Objective 3

9. Jennifer is three years older than her cousin Kyle. In one year, Kyle's age will be two-thirds of Jennifer's age at that time. Find their present ages.

10. The sum of the present ages of Beth and her aunt Greta is 40 years. In 10 years, Greta's age will be twice Beth's age. Find their present ages.

Answers to the Concept Quiz
1. True 2. False 3. True 4. True 5. False 6. True 7. False

Chapter 3 Review Problem Set

For Problems 1–10, solve each proportion.

1. $\dfrac{x}{8} = \dfrac{5}{2}$

2. $\dfrac{x}{6} = \dfrac{7}{2}$

3. $\dfrac{3}{2} = \dfrac{9}{x}$

4. $\dfrac{10}{x} = \dfrac{5}{4}$

5. $\dfrac{x-1}{3} = \dfrac{6}{5}$

6. $\dfrac{x+2}{4} = \dfrac{5}{8}$

7. $\dfrac{x+1}{3} = \dfrac{x-6}{2}$

8. $\dfrac{x+3}{6} = \dfrac{x-2}{4}$

9. $\dfrac{7}{2x-2} = \dfrac{3}{x+4}$

10. $\dfrac{-2}{x-2} = \dfrac{6}{3x+4}$

For Problems 11–18, set up a proportion and solve.

11. A blueprint has a scale in which 1 inch represents 4 feet. Find the dimensions of a rectangular room that measures 3.5 inches by 4.25 inches on the blueprint.

12. On a certain map, 1 inch represents 20 miles. If two cities are 6.5 inches apart on the map, find the number of miles between the cities.

13. If a car travels 200 miles using 10 gallons of gasoline, how far will it travel on 15 gallons of gasoline?

14. A home valued at $150,000 is assessed $3000 in real estate taxes. At the same rate, how much are the taxes on a home with a value of $200,000?

15. If 20 pounds of fertilizer will cover 1400 square feet of grass, how many pounds of fertilizer are needed for 1750 square feet of grass?

16. A board 24 feet long is cut into two pieces whose lengths are in the ratio of 2 to 3. Find the lengths of the two pieces.

17. A sum of $750 is to be divided between two people in the ratio of 3 to 2. How much does each person receive?

18. The ratio of female students to male students at a certain junior college is 5 to 3. If there is a total of 4400 students, find the number of female students and the number of male students.

For Problems 19–26, use proportions to change each common fraction to a percent.

19. $\dfrac{13}{20}$

20. $\dfrac{8}{25}$

21. $\dfrac{7}{8}$

22. $\dfrac{1}{8}$

23. $\dfrac{7}{4}$

24. $\dfrac{3}{16}$

25. $\dfrac{2}{3}$

26. $\dfrac{7}{3}$

For Problems 27–32, set up an equation and solve.

27. Eighteen is what percent of 30?

28. Twenty-four is what percent of 150?

29. Fifteen percent of a number is 6. Find the number.

30. Ten percent of a number is 56. Find the number.

31. What is 120% of 500?

32. What is 42% of 20?

For Problems 33–38, solve each equation.

33. $0.5x + 0.7x = 6$

34. $0.02x + 1.8x = 6.37$

35. $0.07t + 0.12(t - 3) = 0.59$

36. $0.1x + 0.12(1700 - x) = 188$

37. $x + 0.25x = 20$

38. $0.2(x - 3) = 14$

For Problems 39–46, set up an equation and solve.

39. Louise bought a pair of jeans for $36.40, which represents a 30% discount of the original price. Find the original price of the jeans.

40. Find the cost of a $75 shirt that is on sale for 20% off.

41. Ely bought a putter for $72 that was originally listed for $120. What rate of discount did he receive?

42. Dominic bought a computer monitor for $162.50 that was originally listed for $250. What rate of discount did he receive?

43. A retailer has some candles that cost him $5 each. He wants to sell them at a profit of 60% of the cost. What should he charge for the candles?

44. A retailer has some shoes that cost her $25 a pair. She wants to sell them at a profit of 45% of the cost. What should she charge for the shoes?

45. A bookstore manager buys some textbooks for $88 each. He wants to sell them at a profit of 20% based on the selling price. What should he charge for the books?

46. A supermarket manager buys some apples at $0.80 per pound. He wants to sell them at a profit of 50% based on the selling price. What should he charge for the apples?

For Problems 47–50, use the formula $i = Prt$.

47. How much interest will be charged on a 4-year student loan of $8500 at an 8% annual interest rate?

48. How long will $2500 need to be invested at a 6% annual interest rate to earn $600?

49. How much principal, invested at 7% annual interest rate for 3 years, is needed to earn $630?

50. Find the annual interest rate if $820 in interest is earned when $2500 is invested for 4 years.

51. Solve $P = 2l + 2w$ for w if $P = 50$ and $l = 19$.

52. Solve $F = \frac{9}{5}C + 32$ for C if $F = 77$.

53. Solve $A = P + Prt$ for r if $A = 450$, $P = 400$, and $t = 2$.

54. Solve $d = rt$ for t if $d = 150$ and $r = 60$.

55. Find the area of a trapezoid with one base 8 inches long, the other base 14 inches long, and the altitude between the two bases 7 inches.

56. If the area of a triangle is 27 square centimeters, and the length of one side is 9 centimeters, find the length of the altitude to that side.

57. How many yards of fencing are required to enclose a rectangular field that is 120 yards long and 57 yards wide?

58. If the total surface area of a right circular cylinder is 152π square feet, and a radius of a base is 4 feet long, find the height of the cylinder.

59. Solve $V = lwh$ for w.

60. Solve $V = \pi r^2 h$ for h.

61. Solve $A = \frac{1}{2}bh$ for b.

62. Solve $V = \frac{1}{3}Bh$ for B.

63. Solve $2x + 5y = -12$ for y.

64. Solve $3x - y = 2$ for y.

65. Solve $6x - 2y = 5$ for x.

66. Solve $3x + 4y = 7$ for x.

67. Suppose that the length of a certain rectangle is 5 meters longer than twice the width. The perimeter of the rectangle is 46 meters. Find the length and width of the rectangle.

68. A copper wire 110 centimeters long was bent in the shape of a rectangle. The length of the rectangle was 10 centimeters longer than twice the width. Find the dimensions of the rectangle.

69. Seventy-eight yards of fencing were purchased to enclose a rectangular garden. The length of the garden is 1 yard shorter than three times its width. Find the length and width of the garden.

70. One angle of a triangle has a measure of 47°. Of the other two angles, one of them is 3° smaller than three times the other angle. Find the measures of the two remaining angles.

71. Two airplanes leave Chicago at the same time and fly in opposite directions. If one travels at 350 miles per hour and the other at 400 miles per hour, how long will it take them to be 1125 miles apart?

72. Two cars start from the same place traveling in opposite directions. One car travels 7 miles per hour faster than the other car. If at the end of 3 hours they are 369 miles apart, find the speed of each car.

73. Connie rides out into the country on her bicycle at a rate of 10 miles per hour. An hour later Zak leaves from the same place that Connie did and rides his bicycle along the same route at 12 miles per hour. How long will it take Zak to catch Connie?

74. Two cities, A and B, are 406 miles apart. Billie starts at city A in her car traveling toward city B at 52 miles per hour. At the same time, using the same route, Zorka leaves in her car from city B traveling toward city A at 64 miles per hour. How long will it be before the cars meet?

75. How many liters of pure alcohol must be added to 10 liters of a 70% solution to obtain a 90% solution?

76. How many gallons of a 10% salt solution must be mixed with 12 gallons of a 15% salt solution to obtain a 12% salt solution?

77. Suppose that 20 ounces of a punch containing 20% orange juice is added to 30 ounces of a punch containing 30% orange juice. Find the percent of orange juice in the resulting mixture.

78. How many liters of pure alcohol must be added to 10 liters of a 30% solution to obtain a 50% solution?

79. How many cups of grapefruit juice must be added to 40 cups of punch that contains 5% grapefruit juice to obtain a punch that is 10% grapefruit juice?

80. How many milliliters of pure acid must be added to 150 milliliters of a 30% solution of acid to obtain a 40% solution?

81. A sum of $2100 is invested, part of it at 3% interest and the remainder at 5%. If the interest earned by the 5% investment is $51 more than the interest from the 3% investment, find the amount invested at each rate.

82. Eva invested a certain amount of money at 5% interest, and $1500 more than that amount at 6%. Her total yearly interest income was $420. How much did she invest at each rate?

83. A total of $4000 was invested, part of it at 8% and the remainder at 9%. If the total interest income amounted to $350, how much was invested at each rate?

84. If $500 is invested at 6% interest, how much additional money must be invested at 9% so that the total return for both investments averages 8%?

85. Shane is 8 years older than his brother Caleb. In 4 years, Caleb's age will be two-thirds of Shane's age. Find their present ages.

86. The sum of the present ages of Melinda and her mother is 64 years. In 8 years, Melinda will be three-fifths as old as her mother at that time. Find the present ages of Melinda and her mother.

87. At the present time, Nikki is one-third as old as Kaitlin. In 10 years, Kaitlin's age will be 6 years less than twice Nikki's age at that time. Find the present age of both girls.

88. Annilee's present age is two-thirds of Jessie's present age. In 12 years, the sum of their ages will be 54 years. Find their present ages.

Chapter 3 Practice Test

For Problems 1–10, solve each equation.

1. $\dfrac{x+2}{4} = \dfrac{x-3}{5}$

2. $\dfrac{-4}{2x-1} = \dfrac{3}{3x+5}$

3. $\dfrac{x-1}{6} - \dfrac{x+2}{5} = 2$

4. $\dfrac{x+8}{7} - 2 = \dfrac{x-4}{4}$

5. $\dfrac{n}{20-n} = \dfrac{7}{3}$

6. $\dfrac{h}{4} + \dfrac{h}{6} = 1$

7. $0.05n + 0.06(400 - n) = 23$

8. $s = 35 + 0.5s$

9. $0.07n = 45.5 - 0.08(600 - n)$

10. $12t + 8\left(\dfrac{7}{2} - t\right) = 50$

11. Solve $F = \dfrac{9C + 160}{5}$ for C.

12. Solve $y = 2(x - 4)$ for x.

13. Solve $\dfrac{x+3}{4} = \dfrac{y-5}{9}$ for y.

For Problems 14–16, use the geometric formulas given in this chapter to help you find the solution.

14. Find the area of a circular region if the circumference is 16π centimeters. Express the answer in terms of π.

15. If the perimeter of a rectangle is 100 inches and its length is 32 inches, find the area of the rectangle.

16. The area of a triangular plot of ground is 133 square yards. If the length of one side of the plot is 19 yards, find the length of the altitude to that side.

For Problems 17–25, set up an equation and solve.

17. Express $\dfrac{5}{4}$ as a percent.

18. Thirty-five percent of what number is 24.5?

19. Cora bought a digital camera for $132.30, which represented a 30% discount off the original price. What was the original price of the camera?

20. A retailer has some lamps that cost her $40 each. She wants to sell them at a profit of 30% of the cost. What price should she charge for the lamps?

21. _____

22. _____

23. _____

24. _____

25. _____

21. Hugh paid $48 for a pair of golf shoes that were listed for $80. What rate of discount did he receive?

22. The election results in a certain precinct indicated that the ratio of female voters to male voters was 7 to 5. If a total of 1500 people voted, how many women voted?

23. A car leaves a city traveling at 50 miles per hour. One hour later a second car leaves the same city traveling on the same route at 55 miles per hour. How long will it take the second car to overtake the first car?

24. How many centiliters of pure acid must be added to 6 centiliters of a 50% acid solution to obtain a 70% acid solution?

25. How long will it take $4000 to double if it is invested at 9% simple interest?

Chapters 1–3   Cumulative Review Problem Set

For Problems 1–10, simplify each algebraic expression by combining similar terms.

1. $7x - 9x - 14x$
2. $-10a - 4 + 13a + a - 2$
3. $5(x - 3) + 7(x + 6)$
4. $3(x - 1) - 4(2x - 1)$
5. $-3n - 2(n - 1) + 5(3n - 2) - n$
6. $6n + 3(4n - 2) - 2(2n - 3) - 5$
7. $\frac{1}{2}x - \frac{3}{4}x + \frac{2}{3}x - \frac{1}{6}x$
8. $\frac{1}{3}n - \frac{4}{15}n + \frac{5}{6}n - n$
9. $0.4x - 0.7x - 0.8x + x$
10. $0.5(x - 2) + 0.4(x + 3) - 0.2x$

For Problems 11–20, evaluate each algebraic expression for the given values of the variables.

11. $5x - 7y + 2xy$ for $x = -2$ and $y = 5$
12. $2ab - a + 6b$ for $a = 3$ and $b = -4$
13. $-3(x - 1) + 2(x + 6)$ for $x = -5$
14. $5(n + 3) - (n + 4) - n$ for $n = 7$
15. $\frac{3x - 2y}{2x - 3y}$ for $x = 3$ and $y = -6$
16. $\frac{3}{4}n - \frac{1}{3}n + \frac{5}{6}n$ for $n = -\frac{2}{3}$
17. $2a^2 - 4b^2$ for $a = 0.2$ and $b = -0.3$
18. $x^2 - 3xy - 2y^2$ for $x = \frac{1}{2}$ and $y = \frac{1}{4}$
19. $5x - 7y - 8x + 3y$ for $x = 9$ and $y = -8$
20. $\frac{3a - b - 4a + 3b}{a - 6b - 4b - 3a}$ for $a = -1$ and $b = 3$

For Problems 21–26, evaluate each expression.

21. 3^4
22. -2^6
23. $(0.4)^3$
24. $\left(-\frac{1}{2}\right)^5$
25. $\left(\frac{1}{2} + \frac{1}{3}\right)^2$
26. $\left(\frac{3}{4} - \frac{7}{8}\right)^3$

For Problems 27–38, solve each equation.

27. $-5x + 2 = 22$
28. $3x - 4 = 7x + 4$
29. $3(4x - 1) = 6(2x - 1)$
30. $2(x - 1) - 3(x - 2) = 12$
31. $\frac{2}{5}x - \frac{1}{3} = \frac{1}{3}x + \frac{1}{2}$
32. $\frac{t - 2}{4} + \frac{t + 3}{3} = \frac{1}{6}$
33. $\frac{2n - 1}{5} - \frac{n + 2}{4} = 1$
34. $0.09x + 0.12(500 - x) = 54$
35. $-5(n - 1) - (n - 2) = 3(-2n - 1)$
36. $\frac{-2}{x - 1} = \frac{-3}{x + 4}$
37. $0.2x + 0.1(x - 4) = 0.7x - 1$
38. $-(t - 2) + (t - 4) = 2\left(t - \frac{1}{2}\right) - 3\left(t + \frac{1}{3}\right)$

For Problems 39–46, solve each inequality.

39. $4x - 6 > 3x + 1$
40. $-3x - 6 < 12$
41. $-2(n - 1) \leq 3(n - 2) + 1$
42. $\frac{2}{7}x - \frac{1}{4} \geq \frac{1}{4}x + \frac{1}{2}$
43. $0.08t + 0.1(300 - t) > 28$
44. $-4 > 5x - 2 - 3x$
45. $\frac{2}{3}n - 2 \geq \frac{1}{2}n + 1$
46. $-3 < -2(x - 1) - x$

For Problems 47–54, set up an equation or an inequality and solve each problem.

47. Erin's salary this year is $32,000. This represents $2000 more than twice her salary 5 years ago. Find her salary 5 years ago.

48. One of two supplementary angles is 45° less than four times the other angle. Find the measure of each angle.

141

49. Jaamal has 25 coins, consisting of nickels and dimes, that amount to $2.10. How many coins of each kind does he have?

50. Hana bowled 144 and 176 in her first two games. What must she bowl in the third game to have an average of at least 150 for the three games?

51. A board 30 feet long is cut into two pieces, and the lengths are in the ratio of 2 to 3. Find the lengths of the two pieces.

52. A retailer has some shoes that cost him $32 per pair. He wants to sell them at a profit of 20% of the selling price. What price should he charge for the shoes?

53. Two cars start from the same place traveling in opposite directions. One car travels 5 miles per hour faster than the other car. Find their speeds if after 6 hours they are 570 miles apart.

54. How many liters of pure alcohol must be added to 15 liters of a 20% solution to obtain a 40% solution?

Exponents and Polynomials

Chapter 4 Warm-Up Problems

1. Simplify. (a) $6 - (-7)$ (b) $-4 - 3$ (c) $-1 - (-5)$

2. Divide. (a) $\dfrac{60}{4}$ (b) $\dfrac{48}{-6}$ (c) $\dfrac{-24}{-4}$

3. Multiply. (a) $\left(\dfrac{3}{4}\right)\left(\dfrac{3}{4}\right)\left(\dfrac{3}{4}\right)$ (b) $(0.4)^2$ (c) $\left(-\dfrac{2}{5}\right)^3$

4. Simplify. (a) $\dfrac{1.21}{1.1}$ (b) $\dfrac{-5.6}{0.28}$ (c) $6(3.5)$

5. Divide. Express remainders as fractions.
   (a) $12\overline{)108}$ (b) $9\overline{)39}$ (c) $16\overline{)456}$

6. Use the distributive property to simplify.
   (a) $3(4x^2 + 6x - 1)$ (b) $-(y^3 - 2)$ (c) $-(-3n^2 + 8n - 2)$

7. Simplify.
   (a) $2x + 3 - 7x + 4$ (b) $8x^2 - 3x^2 + 4y + y$ (c) $6m^2 - (-3m^2)$

8. Simplify. (a) $x^2 + 6x + 4x + 24$ (b) $m^2 - 3m + 5m - 15$
   (c) $12n^2 - 18n - 8n + 12$ (d) $x^2 - 4x - 4x + 16$

9. Find the area of a rectangle with
   (a) length, 17; width, 12 (b) length, x; width, 15

4.1 Addition and Subtraction of Polynomials
4.2 Multiplying Monomials
4.3 Multiplying Polynomials
4.4 Dividing by Monomials
4.5 Dividing by Binomials
4.6 Zero and Negative Integers as Exponents

Quadratic equations belong to a larger classification called *polynomial equations*. To solve problems involving polynomial equations, we need to develop some basic skills that pertain to polynomials. That is to say, we need to be able to add, subtract, multiply, divide, and factor polynomials. Chapters 4 and 5 will help you develop those skills as you work through problems that involve quadratic equations.

Video tutorials for all section learning objectives are available in a variety of delivery modes.

INTERNET PROJECT

For ease of operations, many of the numbers you find in science courses are written in scientific notation. In scientific notation, numbers are written in the form:

$a \times 10^b$

("*a* times 10 to the power of *b*"), where the exponent *b* is an integer, and $1 \leq |a| < 10$. For example, 7900 can be written as 7.9×10^3 and 0.00062 can be written as 6.2 3 10^{24}. Do an Internet search to find the mass of the Earth. Is the mass expressed in scientific notation? Engineering notation is similar to scientific notation. Do an Internet search on engineering notation and determine how it differs from scientific notation.

4.1 Addition and Subtraction of Polynomials

OBJECTIVES

1. Classify Polynomials by Size and Degree
2. Add Polynomials
3. Subtract Polynomials Using a Horizontal Format
4. Subtract Polynomials Using a Vertical Format
5. Perform Operations on Polynomials Involving Both Addition and Subtraction
6. Use Polynomials to Represent the Area or Perimeter of Geometric Figures

1 Classify Polynomials by Size and Degree

In earlier chapters, we called algebraic expressions such as $4x$, $5y$, $-6ab$, $7x^2$, and $-9xy^2z^3$ "terms." Recall that a term is an indicated product that may contain any number of factors. The variables in a term are called "literal factors," and the numerical factor is called the "numerical coefficient" of the term. Thus, in $-6ab$, a and b are literal factors and the numerical coefficient is -6. Terms that have the same literal factors are called "similar" or "like" terms.

Terms that contain variables with only whole numbers as exponents are called **monomials**. The previously listed terms, $4x$, $5y$, $-6ab$, $7x^2$, and $-9xy^2z^3$, are all monomials. (We will work with some algebraic expressions later, such as $7x^{-1}y^{-1}$ and $4a^{-2}b^{-3}$, which are not monomials.) The **degree of a monomial** is the sum of the exponents of the literal factors. Here are some examples:

$4xy$ is of degree 2.

$5x$ is of degree 1.

$14a^2b$ is of degree 3.

$-17xy^2z^3$ is of degree 6.

$-9y^4$ is of degree 4.

If the monomial contains only one variable, then the exponent of the variable is the degree of the monomial. Any nonzero constant term is said to be of degree zero.

A **polynomial** is a monomial or a finite sum (or difference) of monomials. The **degree of a polynomial** is the degree of the term with the highest degree in the polynomial. Some special classifications of polynomials are made according to the number of terms. We call a one-term polynomial a **monomial**, a two-term polynomial a **binomial**, and a three-term polynomial a **trinomial**. The following examples illustrate some of this terminology:

The polynomial $5x^3y^4$ is a monomial of degree 7.
The polynomial $4x^2y - 3xy$ is a binomial of degree 3.
The polynomial $5x^2 - 6x + 4$ is a trinomial of degree 2.
The polynomial $9x^4 - 7x^3 + 6x^2 + x - 2$ is given no special name, but it is of degree 4.

2 Add Polynomials

In the preceding chapters, you have worked many problems involving the addition and subtraction of polynomials. For example, simplifying $4x^2 + 6x + 7x^2 - 2x$ to $11x^2 + 4x$ by combining similar terms can actually be considered the addition problem $(4x^2 + 6x) + (7x^2 - 2x)$. At this time we will simply review and expand some of those ideas.

EXAMPLE 1

Add $5x^2 + 7x - 2$ and $9x^2 - 12x + 13$.

Solution

We commonly use the horizontal format for such work. Thus,

$$(5x^2 + 7x - 2) + (9x^2 - 12x + 13) = (5x^2 + 9x^2) + (7x - 12x) + (-2 + 13)$$
$$= 14x^2 - 5x + 11$$

The commutative, associative, and distributive properties provide the basis for rearranging, regrouping, and combining similar terms.

EXAMPLE 2

Add $5x - 1$, $3x + 4$, and $9x - 7$.

Solution

$$(5x - 1) + (3x + 4) + (9x - 7) = (5x + 3x + 9x) + [-1 + 4 + (-7)]$$
$$= 17x - 4$$

EXAMPLE 3

Add $-x^2 + 2x - 1$, $2x^3 - x + 4$, and $-5x + 6$.

Solution

$$(-x^2 + 2x - 1) + (2x^3 - x + 4) + (-5x + 6)$$
$$= (2x^3) + (-x^2) + (2x - x - 5x) + (-1 + 4 + 6)$$
$$= 2x^3 - x^2 - 4x + 9$$

3 Subtract Polynomials Using a Horizontal Format

Recall from Chapter 1 that $a - b = a + (-b)$. We define subtraction as *adding the opposite*. This same idea extends to polynomials in general. The opposite of a polynomial is formed by taking the opposite of each term. For example, the opposite of $(2x^2 - 7x + 3)$ is $-2x^2 + 7x - 3$. Symbolically, we express this as

$$-(2x^2 - 7x + 3) = -2x^2 + 7x - 3$$

Now consider some subtraction problems.

EXAMPLE 4 Subtract $2x^2 + 9x - 3$ from $5x^2 - 7x - 1$.

Solution

Use the horizontal format:

$$(5x^2 - 7x - 1) - (2x^2 + 9x - 3) = (5x^2 - 7x - 1) + (-2x^2 - 9x + 3)$$
$$= (5x^2 - 2x^2) + (-7x - 9x) + (-1 + 3)$$
$$= 3x^2 - 16x + 2$$

EXAMPLE 5 Subtract $-8y^2 - y + 5$ from $2y^2 + 9$.

Solution

$$(2y^2 + 9) - (-8y^2 - y + 5) = (2y^2 + 9) + (8y^2 + y - 5)$$
$$= (2y^2 + 8y^2) + (y) + (9 - 5)$$
$$= 10y^2 + y + 4$$

4 Subtract Polynomials Using a Vertical Format

Later, when dividing polynomials, you will need to use a vertical format to subtract polynomials. Let's consider two such examples.

EXAMPLE 6 Subtract $3x^2 + 5x - 2$ from $9x^2 - 7x - 1$.

Solution

$$9x^2 - 7x - 1$$
$$3x^2 + 5x - 2$$

Notice which polynomial goes on the bottom and the alignment of similar terms in columns

Now we can mentally form the opposite of the bottom polynomial and add.

$$9x^2 - 7x - 1$$
$$\underline{3x^2 + 5x - 2}$$
$$6x^2 - 12x + 1$$

The opposite of $3x^2 + 5x - 2$ is $-3x^2 - 5x + 2$

EXAMPLE 7 Subtract $15y^3 + 5y^2 + 3$ from $13y^3 + 7y - 1$.

Solution

$$13y^3 \qquad\quad + 7y - 1$$
$$\underline{15y^3 + 5y^2 \qquad + 3}$$

Similar terms are arranged in columns

$$-2y^3 - 5y^2 + 7y - 4$$

We mentally formed the opposite of the bottom polynomial and added

5 Perform Operations on Polynomials Involving Both Addition and Subtraction

We can use the distributive property along with the properties $a = 1(a)$ and $-a = -1(a)$ when adding and subtracting polynomials. The next examples illustrate this approach.

EXAMPLE 8

Perform the indicated operations.

$$(3x - 4) + (2x - 5) - (7x - 1)$$

Solution

$$(3x - 4) + (2x - 5) - (7x - 1)$$
$$= 1(3x - 4) + 1(2x - 5) - 1(7x - 1)$$
$$= 1(3x) - 1(4) + 1(2x) - 1(5) - 1(7x) - 1(-1)$$
$$= 3x - 4 + 2x - 5 - 7x + 1$$
$$= 3x + 2x - 7x - 4 - 5 + 1$$
$$= -2x - 8$$

Certainly we can do some of the steps mentally; Example 9 gives a possible format.

EXAMPLE 9

Perform the indicated operations.

$$(-y^2 + 5y - 2) - (-2y^2 + 8y + 6) + (4y^2 - 2y - 5)$$

Solution

$$(-y^2 + 5y - 2) - (-2y^2 + 8y + 6) + (4y^2 - 2y - 5)$$
$$= -y^2 + 5y - 2 + 2y^2 - 8y - 6 + 4y^2 - 2y - 5$$
$$= -y^2 + 2y^2 + 4y^2 + 5y - 8y - 2y - 2 - 6 - 5$$
$$= 5y^2 - 5y - 13$$

When we use the horizontal format, as in Examples 8 and 9, we use parentheses to indicate a quantity. In Example 8, the quantities $(3x - 4)$ and $(2x - 5)$ are to be added; from this result we are to subtract the quantity $(7x - 1)$. Brackets, [ ], are also sometimes used as grouping symbols, especially if there is a need to indicate quantities within quantities. To remove the grouping symbols, perform the indicated operations, starting with the innermost set of symbols. Let's consider two examples of this type.

EXAMPLE 10

Perform the indicated operations.

$$3x - [2x + (3x - 1)]$$

Solution

First we need to add the quantities $2x$ and $(3x - 1)$.

$$3x - [2x + (3x - 1)] = 3x - [2x + 3x - 1]$$
$$= 3x - [5x - 1]$$

Now we need to subtract the quantity $[5x - 1]$ from $3x$.

$$3x - [5x - 1] = 3x - 5x + 1$$
$$= -2x + 1$$

EXAMPLE 11

Perform the indicated operations.

$$8 - \{7x - [2 + (x - 1)] + 4x\}$$

Solution

Start with the innermost set of grouping symbols (the parentheses) and proceed as follows:

$$8 - \{7x - [2 + (x - 1)] + 4x\} = 8 - \{7x - [x + 1] + 4x\}$$
$$= 8 - \{7x - x - 1 + 4x\}$$

$$= 8 - \{10x - 1\}$$
$$= 8 - 10x + 1$$
$$= -10x + 9$$

6 Use Polynomials to Represent the Area or Perimeter of Geometric Figures

For a final example in this section, we look at polynomials in a geometric setting.

EXAMPLE 12 Suppose that a parallelogram and a rectangle have dimensions as indicated in Figure 4.1. Find a polynomial that represents the sum of the areas of the two figures.

 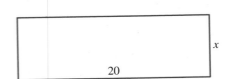

Figure 4.1

Solution

Using the area formulas $A = bh$ and $A = lw$ for parallelograms and rectangles, respectively, we can represent the sum of the areas of the two figures as follows:

Area of the parallelogram $x(x) = x^2$
Area of the rectangle $20(x) = 20x$

We can represent the total area by $x^2 + 20x$.

CONCEPT QUIZ 4.1

For Problems 1–5, answer true or false.

1. The degree of the monomial $4x^2y$ is 3.
2. The degree of the polynomial $2x^4 - 5x^3 + 7x^2 - 4x + 6$ is 10.
3. A three-term polynomial is called a binomial.
4. A polynomial is a monomial or a finite sum of monomials.
5. Monomial terms must have whole number exponents for each variable.

Section 4.1 Classroom Problem Set

Objective 2

1. Add $8x^2 - 2x + 6$ and $3x^2 + 5x - 10$.
2. Add $-2x^2 + 7x - 9$ and $4x^2 - 9x - 14$.
3. Add $6x + 4$, $2x - 3$, and $5x - 8$.
4. Add $-x - 4$, $8x + 9$, and $-7x - 8$.
5. Add $2x^2 - 3x + 4$, $5x^3 + 7x + 2$, and $-2x - 6$.
6. Add $-3x^2 + 2x - 6$, $6x^2 + 7x + 3$, and $-4x^2 - 9$.

Objective 3

7. Subtract $3x^2 - 2x + 1$ from $9x^2 - 8x - 2$.
8. Subtract $-2n^2 - 3n + 4$ from $3n^2 - n + 7$.
9. Subtract $2a^2 - 3a - 1$ from $6a + 4$.
10. Subtract $-4x^2 + 6x - 2$ from $-3x^3 + 2x^2 + 7x - 1$.

Objective 4

11. Subtract $4x^2 - x + 3$ from $11x^2 + 8x - 2$.
12. Subtract $8x^2 - x + 6$ from $6x^2 - x + 11$.
13. Subtract $8y^3 + 6y - 4$ from $5y^3 - 2y^2 + 8$.
14. Subtract $4x^3 + x - 10$ from $3x^2 - 6$.

Objective 5

15. Perform the indicated operations.
    $(6x + 1) + (3x - 2) - (5x - 8)$

16. Perform the indicated operations.

    $(3x - 4) + (9x - 1) - (14x - 7)$

17. Perform the indicated operations.

    $(x^2 + 3x - 4) - (4x^2 + 5x - 6) + (2x + 7)$

18. Perform the indicated operations.

    $(-3x^2 - 2) + (7x^2 - 8) - (9x^2 - 2x - 4)$

19. Perform the indicated operations.

    $10y - [7y + (y - 4)]$

20. Perform the indicated operations.

    $(-3a + 4) - [-7a + (9a - 1)]$

21. Perform the indicated operations.

    $12 - \{4y - [6 + (3y - 2)] + 10\}$

22. Perform the indicated operations.

    $-10x - \{7x - [3x - (2x - 3)]\}$

Objective 6

23. Suppose that a circle and a rectangle have the dimensions as indicated in Figure 4.2. Find a polynomial that represents the sum of the areas of the two figures.

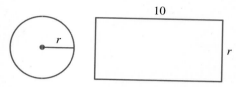

Figure 4.2

24. Suppose that a square has a side of length x and a circle has a radius of the same length x. Find the polynomial that represents the sum of the areas of both figures (see Figure 4.3).

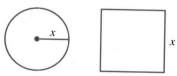

Figure 4.3

THOUGHTS INTO WORDS

1. Explain how to subtract the polynomial $3x^2 + 6x - 2$ from $4x^2 + 7$.

2. Is the sum of two binomials always another binomial? Defend your answer.

3. Is the sum of two binomials ever a trinomial? Defend your answer.

Answers to the Concept Quiz

1. True **2.** False **3.** False **4.** True **5.** True

4.2 Multiplying Monomials

OBJECTIVES

1. Apply the Properties of Exponents to Multiply Monomials
2. Apply the Properties of Exponents to Raise a Monomial to a Power
3. Multiply a Polynomial by a Monomial
4. Apply Multiplication of Monomials to Geometric Problems

1 Apply the Properties of Exponents to Multiply Monomials

The basic idea of using exponents to indicate repeated multiplication was introduced in Section 1.1. Expanding on that idea with some properties of exponents can make multiplying monomials easier. These properties are the direct result of the definition of an exponent. The following examples lead to Property 4.1.

$$x^2 \cdot x^3 = (x \cdot x)(x \cdot x \cdot x) = x^5$$
$$a^3 \cdot a^4 = (a \cdot a \cdot a)(a \cdot a \cdot a \cdot a) = a^7$$
$$b \cdot b^2 = (b)(b \cdot b) = b^3$$

In general,

$$b^n \cdot b^m = \underbrace{(b \cdot b \cdot b \cdot \cdots \cdot b)}_{n \text{ factors of } b}\underbrace{(b \cdot b \cdot b \cdot \cdots \cdot b)}_{m \text{ factors of } b}$$

$$= \underbrace{b \cdot b \cdot b \cdot \cdots \cdot b}_{(n+m) \text{ factors of } b}$$

$$= b^{n+m}$$

Property 4.1

If b is any real number, and n and m are positive integers, then

$$b^n \cdot b^m = b^{n+m}$$

Property 4.1 states that when multiplying powers with the same base, add exponents.

EXAMPLE 1   Multiply the following:

(a) $x^4 \cdot x^3$ (b) $a^8 \cdot a^7$

Solution

(a) $x^4 \cdot x^3 = x^{4+3} = x^7$ (b) $a^8 \cdot a^7 = a^{8+7} = a^{15}$

Consider the following example in which we use the properties of exponents to help simplify the process of multiplying monomials.

EXAMPLE 2   Multiply the following:

(a) $(3x^3)(5x^4)$ (b) $(-4a^2b^3)(6ab^2)$ (c) $(xy)(7xy^5)$ (d) $\left(\dfrac{3}{4}x^2y^3\right)\left(\dfrac{1}{2}x^3y^5\right)$

Solution

(a) $(3x^3)(5x^4) = 3 \cdot 5 \cdot x^3 \cdot x^4$
$\qquad\qquad\quad= 15x^7$

(b) $(-4a^2b^3)(6ab^2) = -4 \cdot 6 \cdot a^2 \cdot a \cdot b^3 \cdot b^2$
$\qquad\qquad\qquad\quad= -24a^3b^5$

(c) $(xy)(7xy^5) = 1 \cdot 7 \cdot x \cdot x \cdot y \cdot y^5$ The numerical coefficient of xy is 1
$\qquad\qquad\quad= 7x^2y^6$

(d) $\left(\dfrac{3}{4}x^2y^3\right)\left(\dfrac{1}{2}x^3y^5\right) = \dfrac{3}{4} \cdot \dfrac{1}{2} \cdot x^2 \cdot x^3 \cdot y^3 \cdot y^5$

$\qquad\qquad\qquad\qquad= \dfrac{3}{8}x^5y^8$

2 Apply the Properties of Exponents to Raise a Monomial to a Power

Another property of exponents is demonstrated by these examples.

$$(x^2)^3 = x^2 \cdot x^2 \cdot x^2 = x^{2+2+2} = x^6$$
$$(a^3)^2 = a^3 \cdot a^3 = a^{3+3} = a^6$$
$$(b^3)^4 = b^3 \cdot b^3 \cdot b^3 \cdot b^3 = b^{3+3+3+3} = b^{12}$$

In general,

$$(b^n)^m = \underbrace{b^n \cdot b^n \cdot b^n \cdot \cdots \cdot b^n}_{\substack{m \text{ factors of } b^n \\ m \text{ of these } ns}}$$
$$= b^{n+n+n+\cdots+n}$$
$$= b^{mn}$$

Property 4.2

If b is any real number, and m and n are positive integers, then

$$(b^n)^m = b^{mn}$$

Property 4.2 states that when raising a power to a power, multiply exponents.

EXAMPLE 3

Raise each to the indicated power.

(a) $(x^4)^3$ **(b)** $(a^5)^6$

Solution

(a) $(x^4)^3 = x^{3 \cdot 4} = x^{12}$ **(b)** $(a^5)^6 = a^{6 \cdot 5} = a^{30}$ ∎

The third property of exponents we use in this section raises a monomial to a power.

$$(2x)^3 = (2x)(2x)(2x) = 2 \cdot 2 \cdot 2 \cdot x \cdot x \cdot x = 2^3 \cdot x^3$$
$$(3a^4)^2 = (3a^4)(3a^4) = 3 \cdot 3 \cdot a^4 \cdot a^4 = (3)^2(a^4)^2$$
$$(-2xy^5)^2 = (-2xy^5)(-2xy^5) = (-2)(-2)(x)(x)(y^5)(y^5) = (-2)^2(x)^2(y^5)^2$$

In general,

$$(ab)^n = \underbrace{ab \cdot ab \cdot ab \cdot \cdots \cdot ab}_{n \text{ factors of } ab}$$
$$= (\underbrace{a \cdot a \cdot a \cdot \cdots \cdot a}_{n \text{ factors of } a})(\underbrace{b \cdot b \cdot b \cdot \cdots \cdot b}_{n \text{ factors of } b})$$
$$= a^n b^n$$

Property 4.3

If a and b are real numbers, and n is a positive integer, then

$$(ab)^n = a^n b^n$$

Property 4.3 states that when raising a monomial to a power, raise each factor to that power.

It is a simple process to raise a monomial to a power when using the properties of exponents. Study the next example.

EXAMPLE 4 Raise each to the indicated power.

(a) $(2x^2y^3)^4$ (b) $(-3ab^5)^3$ (c) $(-2a^4)^5$ (d) $\left(\dfrac{2}{5}x^2y^3\right)^3$

Solution

(a) $(2x^2y^3)^4 = (2)^4(x^2)^4(y^3)^4 = 16x^8y^{12}$

(b) $(-3ab^5)^3 = (-3)^3(a^1)^3(b^5)^3 = -27a^3b^{15}$

(c) $(-2a^4)^5 = (-2)^5(a^4)^5$
$= -32a^{20}$

(d) $\left(\dfrac{2}{5}x^2y^3\right)^3 = \left(\dfrac{2}{5}\right)^3(x^2)^3(y^3)^3$
$= \dfrac{8}{125}x^6y^9$ ∎

Sometimes problems involve first raising monomials to a power and then multiplying the resulting monomials, as in the following example.

EXAMPLE 5 Multiply the following:

(a) $(3x^2)^3(2x^3)^2$ (b) $(-x^2y^3)^5(-2x^2y)^2$

Solution

(a) $(3x^2)^3(2x^3)^2 = (3)^3(x^2)^3(2)^2(x^3)^2$
$= (27)(x^6)(4)(x^6)$
$= 108x^{12}$

(b) $(-x^2y^3)^5(-2x^2y)^2 = (-1)^5(x^2)^5(y^3)^5(-2)^2(x^2)^2(y)^2$
$= (-1)(x^{10})(y^{15})(4)(x^4)(y^2)$
$= -4x^{14}y^{17}$ ∎

3 Multiply a Polynomial by a Monomial

The distributive property along with the properties of exponents forms a basis for finding the product of a monomial and a polynomial. The next example illustrates these ideas.

EXAMPLE 6 Multiply the following:

(a) $(3x)(2x^2 + 6x + 1)$ (b) $(5a^2)(a^3 - 2a^2 - 1)$
(c) $(-2xy)(6x^2y - 3xy^2 - 4y^3)$

Solution

(a) $(3x)(2x^2 + 6x + 1) = (3x)(2x^2) + (3x)(6x) + (3x)(1)$ Apply distributive property
$\qquad = 6x^3 + 18x^2 + 3x$

(b) $(5a^2)(a^3 - 2a^2 - 1) = (5a^2)(a^3) - (5a^2)(2a^2) - (5a^2)(1)$
$\qquad = 5a^5 - 10a^4 - 5a^2$

(c) $(-2xy)(6x^2y - 3xy^2 - 4y^3)$
$\qquad = (-2xy)(6x^2y) - (-2xy)(3xy^2) - (-2xy)(4y^3)$
$\qquad = -12x^3y^2 + 6x^2y^3 + 8xy^4$

■

Once you feel comfortable with this process, you may want to perform most of the work mentally and then simply write down the final result. See whether you understand the following examples.

1. $3x(2x + 3) = 6x^2 + 9x$
2. $-4x(2x^2 - 3x - 1) = -8x^3 + 12x^2 + 4x$
3. $ab(3a^2b - 2ab^2 - b^3) = 3a^3b^2 - 2a^2b^3 - ab^4$

4 Apply Multiplication of Monomials to Geometric Problems

We conclude this section by making a connection between algebra and geometry.

EXAMPLE 7

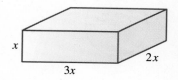

Figure 4.4

Suppose that the dimensions of a rectangular solid are represented by x, $2x$, and $3x$ as shown in Figure 4.4. Express the volume and total surface area of the figure.

Solution

Using the formula $V = lwh$, we can express the volume of the rectangular solid as $(2x)(3x)(x)$, which equals $6x^3$. The total surface area can be described as follows:

Area of front and back faces: $2(x)(3x) = 6x^2$
Area of left and right faces: $2(2x)(x) = 4x^2$
Area of top and bottom faces: $2(2x)(3x) = 12x^2$

We can represent the total surface area by $6x^2 + 4x^2 + 12x^2$, or $22x^2$. ■

CONCEPT QUIZ 4.2

For Problems 1–6, answer true or false.

1. When multiplying factors with the same base, add the exponents.
2. $3^2 \cdot 3^2 = 9^4$
3. $2x^2 \cdot 3x^3 = 6x^6$
4. $(x^2)^3 = x^5$
5. $(-4x^3)^2 = -4x^6$
6. To simplify $(3x^2y)(2x^3y^2)^4$, use the order of operations to first raise $2x^3y^2$ to the fourth power, and then multiply the monomials.

Section 4.2 Classroom Problem Set

Objective 1

1. Multiply. (a) $y^3 \cdot y^5$ (b) $a^2 \cdot a$
2. Multiply. (a) $x^4 \cdot x \cdot x^3$ (b) $a \cdot a \cdot a$
3. Multiply.
   (a) $(-4x^2)(6x^3)$ (b) $\left(\frac{2}{3}a^5 b\right)\left(\frac{6}{7}a^4 b^2\right)$
4. Multiply.
   (a) $(9abc^3)(14bc^2)$ (b) $\left(-\frac{5}{6}x\right)\left(\frac{8}{3}x^2 y\right)$

Objective 2

5. Raise each to the indicated power.
   (a) $(y^3)^2$ (b) $(m^2)^4$
6. Raise each to the indicated power.
   (a) $(n^4)^3$ (b) $(t^5)^5$
7. Raise each to the indicated power.
   (a) $(2x^3)^4$ (b) $(-3a^3 b^5)^2$
8. Raise each to the indicated power.
   (a) $(-3x^3)^3$ (b) $(-x^2 y^3)^7$
9. Multiply.
   (a) $(5a^3)^2(-2a^4)^3$ (b) $(3m^4 n^2)^3(-mn)^2$
10. Multiply.
    (a) $(-x^2 y)^3(6xy)^2$ (b) $(ab^2 c^3)^4(-a^2 b)^3$

Objective 3

11. Multiply.
    (a) $(4a^3)(a^2 - 3ab - 5b^2)$
    (b) $(-2y)(5x^2 y + 6xy + y^3)$
12. Multiply.
    (a) $-8a(4a^2 - 9a - 6)$
    (b) $(5x^2 y)(3x^2 + 7x - 9)$

Objective 4

13. The dimensions of some rectangular regions of a floor plan are shown in Figure 4.5. Express the area of the regions of the floor plan.

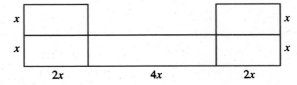

Figure 4.5

14. The dimensions of the rectangular region in Figure 4.6 are represented by $6x$ and $5x$. The radius of the circular region is represented by $2x$. Express the sum of the area of the regions.

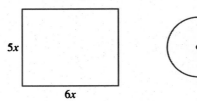

Figure 4.6

THOUGHTS INTO WORDS

1. How would you explain to someone why the product of x^3 and x^4 is x^7 and not x^{12}?

2. Suppose your friend was absent from class the day that this section was discussed. How would you help her understand why the property $(b^n)^m = b^{mn}$ is true?

3. How can Figure 4.7 be used to demonstrate geometrically that $x(x + 2) = x^2 + 2x$?

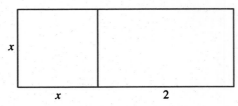

Figure 4.7

Answers to the Concept Quiz
1. True 2. False 3. False 4. False 5. False 6. True

4.3 Multiplying Polynomials

OBJECTIVES

1. Use the Distributive Property to Find the Product of Two Polynomials
2. Use the Shortcut Pattern to Find the Product of Two Binomials
3. Raise a Binomial to a Power
4. Use Special Product Patterns to Find Products
5. Apply Polynomials to Geometric Problems

1 Use the Distributive Property to Find the Product of Two Polynomials

In general, to go from multiplying a monomial times a polynomial to multiplying two polynomials requires the use of the distributive property. Consider some examples.

EXAMPLE 1 Find the product of $(x + 3)$ and $(y + 4)$.

Solution

$$\begin{aligned}(x + 3)(y + 4) &= x(y + 4) + 3(y + 4) \\ &= x(y) + x(4) + 3(y) + 3(4) \\ &= xy + 4x + 3y + 12\end{aligned}$$

Notice that each term of the first polynomial is multiplied times each term of the second polynomial.

EXAMPLE 2 Find the product of $(x - 2)$ and $(y + z + 5)$.

Solution

$$\begin{aligned}(x - 2)(y + z + 5) &= x(y + z + 5) - 2(y + z + 5) \\ &= x(y) + x(z) + x(5) - 2(y) - 2(z) - 2(5) \\ &= xy + xz + 5x - 2y - 2z - 10\end{aligned}$$

Usually, multiplying polynomials will produce similar terms that can be combined to simplify the resulting polynomial.

EXAMPLE 3 Multiply $(x + 3)(x + 2)$.

Solution

$$\begin{aligned}(x + 3)(x + 2) &= x(x + 2) + 3(x + 2) \\ &= x^2 + 2x + 3x + 6 \\ &= x^2 + 5x + 6 \quad \text{Combine like terms}\end{aligned}$$

EXAMPLE 4

Multiply $(x - 4)(x + 9)$.

Solution

$$(x - 4)(x + 9) = x(x + 9) - 4(x + 9)$$
$$= x^2 + 9x - 4x - 36$$
$$= x^2 + 5x - 36 \qquad \text{Combine like terms}$$

EXAMPLE 5

Multiply $(x + 4)(x^2 + 3x + 2)$.

Solution

$$(x + 4)(x^2 + 3x + 2) = x(x^2 + 3x + 2) + 4(x^2 + 3x + 2)$$
$$= x^3 + 3x^2 + 2x + 4x^2 + 12x + 8$$
$$= x^3 + 7x^2 + 14x + 8$$

EXAMPLE 6

Multiply $(2x - y)(3x^2 - 2xy + 4y^2)$.

Solution

$$(2x - y)(3x^2 - 2xy + 4y^2) = 2x(3x^2 - 2xy + 4y^2) - y(3x^2 - 2xy + 4y^2)$$
$$= 6x^3 - 4x^2y + 8xy^2 - 3x^2y + 2xy^2 - 4y^3$$
$$= 6x^3 - 7x^2y + 10xy^2 - 4y^3$$

2 Use the Shortcut Pattern to Find the Product of Two Binomials

Perhaps the most frequently used type of multiplication problem is the product of two binomials. It will be a big help later if you can become proficient at multiplying binomials without showing all of the intermediate steps. This is quite easy to do if you use a three-step shortcut pattern demonstrated by the following examples.

EXAMPLE 7

Multiply $(x + 5)(x + 7)$.

Solution

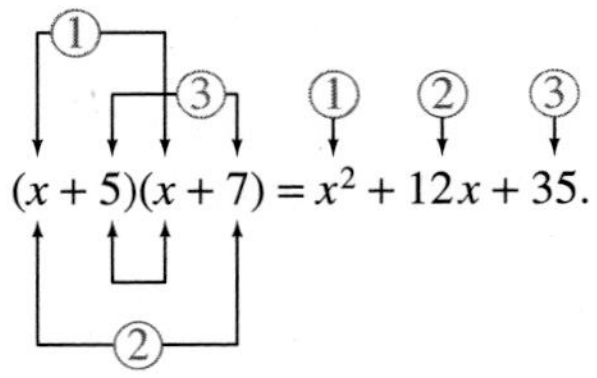

$$(x + 5)(x + 7) = x^2 + 12x + 35.$$

Figure 4.8

Step 1 Multiply $x \cdot x$.

Step 2 Multiply $5 \cdot x$ and $7 \cdot x$ and combine them.

Step 3 Multiply $5 \cdot 7$.

EXAMPLE 8

Multiply $(x - 8)(x + 3)$.

Solution

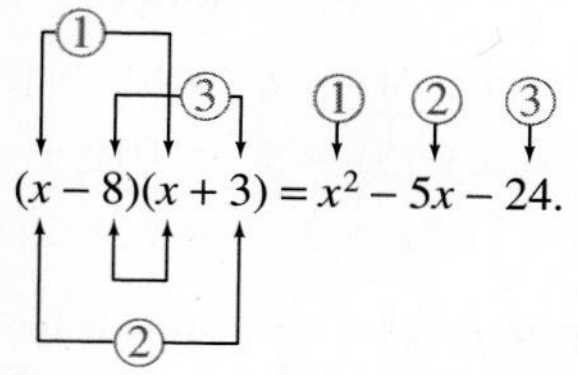

$$(x - 8)(x + 3) = x^2 - 5x - 24.$$

Figure 4.9

EXAMPLE 9

Multiply $(3x + 2)(2x - 5)$.

Solution

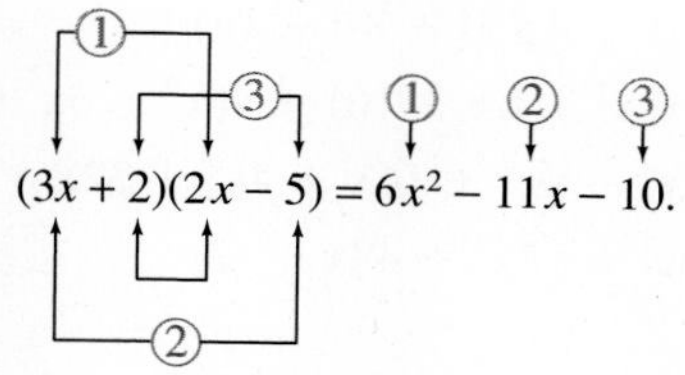

$$(3x + 2)(2x - 5) = 6x^2 - 11x - 10.$$

Figure 4.10

The mnemonic device FOIL is often used to remember the pattern for multiplying binomials. The letters in FOIL represent First, Outside, Inside, and Last. If you look back at Examples 7 through 9, step 1 is to find the product of the first terms in each binomial; step 2 is to find the product of the outside terms and the inside terms; and step 3 is to find the product of the last terms in each binomial. Now see whether you can use the pattern to find these products:

$(x + 3)(x + 7)$
$(3x + 1)(2x + 5)$
$(x - 2)(x - 3)$
$(4x + 5)(x - 2)$

Your answers should be $x^2 + 10x + 21, 6x^2 + 17x + 5, x^2 - 5x + 6,$ and $4x^2 - 3x - 10$.

Keep in mind that the shortcut pattern applies only to finding the product of two binomials. For other situations, such as finding the product of a binomial and a trinomial, we suggest showing the intermediate steps as follows:

$$(x + 3)(x^2 + 6x - 7) = x(x^2) + x(6x) - x(7) + 3(x^2) + 3(6x) - 3(7)$$
$$= x^3 + 6x^2 - 7x + 3x^2 + 18x - 21$$
$$= x^3 + 9x^2 + 11x - 21$$

Perhaps you could omit the first step, and shorten the form as follows:

$$(x - 4)(x^2 - 5x - 6) = x^3 - 5x^2 - 6x - 4x^2 + 20x + 24$$
$$= x^3 - 9x^2 + 14x + 24$$

Remember that you are multiplying each term of the first polynomial times each term of the second polynomial and combining similar terms.

3 Raise a Binomial to a Power

Exponents are also used to indicate repeated multiplication of polynomials. For example, we can write $(x + 4)(x + 4)$ as $(x + 4)^2$. Thus to square a binomial we simply write it as the product of two equal binomials and apply the shortcut pattern.

$$(x + 4)^2 = (x + 4)(x + 4) = x^2 + 8x + 16$$
$$(x - 5)^2 = (x - 5)(x - 5) = x^2 - 10x + 25$$
$$(2x + 3)^2 = (2x + 3)(2x + 3) = 4x^2 + 12x + 9$$

When you square binomials, be careful not to forget the middle term. That is to say, $(x + 3)^2 \neq x^2 + 3^2$; instead, $(x + 3)^2 = (x + 3)(x + 3) = x^2 + 6x + 9$.

The next example suggests a format to use when cubing a binomial.

EXAMPLE 10 Find the indicated product of $(x + 4)^3$.

Solution

$$(x + 4)^3 = (x + 4)(x + 4)(x + 4)$$
$$= (x + 4)(x^2 + 8x + 16) \quad \text{Use FOIL to multiply two of the factors}$$
$$= x(x^2 + 8x + 16) + 4(x^2 + 8x + 16) \quad \text{Use the distributive property}$$
$$= x^3 + 8x^2 + 16x + 4x^2 + 32x + 64$$
$$= x^3 + 12x^2 + 48x + 64$$ ∎

4 Use Special Product Patterns to Find Products

When we multiply binomials, some special patterns occur that you should recognize. We can use these patterns to find products and later to factor polynomials. We will state each of the patterns in general terms followed by examples to illustrate the use of each pattern.

PATTERN 1

$$(a + b)^2 = (a + b)(a + b) = a^2 \; + \; 2ab \; + \; b^2$$

$\quad$ Square of the first term of the binomial $\;+\;$ Twice the product of the two terms of the binomial $\;+\;$ Square of the second term of the binomial

Examples

$$(x + 4)^2 = x^2 + 8x + 16$$
$$(2x + 3y)^2 = 4x^2 + 12xy + 9y^2$$
$$(5a + 7b)^2 = 25a^2 + 70ab + 49b^2$$ ∎

PATTERN 2

$$(a - b)^2 = (a - b)(a - b) = a^2 \; - \; 2ab \; + \; b^2$$

$\quad$ Square of the first term of the binomial $\;-\;$ Twice the product of the two terms of the binomial $\;+\;$ Square of the second term of the binomial

Examples

$$(x - 8)^2 = x^2 - 16x + 64$$
$$(3x - 4y)^2 = 9x^2 - 24xy + 16y^2$$
$$(4a - 9b)^2 = 16a^2 - 72ab + 81b^2$$ ∎

PATTERN 3

$(a + b)(a - b) = a^2 - b^2$

where a^2 = Square of the first term of the binomial, and b^2 = Square of the second term of the binomial.

Examples

$(x + 7)(x - 7) = x^2 - 49$

$(2x + y)(2x - y) = 4x^2 - y^2$

$(3a - 2b)(3a + 2b) = 9a^2 - 4b^2$

5 Apply Polynomials to Geometric Problems

As you might expect, there are geometric interpretations for many of the algebraic concepts presented in this section. We will give you the opportunity to make some of these connections between algebra and geometry in the next problem set. We conclude this section with a problem that allows us to use some algebra and geometry.

EXAMPLE 11   Apply Your Skill

A rectangular piece of tin is 16 inches long and 12 inches wide as shown in Figure 4.11. From each corner a square piece x inches on a side is cut out. The flaps are then turned up to form an open box. Find polynomials that represent the volume and the exterior surface area of the box.

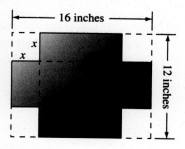

Figure 4.11

Solution

The length of the box is $16 - 2x$, the width is $12 - 2x$, and the height is x. From the volume formula $V = lwh$, the polynomial $(16 - 2x)(12 - 2x)(x)$, which simplifies to $4x^3 - 56x^2 + 192x$, represents the volume.

The outside surface area of the box is the area of the original piece of tin minus the four corners that were cut off. Therefore, the polynomial $16(12) - 4x^2$, or $192 - 4x^2$, represents the outside surface area of the box.

CONCEPT QUIZ 4.3

For Problems 1–5, answer true or false.

1. The algebraic expression $(x + y)^2$ is called the square of a binomial.
2. The algebraic expression $(x + y)(x + 2xy + y)$ is called the product of two binomials.
3. The mnemonic device FOIL stands for first, outside, inside, and last.
4. $(a + 2)^2 = a^2 + 4$
5. $(y + 3)(y - 3) = y^2 + 9$

Section 4.3 Classroom Problem Set

Objective 1

1. Find the product of $(x + 5)$ and $(y + 2)$.
2. Find the product of $(x - 4)$ and $(y + 1)$.
3. Find the product of $(x - 4)$ and $(y + z - 3)$.
4. Find the product of $(a + b)$ and $(x - y - z)$.
5. Multiply $(x + 6)(x + 5)$.
6. Multiply $(x + 7)(x + 1)$.
7. Multiply $(x + 2)(x - 8)$.
8. Multiply $(x - 10)(x + 8)$.
9. Multiply $(x + 3)(x^2 + 2x + 6)$.
10. Multiply $(x - 5)(2x^2 + 3x - 7)$.
11. Multiply $(x - 2y)(5x^2 + 3xy + 4y^2)$.
12. Multiply $(2a - 1)(4a^2 - 5a + 9)$.

Objective 2

13. Multiply $(x + 1)(x + 6)$.
14. Multiply $(n - 4)(n - 3)$.
15. Multiply $(x + 3)(x - 7)$.
16. Multiply $(x - 1)(x + 9)$.
17. Multiply $(4x + 1)(2x - 3)$.
18. Multiply $(5a + 4)(4a - 5)$.

Objective 3

19. Find the indicated product. $(x + 5)^3$
20. Find the indicated product. $(x - 1)^3$

Objective 4

21. Use the pattern $(a + b)^2 = a^2 + 2ab + b^2$ to find $(x + 8)^2$.
22. Use the pattern $(a + b)^2 = a^2 + 2ab + b^2$ to find $(3x + 7)^2$.
23. Use the pattern $(a - b)^2 = a^2 - 2ab + b^2$ to find $(x - 1)^2$.
24. Use the pattern $(a - b)^2 = a^2 - 2ab + b^2$ to find $(4x - 5)^2$.
25. Use the pattern $(a + b)(a - b) = a^2 - b^2$ to find $(3x + 5)(3x - 5)$.
26. Use the pattern $(a + b)(a - b) = a^2 - b^2$ to find $(7x - 9)(7x + 9)$.

Objective 5

27. A rectangular piece of cardboard is 10 inches long and 6 inches wide. From each corner a square piece y inches on a side is cut out. The flaps are then turned up to form an open box. Find the polynomials that represent the volume and the exterior surface area of the box.
28. A square piece of aluminum is 8 inches long on each side. From each corner a square piece x inches on a side is cut out. The flaps are then turned up to form an open box. Find the polynomials that represent the volume and the exterior surface area of the box.

THOUGHTS INTO WORDS

1. Describe the process of multiplying two polynomials.
2. Illustrate as many uses of the distributive property as you can.
3. Determine the number of terms in the product of $(x + y + z)$ and $(a + b + c)$ without doing the multiplication. Explain how you arrived at your answer.

Answers to the Concept Quiz

1. True   2. False   3. True   4. False   5. False

4.4 Dividing by Monomials

OBJECTIVES

1. Apply the Properties of Exponents to Divide Monomials
2. Divide Polynomials by Monomials

1 Apply the Properties of Exponents to Divide Monomials

To develop an effective process for dividing by a monomial we must rely on yet another property of exponents. This property is also a direct consequence of the definition of exponents and is illustrated by the following examples.

$$\frac{x^5}{x^2} = \frac{x \cdot x \cdot x \cdot x \cdot x}{x \cdot x} = x^3$$

$$\frac{a^4}{a^3} = \frac{a \cdot a \cdot a \cdot a}{a \cdot a \cdot a} = a$$

$$\frac{y^7}{y^3} = \frac{y \cdot y \cdot y \cdot y \cdot y \cdot y \cdot y}{y \cdot y \cdot y} = y^4$$

$$\frac{x^4}{x^4} = \frac{x \cdot x \cdot x \cdot x}{x \cdot x \cdot x \cdot x} = 1$$

$$\frac{y^3}{y^3} = \frac{y \cdot y \cdot y}{y \cdot y \cdot y} = 1$$

Property 4.4

If b is any nonzero real number, and n and m are positive integers, then

1. $\dfrac{b^n}{b^m} = b^{n-m}$ when $n > m$

2. $\dfrac{b^n}{b^m} = 1$ when $n = m$

(The situation $n < m$ will be discussed in a later section.)

Applying Property 4.4 to the previous examples yields these results:

$$\frac{x^5}{x^2} = x^{5-2} = x^3$$

$$\frac{a^4}{a^3} = a^{4-3} = a^1 \quad \text{Usually written as } a$$

$$\frac{y^7}{y^3} = y^{7-3} = y^4$$

$$\frac{x^4}{x^4} = 1$$

$$\frac{y^3}{y^3} = 1$$

Property 4.4 along with our knowledge of dividing integers provides the basis for dividing a monomial by another monomial. Consider the next example.

EXAMPLE 1 Divide the monomials.

(a) $\dfrac{16x^5}{2x^3}$ (b) $\dfrac{-81a^{12}}{-9a^4}$ (c) $\dfrac{45x^4}{9x^4}$ (d) $\dfrac{54x^3y^7}{-6xy^5}$

Solution

(a) $\dfrac{16x^5}{2x^3} = 8x^{5-3} = 8x^2$

(b) $\dfrac{-81a^{12}}{-9a^4} = 9a^{12-4} = 9a^8$

(c) $\dfrac{45x^4}{9x^4} = 5 \quad \dfrac{x^4}{x^4} = 1$

(d) $\dfrac{54x^3y^7}{-6xy^5} = -9x^{3-1}y^{7-5} = -9x^2y^2$ ∎

2 Divide Polynomials by Monomials

Recall that $\dfrac{a+b}{c} = \dfrac{a}{c} + \dfrac{b}{c}$; this property serves as the basis for dividing a polynomial by a monomial. Consider these examples.

$$\dfrac{25x^3 + 10x^2}{5x} = \dfrac{25x^3}{5x} + \dfrac{10x^2}{5x} = 5x^2 + 2x$$

$$\dfrac{-35x^8 - 28x^6}{7x^3} = \dfrac{-35x^8}{7x^3} - \dfrac{28x^6}{7x^3} = -5x^5 - 4x^3 \qquad \dfrac{a-b}{c} = \dfrac{a}{c} - \dfrac{b}{c}$$

> To divide a polynomial by a monomial, we simply divide each term of the polynomial by the monomial. Here are some additional examples.

EXAMPLE 2 Perform the division.

(a) $\dfrac{12x^3y^2 - 14x^2y^5}{-2xy}$ (b) $\dfrac{48ab^5 + 64a^2b}{-16ab}$ (c) $\dfrac{33x^6 - 24x^5 - 18x^4}{3x}$

Solution

(a) $\dfrac{12x^3y^2 - 14x^2y^5}{-2xy} = \dfrac{12x^3y^2}{-2xy} - \dfrac{14x^2y^5}{-2xy} = -6x^2y + 7xy^4$

(b) $\dfrac{48ab^5 + 64a^2b}{-16ab} = \dfrac{48ab^5}{-16ab} + \dfrac{64a^2b}{-16ab} = -3b^4 - 4a$

(c) $\dfrac{33x^6 - 24x^5 - 18x^4}{3x} = \dfrac{33x^6}{3x} - \dfrac{24x^5}{3x} - \dfrac{18x^4}{3x}$

$\qquad = 11x^5 - 8x^4 - 6x^3$ ∎

As with many skills, once you feel comfortable with the process, you may want to perform some of the steps mentally. Your work could take on the following format.

$$\dfrac{24x^4y^5 - 56x^3y^9}{8x^2y^3} = 3x^2y^2 - 7xy^6$$

$$\dfrac{13a^2b - 12ab^2}{-ab} = -13a + 12b$$

CONCEPT QUIZ 4.4

For Problems 1–5, answer true or false.

1. When dividing factors that have the same base, add the exponents.
2. $\dfrac{10a^6}{2a^2} = 8a^4$
3. $\dfrac{y^8}{y^4} = y^2$
4. $\dfrac{6x^5 + 3x}{3x} = 2x^4$
5. $\dfrac{x^3}{x^3} = 0$

Section 4.4 Classroom Problem Set

Objective 1

1. Divide the monomials.

   (a) $\dfrac{-32b^8}{4b^2}$ (b) $\dfrac{6a^4b^7}{-6ab^2}$

2. Divide the monomials.

   (a) $\dfrac{72x^3}{-9x^3}$ (b) $\dfrac{-10x^4y^6}{-x^4y}$

Objective 2

3. Perform the division.

   (a) $\dfrac{21m^4n^5 + 14m^2n^3}{-7m^2n}$ (b) $\dfrac{12a^5 - 18a^4 - 36a^3}{6a^3}$

4. Perform the division.

   (a) $\dfrac{-65a^8 - 78a^4}{-13a^2}$ (b) $\dfrac{-12a^2c - 48ac^2 + 12ac}{-12ac}$

THOUGHTS INTO WORDS

1. How would you explain to someone why the quotient of x^8 and x^2 is x^6 and not x^4?

2. Your friend is having difficulty with problems such as $\dfrac{12x^2y}{xy}$ and $\dfrac{36x^3y^2}{-xy}$ where there appears to be no numerical coefficient in the denominator. What can you tell him that might help?

Answers to the Concept Quiz

1. False   2. False   3. False   4. False   5. False

4.5 Dividing by Binomials

OBJECTIVE

1. Divide Polynomials by Binomials

1 Divide Polynomials by Binomials

Perhaps the easiest way to explain the process of dividing a polynomial by a binomial is to work a few examples and describe the step-by-step procedure as we go along.

Chapter 4 Exponents and Polynomials

EXAMPLE 1 Divide $x^2 + 5x + 6$ by $x + 2$.

Solution

Step 1 Use the conventional long division format from arithmetic, and arrange both the dividend and the divisor in descending powers of the variable.

$$x + 2 \overline{\smash{)}x^2 + 5x + 6}$$

Step 2 Find the first term of the quotient by dividing the first term of the dividend by the first term of the divisor.

$$\begin{array}{r} x \phantom{{}+5x+6} \\ x+2\overline{\smash{)}x^2+5x+6} \end{array} \qquad \dfrac{x^2}{x} = x$$

Step 3 Multiply the entire divisor by the term of the quotient found in step 2, and position this product to be subtracted from the dividend.

$$\begin{array}{r} x \phantom{{}+5x+6} \\ x+2\overline{\smash{)}x^2+5x+6} \\ x^2+2x \phantom{{}+6} \end{array} \qquad \begin{array}{l} x(x+2) = \\ x^2+2x \end{array}$$

Step 4 Subtract.
Remember to add the opposite!

$$\begin{array}{r} x \phantom{{}+5x+6} \\ x+2\overline{\smash{)}x^2+5x+6} \\ \underline{x^2+2x \phantom{{}+6}} \\ 3x+6 \end{array}$$

Step 5 Repeat the process beginning with step 2; use the polynomial that resulted from the subtraction in step 4 as a new dividend.

$$\begin{array}{r} x+3 \phantom{{}6} \\ x+2\overline{\smash{)}x^2+5x+6} \\ \underline{x^2+2x \phantom{{}+6}} \\ 3x+6 \\ \underline{3x+6} \\ 0 \end{array} \qquad \begin{array}{l} \dfrac{3x}{x} = 3 \\ \\ 3(x+2) = \\ 3x+6 \end{array}$$

Thus $(x^2 + 5x + 6) \div (x + 2) = x + 3$, which can be checked by multiplying $(x + 2)$ and $(x + 3)$.

$$(x + 2)(x + 3) = x^2 + 5x + 6 \qquad \blacksquare$$

A division problem such as $(x^2 + 5x + 6) \div (x + 2)$ can also be written as $\dfrac{x^2 + 5x + 6}{x + 2}$. Using this format, we can express the final result for Example 1 as $\dfrac{x^2 + 5x + 6}{x + 2} = x + 3$. (Technically, the restriction $x \neq -2$ should be made to avoid division by zero.)

In general, to check a division problem we can multiply the divisor times the quotient and add the remainder, which can be expressed as

Dividend = (Divisor)(Quotient) + Remainder

Sometimes the remainder is expressed as a fractional part of the divisor. The relationship then becomes

$$\dfrac{\text{Dividend}}{\text{Divisor}} = \text{Quotient} + \dfrac{\text{Remainder}}{\text{Divisor}}$$

EXAMPLE 2 Divide $2x^2 - 3x - 20$ by $x - 4$.

Solution

Step 1 $x - 4 \overline{\smash{)}2x^2 - 3x - 20}$

Step 2 $\begin{array}{r} 2x \phantom{{}-3x-20} \\ x-4\overline{\smash{)}2x^2-3x-20} \end{array} \qquad \dfrac{2x^2}{x} = 2x$

Step 3
$$\begin{array}{r} 2x \\ x-4\overline{\smash{)}2x^2 - 3x - 20} \\ 2x^2 - 8x \end{array}$$
$$2x(x-4) = 2x^2 - 8x$$

Step 4
$$\begin{array}{r} 2x \\ x-4\overline{\smash{)}2x^2 - 3x - 20} \\ \underline{2x^2 - 8x} \\ 5x - 20 \end{array}$$

Step 5
$$\begin{array}{r} 2x + 5 \\ x-4\overline{\smash{)}2x^2 - 3x - 20} \\ \underline{2x^2 - 8x} \\ 5x - 20 \\ \underline{5x - 20} \end{array}$$

$$\frac{5x}{x} = 5$$

$$5(x-4) = 5x - 20$$

✔ **Check**

$$(x - 4)(2x + 5) = 2x^2 - 3x - 20$$

Therefore, $\dfrac{2x^2 - 3x - 20}{x - 4} = 2x + 5.$ ∎

Now let's continue to think in terms of the step-by-step division process but organize our work in the typical long division format.

EXAMPLE 3

Divide $12x^2 + x - 6$ by $3x - 2$.

Solution

$$\begin{array}{r} 4x + 3 \\ 3x-2\overline{\smash{)}12x^2 + 1x - 6} \\ \underline{12x^2 - 8x} \\ 9x - 6 \\ \underline{9x - 6} \end{array}$$

✔ **Check**

$$(3x - 2)(4x + 3) = 12x^2 + x - 6$$

Therefore, $\dfrac{12x^2 + x - 6}{3x - 2} = 4x + 3.$ ∎

Each of the next three examples illustrates another aspect of the division process. Study them carefully; then you should be ready to work the exercises in the next problem set.

EXAMPLE 4

Perform the division $(7x^2 - 3x - 4) \div (x - 2)$.

Solution

$$\begin{array}{r} 7x + 11 \\ x-2\overline{\smash{)}7x^2 - 3x - 4} \\ \underline{7x^2 - 14x} \\ 11x - 4 \\ \underline{11x - 22} \\ 18 \end{array}$$

⟵ A remainder of 18

✔ Check

Just as in arithmetic, we check by *adding* the remainder to the product of the divisor and quotient.

$$(x - 2)(7x + 11) + 18 \stackrel{?}{=} 7x^2 - 3x - 4$$
$$7x^2 - 3x - 22 + 18 \stackrel{?}{=} 7x^2 - 3x - 4$$
$$7x^2 - 3x - 4 = 7x^2 - 3x - 4$$

Therefore, $\dfrac{7x^2 - 3x - 4}{x - 2} = 7x + 11 + \dfrac{18}{x - 2}$. ∎

EXAMPLE 5

Perform the division $\dfrac{x^3 - 8}{x - 2}$.

Solution

$$\begin{array}{r}
x^2 + 2x + 4 \\
x - 2 \overline{\smash{)}x^3 + 0x^2 + 0x - 8} \\
\underline{x^3 - 2x^2} \\
2x^2 + 0x - 8 \\
\underline{2x^2 - 4x} \\
4x - 8 \\
\underline{4x - 8} \\
\end{array}$$

← Notice the insertion of x^2 and x terms with zero coefficients

✔ Check

$$(x - 2)(x^2 + 2x + 4) \stackrel{?}{=} x^3 - 8$$
$$x^3 + 2x^2 + 4x - 2x^2 - 4x - 8 \stackrel{?}{=} x^3 - 8$$
$$x^3 - 8 = x^3 - 8$$

Therefore, $\dfrac{x^3 - 8}{x - 2} = x^2 + 2x + 4$. ∎

EXAMPLE 6

Perform the division $\dfrac{x^3 + 5x^2 - 3x - 4}{x^2 + 2x}$.

Solution

$$\begin{array}{r}
x + 3 \\
x^2 + 2x \overline{\smash{)}x^3 + 5x^2 - 3x - 4} \\
\underline{x^3 + 2x^2} \\
3x^2 - 3x - 4 \\
\underline{3x^2 + 6x} \\
-9x - 4 \\
\end{array}$$

← A remainder of $-9x - 4$

We stop the division process when the degree of the remainder is less than the degree of the divisor.

✔ **Check**

$$(x^2 + 2x)(x + 3) + (-9x - 4) \stackrel{?}{=} x^3 + 5x^2 - 3x - 4$$
$$x^3 + 3x^2 + 2x^2 + 6x - 9x - 4 \stackrel{?}{=} x^3 + 5x^2 - 3x - 4$$
$$x^3 + 5x^2 - 3x - 4 = x^3 + 5x^2 - 3x - 4$$

Therefore, $\dfrac{x^3 + 5x^2 - 3x - 4}{x^2 + 2x} = x + 3 + \dfrac{-9x - 4}{x^2 + 2x}$. ∎

CONCEPT QUIZ 4.5

For Problems 1–6, answer true or false.

1. A division problem written as $(x^2 - x - 6) \div (x - 1)$ could also be written as $\dfrac{x^2 - x - 6}{x - 1}$.
2. The division of $\dfrac{x^2 + 7x + 12}{x + 3} = x + 4$ could be checked by multiplying $(x + 4)$ by $(x + 3)$.
3. For the division problem $(2x^2 + 5x + 9) \div (2x + 1)$, the remainder is 7. The remainder for the division problem is sometimes expressed as $\dfrac{7}{2x + 1}$.
4. In general, to check a division problem we can multiply the divisor times the quotient and subtract the remainder.
5. If a term is inserted to act as a placeholder, then the coefficient of the term must be zero.
6. When performing division, the process ends when the degree of the remainder is less than the degree of the divisor.

Section 4.5 Classroom Problem Set

Objective 1

1. Divide $(x^2 + 11x + 28)$ by $(x + 4)$.
2. Divide $(x^2 + 15x + 54)$ by $(x + 6)$.
3. Divide $(3x^2 - 5x - 2)$ by $(x - 2)$.
4. Divide $(7n^2 - 61n - 90)$ by $(n - 10)$.
5. Divide $(8x^2 + 14x - 15)$ by $(4x - 3)$.
6. Divide $(24x^2 + 10x - 21)$ by $(4x - 3)$.
7. Perform the division. $(4x^2 + 6x - 4) \div (x + 3)$
8. Perform the division. $(12x^2 + 28x + 27) \div (6x + 5)$
9. Perform the division. $\dfrac{a^3 + 27}{a + 3}$
10. Perform the division. $\dfrac{8x^3 + 27}{2x + 3}$
11. Perform the division. $\dfrac{a^3 + 5a^2 - 2a + 3}{a^2 + a}$
12. Perform the division. $\dfrac{2x^3 - x^2 - 3x + 5}{x^2 + x}$

THOUGHTS INTO WORDS

1. Give a step-by-step description of how you would do the division problem $(2x^3 + 8x^2 - 29x - 30) \div (x + 6)$.

2. How do you know by inspection that the answer to the following division problem is incorrect?
$$(3x^3 - 7x^2 - 22x + 8) \div (x - 4)$$
$$= 3x^2 + 5x + 1$$

Answers to the Concept Quiz
1. True 2. True 3. True 4. False 5. True 6. True

4.6 Zero and Negative Integers as Exponents

OBJECTIVES

1. Apply the Properties of Exponents Including Negative and Zero Exponents to Simplify Expressions
2. Write Numbers in Scientific Notation
3. Write Numbers Expressed in Scientific Notation in Standard Decimal Notation
4. Use Scientific Notation to Evaluate Numerical Expressions

1 Apply the Properties of Exponents Including Negative and Zero Exponents to Simplify Expressions

Thus far in this text we have used only positive integers as exponents. The next definitions and properties serve as a basis for our work with exponents.

Definition 4.1

If n is a positive integer and b is any real number, then
$$b^n = \underbrace{bbb \cdots b}_{n \text{ factors of } b}$$

Property 4.5

If m and n are positive integers and a and b are real numbers, except $b \neq 0$ whenever it appears in a denominator, then

1. $b^n \cdot b^m = b^{n+m}$
2. $(b^n)^m = b^{mn}$
3. $(ab)^n = a^n b^n$
4. $\left(\dfrac{a}{b}\right)^n = \dfrac{a^n}{b^n}$ Part 4 has not been stated previously
5. $\dfrac{b^n}{b^m} = b^{n-m}$ When $n > m$

   $\dfrac{b^n}{b^m} = 1$ When $n = m$

Property 4.5 pertains to the use of positive integers as exponents. Zero and the negative integers can also be used as exponents. First let's consider the use of 0 as an exponent. We want to use 0 as an exponent in such a way that the basic properties of exponents will continue to hold. Consider the example $x^4 \cdot x^0$. If part 1 of Property 4.5 is to hold, then

$$x^4 \cdot x^0 = x^{4+0} = x^4$$

Note that x^0 acts like 1 because $x^4 \cdot x^0 = x^4$. This suggests the following definition.

4.6 Zero and Negative Integers as Exponents

Definition 4.2

If b is a nonzero real number, then
$$b^0 = 1$$

According to Definition 4.2, the following statements are all true.

$4^0 = 1$

$(-628)^0 = 1$

$\left(\dfrac{4}{7}\right)^0 = 1$

$n^0 = 1, \quad n \neq 0$

$(x^2 y^5)^0 = 1, \quad x \neq 0 \text{ and } y \neq 0$

A similar line of reasoning indicates how negative integers should be used as exponents. Consider the example $x^3 \cdot x^{-3}$. If part 1 of Property 4.5 is to hold, then

$$x^3 \cdot x^{-3} = x^{3+(-3)} = x^0 = 1$$

Thus x^{-3} must be the reciprocal of x^3 because their product is 1; that is,

$$x^{-3} = \dfrac{1}{x^3}$$

This process suggests the following definition.

Definition 4.3

If n is a positive integer, and b is a nonzero real number, then
$$b^{-n} = \dfrac{1}{b^n}$$

According to Definition 4.3, the following statements are all true.

$x^{-6} = \dfrac{1}{x^6}$

$2^{-3} = \dfrac{1}{2^3} = \dfrac{1}{8}$

$10^{-2} = \dfrac{1}{10^2} = \dfrac{1}{100} \quad \text{or} \quad 0.01$

$\dfrac{1}{x^{-4}} = \dfrac{1}{\dfrac{1}{x^4}} = 1 \cdot \dfrac{x^4}{1} = x^4$

$\left(\dfrac{2}{3}\right)^{-2} = \dfrac{1}{\left(\dfrac{2}{3}\right)^2} = \dfrac{1}{\dfrac{4}{9}} = 1 \cdot \dfrac{9}{4} = \dfrac{9}{4}$

Remark: Note in the last example that $\left(\dfrac{2}{3}\right)^{-2} = \left(\dfrac{3}{2}\right)^2$. In other words, to raise a fraction to a negative power, we can invert the fraction and raise it to the corresponding positive power.

We can verify (we will not do so in this text) that all parts of Property 4.5 hold for *all integers*. In fact, we can replace part 5 with this statement.

> **Replacement for Part 5 of Property 4.5**
>
> $$\frac{b^n}{b^m} = b^{n-m} \quad \text{for all integers } n \text{ and } m$$

The next example illustrates the use of this new property. In each part of the example, we simplify the original expression and use only positive exponents in the final result.

EXAMPLE 1 Simplify the expressions and use only positive exponents in the answer.

(a) $\dfrac{x^2}{x^5}$ (b) $\dfrac{a^{-3}}{a^{-7}}$ (c) $\dfrac{y^{-5}}{y^{-2}}$ (d) $\dfrac{x^{-6}}{x^{-6}}$

Solution

(a) $\dfrac{x^2}{x^5} = x^{2-5} = x^{-3} = \dfrac{1}{x^3}$

(b) $\dfrac{a^{-3}}{a^{-7}} = a^{-3-(-7)} = a^{-3+7} = a^4$

(c) $\dfrac{y^{-5}}{y^{-2}} = y^{-5-(-2)} = y^{-5+2} = y^{-3} = \dfrac{1}{y^3}$

(d) $\dfrac{x^{-6}}{x^{-6}} = x^{-6-(-6)} = x^{-6+6} = x^0 = 1$ ∎

The properties of exponents provide a basis for simplifying certain types of numerical expressions, as the following example illustrates.

EXAMPLE 2 Simplify the expressions and use only positive exponents in the answer.

(a) $2^{-4} \cdot 2^6$ (b) $10^5 \cdot 10^{-6}$ (c) $\dfrac{10^2}{10^{-2}}$ (d) $(2^{-3})^{-2}$

Solution

(a) $2^{-4} \cdot 2^6 = 2^{-4+6} = 2^2 = 4$

(b) $10^5 \cdot 10^{-6} = 10^{5+(-6)} = 10^{-1} = \dfrac{1}{10}$ or 0.1

(c) $\dfrac{10^2}{10^{-2}} = 10^{2-(-2)} = 10^{2+2} = 10^4 = 10{,}000$

(d) $(2^{-3})^{-2} = 2^{-3(-2)} = 2^6 = 64$ ∎

Having the use of all integers as exponents also expands the type of work that we can do with algebraic expressions. In the following example, we simplify each given expression and use only positive exponents in the final result.

EXAMPLE 3 Simplify the expressions and use only positive exponents in the answer.

(a) $x^8 \cdot x^{-2}$ (b) $a^{-4} \cdot a^{-3}$ (c) $(y^{-3})^4$ (d) $(x^{-2}y^4)^{-3}$ (e) $\left(\dfrac{x^{-1}}{y^2}\right)^{-2}$

(f) $(4x^{-2})(3x^{-1})$ (g) $\left(\dfrac{12x^{-6}}{6x^{-2}}\right)^{-2}$

Solution

(a) $x^8 x^{-2} = x^{8+(-2)} = x^6$

(b) $a^{-4} a^{-3} = a^{-4+(-3)} = a^{-7} = \dfrac{1}{a^7}$

(c) $(y^{-3})^4 = y^{-3(4)} = y^{-12} = \dfrac{1}{y^{12}}$

(d) $(x^{-2} y^4)^{-3} = (x^{-2})^{-3}(y^4)^{-3} = x^6 y^{-12} = \dfrac{x^6}{y^{12}}$

(e) $\left(\dfrac{x^{-1}}{y^2}\right)^{-2} = \dfrac{(x^{-1})^{-2}}{(y^2)^{-2}} = \dfrac{x^2}{y^{-4}} = x^2 y^4$

(f) $(-4x^{-2})(3x^{-1}) = -12 x^{-2+(-1)} = -12 x^{-3} = -\dfrac{12}{x^3}$

(g) $\left(\dfrac{12 x^{-6}}{6 x^{-2}}\right)^{-2} = (2 x^{-6-(-2)})^{-2} = (2 x^{-4})^{-2}$ Divide the coefficients $\dfrac{12}{6} = 2$

$\qquad = (2)^{-2}(x^{-4})^{-2}$

$\qquad = \left(\dfrac{1}{2^2}\right)(x^8) = \dfrac{x^8}{4}$ ∎

2 Write Numbers in Scientific Notation

Many scientific applications of mathematics involve the measurement of very large and very small quantities. For example:

> The speed of light is approximately 29,979,200,000 centimeters per second.
>
> A light year—the distance light travels in 1 year—is approximately 5,865,696,000,000 miles.
>
> A gigahertz equals 1,000,000,000 hertz.
>
> The length of a typical virus cell equals 0.000000075 of a meter.
>
> The length of the diameter of a water molecule is 0.0000000003 of a meter.

Working with numbers of this type in standard form is quite cumbersome. It is much more convenient to represent very small and very large numbers in scientific notation, sometimes called scientific form. Although negative numbers can be written in scientific form, we will restrict our discussion to positive numbers. A number is in **scientific notation** when it is written as the product of a number between 1 and 10 (including 1) and an integral power of 10. Symbolically, a number in scientific notation has the form $(N)(10^k)$, where $1 \leq N < 10$, and k is an integer. For example, 621 can be written as $(6.21)(10^2)$, and 0.0023 can be written as $(2.3)(10^{-3})$.

To switch from ordinary notation to scientific notation, you can use the following procedure.

> Write the given number as the product of a number greater than or equal to 1 (and less than 10) and an integral power of 10. To determine the exponent of 10, count the number of places that the decimal point moved when going from the original number to the number greater than or equal to 1 and less than 10. This exponent is (a) negative if the original number is less than 1, (b) positive if the original number is greater than 10, and (c) zero if the original number itself is between 1 and 10.

172 Chapter 4 Exponents and Polynomials

EXAMPLE 4

Write the following numbers in scientific notation.

(a) 0.000179   (b) 8175   (c) 3.14

Solution

We can write

(a) $0.000179 = (1.79)(10^{-4})$ According to part (a) of the procedure
(b) $8175 = (8.175)(10^3)$ According to part (b)
(c) $3.14 = (3.14)(10^0)$ According to part (c)   ∎

We can express the applications given earlier in scientific notation as follows:

Speed of light: $29{,}979{,}200{,}000 = (2.99792)(10^{10})$ centimeters per second
Light year: $5{,}865{,}696{,}000{,}000 = (5.865696)(10^{12})$ miles
Gigahertz: $1{,}000{,}000{,}000 = (1)(10^9)$ hertz
Length of a virus cell: $0.000000075 = (7.5)(10^{-8})$ meter
Length of the diameter of a water molecule: $0.0000000003 = (3)(10^{-10})$ meter

3 Write Numbers Expressed in Scientific Notation in Standard Decimal Notation

To switch from scientific notation to ordinary decimal notation you can use the following procedure.

> Move the decimal point the number of places indicated by the exponent of 10. The decimal point is moved to the right if the exponent is positive and to the left if it is negative.

EXAMPLE 5

Write the following numbers in standard decimal notation.

(a) $(4.71)(10^4)$ (b) $(1.78)(10^{-2})$

Solution

Thus we can write

(a) $(4.71)(10^4) = 47{,}100$ Two zeros are needed for place value purposes
(b) $(1.78)(10^{-2}) = 0.0178$ One zero is needed for place value purposes   ∎

4 Use Scientific Notation to Evaluate Numerical Expressions

The use of scientific notation along with the properties of exponents can make some arithmetic problems much easier to evaluate. The next examples illustrate this point.

EXAMPLE 6

Evaluate $(4000)(0.000012)$.

Solution

$$(4000)(0.000012) = (4)(10^3)(1.2)(10^{-5})$$
$$= (4)(1.2)(10^3)(10^{-5})$$
$$= (4.8)(10^{-2})$$
$$= 0.048$$ ∎

EXAMPLE 7

Evaluate $\dfrac{960{,}000}{0.032}$.

Solution

$$\dfrac{960{,}000}{0.032} = \dfrac{(9.6)(10^5)}{(3.2)(10^{-2})}$$

$$= (3)(10^7) \qquad \dfrac{10^5}{10^{-2}} = 10^{5-(-2)} = 10^7$$

$$= 30{,}000{,}000$$

EXAMPLE 8

Evaluate $\dfrac{(6000)(0.00008)}{(40{,}000)(0.006)}$.

Solution

$$\dfrac{(6000)(0.00008)}{(40{,}000)(0.006)} = \dfrac{(6)(10^3)(8)(10^{-5})}{(4)(10^4)(6)(10^{-3})}$$

$$= \dfrac{(48)(10^{-2})}{(24)(10^1)}$$

$$= (2)(10^{-3}) \qquad \dfrac{10^{-2}}{10^1} = 10^{-2-1} = 10^{-3}$$

$$= 0.002$$

CONCEPT QUIZ 4.6

For Problems 1–10, answer true or false.

1. Any nonzero number raised to the zero power is equal to one.
2. The algebraic expression x^{-2} is the reciprocal of x^2 if $x \neq 0$.
3. To raise a fraction to a negative exponent, we can invert the fraction and raise it to the corresponding positive exponent.
4. $\dfrac{1}{y^{-3}} = y^{-3}$
5. A number in scientific notation has the form $(N)(10^k)$, where $1 \leq N < 10$ and k is any real number.
6. A number is less than zero if the exponent is negative when the number is written in scientific notation.
7. $\dfrac{1}{x^{-2}} = x^2$
8. $\dfrac{10^{-2}}{10^{-4}} = 100$
9. $(3.11)(10^{-2}) = 311$
10. $(5.24)(10^{-1}) = 0.524$

Section 4.6 Classroom Problem Set

Objective 1

1. Simplify the expressions and use only positive exponents in the answer.

   (a) $\dfrac{m^4}{m^7}$ (b) $\dfrac{y^{-2}}{y^{-6}}$ (c) $\dfrac{z^{-3}}{z^2}$ (d) $\dfrac{k^5}{k^5}$

2. Simplify the expressions and use only positive exponents in the answer.

   (a) $\dfrac{x^{-1}}{x^5}$ (b) $\dfrac{c^{-1}}{c^{-9}}$ (c) $\dfrac{a^{-4}}{a^{-4}}$ (d) $\dfrac{y^{-2}}{y^4}$

3. Simplify the expressions and use only positive exponents in the answer.

   (a) $3^5 \cdot 3^{-8}$ (b) $10^{-2} \cdot 10^{-1}$

   (c) $\dfrac{10^{-1}}{10^{-4}}$ (d) $(2^{-5})^{-1}$

4. Simplify the expressions and use only positive exponents in the answer.

   (a) $2^{-7} \cdot 2^3$ (b) $4^{-1} \cdot 4^4$

   (c) $\dfrac{3^{-2}}{3^{-5}}$ (d) $(2^{-1} \cdot 3^{-1})^{-1}$

5. Simplify the expressions and use only positive exponents in the answer.

   (a) $z^{-1} \cdot z^{-4}$ (b) $(a^5 b^{-1})^{-2}$

   (c) $\left(\dfrac{m^2}{n^{-4}}\right)^{-1}$ (d) $(2a^{-5}b^2)(-5a^3b)$

6. Simplify the expressions and use only positive exponents in the answer.

   (a) $d^{-3} \cdot d^{-10}$ (b) $(x^{-2}y^{-1})^3$

   (c) $\left(\dfrac{a^{-1}}{b^{-3}}\right)^{-2}$ (d) $(3x^{-2}y^3)(-2x^2y^{-7})$

Objective 2

7. Write the following numbers in scientific notation.

   (a) 0.00000683 (b) 5,600,000 (c) 9.54

8. Write the following numbers in scientific notation.

   (a) 26,980,000 (b) 0.0016582 (c) 63.7

Objective 3

9. Practice your skill. Write the following numbers in standard decimal notation.

   (a) $(8.2)(10^7)$ (b) $(3.46)(10^{-4})$

10. Practice your skill. Write the following numbers in standard decimal notation.

    (a) $(1.14)(10^7)$ (b) $(8.64)(10^{-6})$

Objective 4

11. Use scientific notation and properties of exponents to evaluate $(50{,}000)(0.0000013)$.

12. Use scientific notation and properties of exponents to evaluate $(5{,}000{,}000)(0.00009)$.

13. Use scientific notation and properties of exponents to evaluate $\dfrac{8{,}400}{0.042}$.

14. Use scientific notation and properties of exponents to evaluate $\dfrac{0.0057}{30{,}000}$.

15. Use scientific notation and properties of exponents to evaluate $\dfrac{(900)(0.0006)}{(30{,}000)(0.02)}$.

16. Use scientific notation and properties of exponents to evaluate $\dfrac{(0.0008)(0.07)}{(20{,}000)(0.0004)}$.

THOUGHTS INTO WORDS

1. Is the following simplification process correct?

   $$(2^{-2})^{-1} = \left(\dfrac{1}{2^2}\right)^{-1} = \left(\dfrac{1}{4}\right)^{-1} = \dfrac{1}{\left(\dfrac{1}{4}\right)^1} = 4$$

   Can you suggest a better way to do the problem?

2. Explain the importance of scientific notation.

Answers to the Concept Quiz

1. True **2.** True **3.** True **4.** False **5.** False **6.** False **7.** True **8.** True **9.** False **10.** True

Chapter 4 Review Problem Set

For Problems 1–6, match each polynomial with its description.

1. $-3xy^4$
2. $a^2 + 2ab - b^2$
3. $4x^2y + 8xy$
4. $2x^4 - x^3 + 5x^2$
5. $5a^3$
6. $3x^2y^2 + 4y^3$

A. Binomial of degree 4
B. Monomial of degree 3
C. Monomial of degree 5
D. Trinomial of degree 4
E. Binomial of degree 3
F. Trinomial of degree 2

For Problems 7–10, perform the additions and subtractions.

7. $(5x^2 - 6x + 4) + (3x^2 - 7x - 2)$
8. $(7y^2 + 9y - 3) - (4y^2 - 2y + 6)$
9. $(2x^2 + 3x - 4) + (4x^2 - 3x - 6) - (3x^2 - 2x - 1)$
10. $(-3x^2 - 2x + 4) - (x^2 - 5x - 6) - (4x^2 + 3x - 8)$

For Problems 11–18, remove parentheses and combine similar terms.

11. $5(2x - 1) + 7(x + 3) - 2(3x + 4)$
12. $3(2x^2 - 4x - 5) - 5(3x^2 - 4x + 1)$
13. $6(y^2 - 7y - 3) - 4(y^2 + 3y - 9)$
14. $3(a - 1) - 2(3a - 4) - 5(2a + 7)$
15. $-(a + 4) + 5(-a - 2) - 7(3a - 1)$
16. $-2(3n - 1) - 4(2n + 6) + 5(3n + 4)$
17. $3(n^2 - 2n - 4) - 4(2n^2 - n - 3)$
18. $-5(-n^2 + n - 1) + 3(4n^2 - 3n - 7)$

For Problems 19–26, find the indicated products.

19. $(5x^2)(7x^4)$
20. $(-6x^3)(9x^5)$
21. $(-4xy^2)(-6x^2y^3)$
22. $(2a^3b^4)(-3ab^5)$
23. $(2a^2b^3)^3$
24. $(-3xy^2)^2$
25. $(-4xy^3)^2$
26. $(5a^2b)^3$

For Problems 27–34, find the indicated products. Be sure to simplify your answers.

27. $5x(7x + 3)$
28. $(-3x^2)(8x - 1)$
29. $(x - 2)(3x^2 + x - 1)$
30. $(x + 6)(2x^2 + 5x - 4)$
31. $(x - 2)(x^2 - x + 6)$
32. $(2x - 1)(x^2 + 4x + 7)$
33. $(x^2 - x - 1)(x^2 + 2x + 5)$
34. $(n^2 + 2n + 4)(n^2 - 7n - 1)$

For Problems 35–50, find the indicated products. Be sure to simplify your answers.

35. $(x + 9)(x + 8)$
36. $(x + 4)(x + 6)$
37. $(x - 6)(x - 2)$
38. $(x - 5)(x - 6)$
39. $(x - 7)(x + 2)$
40. $(x + 10)(x - 4)$
41. $(2x + 1)(x + 5)$
42. $(3x + 2)(x + 4)$
43. $(5x - 1)(x + 3)$
44. $(4x + 3)(x - 5)$
45. $(2x - 1)(7x + 3)$
46. $(4a - 7)(5a + 8)$
47. $(5n - 1)(6n + 5)$
48. $(3n + 4)(4n - 1)$
49. $(a + 5)^3$
50. $(a - 6)^3$

For Problems 51–56, use one of the appropriate patterns, $(a + b)^2 = a^2 + 2ab^2 + b^2$, $(a - b)^2 = a^2 - 2ab + b^2$, or $(a + b)(a - b) = a^2 - b^2$, to find the indicated products.

51. $(y + 8)^2$
52. $(m - 4)^2$
53. $(3a - 5)^2$
54. $(2x + 7)^2$
55. $(2n + 1)(2n - 1)$
56. $(4n + 5)(4n - 5)$

57. A piece of vinyl flooring that is 20 feet by 30 feet has a square piece x feet on a side cut out from each corner. Find the area of the vinyl flooring piece after the corners are removed.

58. A piece of cardboard that is 12 inches by 18 inches has a square piece x inches on a side cut out from each corner. After the corners are removed, the flaps are turned up to form an open box. Find the polynomial that represents the volume of the box.

For Problems 59–68, perform the divisions.

59. $\dfrac{36x^4y^5}{-3xy^2}$
60. $\dfrac{-56a^5b^7}{-8a^2b^3}$
61. $\dfrac{-18x^4y^3 - 54x^6y^2}{6x^2y^2}$
62. $\dfrac{-30a^5b^{10} + 39a^4b^8}{-3ab}$
63. $\dfrac{56x^4 - 40x^3 - 32x^2}{4x^2}$
64. $\dfrac{12a^3 - 15a^2 - 3a}{3a}$
65. $(x^2 + 9x - 1) \div (x + 5)$

66. $(21x^2 - 4x - 12) \div (3x + 2)$

67. $(2x^3 - 3x^2 + 2x - 4) \div (x - 2)$

68. $(6x^2 + 7x - 9) \div (2x - 1)$

For Problems 69–80, evaluate each expression.

69. $3^2 + 2^2$

70. $(3 + 2)^2$

71. 2^{-4}

72. $(-5)^0$

73. -5^0

74. $\dfrac{1}{3^{-2}}$

75. $\left(\dfrac{3}{4}\right)^{-2}$

76. $\dfrac{1}{\left(\dfrac{1}{4}\right)^{-1}}$

77. $\dfrac{1}{(-2)^{-3}}$

78. $2^{-1} + 3^{-2}$

79. $3^0 + 2^{-2}$

80. $(2 + 3)^{-2}$

For Problems 81–94, simplify each of the following, and express your answers using positive exponents only.

81. $x^5 x^{-8}$

82. $(3x^5)(4x^{-2})$

83. $\dfrac{x^{-4}}{x^{-6}}$

84. $\dfrac{x^{-6}}{x^{-4}}$

85. $\dfrac{-18xy^4}{2x^3y}$

86. $\dfrac{-20x^2y}{-5x^4y^3}$

87. $\dfrac{24a^5}{3a^{-1}}$

88. $\dfrac{48n^{-2}}{12n^{-1}}$

89. $(x^{-2}y)^{-1}$

90. $(a^2b^{-3})^{-2}$

91. $(2x)^{-1}$

92. $(3n^2)^{-2}$

93. $(2n^{-1})^{-3}$

94. $(4ab^{-1})(-3a^{-1}b^2)$

For Problems 95–98, write each expression in standard decimal form.

95. $(6.1)(10^2)$

96. $(5.6)(10^4)$

97. $(8)(10^{-2})$

98. $(9.2)(10^{-4})$

For Problems 99–102, write each number in scientific notation.

99. 9000

100. 47

101. 0.047

102. 0.00021

For Problems 103–106, use scientific notation and the properties of exponents to evaluate each expression.

103. $(0.00004)(12{,}000)$

104. $(0.0021)(2000)$

105. $\dfrac{0.0056}{0.0000028}$

106. $\dfrac{0.00078}{39{,}000}$

Chapter 4   Practice Test

1. Find the sum of $-7x^2 + 6x - 2$ and $5x^2 - 8x + 7$.
2. Subtract $-x^2 + 9x - 14$ from $-4x^2 + 3x + 6$.
3. Remove parentheses and combine similar terms for the expression $3(2x - 1) - 6(3x - 2) - (x + 7)$.
4. Find the product $(-4xy^2)(7x^2y^3)$.
5. Find the product $(2x^2y)^2(3xy^3)$.

For Problems 6–12, find the indicated products and express answers in simplest form.

6. $(x - 9)(x + 2)$
7. $(n + 14)(n - 7)$
8. $(5a + 3)(8a + 7)$
9. $(3x - 7y)^2$
10. $(x + 3)(2x^2 - 4x - 7)$
11. $(9x - 5y)(9x + 5y)$
12. $(3x - 7)(5x - 11)$
13. Find the indicated quotient: $\dfrac{-96x^4y^5}{-12x^2y}$.
14. Find the indicated quotient: $\dfrac{56x^2y - 72xy^2}{-8xy}$.
15. Find the indicated quotient: $(2x^3 + 5x^2 - 22x + 15) \div (2x - 3)$.
16. Find the indicated quotient: $(4x^3 + 23x^2 + 36) \div (x + 6)$.
17. Evaluate $\left(\dfrac{2}{3}\right)^{-3}$.
18. Evaluate $4^{-2} + 4^{-1} + 4^0$.
19. Evaluate $\dfrac{1}{2^{-4}}$.
20. Find the product $(-6x^{-4})(4x^2)$ and express the answer using a positive exponent.
21. Simplify $\left(\dfrac{8x^{-1}}{2x^2}\right)^{-1}$ and express the answer using a positive exponent.
22. Simplify $(x^{-3}y^5)^{-2}$ and express the answer using positive exponents.
23. Write 0.00027 in scientific notation.
24. Express $(9.2)(10^6)$ in standard decimal form.
25. Evaluate $(0.000002)(3000)$.

1. _____
2. _____
3. _____
4. _____
5. _____
6. _____
7. _____
8. _____
9. _____
10. _____
11. _____
12. _____
13. _____
14. _____
15. _____
16. _____
17. _____
18. _____
19. _____
20. _____
21. _____
22. _____
23. _____
24. _____
25. _____

Chapters 1–4 Cumulative Review Problem Set

For Problems 1–10, evaluate each of the numerical expressions.

1. $5 + 3(2 - 7)^2 \div 3 \cdot 5$
2. $8 \div 2 \cdot (-1) + 3$
3. $7 - 2^2 \cdot 5 \div (-1)$
4. $4 + (-2) - 3(6)$
5. $(-3)^4$
6. -2^5
7. $\left(\dfrac{2}{3}\right)^{-1}$
8. $\dfrac{1}{4^{-2}}$
9. $\left(\dfrac{1}{2} - \dfrac{1}{3}\right)^{-2}$
10. $2^0 + 2^{-1} + 2^{-2}$

For Problems 11–16, evaluate each algebraic expression for the given values of the variables.

11. $\dfrac{2x + 3y}{x - y}$ for $x = \dfrac{1}{2}$ and $y = -\dfrac{1}{3}$
12. $\dfrac{2}{5}n - \dfrac{1}{3}n - n + \dfrac{1}{2}n$ for $n = -\dfrac{3}{4}$
13. $\dfrac{3a - 2b - 4a + 7b}{-a - 3a + b - 2b}$ for $a = -1$ and $b = -\dfrac{1}{3}$
14. $-2(x - 4) + 3(2x - 1) - (3x - 2)$ for $x = -2$
15. $(x^2 + 2x - 4) - (x^2 - x - 2) + (2x^2 - 3x - 1)$ for $x = -1$
16. $2(n^2 - 3n - 1) - (n^2 + n + 4) - 3(2n - 1)$ for $n = 3$

For Problems 17–29, find the indicated products.

17. $(3x^2y^3)(-5xy^4)$
18. $(-6ab^4)(-2b^3)$
19. $(-2x^2y^5)^3$
20. $-3xy(2x - 5y)$
21. $(5x - 2)(3x - 1)$
22. $(7x - 1)(3x + 4)$
23. $(-x - 2)(2x + 3)$
24. $(7 - 2y)(7 + 2y)$
25. $(x - 2)(3x^2 - x - 4)$
26. $(2x - 5)(x^2 + x - 4)$
27. $(2n + 3)^2$
28. $(1 - 2n)^3$
29. $(x^2 - 2x + 6)(2x^2 + 5x - 6)$

For Problems 30–34, perform the indicated divisions.

30. $\dfrac{-52x^3y^4}{13xy^2}$
31. $\dfrac{-126a^3b^5}{-9a^2b^3}$
32. $\dfrac{56xy^2 - 64x^3y - 72x^4y^4}{8xy}$
33. $(2x^3 + 2x^2 - 19x - 21) \div (x + 3)$
34. $(3x^3 + 17x^2 + 6x - 4) \div (3x - 1)$

For Problems 35–38, simplify each expression and express your answers using positive exponents only.

35. $(-2x^3)(3x^{-4})$
36. $\dfrac{4x^{-2}}{2x^{-1}}$
37. $(3x^{-1}y^{-2})^{-1}$
38. $(xy^2z^{-1})^{-2}$

For Problems 39–41, use scientific notation and the properties of exponents to help evaluate each numerical expression.

39. $(0.00003)(4000)$
40. $(0.0002)(0.003)^2$
41. $\dfrac{0.00034}{0.0000017}$

For Problems 42–49, solve each of the equations.

42. $5x + 8 = 6x - 3$
43. $-2(4x - 1) = -5x + 3 - 2x$
44. $\dfrac{y}{2} - \dfrac{y}{3} = 8$
45. $6x + 8 - 4x = 10(3x + 2)$
46. $1.6 - 2.4x = 5x - 65$
47. $-3(x - 1) + 2(x + 3) = -4$
48. $\dfrac{3n + 1}{5} + \dfrac{n - 2}{3} = \dfrac{2}{15}$
49. $0.06x + 0.08(1500 - x) = 110$

For Problems 50–55, solve each of the inequalities.

50. $2x - 7 \le -3(x + 4)$
51. $6x + 5 - 3x > 5$
52. $4(x - 5) + 2(3x + 6) < 0$
53. $-5x + 3 > -4x + 5$
54. $\dfrac{3x}{4} - \dfrac{x}{2} \le \dfrac{5x}{6} - 1$
55. $0.08(700 - x) + 0.11x \ge 65$

For Problems 56–62, set up an equation and solve each problem.

56. The sum of 4 and three times a certain number is the same as the sum of the number and 10. Find the number.
57. Fifteen percent of some number is 6. Find the number.
58. Lou has 18 coins consisting of dimes and quarters. If the total value of the coins is $3.30, how many coins of each denomination does he have?
59. A sum of $1500 is invested, part of it at 8% interest and the remainder at 9%. If the total interest amounts to $128, find the amount invested at each rate.
60. How many gallons of water must be added to 15 gallons of a 12% salt solution to change it to a 10% salt solution?
61. Two airplanes leave Atlanta at the same time and fly in opposite directions. If one travels at 400 miles per hour and the other at 450 miles per hour, how long will it take them to be 2975 miles apart?
62. The length of a rectangle is 1 meter more than twice its width. If the perimeter of the rectangle is 44 meters, find the length and width.

Factoring, Solving Equations, and Problem Solving

5

Chapter 5 Warm-Up Problems

1. Find the greatest common factor.
   - (a) 8 and 18
   - (b) 36 and 42
   - (c) 16, 24, and 60
2. Simplify each of the expressions.
   - (a) $3(8); 3 + 8$
   - (b) $-11(6); -11 + 6$
   - (c) $(-2)(-18); -2 + (-18)$
3. Evaluate.
   - (a) $y^2 - 7y$ if $y = -2$
   - (b) $x(x + 8)$ if $x = -10$
   - (c) $(m + 3)(m - 12)$ if $m = 12$
4. Simplify.
   - (a) $x^2 + (x + 2)^2$
   - (b) $5m(2n - 1) - 3m(2n - 1)$
   - (c) $-5xy(x^2 - 1)$
5. Simplify.
   - (a) $x^2 + 7x - 8x - 56$
   - (b) $m^2 - 9m + 9m - 81$
   - (c) $4y^2 + 14y + 14y + 49$
6. Multiply.
   - (a) $(6x - 7)(3x + 1)$
   - (b) $(2m - 3)^2$
   - (c) $(x - 4y)(x + 4y)$
7. Solve for the variable.
   - (a) $m - 8 = 0$
   - (b) $4x - 12 = 0$
   - (c) $3 - 6n = 0$
8. Find the area of a triangle given
   - (a) base, 16; height, 5
   - (b) base, $3x$; height, 9
   - (c) base, m; height, $m + 7$

5.1 Factoring by Using the Distributive Property

5.2 Factoring the Difference of Two Squares

5.3 Factoring Trinomials of the Form $x^2 + bx + c$

5.4 Factoring Trinomials of the Form $ax^2 + bx + c$

5.5 Factoring, Solving Equations, and Problem Solving

T he distributive property has allowed us to combine similar terms and multiply polynomials. In this chapter, we will see yet another use of the distributive property as we learn how to **factor polynomials**. Factoring polynomials will allow us to solve other kinds of equations, which will, in turn, help us to solve a greater variety of word problems.

Video tutorials for all section learning objectives are available in a variety of delivery modes.

INTERNET PROJECT

Pythagoras is widely known for the Pythagorean theorem pertaining to right triangles. Do an Internet search to determine at least two other fields where Pythagoras made significant contributions. Pythagoras also founded a school. While conducting your search, find the name given to the students attending Pythagoras' school and some of the school rules for students. Can you think of any modern-day schools that might have the same requirements?

5.1 Factoring by Using the Distributive Property

OBJECTIVES

1. Find the Greatest Common Factor
2. Factor Out the Greatest Common Factor
3. Factor by Grouping
4. Solve Equations by Factoring
5. Solve Word Problems Using Factoring

1 Find the Greatest Common Factor

In Chapter 1, we found the *greatest common factor* of two or more whole numbers by inspection or by using the prime factored form of the numbers. For example, by inspection we see that the greatest common factor of 8 and 12 is 4. This means that 4 is the largest whole number that is a factor of both 8 and 12. If it is difficult to determine the greatest common factor by inspection, then we can use the prime factorization technique as follows:

$$42 = 2 \cdot 3 \cdot 7$$
$$70 = 2 \cdot 5 \cdot 7$$

We see that $2 \cdot 7 = 14$ is the greatest common factor of 42 and 70.

It is meaningful to extend the concept of *greatest common factor* to monomials. Consider the next example.

EXAMPLE 1 Find the greatest common factor of $8x^2$ and $12x^3$.

Solution

$$8x^2 = 2 \cdot 2 \cdot 2 \cdot x \cdot x$$
$$12x^3 = 2 \cdot 2 \cdot 3 \cdot x \cdot x \cdot x$$

Therefore, the greatest common factor is $2 \cdot 2 \cdot x \cdot x = 4x^2$. ∎

By *the greatest common factor of two or more monomials* we mean the monomial with the largest numerical coefficient and highest power of the variables that is a factor of the given monomials.

5.1 Factoring by Using the Distributive Property 181

EXAMPLE 2

Find the greatest common factor of $16x^2y$, $24x^3y^2$, and $32xy$.

Solution

$$16x^2y = 2 \cdot 2 \cdot 2 \cdot 2 \cdot x \cdot x \cdot y$$
$$24x^3y^2 = 2 \cdot 2 \cdot 2 \cdot 3 \cdot x \cdot x \cdot x \cdot y \cdot y$$
$$32xy = 2 \cdot 2 \cdot 2 \cdot 2 \cdot 2 \cdot x \cdot y$$

Therefore, the greatest common factor is $2 \cdot 2 \cdot 2 \cdot x \cdot y = 8xy$. ∎

2 Factor Out the Greatest Common Factor

We have used the distributive property to multiply a polynomial by a monomial; for example,

$$3x(x + 2) = 3x^2 + 6x$$

Suppose we start with $3x^2 + 6x$ and want to express it in factored form. We use the distributive property in the form $ab + ac = a(b + c)$.

$$3x^2 + 6x = 3x(x) + 3x(2) \quad \text{$3x$ is the greatest common factor of $3x^2$ and $6x$}$$
$$= 3x(x + 2) \quad \text{Use the distributive property}$$

The next four examples further illustrate this process of *factoring out the greatest common monomial factor*.

EXAMPLE 3

Factor $12x^3 - 8x^2$.

Solution

$$12x^3 - 8x^2 = 4x^2(3x) - 4x^2(2)$$
$$= 4x^2(3x - 2) \quad ab - ac = a(b - c)$$ ∎

EXAMPLE 4

Factor $12x^2y + 18xy^2$.

Solution

$$12x^2y + 18xy^2 = 6xy(2x) + 6xy(3y)$$
$$= 6xy(2x + 3y)$$ ∎

EXAMPLE 5

Factor $24x^3 + 30x^4 - 42x^5$.

Solution

$$24x^3 + 30x^4 - 42x^5 = 6x^3(4) + 6x^3(5x) - 6x^3(7x^2)$$
$$= 6x^3(4 + 5x - 7x^2)$$ ∎

EXAMPLE 6

Factor $9x^2 + 9x$.

Solution

$$9x^2 + 9x = 9x(x) + 9x(1)$$
$$= 9x(x + 1)$$ ∎

We want to emphasize the point made just before Example 3. It is important to realize that we are factoring out the *greatest* common monomial factor. We could factor an expression such as $9x^2 + 9x$ in Example 6 as $9(x^2 + x)$, $3(3x^2 + 3x)$, $3x(3x + 3)$, or even $\frac{1}{2}(18x^2 + 18x)$, but it is the form $9x(x + 1)$ that we want. We can accomplish this by factoring out the greatest common monomial factor; we sometimes refer to this process as **factoring completely**. A polynomial with integral coefficients is in completely factored form if these conditions are met:

1. It is expressed as a product of polynomials with integral coefficients.
2. No polynomial, other than a monomial, within the factored form can be further factored into polynomials with integral coefficients.

Thus $9(x^2 + x)$, $3(3x^2 + 3x)$, and $3x(3x + 2)$ are not completely factored because they violate condition 2. The form $\frac{1}{2}(18x^2 + 18x)$ violates both conditions 1 and 2.

3 Factor by Grouping

Sometimes there may be a **common binomial factor** rather than a common monomial factor. For example, each of the two terms of $x(y + 2) + z(y + 2)$ has a common binomial factor of $(y + 2)$. Thus we can factor $(y + 2)$ from each term and get

$$x(y + 2) + z(y + 2) = (y + 2)(x + z)$$

Consider a few more examples involving a common binomial factor:

$$a(b + c) - d(b + c) = (b + c)(a - d)$$
$$x(x + 2) + 3(x + 2) = (x + 2)(x + 3)$$
$$x(x + 5) - 4(x + 5) = (x + 5)(x - 4)$$

It may be that the original polynomial exhibits no apparent common monomial or binomial factor, which is the case with

$$ab + 3a + bc + 3c$$

However, by factoring a from the first two terms and c from the last two terms, we see that

$$ab + 3a + bc + 3c = a(b + 3) + c(b + 3)$$

Now a common binomial factor of $(b + 3)$ is obvious, and we can proceed as before:

$$a(b + 3) + c(b + 3) = (b + 3)(a + c)$$

This factoring process is called **factoring by grouping**. Let's consider two more examples of factoring by grouping.

EXAMPLE 7

Factor each polynomial completely.

(a) $x^2 - x + 5x - 5$ (b) $6x^2 - 4x - 3x + 2$

Solution

(a) $x^2 - x + 5x - 5 = x(x - 1) + 5(x - 1)$ Factor x from first two terms and 5 from last two terms

$\qquad\qquad\qquad\quad = (x - 1)(x + 5)$ Factor common binomial factor of $(x - 1)$ from both terms

(b) $6x^2 - 4x - 3x + 2 = 2x(3x - 2) - 1(3x - 2)$ Factor $2x$ from first two terms and -1 from last two terms

$\qquad\qquad\qquad\quad = (3x - 2)(2x - 1)$ Factor common binomial factor of $(3x - 2)$ from both terms

4 Solve Equations by Factoring

Suppose we are told that the product of two numbers is 0. What do we know about the numbers? Do you agree we can conclude that at least one of the numbers must be 0? The next property formalizes this idea.

> **Property 5.1**
>
> For all real numbers a and b,
>
> $ab = 0$ if and only if $a = 0$ or $b = 0$

Property 5.1 provides us with another technique for solving equations.

EXAMPLE 8

Solve $x^2 + 6x = 0$.

Solution

To solve equations by applying Property 5.1, one side of the equation must be a product, and the other side of the equation must be zero. This equation already has zero on the right-hand side of the equation, but the left-hand side of this equation is a sum. We will factor the left-hand side, $x^2 + 6x$, to change the sum into a product.

$$x^2 + 6x = 0$$
$$x(x + 6) = 0$$
$$x = 0 \quad \text{or} \quad x + 6 = 0 \qquad ab = 0 \text{ if and only if } a = 0 \text{ or } b = 0$$
$$x = 0 \quad \text{or} \quad x = -6$$

The solution set is $\{-6, 0\}$. (Be sure to check both values in the original equation.) ∎

EXAMPLE 9

Solve $x^2 = 12x$.

Solution

In order to solve this equation by Property 5.1, we will first get zero on the right-hand side of the equation by adding $-12x$ to each side. Then we factor the expression on the left-hand side of the equation.

$$x^2 = 12x$$
$$x^2 - 12x = 0 \qquad \text{Added } -12x \text{ to both sides}$$
$$x(x - 12) = 0$$
$$x = 0 \quad \text{or} \quad x - 12 = 0 \qquad ab = 0 \text{ if and only if } a = 0 \text{ or } b = 0$$
$$x = 0 \quad \text{or} \quad x = 12$$

The solution set is $\{0, 12\}$. ∎

Remark: Note in Example 9 that we *did not* divide both sides of the original equation by x. Doing so would cause us to lose the solution of 0.

EXAMPLE 10

Solve $4x^2 - 3x = 0$.

Solution

$$4x^2 - 3x = 0$$
$$x(4x - 3) = 0$$

$$x = 0 \quad \text{or} \quad 4x - 3 = 0 \qquad ab = 0 \text{ if and only if } a = 0 \text{ or } b = 0$$
$$x = 0 \quad \text{or} \quad 4x = 3$$
$$x = 0 \quad \text{or} \quad x = \frac{3}{4}$$

The solution set is $\left\{0, \dfrac{3}{4}\right\}$. ∎

EXAMPLE 11

Solve $x(x + 2) + 3(x + 2) = 0$.

Solution

In order to solve this equation by Property 5.1, we will factor the left-hand side of the equation. The greatest common factor of the terms is $(x + 2)$.

$$x(x + 2) + 3(x + 2) = 0$$
$$(x + 2)(x + 3) = 0$$
$$x + 2 = 0 \quad \text{or} \quad x + 3 = 0 \qquad ab = 0 \text{ if and only if } a = 0 \text{ or } b = 0$$
$$x = -2 \quad \text{or} \quad x = -3$$

The solution set is $\{-3, -2\}$. ∎

5 Solve Word Problems Using Factoring

Each time we expand our equation-solving capabilities, we gain more techniques for solving word problems. Let's solve a geometric problem with the ideas we learned in this section.

EXAMPLE 12

Apply Your Skill

The area of a square is numerically equal to twice its perimeter. Find the length of a side of the square.

Solution

Sketch a square and let s represent the length of each side (see Figure 5.1). Then the area is represented by s^2 and the perimeter by $4s$. Thus

$$s^2 = 2(4s)$$
$$s^2 = 8s$$
$$s^2 - 8s = 0$$
$$s(s - 8) = 0$$
$$s = 0 \quad \text{or} \quad s - 8 = 0$$
$$s = 0 \quad \text{or} \quad s = 8$$

Because 0 is not a reasonable answer to the problem, the solution is 8. (Be sure to check this solution in the original statement of the example!) ∎

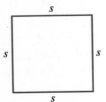

Figure 5.1

CONCEPT QUIZ 5.1

For Problems 1–10, answer true or false.

1. The greatest common factor of $6x^2y^3 - 12x^3y^2 + 18x^4y$ is $2x^2y$.
2. If the factored form of a polynomial can be factored further, then it has not met the conditions for being considered "factored completely."
3. Common factors are always monomials.
4. If the product of x and y is zero, then x is zero or y is zero.

5. The factored form $3a(2a^2 + 4)$ is factored completely.
6. The solutions for the equation $x(x + 2) = 7$ are 7 and 5.
7. The solution set for $x^2 = 7x$ is $\{7\}$.
8. The solution set for $x(x - 2) - 3(x - 2) = 0$ is $\{2, 3\}$.
9. The solution set for $-3x = x^2$ is $\{-3, 0\}$.
10. The solution set for $x(x + 6) = 2(x + 6)$ is $\{-6\}$.

Section 5.1   Classroom Problem Set

Objective 1

1. Find the greatest common factor of $14a^2$ and $7a^5$.
2. Find the greatest common factor of $42ab^3$ and $70a^2b^2$.
3. Find the greatest common factor of $18m^2n^4$, $4m^3n^5$, and $10m^4n^3$.
4. Find the greatest common factor of $16x^2y^2$, $40x^2y^3$, and $56x^3y^4$.

Objective 2

5. Factor $15a^2 + 21a^6$.
6. Factor $24x - 40xy$.
7. Factor $8m^3n^2 + 2m^6n$.
8. Factor $84x^2y^3 - 12xy^3$.
9. Factor $48y^8 - 16y^6 + 24y^4$.
10. Factor $44a^5 - 24a^3 - 20a^2$.
11. Factor $8b^3 + 8b^2$.
12. Factor $4a^3 - 6a^5$.

Objective 3

13. Factor each polynomial completely.
    (a) $ab + 5a + 3b + 15$ (b) $xy - 2x + 4y - 8$

14. Factor each polynomial completely.
    (a) $ac + bc + a + b$ (b) $2x - 2y - ax + ay$

Objective 4

15. Solve $y^2 - 7y = 0$.
16. Solve $x^2 + x = 0$.
17. Solve $a^2 = 15a$.
18. Solve $7x^2 = -3x$.
19. Solve $5y^2 + 2y = 0$.
20. Solve $6n^2 - 24n = 0$.
21. Solve $m(m - 2) + 5(m - 2) = 0$.
22. Solve $x(x + 9) - 2(x + 9) = 0$.

Objective 5

23. The area of a square is numerically equal to three times its perimeter. Find the length of a side of the square.
24. Four times the square of a number equals 20 times that number. What is the number?

THOUGHTS INTO WORDS

1. Suppose that your friend factors $24x^2y + 36xy$ like this:
$$24x^2y + 36xy = 4xy(6x + 9)$$
$$= (4xy)(3)(2x + 3)$$
$$= 12xy(2x + 3)$$
Is this correct? Would you suggest any changes?

2. The following solution is given for the equation $x(x - 10) = 0$.
$$x(x - 10) = 0$$
$$x^2 - 10x = 0$$
$$x(x - 10) = 0$$
$$x = 0 \quad \text{or} \quad x - 10 = 0$$
$$x = 0 \quad \text{or} \quad x = 10$$

The solution set is $\{0, 10\}$. Is this solution correct? Would you suggest any changes?

Answers to the Concept Quiz

1. False   2. True   3. False   4. True   5. False   6. False   7. False   8. True   9. True   10. False

5.2 Factoring the Difference of Two Squares

OBJECTIVES

1. Factor the Difference of Two Squares
2. Solve Equations by Factoring the Difference of Two Squares
3. Solve Word Problems Using Factoring

1 Factor the Difference of Two Squares

In Section 4.3, we noted some special multiplication patterns. One of these patterns was

$$(a - b)(a + b) = a^2 - b^2$$

Here is another version of that pattern:

Difference of Two Squares

$$a^2 - b^2 = (a - b)(a + b)$$

To apply the difference-of-two-squares pattern is a fairly simple process, as these next examples illustrate. The steps inside the box are often performed mentally.

$$x^2 - 36 = (x)^2 - (6)^2 = (x - 6)(x + 6)$$
$$4x^2 - 25 = (2x)^2 - (5)^2 = (2x - 5)(2x + 5)$$
$$9x^2 - 16y^2 = (3x)^2 - (4y)^2 = (3x - 4y)(3x + 4y)$$
$$64 - y^2 = (8)^2 - (y)^2 = (8 - y)(8 + y)$$

Because multiplication is commutative, the order of writing the factors is not important. For example, $(x - 6)(x + 6)$ can also be written as $(x + 6)(x - 6)$.

Remark: You must be careful not to assume an analogous factoring pattern for the *sum* of two squares; it does not exist. For example, $x^2 + 4 \neq (x + 2)(x + 2)$ because $(x + 2)(x + 2) = x^2 + 4x + 4$. We say that the *sum of two squares is not factorable using integers*. The phrase "using integers" is necessary because $x^2 + 4$ could be written as $\frac{1}{2}(2x^2 + 8)$, but such *factoring* is of no help. Furthermore, we do not consider $(1)(x^2 + 4)$ as factoring $x^2 + 4$.

It is possible that both the technique of *factoring out a common monomial factor* and the *difference-of-two-squares* pattern can be applied to the same polynomial. In general, it is best to look for a common monomial factor first.

EXAMPLE 1   Factor $2x^2 - 50$.

Solution

$$2x^2 - 50 = 2(x^2 - 25) \quad \text{Common factor of 2}$$
$$= 2(x - 5)(x + 5) \quad \text{Difference of squares}$$

In Example 1, by expressing $2x^2 - 50$ as $2(x - 5)(x + 5)$, we say that it has been *factored completely*. That means the factors 2, $x - 5$, and $x + 5$ cannot be factored any further using integers.

EXAMPLE 2

Factor completely $18y^3 - 8y$.

Solution

$$18y^3 - 8y = 2y(9y^2 - 4) \quad \text{Common factor of } 2y$$
$$= 2y(3y - 2)(3y + 2) \quad \text{Difference of squares}$$

Sometimes it is possible to apply the difference-of-two-squares pattern more than once. Consider the next example.

EXAMPLE 3

Factor completely $x^4 - 16$.

Solution

$$x^4 - 16 = (x^2 + 4)(x^2 - 4)$$
$$= (x^2 + 4)(x + 2)(x - 2)$$

The following examples should help you to summarize the factoring ideas presented thus far.

$$5x^2 + 20 = 5(x^2 + 4)$$
$$25 - y^2 = (5 - y)(5 + y)$$
$$3 - 3x^2 = 3(1 - x^2) = 3(1 + x)(1 - x)$$
$$36x^2 - 49y^2 = (6x - 7y)(6x + 7y)$$

$a^2 + 9$ is not factorable using integers

$9x + 17y$ is not factorable using integers

2 Solve Equations by Factoring the Difference of Two Squares

Each time we learn a new factoring technique, we also develop more power for solving equations. Let's consider how we can use the difference-of-squares factoring pattern to help solve certain kinds of equations.

EXAMPLE 4

Solve $x^2 = 25$.

Solution

$$x^2 = 25$$
$$x^2 - 25 = 0$$
$$(x + 5)(x - 5) = 0$$

$x + 5 = 0 \quad \text{or} \quad x - 5 = 0 \quad$ Remember: $ab = 0$ if and only if
$x = -5 \quad \text{or} \quad x = 5 \quad\quad a = 0$ or $b = 0$

The solution set is $\{-5, 5\}$. Check these answers!

EXAMPLE 5

Solve $9x^2 = 25$.

Solution

$$9x^2 = 25$$
$$9x^2 - 25 = 0$$
$$(3x + 5)(3x - 5) = 0$$
$$3x + 5 = 0 \quad \text{or} \quad 3x - 5 = 0$$
$$3x = -5 \quad \text{or} \quad 3x = 5$$
$$x = -\frac{5}{3} \quad \text{or} \quad x = \frac{5}{3}$$

The solution set is $\left\{-\frac{5}{3}, \frac{5}{3}\right\}$.

EXAMPLE 6

Solve $5y^2 = 20$.

Solution

$$5y^2 = 20$$
$$\frac{5y^2}{5} = \frac{20}{5} \quad \text{Divide both sides by 5}$$
$$y^2 = 4$$
$$y^2 - 4 = 0$$
$$(y + 2)(y - 2) = 0$$
$$y + 2 = 0 \quad \text{or} \quad y - 2 = 0$$
$$y = -2 \quad \text{or} \quad y = 2$$

The solution set is $\{-2, 2\}$. Check it!

EXAMPLE 7

Solve $x^3 - 9x = 0$.

Solution

$$x^3 - 9x = 0$$
$$x(x^2 - 9) = 0$$
$$x(x - 3)(x + 3) = 0$$
$$x = 0 \quad \text{or} \quad x - 3 = 0 \quad \text{or} \quad x + 3 = 0$$
$$x = 0 \quad \text{or} \quad x = 3 \quad \text{or} \quad x = -3$$

The solution set is $\{-3, 0, 3\}$.

The more we know about solving equations, the more easily we can solve word problems.

3 Solve Word Problems Using Factoring

EXAMPLE 8

Apply Your Skill

The combined area of two squares is 20 square centimeters. Each side of one square is twice as long as a side of the other square. Find the lengths of the sides of each square.

Solution

We can sketch two squares and label the sides of the smaller square s (see Figure 5.2). Then the sides of the larger square are $2s$. The sum of the areas of the two squares is 20 square centimeters, so we set up and solve the following equation:

$$s^2 + (2s)^2 = 20$$
$$s^2 + 4s^2 = 20$$
$$5s^2 = 20$$
$$s^2 = 4$$
$$s^2 - 4 = 0$$
$$(s + 2)(s - 2) = 0$$
$$s + 2 = 0 \quad \text{or} \quad s - 2 = 0$$
$$s = -2 \quad \text{or} \quad s = 2$$

Figure 5.2

Because s represents the length of a side of a square, we must disregard the solution -2. Thus one square has sides of length 2 centimeters, and the other square has sides of length $2(2) = 4$ centimeters. ∎

CONCEPT QUIZ 5.2

For Problems 1–8, answer true or false.

1. A binomial that has two perfect square terms that are subtracted is called the difference of two squares.
2. The sum of two squares is factorable using integers.
3. When factoring it is usually best to look for a common factor first.
4. The polynomial $4x^2 + y^2$ factors into $(2x + y)(2x + y)$.
5. The completely factored form of $y^4 - 81$ is $(y^2 + 9)(y^2 - 9)$.
6. The solution set for $x^2 = -16$ is $\{-4\}$.
7. The solution set for $5x^3 - 5x = 0$ is $\{-1, 0, 1\}$.
8. The solution set for $x^4 - 9x^2 = 0$ is $\{-3, 0, 3\}$.

Section 5.2   Classroom Problem Set

Objective 1

1. Factor $3y^2 - 12$.
2. Factor $5a^2 - 45$.
3. Factor completely $32x^3 - 2x$.
4. Factor completely $4x - 36x^3$.
5. Factor completely $a^4 - 81$.
6. Factor completely $16 - x^4$.

Objective 2

7. Solve $x^2 = 64$.
8. Solve $144 = n^2$.
9. Solve $4x^2 = 9$.
10. Solve $25x^2 = 4$.
11. Solve $3x^2 = 75$.
12. Solve $5 = 45x^2$.
13. Solve $x^3 - 49x = 0$.
14. Solve $2n^3 = 8n$.

Objective 3

15. The combined area of two squares is 250 square feet. Each side of one square is three times as long as a side of the other square. Find the lengths of the sides of each square.

16. The difference of the areas of two squares is 192 square inches. Each side of the larger square is twice the length of a side of the smaller square. Find the length of a side of each square.

THOUGHTS INTO WORDS

1. How do we know that the equation $x^2 + 1 = 0$ has no solutions in the set of real numbers?

2. Consider the following solution:

$$4x^2 - 36 = 0$$
$$4(x^2 - 9) = 0$$
$$4(x + 3)(x - 3) = 0$$
$$4 = 0 \quad \text{or} \quad x + 3 = 0 \quad \text{or} \quad x - 3 = 0$$
$$4 = 0 \quad \text{or} \quad x = -3 \quad \text{or} \quad x = 3$$

The solution set is $\{-3, 3\}$. Is this a correct solution? Do you have any suggestion to offer the person who worked on this problem?

3. Why is the following factoring process incomplete?

$$16x^2 - 64 = (4x + 8)(4x - 8)$$

How should the factoring be done?

Answers to the Concept Quiz

1. True 2. False 3. True 4. False 5. False 6. False 7. True 8. True

5.3 Factoring Trinomials of the Form $x^2 + bx + c$

OBJECTIVES

1. Factor Trinomials of the Form $x^2 + bx + c$
2. Use Factoring of Trinomials to Solve Equations
3. Solve Word Problems Including Consecutive Number Problems
4. Use the Pythagorean Theorem to Solve Problems

1 Factor Trinomials of the Form $x^2 + bx + c$

One of the most common types of factoring used in algebra is to express a trinomial as the product of two binomials. In this section, we will consider trinomials where the coefficient of the squared term is 1, that is, trinomials of the form $x^2 + bx + c$.

Again, to develop a factoring technique we first look at some multiplication ideas. Consider the product $(x + r)(x + s)$, and use the distributive property to show how each term of the resulting trinomial is formed.

$$(x + r)(x + s) = x(x) + x(s) + r(x) + r(s)$$
$$\qquad\qquad\qquad\quad \downarrow \qquad \underbrace{\qquad\qquad} \qquad \downarrow$$
$$\qquad\qquad\qquad\quad x^2 + \quad (s + r)x \quad + \quad rs$$

Note that the coefficient of the middle term is the *sum* of r and s and that the last term is the *product* of r and s. These two relationships are used in the next examples.

EXAMPLE 1   Factor $x^2 + 7x + 12$.

Solution

We need to fill in the blanks with two numbers whose sum is 7 and whose product is 12.

$$x^2 + 7x + 12 = (x + \underline{\quad})(x + \underline{\quad})$$

5.3 Factoring Trinomials of the Form $x^2 + bx + c$

This can be done by setting up a table showing possible numbers.

Product	Sum
1(12) = 12	1 + 12 = 13
2(6) = 12	2 + 6 = 8
3(4) = 12	3 + 4 = 7

The bottom line contains the numbers that we need. Thus
$$x^2 + 7x + 12 = (x + 3)(x + 4)$$

EXAMPLE 2

Factor $x^2 - 11x + 24$.

Solution

We need two numbers whose product is 24 and whose sum is -11.

Product	Sum
$(-1)(-24) = 24$	$-1 + (-24) = -25$
$(-2)(-12) = 24$	$-2 + (-12) = -14$
$(-3)(-8) = 24$	$-3 + (-8) = -11$
$(-4)(-6) = 24$	$-4 + (-6) = -10$

The third line contains the numbers that we want. Thus
$$x^2 - 11x + 24 = (x - 3)(x - 8)$$

EXAMPLE 3

Factor $x^2 + 3x - 10$.

Solution

We need two numbers whose product is -10 and whose sum is 3.

Product	Sum
$1(-10) = -10$	$1 + (-10) = -9$
$-1(10) = -10$	$-1 + 10 = 9$
$2(-5) = -10$	$2 + (-5) = -3$
$-2(5) = -10$	$-2 + 5 = 3$

The bottom line is the key line. Thus
$$x^2 + 3x - 10 = (x + 5)(x - 2)$$

EXAMPLE 4

Factor $x^2 - 2x - 8$.

Solution

We need two numbers whose product is -8 and whose sum is -2.

Product	Sum
$1(-8) = -8$	$1 + (-8) = -7$
$-1(8) = -8$	$-1 + 8 = 7$
$2(-4) = -8$	$2 + (-4) = -2$
$-2(4) = -8$	$-2 + 4 = 2$

The third line has the information we want.

$$x^2 - 2x - 8 = (x - 4)(x + 2)$$

The tables in the last four examples illustrate one way of organizing your thoughts for such problems. We showed complete tables; that is, for Example 4, we included the last line even though the desired numbers were obtained in the third line. If you use such tables, keep in mind that as soon as you get the desired numbers, the table need not be continued beyond that point. Furthermore, there will be times that you will be able to find the numbers without using a table. The key ideas are the product and sum relationships.

EXAMPLE 5

Factor $x^2 - 13x + 12$.

Solution

Product	Sum
$(-1)(-12) = 12$	$(-1) + (-12) = -13$

We need not complete the table.

$$x^2 - 13x + 12 = (x - 1)(x - 12)$$

In the next example, we refer to the concept of absolute value. Recall that the absolute value is the number without regard for the sign. For example,

$$|4| = 4 \quad \text{and} \quad |-4| = 4$$

EXAMPLE 6

Factor $x^2 - x - 56$.

Solution

Note that the coefficient of the middle term is -1. Therefore, we are looking for two numbers whose product is -56; their sum is -1, so the absolute value of the negative number must be 1 larger than the absolute value of the positive number. The numbers are -8 and 7, and we have

$$x^2 - x - 56 = (x - 8)(x + 7)$$

EXAMPLE 7

Factor $x^2 + 10x + 12$.

Solution

Product	Sum
$1(12) = 12$	$1 + 12 = 13$
$2(6) = 12$	$2 + 6 = 8$
$3(4) = 12$	$3 + 4 = 7$

Because the table is complete and no two factors of 12 produce a sum of 10, we conclude that

$$x^2 + 10x + 12$$

is not factorable using integers. ∎

In a problem such as Example 7, we need to be sure that we have tried all possibilities before we conclude that the trinomial is not factorable.

2 Use Factoring of Trinomials to Solve Equations

The property $ab = 0$ if and only if $a = 0$ or $b = 0$ continues to play an important role as we solve equations that involve the factoring ideas of this section. Consider the following examples.

EXAMPLE 8   Solve $x^2 + 8x + 15 = 0$.

Solution

$$x^2 + 8x + 15 = 0$$
$$(x + 3)(x + 5) = 0 \quad \text{Factor the left side}$$
$$x + 3 = 0 \quad \text{or} \quad x + 5 = 0 \quad \text{Use } ab = 0 \text{ if and only if } a = 0 \text{ or } b = 0$$
$$x = -3 \quad \text{or} \quad x = -5$$

The solution set is $\{-5, -3\}$. ∎

EXAMPLE 9   Solve $x^2 + 5x - 6 = 0$.

Solution

$$x^2 + 5x - 6 = 0$$
$$(x + 6)(x - 1) = 0$$
$$x + 6 = 0 \quad \text{or} \quad x - 1 = 0$$
$$x = -6 \quad \text{or} \quad x = 1$$

The solution set is $\{-6, 1\}$. ∎

EXAMPLE 10   Solve $y^2 - 4y = 45$.

Solution

$$y^2 - 4y = 45$$
$$y^2 - 4y - 45 = 0$$
$$(y - 9)(y + 5) = 0$$
$$y - 9 = 0 \quad \text{or} \quad y + 5 = 0$$
$$y = 9 \quad \text{or} \quad y = -5$$

The solution set is $\{-5, 9\}$. ∎

Don't forget that we can always check to be absolutely sure of our solutions. Let's check the solutions for Example 10. If $y = 9$, then $y^2 - 4y = 45$ becomes

$$9^2 - 4(9) \stackrel{?}{=} 45$$
$$81 - 36 \stackrel{?}{=} 45$$
$$45 = 45$$

If $y = -5$, then $y^2 - 4y = 45$ becomes

$$(-5)^2 - 4(-5) \stackrel{?}{=} 45$$
$$25 + 20 \stackrel{?}{=} 45$$
$$45 = 45$$

3 Solve Word Problems Including Consecutive Number Problems

The more we know about factoring and solving equations, the more easily we can solve word problems.

EXAMPLE 11 Apply Your Skill

Find two consecutive integers whose product is 72.

Solution

Let n represent one integer. Then $n + 1$ represents the next integer.

$$n(n + 1) = 72 \quad \text{The product of the two integers is 72}$$
$$n^2 + n = 72$$
$$n^2 + n - 72 = 0$$
$$(n + 9)(n - 8) = 0$$
$$n + 9 = 0 \quad \text{or} \quad n - 8 = 0$$
$$n = -9 \quad \text{or} \quad n = 8$$

If $n = -9$, then $n + 1 = -9 + 1 = -8$. If $n = 8$, then $n + 1 = 8 + 1 = 9$. Thus the consecutive integers are -9 and -8 or 8 and 9. ∎

EXAMPLE 12 Apply Your Skill

A rectangular plot is 6 meters longer than it is wide. The area of the plot is 16 square meters. Find the length and width of the plot.

Solution

We let w represent the width of the plot, and then $w + 6$ represents the length (see Figure 5.3).

Figure 5.3

Using the area formula $A = lw$, we obtain

$$w(w + 6) = 16$$
$$w^2 + 6w = 16$$
$$w^2 + 6w - 16 = 0$$

$$(w + 8)(w - 2) = 0$$
$$w + 8 = 0 \quad \text{or} \quad w - 2 = 0$$
$$w = -8 \quad \text{or} \quad w = 2$$

The solution -8 is not possible for the width of a rectangle, so the plot is 2 meters wide, and its length $(w + 6)$ is 8 meters. ∎

4 Use the Pythagorean Theorem to Solve Problems

The Pythagorean theorem, an important theorem pertaining to right triangles, can also serve as a guideline for solving certain types of problems. The Pythagorean theorem states that **in any right triangle, the square of the longest side** (*called the hypotenuse*) **is equal to the sum of the squares of the other two sides** (*called legs*); see Figure 5.4. We can use this theorem to help solve a problem.

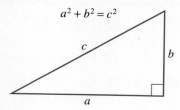

Figure 5.4

EXAMPLE 13   Apply Your Skill

Suppose that the lengths of the three sides of a right triangle are consecutive whole numbers. Find the lengths of the three sides.

Solution

Let s represent the length of the shortest leg. Then $s + 1$ represents the length of the other leg, and $s + 2$ represents the length of the hypotenuse. Using the Pythagorean theorem as a guideline, we obtain the following equation:

$$\underbrace{s^2 + (s + 1)^2}_{\text{Sum of squares of two legs}} = \underbrace{(s + 2)^2}_{\text{Square of hypotenuse}}$$

Solving this equation yields

$$s^2 + s^2 + 2s + 1 = s^2 + 4s + 4$$
$$2s^2 + 2s + 1 = s^2 + 4s + 4$$
$$s^2 + 2s + 1 = 4s + 4 \qquad \text{Add } -s^2 \text{ to both sides}$$
$$s^2 - 2s + 1 = 4 \qquad \text{Add } -4s \text{ to both sides}$$
$$s^2 - 2s - 3 = 0 \qquad \text{Add } -4 \text{ to both sides}$$
$$(s - 3)(s + 1) = 0$$
$$s - 3 = 0 \quad \text{or} \quad s + 1 = 0$$
$$s = 3 \quad \text{or} \quad s = -1$$

The solution of -1 is not possible for the length of a side, so the shortest side (s) is of length 3. The other two sides ($s + 1$ and $s + 2$) have lengths of 4 and 5. ∎

CONCEPT QUIZ 5.3

For Problems 1–10, answer true or false.

1. Any trinomial of the form $x^2 + bx + c$ can be factored (using integers) into the product of two binomials.
2. To factor $x^2 - 4x - 60$ we look for two numbers whose product is -60 and whose sum is -4.
3. A trinomial of the form $x^2 + bx + c$ will never have a common factor other than 1.
4. If n represents an odd integer, then $n + 1$ represents the next consecutive odd integer.
5. The Pythagorean theorem applies only to right triangles.
6. In a right triangle the longest side is called the hypotenuse.
7. The polynomial $x^2 + 25x + 72$ is not factorable.
8. The polynomial $x^2 + 27x + 72$ is not factorable.
9. The solution set of the equation $x^2 + 2x - 63 = 0$ is $\{-9, 7\}$.
10. The solution set of the equation $x^2 - 5x - 66 = 0$ is $\{-11, -6\}$.

Section 5.3 Classroom Problem Set

Objective 1

1. Factor $y^2 + 9y + 20$.
2. Factor $x^2 + 11y + 24$.
3. Factor $m^2 - 8m + 15$.
4. Factor $n^2 - 7n + 10$.
5. Factor $y^2 + 5y - 24$.
6. Factor $x^2 + 5x - 66$.
7. Factor $y^2 - 4y - 12$.
8. Factor $a^2 - a - 30$.
9. Factor $y^2 - 7y + 6$.
10. Factor $c^2 - 11c + 18$.
11. Factor $a^2 + a - 12$.
12. Factor $d^2 + 3d - 88$.
13. Factor $y^2 + 7y + 18$.
14. Factor $x^2 - 14x + 32$.

Objective 2

15. Solve $y^2 - 14y + 24 = 0$.
16. Solve $x^2 + 9x + 20 = 0$.
17. Solve $a^2 + 3a - 28 = 0$.
18. Solve $n^2 + n - 56 = 0$.
19. Solve $x^2 - 18x = 40$.
20. Solve $x(x - 12) = -35$.

Objective 3

21. Find two consecutive integers whose product is 110.
22. Find two consecutive even whole numbers whose product is 168.
23. A rectangular plot is 2 yards longer than it is wide. The area of the plot is 80 square yards. Find the length and width of the plot.
24. The area of the floor of a rectangular room is 84 square feet. The length of the room is 5 feet more than its width. Find the length and width of the room.

Objective 4

25. The length of one leg of a right triangle is 7 inches more than the length of the other leg. The length of the hypotenuse is 13 inches. Find the length of the two legs.
26. One leg of a right triangle is 7 meters longer than the other leg. The hypotenuse is 1 meter longer than the longer leg. Find the lengths of all three sides of the right triangle.

THOUGHTS INTO WORDS

1. What does the expression "not factorable using integers" mean to you?
2. Discuss the role that factoring plays in solving equations.
3. Give an explanation of how you would solve the equation $(x - 3)(x + 4) = 0$, and also how you would solve $(x - 3)(x + 4) = 8$.

Answers to the Concept Quiz

1. False 2. True 3. True 4. False 5. True 6. True 7. True 8. False 9. True 10. False

5.4 Factoring Trinomials of the Form $ax^2 + bx + c$

OBJECTIVES

1. Factor Trinomials Where the Leading Coefficient Is Not One
2. Solve Equations That Involve Factoring

1 Factor Trinomials Where the Leading Coefficient Is Not One

Now let's consider factoring trinomials where the coefficient of the squared term is not 1. We first illustrate an informal trial-and-error technique that works well for certain types of trinomials. This technique relies on our knowledge of multiplication of binomials.

EXAMPLE 1   Factor $2x^2 + 7x + 3$.

Solution

By looking at the first term, $2x^2$, and the positive signs of the other two terms, we know that the binomials are of the form

$$(2x + \underline{\quad})(x + \underline{\quad})$$

Because the factors of the constant term 3 are 1 and 3, we have only two possibilities to try:

$$(2x + 3)(x + 1) \quad \text{or} \quad (2x + 1)(x + 3)$$

By checking the middle term of both of these products, we find that the second one yields the correct middle term of $7x$. Therefore,

$$2x^2 + 7x + 3 = (2x + 1)(x + 3)$$
■

EXAMPLE 2   Factor $6x^2 - 17x + 5$.

Solution

First, we note that $6x^2$ can be written as $2x \cdot 3x$ or $6x \cdot x$. Second, because the middle term of the trinomial is negative, and the last term is positive, we know that the binomials are of the form

$$(2x - \underline{\quad})(3x - \underline{\quad}) \quad \text{or} \quad (6x - \underline{\quad})(x - \underline{\quad})$$

The factors of the constant term 5 are 1 and 5, so we have the following possibilities:

$$(2x - 5)(3x - 1) \quad (2x - 1)(3x - 5)$$
$$(6x - 5)(x - 1) \quad (6x - 1)(x - 5)$$

By checking the middle term for each of these products, we find that the product $(2x - 5)(3x - 1)$ produces the desired term of $-17x$. Therefore,

$$6x^2 - 17x + 5 = (2x - 5)(3x - 1)$$
■

EXAMPLE 3

Factor $8x^2 - 8x - 30$.

Solution

First, we note that the polynomial $8x^2 - 8x - 30$ has a common factor of 2. Factoring out the common factor gives us $2(4x^2 - 4x - 15)$. Now we need to factor $4x^2 - 4x - 15$.

Now we note that $4x^2$ can be written as $4x \cdot x$ or $2x \cdot 2x$. Also, the last term, -15, can be written as $(1)(-15)$, $(-1)(15)$, $(3)(-5)$, or $(-3)(5)$. Thus we can generate the possibilities for the binomial factors as follows:

Using 1 and −15	Using −1 and 15
$(4x - 15)(x + 1)$	$(4x - 1)(x + 15)$
$(4x + 1)(x - 15)$	$(4x + 15)(x - 1)$
$(2x + 1)(2x - 15)$	$(2x - 1)(2x + 15)$

Using 3 and −5	Using −3 and 5
$(4x + 3)(x - 5)$	$(4x - 3)(x + 5)$
$(4x - 5)(x + 3)$	$(4x + 5)(x - 3)$
$(2x - 5)(2x + 3)$	$(2x + 5)(2x - 3)$

By checking the middle term of each of these products, we find that the product indicated with a check mark produces the desired middle term of $-4x$. Therefore,

$$8x^2 - 8x - 30 = 2(2x - 5)(2x + 3)$$ ■

Let's pause for a moment and look back over Examples 1, 2, and 3. Example 3 clearly created the most difficulty because we had to consider so many possibilities. We have suggested one possible format for considering the possibilities, but as you practice such problems, you may develop a format of your own that works better for you. Whatever format you use, the key idea is to organize your work so that you consider all possibilities. Let's look at another example.

EXAMPLE 4

Factor $4x^2 + 6x + 9$.

Solution

First, we note that $4x^2$ can be written as $4x \cdot x$ or $2x \cdot 2x$. Second, because the middle term is positive and the last term is positive, we know that the binomials are of the form

$$(4x + \underline{})(x + \underline{}) \quad \text{or} \quad (2x + \underline{})(2x + \underline{})$$

Because 9 can be written as $9 \cdot 1$ or $3 \cdot 3$, we have only the following five possibilities to try:

$(4x + 9)(x + 1)$ $(4x + 1)(x + 9)$
$(4x + 3)(x + 3)$ $(2x + 1)(2x + 9)$
$(2x + 3)(2x + 3)$

When we try all of these possibilities, we find that none of them yields a middle term of $6x$. Therefore, $4x^2 + 6x + 9$ is *not factorable* using integers. ■

Remark: Example 4 illustrates the importance of organizing your work so that you try *all* possibilities before you conclude that a particular trinomial is not factorable.

2 Solve Equations That Involve Factoring

The ability to factor certain trinomials of the form $ax^2 + bx + c$ provides us with greater equation-solving capabilities. Consider the next examples.

EXAMPLE 5 Solve $3x^2 + 17x + 10 = 0$.

Solution

$3x^2 + 17x + 10 = 0$
$(x + 5)(3x + 2) = 0$ Factoring $3x^2 + 17x + 10$ as $(x + 5)(3x + 2)$ may require some extra work on scratch paper

$x + 5 = 0$ or $3x + 2 = 0$ $ab = 0$ if and only if $a = 0$ or $b = 0$
$x = -5$ or $3x = -2$
$x = -5$ or $x = -\dfrac{2}{3}$

The solution set is $\left\{-5, -\dfrac{2}{3}\right\}$. Check it!

EXAMPLE 6 Solve $24x^2 + 2x - 15 = 0$.

Solution

$24x^2 + 2x - 15 = 0$
$(4x - 3)(6x + 5) = 0$
$4x - 3 = 0$ or $6x + 5 = 0$
$4x = 3$ or $6x = -5$
$x = \dfrac{3}{4}$ or $x = -\dfrac{5}{6}$

The solution set is $\left\{-\dfrac{5}{6}, \dfrac{3}{4}\right\}$.

CONCEPT QUIZ 5.4

For Problems 1–8, answer true or false.

1. Any trinomial of the form $ax^2 + bx + c$ can be factored (using integers) into the product of two binomials.
2. To factor $2x^2 - x - 3$, we look for two numbers whose product is -3 and whose sum is -1.
3. A trinomial of the form $ax^2 + bx + c$ will never have a common factor other than 1.
4. The factored form $(x + 3)(2x + 4)$ is factored completely.
5. The difference-of-squares polynomial $9x^2 - 25$ could be written as the trinomial $9x^2 + 0x - 25$.
6. The polynomial $12x^2 + 11x - 12$ is not factorable.
7. The solution set of the equation $6x^2 + 13x - 5 = 0$ is $\left\{\dfrac{1}{3}, \dfrac{2}{5}\right\}$.
8. The solution set of the equation $18x^2 - 39x + 20 = 0$ is $\left\{\dfrac{5}{6}, \dfrac{4}{3}\right\}$.

Section 5.4 Classroom Problem Set

Objective 1

1. Factor $3y^2 + 16y + 5$.
2. Factor $7x^2 + 19x + 10$.
3. Factor $4a^2 - 8a + 3$.
4. Factor $6x^2 - 17x + 12$.
5. Factor $6y^2 + 27y + 30$.
6. Factor $6a^2 - 4a - 16$.

Objective 2

7. Solve $2y^2 + 7y + 6 = 0$.
8. Solve $4x^2 - 31x + 21 = 0$.
9. Solve $2a^2 - 5a - 12 = 0$.
10. Solve $12n^2 + 28n - 5 = 0$.

THOUGHTS INTO WORDS

1. Explain your thought process when factoring
$$24x^2 - 17x - 20$$

2. Your friend factors $8x^2 - 32x + 32$ as follows:
$$\begin{aligned} 8x^2 - 32x + 32 &= (4x - 8)(2x - 4) \\ &= 4(x - 2)(2)(x - 2) \\ &= 8(x - 2)(x - 2) \end{aligned}$$
Is she correct? Do you have any suggestions for her?

3. Your friend solves the equation $8x^2 - 32x + 32 = 0$ as follows:
$$8x^2 - 32x + 32 = 0$$
$$(4x - 8)(2x - 4) = 0$$
$$4x - 8 = 0 \quad \text{or} \quad 2x - 4 = 0$$
$$4x = 8 \quad \text{or} \quad 2x = 4$$
$$x = 2 \quad \text{or} \quad x = 2$$
The solution set is {2}. Is your friend correct? Do you have any changes to recommend?

Answers to the Concept Quiz

1. False 2. False 3. False 4. False 5. True 6. True 7. False 8. True

5.5 Factoring, Solving Equations, and Problem Solving

OBJECTIVES

1. Factor Perfect-Square Trinomials
2. Recognize the Different Types of Factoring Patterns
3. Use Factoring to Solve Equations
4. Solve Word Problems That Involve Factoring

1 Factor Perfect-Square Trinomials

Before we summarize our work with factoring techniques, let's look at two more special factoring patterns. These patterns emerge when multiplying binomials. Consider the following examples.

$$(x + 5)^2 = (x + 5)(x + 5) = x^2 + 10x + 25$$
$$(2x + 3)^2 = (2x + 3)(2x + 3) = 4x^2 + 12x + 9$$
$$(4x + 7)^2 = (4x + 7)(4x + 7) = 16x^2 + 56x + 49$$

In general, $(a + b)^2 = (a + b)(a + b) = a^2 + 2ab + b^2$. Also,
$$(x - 6)^2 = (x - 6)(x - 6) = x^2 - 12x + 36$$
$$(3x - 4)^2 = (3x - 4)(3x - 4) = 9x^2 - 24x + 16$$
$$(5x - 2)^2 = (5x - 2)(5x - 2) = 25x^2 - 20x + 4$$

In general, $(a - b)^2 = (a - b)(a - b) = a^2 - 2ab + b^2$. Thus we have the following patterns.

> **Perfect-Square Trinomials**
> $$a^2 + 2ab + b^2 = (a + b)^2$$
> $$a^2 - 2ab + b^2 = (a - b)^2$$

Trinomials of the form $a^2 + 2ab + b^2$ or $a^2 - 2ab + b^2$ are called **perfect-square trinomials**. They are easy to recognize because of the nature of their terms. For example, $9x^2 + 30x + 25$ is a perfect-square trinomial for these reasons:

1. The first term is a square: $(3x)^2$
2. The last term is a square: $(5)^2$
3. The middle term is twice the product of the quantities being squared in the first and last terms: $2(3x)(5)$

Likewise, $25x^2 - 40xy + 16y^2$ is a perfect-square trinomial for these reasons:

1. The first term is a square: $(5x)^2$
2. The last term is a square: $(4y)^2$
3. The middle term is twice the product of the quantities being squared in the first and last terms: $2(5x)(4y)$

Once we know that we have a perfect-square trinomial, the factoring process follows immediately from the two basic patterns.
$$9x^2 + 30x + 25 = (3x + 5)^2$$
$$25x^2 - 40xy + 16y^2 = (5x - 4y)^2$$

Here are some additional examples of perfect-square trinomials and their factored forms.

EXAMPLE 1

Factor the following.

(a) $x^2 - 16x + 64$ (b) $16x^2 - 56x + 49$ (c) $25x^2 + 20xy + 4y^2$
(d) $1 + 6y + 9y^2$ (e) $4m^2 - 4mn + n^2$

Solution

(a) $x^2 - 16x + 64 = (x)^2 - 2(x)(8) + (8)^2 = (x - 8)^2$
(b) $16x^2 - 56x + 49 = (4x)^2 - 2(4x)(7) + (7)^2 = (4x - 7)^2$
(c) $25x^2 + 20xy + 4y^2 = (5x)^2 + 2(5x)(2y) + (2y)^2 = (5x + 2y)^2$
(d) $1 + 6y + 9y^2 = (1)^2 + 2(1)(3y) + (3y)^2 = (1 + 3y)^2$
(e) $4m^2 - 4mn + n^2 = (2m)^2 - 2(2m)(n) + (n)^2 = (2m - n)^2$

You may want to do the middle part mentally, after you feel comfortable with the process. ∎

2 Recognize the Different Types of Factoring Patterns

In this chapter, we have considered some basic factoring techniques one at a time, but you must be able to apply them as needed in a variety of situations. Let's first summarize the techniques and then consider some examples. These are the techniques we have discussed in this chapter:

1. Factoring by using the distributive property to factor out the greatest common monomial or binomial factor
2. Factoring by grouping
3. Factoring by applying the difference-of-squares pattern
4. Factoring by applying the perfect-square-trinomial pattern
5. Factoring trinomials of the form $x^2 + bx + c$ into the product of two binomials
6. Factoring trinomials of the form $ax^2 + bx + c$ into the product of two binomials

As a general guideline, **always look for a greatest common monomial factor first**, and then proceed with the other factoring techniques.

In each of the following examples, we have factored completely whenever possible. Study these examples carefully and note the factoring techniques we have used.

EXAMPLE 2

Factor completely $2x^2 + 12x + 10$.

Solution

$2x^2 + 12x + 10 = 2(x^2 + 6x + 5)$ First, factor out the common factor of 2
$ = 2(x + 1)(x + 5)$

EXAMPLE 3

Factor completely $4x^2 + 36$.

Solution

$4x^2 + 36 = 4(x^2 + 9)$ Remember that the *sum* of two squares is not factorable using integers unless there is a common factor

EXAMPLE 4

Factor completely $4t^2 + 20t + 25$.

Solution

$4t^2 + 20t + 25 = (2t + 5)^2$ If you fail to recognize a perfect-square trinomial, no harm is done; simply proceed to factor into the product of two binomials, and then you will recognize that the two binomials are the same

EXAMPLE 5

Factor completely $x^2 - 3x - 8$.

Solution

$x^2 - 3x - 8$ is not factorable using integers. This becomes obvious from the table.

5.5 Factoring, Solving Equations, and Problem Solving

Product	Sum
$1(-8) = -8$	$1 + (-8) = -7$
$-1(8) = -8$	$-1 + 8 = 7$
$2(-4) = -8$	$2 + (-4) = -2$
$-2(4) = -8$	$-2 + 4 = 2$

No two factors of -8 produce a sum of -3.

EXAMPLE 6

Factor completely $6y^2 - 13y - 28$.

Solution

$$6y^2 - 13y - 28 = (2y - 7)(3y + 4).$$

We found the binomial factors as follows:

$(y + \underline{})(6y - \underline{})$

or

$(y - \underline{})(6y + \underline{})$

or

$(2y - \underline{})(3y + \underline{})$ ←

or

$(2y + \underline{})(3y - \underline{})$

$1 \cdot 28$ or $28 \cdot 1$
$2 \cdot 14$ or $14 \cdot 2$
$4 \cdot 7$ or $\boxed{7 \cdot 4}$

EXAMPLE 7

Factor completely $32x^2 - 50y^2$.

Solution

$32x^2 - 50y^2 = 2(16x^2 - 25y^2)$ First, factor out the common factor of 2
$= 2(4x + 5y)(4x - 5y)$ Factor the difference of squares

3 Use Factoring to Solve Equations

Each time we considered a new factoring technique in this chapter, we used that technique to help solve some equations. It is important that you be able to recognize which technique works for a particular type of equation.

EXAMPLE 8

Solve $x^2 = 25x$.

Solution

$$x^2 = 25x$$
$$x^2 - 25x = 0$$
$$x(x - 25) = 0$$
$$x = 0 \quad \text{or} \quad x - 25 = 0$$
$$x = 0 \quad \text{or} \quad x = 25$$

The solution set is $\{0, 25\}$. Check it!

EXAMPLE 9

Solve $x^3 - 36x = 0$.

Solution

$$x^3 - 36x = 0$$
$$x(x^2 - 36) = 0$$
$$x(x + 6)(x - 6) = 0$$
$$x = 0 \quad \text{or} \quad x + 6 = 0 \quad \text{or} \quad x - 6 = 0 \quad \text{If } abc = 0, \text{ then } a = 0$$
$$\text{or } b = 0 \text{ or } c = 0$$
$$x = 0 \quad \text{or} \quad x = -6 \quad \text{or} \quad x = 6$$

The solution set is $\{-6, 0, 6\}$. Does it check? ∎

EXAMPLE 10

Solve $10x^2 - 13x - 3 = 0$.

Solution

$$10x^2 - 13x - 3 = 0$$
$$(5x + 1)(2x - 3) = 0$$
$$5x + 1 = 0 \quad \text{or} \quad 2x - 3 = 0$$
$$5x = -1 \quad \text{or} \quad 2x = 3$$
$$x = -\frac{1}{5} \quad \text{or} \quad x = \frac{3}{2}$$

The solution set is $\left\{-\frac{1}{5}, \frac{3}{2}\right\}$. Does it check? ∎

EXAMPLE 11

Solve $4x^2 - 28x + 49 = 0$.

Solution

$$4x^2 - 28x + 49 = 0$$
$$(2x - 7)^2 = 0$$
$$(2x - 7)(2x - 7) = 0$$
$$2x - 7 = 0 \quad \text{or} \quad 2x - 7 = 0$$
$$2x = 7 \quad \text{or} \quad 2x = 7$$
$$x = \frac{7}{2} \quad \text{or} \quad x = \frac{7}{2}$$

The solution set is $\left\{\frac{7}{2}\right\}$. ∎

Pay special attention to the next example. We need to change the form of the original equation before we can apply the property $ab = 0$ if and only if $a = 0$ or $b = 0$. The unique feature of this property is that an indicated product is set equal to zero.

EXAMPLE 12

Solve $(x + 1)(x + 4) = 40$.

Solution

$$(x + 1)(x + 4) = 40$$
$$x^2 + 5x + 4 = 40 \quad \text{Multiply the binomials}$$

$$x^2 + 5x - 36 = 0 \quad \text{Added } -40 \text{ to each side}$$
$$(x + 9)(x - 4) = 0$$
$$x + 9 = 0 \quad \text{or} \quad x - 4 = 0$$
$$x = -9 \quad \text{or} \quad x = 4$$

The solution set is $\{-9, 4\}$. Check it!

EXAMPLE 13

Solve $2n^2 + 16n - 40 = 0$.

Solution

$$2n^2 + 16n - 40 = 0$$
$$2(n^2 + 8n - 20) = 0$$
$$n^2 + 8n - 20 = 0 \quad \text{Multiplied both sides by } \tfrac{1}{2}$$
$$(n + 10)(n - 2) = 0$$
$$n + 10 = 0 \quad \text{or} \quad n - 2 = 0$$
$$n = -10 \quad \text{or} \quad n = 2$$

The solution set is $\{-10, 2\}$. Does it check?

4 Solve Word Problems That Involve Factoring

The preface to this book states that a common thread throughout the book is *to learn a skill*, *to use that skill to help solve equations*, and then *to use equations to help solve problems*. This approach should be very apparent in this chapter. Our new factoring skills have provided more ways of solving equations, which in turn gives us more power to solve word problems. We conclude the chapter by solving a few more examples.

EXAMPLE 14

Apply Your Skill

Find two numbers whose product is 65 if one of the numbers is 3 more than twice the other number.

Solution

Let n represent one of the numbers; then $2n + 3$ represents the other number. Because their product is 65, we can set up and solve the following equation:

$$n(2n + 3) = 65$$
$$2n^2 + 3n - 65 = 0$$
$$(2n + 13)(n - 5) = 0$$
$$2n + 13 = 0 \quad \text{or} \quad n - 5 = 0$$
$$2n = -13 \quad \text{or} \quad n = 5$$
$$n = -\frac{13}{2} \quad \text{or} \quad n = 5$$

If $n = -\dfrac{13}{2}$, then $2n + 3 = 2\left(-\dfrac{13}{2}\right) + 3 = -10$. If $n = 5$, then $2n + 3 = 2(5) + 3 = 13$. Thus the numbers are $-\dfrac{13}{2}$ and -10, or 5 and 13.

EXAMPLE 15 — Apply Your Skill

The area of a triangular sheet of paper is 14 square inches. One side of the triangle is 3 inches longer than the altitude to that side. Find the length of the one side and the length of the altitude to that side.

Solution

Let h represent the altitude to the side. Then $h + 3$ represents the length of the side of the triangle (see Figure 5.5). Because the formula for finding the area of a triangle is $A = \frac{1}{2}bh$, we have

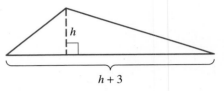

Figure 5.5

$$\frac{1}{2}h(h+3) = 14$$
$$h(h+3) = 28 \quad \text{Multiplied both sides by 2}$$
$$h^2 + 3h = 28$$
$$h^2 + 3h - 28 = 0$$
$$(h+7)(h-4) = 0$$
$$h+7 = 0 \quad \text{or} \quad h-4 = 0$$
$$h = -7 \quad \text{or} \quad h = 4$$

The solution -7 is not reasonable. Thus the altitude is 4 inches, and the length of the side to which that altitude is drawn is 7 inches. ∎

CONCEPT QUIZ 5.5

For Problems 1–7, match each factoring problem with the type of pattern that would be used to factor the problem.

1. $x^2 + 2xy + y^2$
2. $x^2 - y^2$
3. $ax + ay + bx + by$
4. $x^2 + bx + c$
5. $ax^2 + bx + c$
6. $ax^2 + ax + a$
7. $(a+b)x + (a+b)y$

A. Trinomial with an x-squared coefficient of one
B. Common binomial factor
C. Difference of two squares
D. Common factor
E. Factor by grouping
F. Perfect-square trinomial
G. Trinomial with an x-squared coefficient of not one

Section 5.5 Classroom Problem Set

Objective 1

1. Factor the following.
   (a) $a^2 + 10a + 25$
   (b) $36x^2 - 12x + 1$
   (c) $49m^2 + 56mn + 16n^2$

2. Factor the following.
   (a) $x^2 + 18x + 81$
   (b) $36a^2 - 84ax + 49$
   (c) $64x^2 + 16xy + y^2$

Objective 2

3. Factor completely $3x^2 - 6x - 72$.
4. Factor completely $2x^2 + 38x + 36$.
5. Factor completely $4x^2 + 100$.
6. Factor completely $2x^2 - 10$.
7. Factor completely $9x^2 - 30x + 25$.
8. Factor completely $36x^2 - 84x + 49$.
9. Factor completely $x^2 - 8x - 24$.
10. Factor completely $5x^2 - 5x - 6$.

11. Factor completely $3x^2 - 10x - 8$.
12. Factor completely $30x^2 - x - 1$.
13. Factor completely $4x^2 - 100$.
14. Factor completely $7x^3 - 7x$.

Objective 3

15. Solve $y^2 = 4y$.
16. Solve $-3x^2 - 24x = 0$.
17. Solve $z^3 - 25z = 0$.
18. Solve $-2x^3 + 8x = 0$.
19. Solve $3y^2 + 2y - 8 = 0$.
20. Solve $14n^2 - 19n - 3 = 0$.
21. Solve $25y^2 + 20y + 4 = 0$.
22. Solve $9x^2 - 42x + 49 = 0$.
23. Solve $(x + 1)(x + 2) = 20$.
24. Solve $(3n + 1)(n + 2) = 12$.
25. Solve $3y^2 - 12y - 36 = 0$.
26. Solve $x^3 = 6x^2$.

Objective 4

27. Find two numbers whose product is 36 if one of the numbers is 1 more than twice the other number.
28. Find two numbers whose product is -1. One of the numbers is 3 more than twice the other number.
29. The area of a triangular piece of glass is 30 square inches. One side of the triangle is 4 inches longer than the altitude to that side. Find the length of that side and the length of the altitude to that side.
30. The sum of the areas of two circles is 100π square centimeters. The length of a radius of the larger circle is 2 centimeters more than the length of a radius of the smaller circle. Find the length of a radius of each circle.

THOUGHTS INTO WORDS

1. When factoring polynomials, why do you think that it is best to look for a greatest common monomial factor first?
2. Explain how you would solve $(4x - 3)(8x + 5) = 0$ and also how you would solve $(4x - 3)(8x + 5) = -9$.
3. Explain how you would solve
$$(x + 2)(x + 3) = (x + 2)(3x - 1)$$
Do you see more than one approach to this problem?

Answers to the Concept Quiz

1. F or A   2. C   3. E   4. A   5. G   6. D   7. B

Chapter 5 Review Problem Set

For Problems 1–4, find the greatest common factor of the expressions.

1. $6x^2y^2$, $4x^3y$, and $14x^4y^3$
2. $9xy^2$, $21x^3y^2$, and $15x^4y^3$
3. $24a^3b^2$, $8a^2b^3$, and $12ab^3$
4. $36m^5n^3$, $24m^3n^4$, and $60m^4n^2$

For Problems 5–8, factor out the greatest common factor.

5. $15x^4 + 21x^2$
6. $24a^3 - 20a^6$
7. $10x^4y - 50x^3y^2 + 5x^2y^3$
8. $12m^2n + 20m^3n^2 - 24m^4n^3$

For Problems 9–12, factor completely by grouping.

9. $ab - 4a + 3b - 12$
10. $2xy + 12x + 5y + 30$
11. $x^3 + x^2y + xy^2 + y^3$
12. $8m^2 + 2m + 12mn + 3n$

For Problems 13–16, factor the difference of squares. Be sure to factor completely.

13. $25x^2 - 16y^2$
14. $18x^2 - 50$
15. $x^4 - 16$
16. $x^3 - 49x$

For Problems 17–28, factor completely.

17. $x^2 - 11x + 24$
18. $x^2 + 13x + 12$
19. $x^2 - 5x - 24$
20. $x^2 + 8x - 20$
21. $2x^2 + 5x + 3$
22. $5x^2 + 17x + 6$
23. $4x^2 + 5x - 6$
24. $3x^2 - 10x - 8$
25. $9x^2 + 12x + 4$
26. $4x^2 - 20x + 25$
27. $16x^2 - 8x + 1$
28. $x^2 + 12xy + 36y^2$

For Problems 29–52, factor completely. Indicate any polynomials that are not factorable using integers.

29. $x^2 - 9x + 14$
30. $3x^2 + 21x$
31. $9x^2 - 4$
32. $4x^2 + 8x - 5$
33. $25x^2 - 60x + 36$
34. $n^3 + 13n^2 + 40n$
35. $y^2 + 11y - 12$
36. $3xy^2 + 6x^2y$
37. $x^4 - 1$
38. $18n^2 + 9n - 5$
39. $x^2 + 7x + 24$
40. $4x^2 - 3x - 7$
41. $3n^2 + 3n - 90$
42. $x^3 - xy^2$
43. $2x^2 + 3xy - 2y^2$
44. $4n^2 - 6n - 40$
45. $5x + 5y + ax + ay$
46. $21t^2 - 5t - 4$
47. $2x^3 - 2x$
48. $3x^3 - 108x$
49. $16x^2 + 40x + 25$
50. $xy - 3x - 2y + 6$
51. $15x^2 - 7xy - 2y^2$
52. $6n^4 - 5n^3 + n^2$

For Problems 53–72, solve each equation.

53. $x^2 + 4x - 12 = 0$
54. $x^2 = 11x$
55. $2x^2 + 3x - 20 = 0$
56. $9n^2 + 21n - 8 = 0$
57. $6n^2 = 24$
58. $16y^2 + 40y + 25 = 0$
59. $t^3 - t = 0$
60. $28x^2 + 71x + 18 = 0$
61. $x^2 + 3x - 28 = 0$
62. $(x - 2)(x + 2) = 21$
63. $5n^2 + 27n = 18$
64. $4n^2 + 10n = 14$
65. $2x^3 - 8x = 0$
66. $x^2 - 20x + 96 = 0$
67. $4t^2 + 17t - 15 = 0$
68. $3(x + 2) - x(x + 2) = 0$
69. $(2x - 5)(3x + 7) = 0$
70. $(x + 4)(x - 1) = 50$
71. $-7n - 2n^2 = -15$
72. $-23x + 6x^2 = -20$

Set up an equation and solve each of the following problems.

73. The larger of two numbers is one less than twice the smaller number. The difference of their squares is 33. Find the numbers.

74. The length of a rectangle is 2 centimeters less than five times the width of the rectangle. The area of the rectangle is 16 square centimeters. Find the length and width of the rectangle.

75. Suppose that the combined area of two squares is 104 square inches. Each side of the larger square is five times as long as a side of the smaller square. Find the size of each square.

76. The longer leg of a right triangle is one unit shorter than twice the length of the shorter leg. The hypotenuse is one unit longer than twice the length of the shorter leg. Find the lengths of the three sides of the triangle.

77. The product of two numbers is 26, and one of the numbers is one larger than six times the other number. Find the numbers.

78. Find three consecutive positive odd whole numbers such that the sum of the squares of the two smaller numbers is nine more than the square of the largest number.

79. The number of books per shelf in a bookcase is one less than nine times the number of shelves. If the bookcase contains 140 books, find the number of shelves.

80. The combined area of a square and a rectangle is 225 square yards. The length of the rectangle is eight times the width of the rectangle, and the length of a side of the square is the same as the width of the rectangle. Find the dimensions of the square and the rectangle.

81. Suppose that we want to find two consecutive integers such that the sum of their squares is 613. What are they?

82. If numerically the volume of a cube equals the total surface area of the cube, find the length of an edge of the cube.

83. The combined area of two circles is 53π square meters. The length of a radius of the larger circle is 1 meter more than three times the length of a radius of the smaller circle. Find the length of a radius of each circle.

84. The product of two consecutive odd whole numbers is one less than five times their sum. Find the numbers.

Chapter 5 Practice Test

For Problems 1–10, factor each expression completely.

1. $x^2 + 3x - 10$
2. $x^2 - 5x - 24$
3. $2x^3 - 2x$
4. $x^2 + 21x + 108$
5. $18n^2 + 21n + 6$
6. $ax + ay + 2bx + 2by$
7. $4x^2 + 17x - 15$
8. $6x^2 + 24$
9. $30x^3 - 76x^2 + 48x$
10. $28 + 13x - 6x^2$

For Problems 11–21, solve each equation.

11. $7x^2 = 63$
12. $x^2 + 5x - 6 = 0$
13. $4n^2 = 32n$
14. $(3x - 2)(2x + 5) = 0$
15. $(x - 3)(x + 7) = -9$
16. $x^3 + 16x^2 + 48x = 0$
17. $9(x - 5) - x(x - 5) = 0$
18. $3t^2 + 35t = 12$
19. $8 - 10x - 3x^2 = 0$
20. $3x^3 = 75x$
21. $25n^2 - 70n + 49 = 0$

For Problems 22–25, set up an equation and solve each problem.

22. The length of a rectangle is 2 inches less than twice its width. If the area of the rectangle is 112 square inches, find the length of the rectangle.

23. The length of one leg of a right triangle is 4 centimeters more than the length of the other leg. The length of the hypotenuse is 8 centimeters more than the length of the shorter leg. Find the length of the shorter leg.

24. A room contains 112 chairs. The number of chairs per row is five less than three times the number of rows. Find the number of chairs per row.

25. Suppose the sum of the squares of three consecutive integers is 77. Find the integers.

6 Rational Expressions and Rational Equations

- 6.1 Simplifying Rational Expressions
- 6.2 Multiplying and Dividing Rational Expressions
- 6.3 Adding and Subtracting Rational Expressions
- 6.4 More on Addition and Subtraction of Rational Expressions
- 6.5 Rational Equations and Problem Solving
- 6.6 More Rational Equations and Problem Solving

Chapter 6 Warm-Up Problems

1. Simplify.

    (a) $\dfrac{5}{9} + \dfrac{6}{9}$ (b) $\dfrac{3x}{11} - \left(-\dfrac{8x}{11}\right)$ (c) $\dfrac{9}{14} - \dfrac{8}{21}$

2. Find the least common denominators (LCDs).

    (a) $\dfrac{11}{32} + \dfrac{5}{18}$ (b) $\dfrac{1}{6} - \dfrac{2}{21}$ (c) $\dfrac{1}{4x^2y} + \dfrac{1}{6xy}$

3. Reduce the fraction to lowest terms.

    (a) $\dfrac{8x}{26x}$ (b) $\dfrac{-56m}{77mn}$ (c) $\dfrac{-22x^3y^4}{-46xy^5}$

4. Divide.

    (a) $\dfrac{3}{4} \div \dfrac{5}{12}$ (b) $\dfrac{4}{x} \div \dfrac{6}{x^2}$ (c) $\dfrac{-14x^2y}{21xy}$

5. Use the distributive property to simplify.

    (a) $20\left(\dfrac{1}{4}x + \dfrac{3}{5}\right)$ (b) $18\left(\dfrac{3}{2} - \dfrac{1}{3}\right)$ (c) $4x\left(\dfrac{1}{2x} - \dfrac{9}{4x}\right)$

6. Simplify.

    (a) $5x + 3 - 4x + 1$ (b) $(6x + 1) - (2x + 3)$ (c) $\dfrac{3(m-1) - (m-1)}{2}$

7. Factor.

    (a) $x^2 + 7x - 30$ (b) $12x^2 - 5x - 3$ (c) $12x^2y + 8xy^2$

8. (a) What is the reciprocal of $\dfrac{57}{20}$?

    (b) How long would it take a car to travel 260 miles at a rate of 65 miles per hour?

    (c) Write the expression for how fast a train would be traveling if it went 570 miles in x hours.

I n Chapter 1 our study of common fractions led naturally to some work with simple algebraic fractions. Then in Chapters 4 and 5 we discussed the basic operations that pertain to polynomials. Now we can use some ideas about polynomials—specifically the factoring techniques—to expand our study of rational expressions. This, in turn, gives us more techniques for solving equations, which increases our problem-solving capabilities.

Video tutorials for all section learning objectives are available in a variety of delivery modes.

INTERNET PROJECT

Because fractions in this chapter have denominators that contain variables, certain values for the variable are excluded so the denominator does not equal zero. Zero is a very special number. Do an Internet search on the history of zero to discover the two important but different uses of zero. What Indian mathematician was one of the first to discover some of the properties of zero? Was he able to discover properties regarding division by zero?

6.1 Simplifying Rational Expressions

OBJECTIVES

1. Simplify Rational Expressions
2. Simplify Rational Expressions Involving Opposites

1 Simplify Rational Expressions

If the numerator and denominator of a fraction are polynomials, then we call the fraction an **algebraic fraction** or a **rational expression**. Here are some examples of rational expressions.

$$\frac{4}{x-2} \qquad \frac{x^2+2x-4}{x^2-9} \qquad \frac{y+x^2}{xy-3} \qquad \frac{x^3+2x^2-3x-4}{x^2-2x-6}$$

Because we must avoid division by zero, no values can be assigned to variables that create a denominator of zero. Thus the fraction $\frac{4}{x-2}$ is meaningful for all real number values of x except for $x = 2$. Rather than making a restriction for each individual fraction, we will simply assume that all denominators represent nonzero real numbers.

Recall that the **fundamental principle of fractions** $\left(\frac{ak}{bk} = \frac{a}{b}\right)$ provides the basis for expressing fractions in reduced (or simplified) form, as the next examples demonstrate.

$$\frac{18}{24} = \frac{3 \cdot 6}{4 \cdot 6} = \frac{3}{4} \qquad \frac{-42xy}{77y} = -\frac{2 \cdot 3 \cdot 7 \cdot x \cdot y}{7 \cdot 11 \cdot y} = -\frac{6x}{11}$$

$$\frac{15x}{25x} = \frac{3 \cdot 5 \cdot x}{5 \cdot 5 \cdot x} = \frac{3}{5} \qquad \frac{28x^2y^2}{-63x^2y^3} = -\frac{4 \cdot 7 \cdot x^2 \cdot y^2}{9 \cdot 7 \cdot x^2 \cdot y^3} = -\frac{4}{9y}$$

We can use the factoring techniques from Chapter 5 to factor numerators and/or denominators so that we can apply the fundamental principle of fractions. The following examples should clarify this process.

EXAMPLE 1 Simplify $\dfrac{x^2+6x}{x^2-36}$.

Solution

$$\frac{x^2+6x}{x^2-36} = \frac{x(x+6)}{(x-6)(x+6)} = \frac{x}{x-6}$$

∎

EXAMPLE 2

Simplify $\dfrac{a+2}{a^2+4a+4}$.

Solution

$$\dfrac{a+2}{a^2+4a+4} = \dfrac{1(a+2)}{(a+2)(a+2)} = \dfrac{1}{a+2}$$

EXAMPLE 3

Simplify $\dfrac{x^2+4x-21}{2x^2+15x+7}$.

Solution

$$\dfrac{x^2+4x-21}{2x^2+15x+7} = \dfrac{(x-3)(x+7)}{(2x+1)(x+7)} = \dfrac{x-3}{2x+1}$$

EXAMPLE 4

Simplify $\dfrac{a^2b+ab^2}{ab+b^2}$.

Solution

$$\dfrac{a^2b+ab^2}{ab+b^2} = \dfrac{ab(a+b)}{b(a+b)} = a$$

EXAMPLE 5

Simplify $\dfrac{4x^3y-36xy}{2x^2-4x-30}$.

Solution

$$\dfrac{4x^3y-36xy}{2x^2-4x-30} = \dfrac{4xy(x^2-9)}{2(x^2-2x-15)}$$

$$= \dfrac{\overset{2}{4}xy(x+3)(x-3)}{2(x-5)(x+3)}$$

$$= \dfrac{2xy(x-3)}{x-5}$$

Notice in Example 5 that we left the numerator of the final fraction in factored form. We do this when polynomials other than monomials are involved. Either $\dfrac{2xy(x-3)}{x-5}$ or $\dfrac{2x^2y-6xy}{x-5}$ is an acceptable answer.

EXAMPLE 6

Simplify $\dfrac{-3x+9}{x^2-3x}$.

Solution

$$\dfrac{-3x+9}{x^2-3x} = \dfrac{-3(x-3)}{x(x-3)} \quad \text{In the numerator, factor out a common factor of } -3$$

$$= \dfrac{-3(x-3)}{x(x-3)}$$

$$= \dfrac{-3}{x} = -\dfrac{3}{x}$$

2 Simplify Rational Expressions Involving Opposites

Remember that the quotient of any nonzero real number and its opposite is -1. For example, $\dfrac{7}{-7} = -1$ and $\dfrac{-9}{9} = -1$. Likewise, the indicated quotient of any polynomial and its opposite is equal to -1. Consider these examples.

$\dfrac{x}{-x} = -1$ because x and $-x$ are opposites.

$\dfrac{x-y}{y-x} = -1$ because $x-y$ and $y-x$ are opposites.

$\dfrac{a^2-9}{9-a^2} = -1$ because a^2-9 and $9-a^2$ are opposites.

Use this property to simplify rational expressions in the final examples of this section.

EXAMPLE 7

Simplify $\dfrac{14-7n}{n-2}$.

Solution

$$\dfrac{14-7n}{n-2} = \dfrac{7(2-n)}{n-2}$$

$$= 7(-1) \qquad \dfrac{2-n}{n-2} = -1$$

$$= -7$$

EXAMPLE 8

Simplify $\dfrac{x^2+4x-21}{15-2x-x^2}$.

Solution

$$\dfrac{x^2+4x-21}{15-2x-x^2} = \dfrac{(x+7)(x-3)}{(5+x)(3-x)}$$

$$= \left(\dfrac{x+7}{x+5}\right)(-1) \qquad \dfrac{x-3}{3-x} = -1$$

$$= -\dfrac{x+7}{x+5} \quad \text{or} \quad \dfrac{-x-7}{x+5}$$

CONCEPT QUIZ 6.1

For Problems 1–5, answer true or false.

1. A fraction in which both numerator and denominator are polynomials is called a polynomial fraction.
2. For rational expressions, no values can be assigned to variables that create a denominator of zero.
3. The fundamental principle of fractions is applied to reduce rational expressions.
4. The rational expression $\dfrac{x-2}{x+2}$ simplifies to -1.
5. The rational expression $\dfrac{5-a}{a-5}$ simplifies to -1.

Section 6.1 Classroom Problem Set

Objective 1

1. Simplify $\dfrac{2a^2 + 8a}{a^2 - 16}$.

2. Simplify $\dfrac{x^2 - 1}{3x^2 - 3x}$.

3. Simplify $\dfrac{x - 3}{x^2 - 5x + 6}$.

4. Simplify $\dfrac{n^2 + 2n}{n^2 + 3n + 2}$.

5. Simplify $\dfrac{a^2 + a - 20}{2a^2 + 11a + 5}$.

6. Simplify $\dfrac{3n^2 - 10n - 8}{n^2 - 16}$.

7. Simplify $\dfrac{xy^2 + x^2y}{x^2 + xy}$.

8. Simplify $\dfrac{ab^3 + a^2b}{b^2 + a}$.

9. Simplify $\dfrac{5a^2 - 8a - 4}{3a^3b - 12ab}$.

10. Simplify $\dfrac{18(x + 2)^3}{16(x + 2)^2}$.

11. Simplify $\dfrac{-5x - 10}{6x + 12}$.

12. Simplify $\dfrac{6n - 60}{n^2 - 20n + 100}$.

Objective 2

13. Simplify $\dfrac{20 - 5y}{y - 4}$.

14. Simplify $\dfrac{5x - 40}{80 - 10x}$.

15. Simplify $\dfrac{12 + a - a^2}{a^2 - 2a - 8}$.

16. Simplify $\dfrac{x^2 + 7x - 18}{12 - 4x - x^2}$.

THOUGHTS INTO WORDS

1. Explain the role that factoring plays in simplifying rational expressions.

2. Which of the following simplification processes are correct? Explain your answer.

$\dfrac{2x}{x} = 2 \qquad \dfrac{x + 2}{x} = 2 \qquad \dfrac{x(x + 2)}{x} = x + 2$

Answers to the Concept Quiz

1. False   2. True   3. True   4. False   5. True

6.2 Multiplying and Dividing Rational Expressions

OBJECTIVES

1. Multiply Rational Expressions
2. Divide Rational Expressions

1 Multiply Rational Expressions

In Chapter 1 we defined the product of two rational numbers as $\dfrac{a}{b} \cdot \dfrac{c}{d} = \dfrac{ac}{bd}$. This definition extends to rational expressions in general.

6.2 Multiplying and Dividing Rational Expressions

Definition 6.1

If $\dfrac{A}{B}$ and $\dfrac{C}{D}$ are rational expressions, with $B \neq 0$ and $D \neq 0$, then

$$\frac{A}{B} \cdot \frac{C}{D} = \frac{AC}{BD}$$

In other words, to multiply rational expressions we multiply the numerators, multiply the denominators, and *express the product in simplified form*. The following examples illustrate this concept.

1. $\dfrac{2x}{3y} \cdot \dfrac{5y}{4x} = \dfrac{2 \cdot 5 \cdot x \cdot y}{3 \cdot 4 \cdot x \cdot y} = \dfrac{5}{6}$

2. $\dfrac{4a}{6b} \cdot \dfrac{8b}{12a^2} = \dfrac{4 \cdot 8 \cdot a \cdot b}{6 \cdot 12 \cdot a^2 \cdot b} = \dfrac{4}{9a}$

3. $\dfrac{-9x^2}{15xy} \cdot \dfrac{5y^2}{7x^2y^3} = -\dfrac{9 \cdot 5 \cdot x^2 \cdot y^2}{15 \cdot 7 \cdot x^3 \cdot y^4} = -\dfrac{3}{7xy^2}$

Notice that we used the commutative property of multiplication to rearrange factors in a more convenient form for recognizing common factors of the numerator and denominator

When multiplying rational expressions, we sometimes need to factor the numerators and/or denominators so that we can recognize common factors. Consider the next examples.

EXAMPLE 1

Multiply and simplify $\dfrac{x}{x^2 - 9} \cdot \dfrac{x+3}{y}$.

Solution

$$\frac{x}{x^2 - 9} \cdot \frac{x+3}{y} = \frac{x(x+3)}{(x+3)(x-3)(y)}$$

$$= \frac{x}{y(x-3)}$$

$\dfrac{x}{xy - 3y}$ is also an acceptable answer. ∎

Remember, when working with rational expressions, we are assuming that all denominators represent nonzero real numbers. Therefore, in Example 1, we are claiming that $\dfrac{x}{x^2 - 9} \cdot \dfrac{x+3}{y} = \dfrac{x}{y(x-3)}$ for all real numbers except -3 and 3 for x, and 0 for y.

EXAMPLE 2

Multiply and simplify $\dfrac{x}{x^2 + 2x} \cdot \dfrac{x^2 + 10x + 16}{5}$.

Solution

$$\frac{x}{x^2 + 2x} \cdot \frac{x^2 + 10x + 16}{5} = \frac{x(x+2)(x+8)}{x(x+2)(5)} = \frac{x+8}{5}$$ ∎

EXAMPLE 3 Multiply and simplify $\dfrac{a^2 - 3a}{a + 5} \cdot \dfrac{a^2 + 3a - 10}{a^2 - 5a + 6}$.

Solution

$$\dfrac{a^2 - 3a}{a + 5} \cdot \dfrac{a^2 + 3a - 10}{a^2 - 5a + 6} = \dfrac{a(a-3)(a+5)(a-2)}{(a+5)(a-2)(a-3)} = a$$

■

EXAMPLE 4 Multiply and simplify $\dfrac{6n^2 + 7n - 3}{n + 1} \cdot \dfrac{n^2 - 1}{2n^2 + 3n}$.

Solution

$$\dfrac{6n^2 + 7n - 3}{n + 1} \cdot \dfrac{n^2 - 1}{2n^2 + 3n} = \dfrac{(2n+3)(3n-1)(n+1)(n-1)}{(n+1)(n)(2n+3)}$$

$$= \dfrac{(3n-1)(n-1)}{n}$$

■

2 Divide Rational Expressions

Recall that to divide two rational numbers in $\dfrac{a}{b}$ form, we invert the divisor and multiply. Symbolically we express this as $\dfrac{a}{b} \div \dfrac{c}{d} = \dfrac{a}{b} \cdot \dfrac{d}{c}$. Furthermore, we call the numbers $\dfrac{c}{d}$ and $\dfrac{d}{c}$ *reciprocals* of each other because their product is 1. Thus we can also describe division as *to divide by a fraction, multiply by its reciprocal*. We define division of rational expressions in the same way using the same vocabulary.

Definition 6.2

If $\dfrac{A}{B}$ and $\dfrac{C}{D}$ are rational expressions, with $B \neq 0$, $D \neq 0$, and $C \neq 0$, then

$$\dfrac{A}{B} \div \dfrac{C}{D} = \dfrac{A}{B} \cdot \dfrac{D}{C} = \dfrac{AD}{BC}$$

Consider some examples.

1. $\dfrac{4x}{7y} \div \dfrac{6x^2}{14y^2} = \dfrac{4x}{7y} \cdot \dfrac{14y^2}{6x^2} = \dfrac{4 \cdot 14 \cdot x \cdot y^2}{7 \cdot 6 \cdot x^2 \cdot y} = \dfrac{4y}{3x}$

2. $\dfrac{-8ab}{9b} \div \dfrac{18a^3}{15a^2b} = \dfrac{-8ab}{9b} \cdot \dfrac{15a^2b}{18a^3} = -\dfrac{8 \cdot 15 \cdot a^3 \cdot b^2}{9 \cdot 18 \cdot a^3 \cdot b} = -\dfrac{20b}{27}$

3. $\dfrac{x^2y^3}{4ab} \div \dfrac{5xy^2}{-9a^2b} = \dfrac{x^2y^3}{4ab} \cdot \dfrac{-9a^2b}{5xy^2} = -\dfrac{9 \cdot x^2 \cdot y^3 \cdot a^2 \cdot b}{4 \cdot 5 \cdot a \cdot b \cdot x \cdot y^2} = -\dfrac{9axy}{20}$

The key idea when dividing fractions is *first* to convert to an equivalent multiplication problem and *second* to factor numerator and denominator completely and look for common factors.

EXAMPLE 5

Divide and simplify $\dfrac{x^2 - 4x}{xy} \div \dfrac{x^2 - 16}{y^3 + y^2}$.

Solution

$$\frac{x^2 - 4x}{xy} \div \frac{x^2 - 16}{y^3 + y^2} = \frac{x^2 - 4x}{xy} \cdot \frac{y^3 + y^2}{x^2 - 16}$$

$$= \frac{\cancel{x}(x - 4)(\cancel{y^2}^y)(y + 1)}{\cancel{xy}(x + 4)(x - 4)}$$

$$= \frac{y(y + 1)}{x + 4}$$

EXAMPLE 6

Divide and simplify $\dfrac{a^2 + 3a - 18}{a^2 + 4} \div \dfrac{1}{3a^2 + 12}$.

Solution

$$\frac{a^2 + 3a - 18}{a^2 + 4} \div \frac{1}{3a^2 + 12} = \frac{a^2 + 3a - 18}{a^2 + 4} \cdot \frac{3a^2 + 12}{1}$$

$$= \frac{(a + 6)(a - 3)(3)(\cancel{a^2 + 4})}{\cancel{a^2 + 4}}$$

$$= 3(a + 6)(a - 3)$$

EXAMPLE 7

Divide and simplify $\dfrac{2n^2 - 7n - 4}{6n^2 + 7n + 2} \div (n - 4)$.

Solution

$$\frac{2n^2 - 7n - 4}{6n^2 + 7n + 2} \div (n - 4) = \frac{2n^2 - 7n - 4}{6n^2 + 7n + 2} \cdot \frac{1}{n - 4}$$

$$= \frac{\cancel{(2n + 1)}\cancel{(n - 4)}}{\cancel{(2n + 1)}(3n + 2)\cancel{(n - 4)}}$$

$$= \frac{1}{3n + 2}$$

In a problem such as Example 7, it may be helpful to write the divisor with a denominator of 1. Thus we can write $n - 4$ as $\dfrac{n - 4}{1}$; its reciprocal then is obviously $\dfrac{1}{n - 4}$.

CONCEPT QUIZ 6.2

For Problems 1–5, answer true or false.

1. To multiply rational expressions we multiply the numerators, multiply the denominators, and express the result in simplified form.
2. To divide by a fraction, multiply by its reciprocal.
3. The reciprocal of $\dfrac{a}{b}$ is $-\dfrac{b}{a}$.
4. The product of two numbers that are reciprocals is 1.
5. The reciprocal of $\dfrac{1}{x}$ is x.

Section 6.2 Classroom Problem Set

Objective 1

1. Multiply and simplify $\dfrac{a}{a^2 - 5a + 4} \cdot \dfrac{a - 4}{b}$.

2. Multiply and simplify $\dfrac{2x^2 + xy}{xy} \cdot \dfrac{y}{10x + 5y}$.

3. Multiply and simplify $\dfrac{y^2 - 4y - 21}{2} \cdot \dfrac{y}{y^2 + 3y}$.

4. Multiply and simplify $\dfrac{x^2 + 15x + 54}{x^2 + 2} \cdot \dfrac{3x^2 + 6}{x^2 + 10x + 9}$.

5. Multiply and simplify $\dfrac{2x - 6}{3x + 12} \cdot \dfrac{x^2 + 5x + 4}{x^2 - 2x - 3}$.

6. Multiply and simplify $\dfrac{2a^2 - 11a - 21}{3a^2 + a} \cdot \dfrac{3a^2 - 11a - 4}{2a^2 - 5a - 12}$.

7. Multiply and simplify $\dfrac{y^2 - 25}{2y^2 - 7y - 15} \cdot \dfrac{2y^2 + y}{y + 5}$.

8. Multiply and simplify $\dfrac{n^3 - n}{n^2 + 7n + 6} \cdot \dfrac{4n + 24}{n^2 - n}$.

Objective 2

9. Divide and simplify $\dfrac{5a + 15}{20b} \div \dfrac{a^2 - 9}{ab + 4b}$.

10. Divide and simplify $\dfrac{6xy}{4xy + 4y^2} \div \dfrac{7x - 7y}{x^2 - y^2}$.

11. Divide and simplify $\dfrac{x^2 + 4x - 12}{2x + 3} \div \dfrac{1}{4x + 6}$.

12. Divide and simplify $\dfrac{x^2 + 4xy + 4y^2}{x^2} \div \dfrac{x^2 - 4y^2}{x^2 - 2xy}$.

13. Divide and simplify $\dfrac{3x^2 - 5x + 2}{3x^2 + x - 2} \div (x - 1)$.

14. Divide and simplify $\dfrac{2x^2 - xy - 3y^2}{(x + y)^2} \div \dfrac{4x^2 - 12xy + 9y^2}{10x - 15y}$.

THOUGHTS INTO WORDS

1. Give a step-by-step description of how to do the following multiplication problem:
$$\dfrac{x^2 - x}{x^2 - 1} \cdot \dfrac{x^2 + x - 6}{x^2 + 4x - 12}$$

2. Is $\left(\dfrac{x}{x + 1} \div \dfrac{x - 1}{x}\right) \div \dfrac{1}{x} = \dfrac{x}{x + 1} \div \left(\dfrac{x - 1}{x} \div \dfrac{1}{x}\right)$? Justify your answer.

3. Explain why the quotient $\dfrac{x - 2}{x + 1} \div \dfrac{x}{x - 1}$ is undefined for $x = -1$, $x = 1$, and $x = 0$ but is defined for $x = 2$.

Answers to the Concept Quiz

1. True 2. True 3. False 4. True 5. True

6.3 Adding and Subtracting Rational Expressions

OBJECTIVES

1. Add or Subtract Rational Expressions with Like Denominators
2. Add or Subtract Rational Expressions with Unlike Denominators

1 Add or Subtract Rational Expressions with Like Denominators

In Chapter 1 we defined addition and subtraction of rational numbers as $\frac{a}{b} + \frac{c}{b} = \frac{a+c}{b}$ and $\frac{a}{b} - \frac{c}{b} = \frac{a-c}{b}$, respectively. These definitions extend to rational expressions in general.

Definition 6.3

If $\frac{A}{B}$ and $\frac{C}{B}$ are rational expressions, with $B \neq 0$, then

$$\frac{A}{B} + \frac{C}{B} = \frac{A+C}{B} \quad \text{and} \quad \frac{A}{B} - \frac{C}{B} = \frac{A-C}{B}$$

Thus if the denominators of two rational expressions are the same, then we can add or subtract the fractions by adding or subtracting the numerators and placing the result over the common denominator. Here are some examples:

$$\frac{5}{x} + \frac{7}{x} = \frac{5+7}{x} = \frac{12}{x}$$

$$\frac{8}{xy} - \frac{3}{xy} = \frac{8-3}{xy} = \frac{5}{xy}$$

$$\frac{14}{2x+1} + \frac{15}{2x+1} = \frac{14+15}{2x+1} = \frac{29}{2x+1}$$

$$\frac{3}{a-1} - \frac{4}{a-1} = \frac{3-4}{a-1} = \frac{-1}{a-1} \quad \text{or} \quad -\frac{1}{a-1}$$

In the next example, notice how we put to use our previous work with simplifying polynomials.

EXAMPLE 1 Add or subtract as indicated.

(a) $\dfrac{x+3}{4} + \dfrac{2x-3}{4}$ (b) $\dfrac{x+5}{7} - \dfrac{x+2}{7}$ (c) $\dfrac{3x+1}{xy} + \dfrac{2x+3}{xy}$

(d) $\dfrac{2(3n+1)}{n} - \dfrac{3(n-1)}{n}$

Solution

(a) $\dfrac{x+3}{4} + \dfrac{2x-3}{4} = \dfrac{(x+3) + (2x-3)}{4} = \dfrac{3x}{4}$

(b) $\dfrac{x+5}{7} - \dfrac{x+2}{7} = \dfrac{(x+5) - (x+2)}{7} = \dfrac{x+5-x-2}{7} = \dfrac{3}{7}$

(c) $\dfrac{3x+1}{xy} + \dfrac{2x+3}{xy} = \dfrac{(3x+1)+(2x+3)}{xy} = \dfrac{5x+4}{xy}$

(d) $\dfrac{2(3n+1)}{n} - \dfrac{3(n-1)}{n} = \dfrac{2(3n+1)-3(n-1)}{n}$

$= \dfrac{6n+2-3n+3}{n} = \dfrac{3n+5}{n}$ ∎

It may be necessary to simplify the fraction that results from adding or subtracting two fractions.

EXAMPLE 2

Add or subtract as indicated. Express your answer in simplest form.

(a) $\dfrac{4x-3}{8} + \dfrac{2x+3}{8}$ (b) $\dfrac{3n-1}{12} - \dfrac{n-5}{12}$ (c) $\dfrac{-2x+3}{x^2-4} + \dfrac{3x-1}{x^2-4}$

Solution

(a) $\dfrac{4x-3}{8} + \dfrac{2x+3}{8} = \dfrac{(4x-3)+(2x+3)}{8} = \dfrac{6x}{8} = \dfrac{3x}{4}$

(b) $\dfrac{3n-1}{12} - \dfrac{n-5}{12} = \dfrac{(3n-1)-(n-5)}{12} = \dfrac{3n-1-n+5}{12}$

$= \dfrac{2n+4}{12} = \dfrac{2(n+2)}{12} = \dfrac{n+2}{6}$

(c) $\dfrac{-2x+3}{x^2-4} + \dfrac{3x-1}{x^2-4} = \dfrac{(-2x+3)+(3x-1)}{x^2-4} = \dfrac{x+2}{x^2-4}$

$= \dfrac{\cancel{x+2}}{\cancel{(x+2)}(x-2)}$

$= \dfrac{1}{x-2}$ ∎

2 Add or Subtract Rational Expressions with Unlike Denominators

Recall that to add or subtract rational numbers with different denominators, we first change them to equivalent fractions that have a common denominator. In fact, we found that by using the least common denominator (LCD), our work was easier. Let's carefully review the process, because it will also work with algebraic fractions in general.

EXAMPLE 3

Add $\dfrac{3}{5} + \dfrac{1}{4}$.

Solution

By inspection, we see that the LCD is 20. Thus we can change both fractions to equivalent fractions that have a denominator of 20.

$\dfrac{3}{5} + \dfrac{1}{4} = \dfrac{3}{5}\left(\dfrac{4}{4}\right) + \dfrac{1}{4}\left(\dfrac{5}{5}\right) = \dfrac{12}{20} + \dfrac{5}{20} = \dfrac{17}{20}$

↑ ↑
Form Form
of 1 of 1

∎

EXAMPLE 4

Subtract $\dfrac{5}{18} - \dfrac{7}{24}$.

Solution

If we cannot find the LCD by inspection, then we can use the prime factorization forms.

$$\left.\begin{array}{l}18 = 2 \cdot 3 \cdot 3 \\ 24 = 2 \cdot 2 \cdot 2 \cdot 3\end{array}\right\} \longrightarrow \text{LCD} = 2 \cdot 2 \cdot 2 \cdot 3 \cdot 3 = 72$$

$$\dfrac{5}{18} - \dfrac{7}{24} = \dfrac{5}{18}\left(\dfrac{4}{4}\right) - \dfrac{7}{24}\left(\dfrac{3}{3}\right) = \dfrac{20}{72} - \dfrac{21}{72} = -\dfrac{1}{72}$$

Now let's consider adding and subtracting rational expressions with different denominators.

EXAMPLE 5

Add $\dfrac{x-2}{4} + \dfrac{3x+1}{3}$.

Solution

By inspection, we see that the LCD is 12.

$$\dfrac{x-2}{4} + \dfrac{3x+1}{3} = \left(\dfrac{x-2}{4}\right)\left(\dfrac{3}{3}\right) + \left(\dfrac{3x+1}{3}\right)\left(\dfrac{4}{4}\right)$$

$$= \dfrac{3(x-2)}{12} + \dfrac{4(3x+1)}{12}$$

$$= \dfrac{3(x-2) + 4(3x+1)}{12}$$

$$= \dfrac{3x - 6 + 12x + 4}{12}$$

$$= \dfrac{15x - 2}{12}$$

EXAMPLE 6

Subtract $\dfrac{n-2}{2} - \dfrac{n-6}{6}$.

Solution

By inspection, we see that the LCD is 6.

$$\dfrac{n-2}{2} - \dfrac{n-6}{6} = \left(\dfrac{n-2}{2}\right)\left(\dfrac{3}{3}\right) - \dfrac{n-6}{6}$$

$$= \dfrac{3(n-2)}{6} - \dfrac{(n-6)}{6}$$

$$= \dfrac{3(n-2) - (n-6)}{6}$$

$$= \dfrac{3n - 6 - n + 6}{6}$$

$$= \dfrac{2n}{6}$$

$$= \dfrac{n}{3}$$

Don't forget to simplify!

It does not create any serious difficulties when the denominators contain variables; our approach remains basically the same.

EXAMPLE 7

Add $\dfrac{3}{4x} + \dfrac{7}{3x}$.

Solution

By inspection, we see that the LCD is $12x$.

$$\dfrac{3}{4x} + \dfrac{7}{3x} = \dfrac{3}{4x}\left(\dfrac{3}{3}\right) + \dfrac{7}{3x}\left(\dfrac{4}{4}\right) = \dfrac{9}{12x} + \dfrac{28}{12x} = \dfrac{9+28}{12x} = \dfrac{37}{12x}$$

EXAMPLE 8

Subtract $\dfrac{11}{12x} - \dfrac{5}{14x}$.

Solution

$$\left.\begin{array}{l} 12x = 2 \cdot 2 \cdot 3 \cdot x \\ 14x = 2 \cdot 7 \cdot x \end{array}\right\} \longrightarrow \text{LCD} = 2 \cdot 2 \cdot 3 \cdot 7 \cdot x = 84x$$

$$\dfrac{11}{12x} - \dfrac{5}{14x} = \dfrac{11}{12x}\left(\dfrac{7}{7}\right) - \dfrac{5}{14x}\left(\dfrac{6}{6}\right)$$

$$= \dfrac{77}{84x} - \dfrac{30}{84x} = \dfrac{77 - 30}{84x} = \dfrac{47}{84x}$$

EXAMPLE 9

Add $\dfrac{2}{y} + \dfrac{4}{y-2}$.

Solution

By inspection, we see that the LCD is $y(y - 2)$.

$$\dfrac{2}{y} + \dfrac{4}{y-2} = \dfrac{2}{y}\underbrace{\left(\dfrac{y-2}{y-2}\right)}_{\text{Form of 1}} + \dfrac{4}{y-2}\underbrace{\left(\dfrac{y}{y}\right)}_{\text{Form of 1}}$$

$$= \dfrac{2(y-2)}{y(y-2)} + \dfrac{4y}{y(y-2)}$$

$$= \dfrac{2(y-2) + 4y}{y(y-2)}$$

$$= \dfrac{2y - 4 + 4y}{y(y-2)} = \dfrac{6y - 4}{y(y-2)}$$

Notice the final result in Example 9. The numerator, $6y - 4$, can be factored into $2(3y - 2)$. However, because this produces no common factors with the denominator, the fraction cannot be simplified. Thus the final answer can be left as $\dfrac{6y-4}{y(y-2)}$; it is also acceptable to express it as $\dfrac{2(3y-2)}{y(y-2)}$.

EXAMPLE 10

Subtract $\dfrac{4}{x+2} - \dfrac{7}{x+3}$.

Solution

By inspection, we see that the LCD is $(x+2)(x+3)$.

$$\dfrac{4}{x+2} - \dfrac{7}{x+3} = \left(\dfrac{4}{x+2}\right)\left(\dfrac{x+3}{x+3}\right) - \left(\dfrac{7}{x+3}\right)\left(\dfrac{x+2}{x+2}\right)$$

$$= \dfrac{4(x+3)}{(x+2)(x+3)} - \dfrac{7(x+2)}{(x+3)(x+2)}$$

$$= \dfrac{4(x+3) - 7(x+2)}{(x+2)(x+3)}$$

$$= \dfrac{4x + 12 - 7x - 14}{(x+2)(x+3)}$$

$$= \dfrac{-3x - 2}{(x+2)(x+3)}$$

CONCEPT QUIZ 6.3

For Problems 1–5, answer true or false.

1. To add two rational expressions, add the numerators and add the denominators.
2. To add or subtract two rational expressions, the expressions must have a common denominator.
3. After adding rational expressions, the result may need to be simplified.
4. To add or subtract rational expressions with different denominators, the expressions are changed to equivalent expressions that have common denominators.
5. When adding $\dfrac{5}{3n}$ and $\dfrac{7}{2n}$, the least common denominator is $6n^2$.

Section 6.3 Classroom Problem Set

Objective 1

1. Add or subtract as indicated.
   (a) $\dfrac{2x+4}{3} - \dfrac{x+1}{3}$
   (b) $\dfrac{5(y+6)}{3y} + \dfrac{2(4y-1)}{3y}$

2. Add or subtract as indicated.
   (a) $\dfrac{-3a-2}{4} - \dfrac{a+5}{4}$
   (b) $\dfrac{2(n-4)}{3n} + \dfrac{4(n+2)}{3n}$

3. Add or subtract as indicated.
   (a) $\dfrac{8x+4}{9} - \dfrac{5x+4}{9}$
   (b) $\dfrac{-3x+8}{x^2+5x+6} + \dfrac{4x-6}{x^2+5x+6}$

4. Add or subtract as indicated.
   (a) $\dfrac{a^2}{a+2} - \dfrac{4}{a+2}$
   (b) $\dfrac{2x+5}{x^2-2x-8} - \dfrac{x+9}{x^2-2x-8}$

Objective 2

5. Add $\dfrac{4}{7} + \dfrac{1}{3}$.

6. Add $\dfrac{1}{4} + \dfrac{2}{9}$.

7. Add $\dfrac{5}{12} + \dfrac{7}{30}$.

8. Subtract $\dfrac{3}{10} - \dfrac{7}{12}$.

9. Add $\dfrac{2x+7}{5} + \dfrac{x-4}{3}$.

10. Add $\dfrac{2n+3}{4} + \dfrac{4n-1}{7}$.

11. Subtract $\dfrac{5x+3}{6} - \dfrac{x+1}{2}$.

12. Subtract $\dfrac{5n-2}{12} - \dfrac{4n+7}{6}$.

13. Add $\dfrac{5}{2x} + \dfrac{1}{3x}$.

14. Subtract $\dfrac{11}{9y} - \dfrac{8}{15y}$.

15. Subtract $\dfrac{3}{8a} - \dfrac{7}{10a}$.

16. Add $\dfrac{7}{8x} + \dfrac{5}{12x}$.

17. Add $\dfrac{4}{n+1} + \dfrac{14}{n}$.

18. Add $\dfrac{9}{x} + \dfrac{4}{x-8}$.

19. Subtract $\dfrac{2}{a+5} - \dfrac{3}{a+4}$.

20. Subtract $\dfrac{5}{x-1} - \dfrac{4}{x+6}$.

THOUGHTS INTO WORDS

1. Give a step-by-step description of how to do this addition problem:

$$\dfrac{3x-1}{6} + \dfrac{2x+3}{9}$$

2. Why are $\dfrac{3}{x-2}$ and $\dfrac{3}{2-x}$ opposites? What should be the result of adding $\dfrac{3}{x-2}$ and $\dfrac{3}{2-x}$?

3. Suppose that your friend does an addition problem as follows:

$$\dfrac{5}{8} + \dfrac{7}{12} = \dfrac{5(12) + 8(7)}{8(12)} = \dfrac{60 + 56}{96} = \dfrac{116}{96} = \dfrac{29}{24}$$

Is this answer correct? What advice would you offer your friend?

Answers to the Concept Quiz

1. False 2. True 3. True 4. True 5. False

6.4 More on Addition and Subtraction of Rational Expressions

OBJECTIVES

1. Add or Subtract Rational Expressions
2. Simplify Complex Fractions
3. Write Rational Expressions to Express Rates

1 Add or Subtract Rational Expressions

In this section, we expand our work with adding and subtracting rational expressions, and we discuss the process of simplifying complex fractions. Before we begin, however, this seems like an appropriate time to offer a bit of advice regarding your study of algebra. Success in algebra depends on having a good understanding of the concepts as well as being able to perform the various computations. As for the computational work, you should adopt a carefully organized format that shows as many steps as you need in order to minimize the chances of making careless errors. Don't be eager to find shortcuts for certain computations before you have a thorough understanding of the steps involved in the process. This advice is especially appropriate at the beginning of this section.

Study Examples 1–4 very carefully. Note that the same basic procedure is followed when solving each problem:

Step 1 Factor the denominators.

Step 2 Find the LCD.

Step 3 Change each fraction to an equivalent fraction that has the LCD as its denominator.

Step 4 Combine the numerators and place over the LCD.

Step 5 Simplify by performing the addition or subtraction.

Step 6 Look for ways to reduce the resulting fraction.

EXAMPLE 1

Add $\dfrac{3}{x^2 + 2x} + \dfrac{5}{x}$.

Solution

1st denominator: $x^2 + 2x = x(x + 2)$
2nd denominator: x
$\longrightarrow$ LCD is $x(x + 2)$

$$\dfrac{3}{x^2 + 2x} + \dfrac{5}{x} = \underbrace{\dfrac{3}{x(x + 2)}}_{\substack{\text{This fraction} \\ \text{has the LCD} \\ \text{as its} \\ \text{denominator}}} + \dfrac{5}{x}\underbrace{\left(\dfrac{x + 2}{x + 2}\right)}_{\substack{\text{Form} \\ \text{of 1}}}$$

$$= \dfrac{3}{x(x + 2)} + \dfrac{5(x + 2)}{x(x + 2)} = \dfrac{3 + 5(x + 2)}{x(x + 2)}$$

$$= \dfrac{3 + 5x + 10}{x(x + 2)} = \dfrac{5x + 13}{x(x + 2)}$$

EXAMPLE 2

Subtract $\dfrac{4}{x^2 - 4} - \dfrac{1}{x - 2}$.

Solution

$x^2 - 4 = (x + 2)(x - 2)$
$x - 2 = x - 2$
$\longrightarrow$ LCD is $(x + 2)(x - 2)$

$$\dfrac{4}{x^2 - 4} - \dfrac{1}{x - 2} = \dfrac{4}{(x + 2)(x - 2)} - \left(\dfrac{1}{x - 2}\right)\left(\dfrac{x + 2}{x + 2}\right)$$

$$= \dfrac{4}{(x + 2)(x - 2)} - \dfrac{1(x + 2)}{(x + 2)(x - 2)}$$

$$= \dfrac{4 - 1(x + 2)}{(x + 2)(x - 2)} = \dfrac{4 - x - 2}{(x + 2)(x - 2)}$$

$$= \dfrac{-x + 2}{(x + 2)(x - 2)}$$

$$= \dfrac{-1(x - 2)}{(x + 2)(x - 2)} \quad \longrightarrow \quad \begin{array}{l}\text{Note the changing of} \\ -x + 2 \text{ to } -1(x - 2)\end{array}$$

$$= -\dfrac{1}{x + 2}$$

EXAMPLE 3

Add $\dfrac{2}{a^2 - 9} + \dfrac{3}{a^2 + 5a + 6}$.

Solution

$$\left.\begin{array}{l} a^2 - 9 = (a + 3)(a - 3) \\ a^2 + 5a + 6 = (a + 3)(a + 2) \end{array}\right\} \longrightarrow \text{LCD is } (a + 3)(a - 3)(a + 2).$$

$$\dfrac{2}{a^2 - 9} + \dfrac{3}{a^2 + 5a + 6}$$

$$= \left(\dfrac{2}{(a + 3)(a - 3)}\right)\underset{\underset{\text{Form of 1}}{\uparrow}}{\left(\dfrac{a + 2}{a + 2}\right)} + \left(\dfrac{3}{(a + 3)(a + 2)}\right)\underset{\underset{\text{Form of 1}}{\uparrow}}{\left(\dfrac{a - 3}{a - 3}\right)}$$

$$= \dfrac{2(a + 2)}{(a + 3)(a - 3)(a + 2)} + \dfrac{3(a - 3)}{(a + 3)(a - 3)(a + 2)}$$

$$= \dfrac{2(a + 2) + 3(a - 3)}{(a + 3)(a - 3)(a + 2)} = \dfrac{2a + 4 + 3a - 9}{(a + 3)(a - 3)(a + 2)}$$

$$= \dfrac{5a - 5}{(a + 3)(a - 3)(a + 2)} \quad \text{or} \quad \dfrac{5(a - 1)}{(a + 3)(a - 3)(a + 2)} \quad \blacksquare$$

EXAMPLE 4

Perform the indicated operations.

$$\dfrac{2x}{x^2 - y^2} + \dfrac{3}{x + y} - \dfrac{2}{x - y}$$

Solution

$$\left.\begin{array}{l} x^2 - y^2 = (x + y)(x - y) \\ x + y = x + y \\ x - y = x - y \end{array}\right\} \longrightarrow \text{LCD is } (x + y)(x - y).$$

$$\dfrac{2x}{x^2 - y^2} + \dfrac{3}{x + y} - \dfrac{2}{x - y}$$

$$= \dfrac{2x}{(x + y)(x - y)} + \left(\dfrac{3}{x + y}\right)\left(\dfrac{x - y}{x - y}\right) - \left(\dfrac{2}{x - y}\right)\left(\dfrac{x + y}{x + y}\right)$$

$$= \dfrac{2x}{(x + y)(x - y)} + \dfrac{3(x - y)}{(x + y)(x - y)} - \dfrac{2(x + y)}{(x + y)(x - y)}$$

$$= \dfrac{2x + 3(x - y) - 2(x + y)}{(x + y)(x - y)}$$

$$= \dfrac{2x + 3x - 3y - 2x - 2y}{(x + y)(x - y)}$$

$$= \dfrac{3x - 5y}{(x + y)(x - y)} \quad \blacksquare$$

2 Simplify Complex Fractions

Fractional forms that contain fractions in the numerators and/or denominators are called **complex fractions**. Here are some examples of complex fractions:

6.4 More on Addition and Subtraction of Rational Expressions

$$\dfrac{\dfrac{2}{3}}{\dfrac{4}{5}} \qquad \dfrac{\dfrac{1}{x}}{\dfrac{3}{y}} \qquad \dfrac{\dfrac{1}{2}+\dfrac{1}{3}}{\dfrac{5}{6}-\dfrac{1}{4}} \qquad \dfrac{\dfrac{2}{x}+\dfrac{2}{y}}{\dfrac{5}{x}-\dfrac{1}{y^2}}$$

It is often necessary to *simplify* a complex fraction—that is, to express it as a simple fraction. We will illustrate this process with the next four examples.

EXAMPLE 5

Simplify $\dfrac{\dfrac{2}{3}}{\dfrac{4}{5}}$.

Solution

This type of problem creates no difficulty, because it is merely a division problem. Thus,

$$\dfrac{\dfrac{2}{3}}{\dfrac{4}{5}} = \dfrac{2}{3} \div \dfrac{4}{5} = \dfrac{\overset{1}{\cancel{2}}}{3} \cdot \dfrac{5}{\underset{2}{\cancel{4}}} = \dfrac{5}{6}$$ ∎

EXAMPLE 6

Simplify $\dfrac{\dfrac{1}{x}}{\dfrac{3}{y}}$.

Solution

$$\dfrac{\dfrac{1}{x}}{\dfrac{3}{y}} = \dfrac{1}{x} \div \dfrac{3}{y} = \dfrac{1}{x} \cdot \dfrac{y}{3} = \dfrac{y}{3x}$$ ∎

EXAMPLE 7

Simplify $\dfrac{\dfrac{1}{2}+\dfrac{1}{3}}{\dfrac{5}{6}-\dfrac{1}{4}}$.

Let's look at two possible strategies for such a problem.

Solution A

$$\dfrac{\dfrac{1}{2}+\dfrac{1}{3}}{\dfrac{5}{6}-\dfrac{1}{4}} = \dfrac{\dfrac{3}{6}+\dfrac{2}{6}}{\dfrac{10}{12}-\dfrac{3}{12}} = \dfrac{\dfrac{5}{6}}{\dfrac{7}{12}} = \dfrac{5}{\underset{1}{\cancel{6}}} \cdot \dfrac{\overset{2}{\cancel{12}}}{7} = \dfrac{10}{7}$$

← Invert divisor and multiply

Solution B

The least common multiple of all four denominators (2, 3, 6, and 4) is 12. We multiply the entire complex fraction by a form of 1, specifically $\frac{12}{12}$.

$$\frac{\frac{1}{2}+\frac{1}{3}}{\frac{5}{6}-\frac{1}{4}} = \left(\frac{12}{12}\right)\left(\frac{\frac{1}{2}+\frac{1}{3}}{\frac{5}{6}-\frac{1}{4}}\right)$$

$$= \frac{12\left(\frac{1}{2}+\frac{1}{3}\right)}{12\left(\frac{5}{6}-\frac{1}{4}\right)} = \frac{12\left(\frac{1}{2}\right)+12\left(\frac{1}{3}\right)}{12\left(\frac{5}{6}\right)-12\left(\frac{1}{4}\right)}$$

$$= \frac{6+4}{10-3} = \frac{10}{7}$$

EXAMPLE 8

Simplify $\dfrac{\dfrac{2}{x}+\dfrac{3}{y}}{\dfrac{5}{x}-\dfrac{1}{y^2}}$.

Solution A

$$\frac{\frac{2}{x}+\frac{3}{y}}{\frac{5}{x}-\frac{1}{y^2}} = \frac{\frac{2}{x}\left(\frac{y}{y}\right)+\frac{3}{y}\left(\frac{x}{x}\right)}{\frac{5}{x}\left(\frac{y^2}{y^2}\right)-\frac{1}{y^2}\left(\frac{x}{x}\right)} = \frac{\frac{2y}{xy}+\frac{3x}{xy}}{\frac{5y^2}{xy^2}-\frac{x}{xy^2}}$$

$$= \frac{\dfrac{2y+3x}{xy}}{\dfrac{5y^2-x}{xy^2}}$$

$$= \frac{2y+3x}{xy} \cdot \frac{xy^2}{5y^2-x} \qquad \text{Invert divisor and multiply}$$

$$= \frac{y(2y+3x)}{5y^2-x}$$

Solution B

The least common multiple of all four denominators (x, y, x, and y^2) is xy^2. We multiply the entire complex fraction by a form of 1, specifically, $\dfrac{xy^2}{xy^2}$.

$$\frac{\frac{2}{x}+\frac{3}{y}}{\frac{5}{x}-\frac{1}{y^2}} = \left(\frac{xy^2}{xy^2}\right)\left(\frac{\frac{2}{x}+\frac{3}{y}}{\frac{5}{x}-\frac{1}{y^2}}\right)$$

$$= \frac{xy^2\left(\frac{2}{x}+\frac{3}{y}\right)}{xy^2\left(\frac{5}{x}-\frac{1}{y^2}\right)} = \frac{xy^2\left(\frac{2}{x}\right)+xy^2\left(\frac{3}{y}\right)}{xy^2\left(\frac{5}{x}\right)-xy^2\left(\frac{1}{y^2}\right)}$$

$$= \frac{2y^2+3xy}{5y^2-x} \quad \text{or} \quad \frac{y(2y+3x)}{5y^2-x}$$

6.4 More on Addition and Subtraction of Rational Expressions

Certainly, either approach (Solution A or Solution B) will work for problems such as Examples 7 and 8. You should carefully examine Solution B of each example. This approach works very effectively with complex fractions when the least common multiple of all the denominators of the simple fractions is easy to find. We can summarize the two methods for simplifying a complex fraction as follows:

> **Two Methods for Simplifying Complex Fractions**
> 1. Simplify the numerator and denominator of the fraction separately. Then divide the simplified numerator by the simplified denominator.
> 2. Multiply the numerator and denominator of the complex fraction by the least common multiple of all of the denominators that appear in the complex fraction.

EXAMPLE 9   Simplify $\dfrac{\dfrac{2}{x} - 3}{4 + \dfrac{5}{y}}$.

Solution

$$\dfrac{\dfrac{2}{x} - 3}{4 + \dfrac{5}{y}} = \left(\dfrac{xy}{xy}\right)\left(\dfrac{\dfrac{2}{x} - 3}{4 + \dfrac{5}{y}}\right)$$

$$= \dfrac{(xy)\left(\dfrac{2}{x}\right) - (xy)(3)}{(xy)(4) + (xy)\left(\dfrac{5}{y}\right)}$$

$$= \dfrac{2y - 3xy}{4xy + 5x} \quad \text{or} \quad \dfrac{y(2 - 3x)}{x(4y + 5)} \quad \blacksquare$$

3 Write Rational Expressions to Express Rates

We can use our skill at translating expressions to express rates, which are essentially a comparison between two quantities. We see the concept of rates being used every day, and they are used in many fields. The following list shows some examples of rates that you may have encountered.

Rate that an automobile is traveling	Miles per hour
Rate of Internet speed	Megabits per second
Exchange rate of money	Euros per dollar
Price of bananas	Cents per pound
Rate of gasoline consumption	Miles per gallon
Rate of doing homework problems	Problems per hour
Rate of pay	Dollars per hour

For our work in algebra we need to be able to write expressions for rates using variables to represent the quantities. Example 10 uses rational expressions to express rates and quantities related to rates.

EXAMPLE 10

Write a rational expression to answer each question.

(a) If Jorge drove x miles in y hours, what was his rate in miles per hour?

(b) If Megan can clean the entire house in k hours, what fraction of the house has she cleaned in 3 hours?

(c) If m pounds of chocolate cost n dollars, what is the price per pound?

Solution

(a) To express Jorge's rate in miles per hour, we divide the number of miles by the numbers of hours. Therefore Jorge's rate is expressed as $\dfrac{x}{y}$.

(b) If it takes Megan k hours to clean the house, then in 1 hour she gets $\dfrac{1}{k}$ of the job done. So if she works for 3 hours, she will get $\dfrac{3}{k}$ of the job done.

(c) If m pounds of chocolate cost n dollars, then the price per pound can be expressed by dividing the number of pounds by the price. Therefore the price per pound is expressed as $\dfrac{n}{m}$. ∎

CONCEPT QUIZ 6.4

For Problems 1–5, answer true or false.

1. When adding two rational expressions, the result cannot be 0.
2. A complex fraction is a fraction that contains fractions in the numerator or denominator.
3. The complex fraction $\dfrac{\frac{1}{x} + \frac{1}{2}}{\frac{1}{y} + \frac{1}{3}}$ can be simplified by multiplying the numerator by $2x$ and the denominator by $3y$.
4. The complex fraction $\dfrac{\frac{1}{3m} + \frac{1}{2n}}{\frac{1}{2m} - \frac{1}{3n}}$ can be simplified by multiplying the fraction by a form of 1, specifically $\dfrac{6mn}{6mn}$.
5. If Kevin buys four custom wheels for y dollars, then the price per wheel is represented by $\dfrac{4}{y}$.

Section 6.4 Classroom Problem Set

Objective 1

1. Add $\dfrac{4}{y} + \dfrac{2}{y^2 + 3y}$.

2. Add $\dfrac{3}{x^2 + 2x} + \dfrac{7}{x}$.

3. Subtract $\dfrac{7}{a + 2} - \dfrac{8}{a^2 + 5a + 6}$.

4. Subtract $\dfrac{8x}{x^2 - 1} - \dfrac{4}{x - 1}$.

5. Add $\dfrac{3}{x^2 - 4} + \dfrac{5}{x^2 + 6x + 8}$.

6. Add $\dfrac{3}{a^2 + 5a + 4} + \dfrac{a}{a^2 - 1}$.

7. Perform the indicated operations:

$$\dfrac{4y}{y^2 - 2y - 3} + \dfrac{6}{y - 3} - \dfrac{5}{y + 1}$$

8. Perform the indicated operations:

$$\dfrac{5x}{3x^2 + 7x - 20} - \dfrac{1}{3x - 5} - \dfrac{2}{x + 4}$$

Objective 2

9. Simplify $\dfrac{\dfrac{3}{8}}{\dfrac{7}{16}}$.

10. Simplify $\dfrac{-\dfrac{3}{4}}{\dfrac{3}{16}}$.

11. Simplify $\dfrac{\dfrac{2}{a}}{\dfrac{5}{b}}$.

12. Simplify $\dfrac{-\dfrac{6}{x}}{\dfrac{8}{y}}$.

13. Simplify $\dfrac{\dfrac{4}{5}-\dfrac{1}{2}}{\dfrac{3}{10}+\dfrac{1}{5}}$.

14. Simplify $\dfrac{\dfrac{3}{8}+\dfrac{1}{4}}{\dfrac{1}{2}+\dfrac{3}{16}}$.

15. Simplify $\dfrac{\dfrac{3}{a^2}+\dfrac{2}{b}}{\dfrac{1}{a}-\dfrac{5}{b}}$.

16. Simplify $\dfrac{\dfrac{2}{x}-\dfrac{3}{x^2}}{\dfrac{7}{x^2}-\dfrac{4}{x}}$.

17. Simplify $\dfrac{6+\dfrac{2}{m}}{\dfrac{3}{n}-7}$.

18. Simplify $\dfrac{\dfrac{6}{a}+2}{7-\dfrac{3}{c}}$.

Objective 3

19. Write a rational expression to answer each question.
    (a) If Selena jogs x miles in y hours, what is her rate of jogging?
    (b) If Danielle purchased m pounds of apples for n dollars, what is the price (in dollars) per pound of the apples?
    (c) If it takes Taylor a hours to wash b automobiles, what is his rate of washing automobiles?

20. Write a rational expression to answer each question.
    (a) If Roy traveled m miles in h hours, what was his rate in miles per hour?
    (b) If p pounds of candy cost c cents, what is the price per pound?
    (c) If the area of a rectangle is 47 square inches, and the length is x inches, what is the width of the rectangle?

THOUGHTS INTO WORDS

1. Which of the two techniques presented in the text would you use to simplify $\dfrac{\dfrac{1}{4}+\dfrac{1}{3}}{\dfrac{3}{4}-\dfrac{1}{6}}$? Which technique would you use to simplify $\dfrac{\dfrac{3}{8}-\dfrac{5}{7}}{\dfrac{7}{9}+\dfrac{6}{25}}$? Explain your choice for each problem.

Answers to the Concept Quiz
1. False **2.** True **3.** False **4.** True **5.** False

6.5 Rational Equations and Problem Solving

OBJECTIVES
1. Solve Rational Equations with Constants in the Denominator
2. Solve Rational Equations with Variables in the Denominator
3. Solve Word Problems

1 Solve Rational Equations with Constants in the Denominator

We will consider two basic types of rational equations in this text. One type has only constants as denominators, and the other type has some variables in the denominator. In Chapter 2 we considered rational equations that had only constants in the denominator. Let's review our approach to these equations, because we will be using that same basic technique to solve any rational equation.

EXAMPLE 1 Solve $\dfrac{x-2}{3} + \dfrac{x+1}{4} = \dfrac{1}{6}$.

Solution

$$\dfrac{x-2}{3} + \dfrac{x+1}{4} = \dfrac{1}{6}$$

$$12\left(\dfrac{x-2}{3} + \dfrac{x+1}{4}\right) = 12\left(\dfrac{1}{6}\right) \quad \text{Multiply both sides by 12, the LCD of all three denominators}$$

$$12\left(\dfrac{x-2}{3}\right) + 12\left(\dfrac{x+1}{4}\right) = 12\left(\dfrac{1}{6}\right)$$

$$4(x-2) + 3(x+1) = 2$$

$$4x - 8 + 3x + 3 = 2$$

$$7x - 5 = 2$$

$$7x = 7$$

$$x = 1$$

The solution set is {1}. (Check it!) ∎

2 Solve Rational Equations with Variables in the Denominator

If an equation contains a variable in one or more denominators, then we proceed in essentially the same way except that we must avoid any value of the variable that makes a denominator zero. Consider the next example.

EXAMPLE 2 Solve $\dfrac{3}{x} + \dfrac{1}{2} = \dfrac{5}{x}$.

Solution

First, we need to realize that *x cannot equal zero*. Then we can proceed in the usual way.

$$\dfrac{3}{x} + \dfrac{1}{2} = \dfrac{5}{x}$$

$$2x\left(\dfrac{3}{x} + \dfrac{1}{2}\right) = 2x\left(\dfrac{5}{x}\right) \quad \text{Multiply both sides by } 2x, \text{ the LCD of all denominators}$$

$$6 + x = 10$$

$$x = 4$$

✔ **Check**

$\dfrac{3}{x} + \dfrac{1}{2} = \dfrac{5}{x}$ becomes $\dfrac{3}{4} + \dfrac{1}{2} \stackrel{?}{=} \dfrac{5}{4}$ when $x = 4$

$$\dfrac{3}{4} + \dfrac{2}{4} \stackrel{?}{=} \dfrac{5}{4}$$

$$\dfrac{5}{4} = \dfrac{5}{4}$$

The solution set is {4}. ∎

6.5 Rational Equations and Problem Solving 233

EXAMPLE 3 Solve $\dfrac{5}{x+2} = \dfrac{2}{x-1}$.

Solution

Because neither denominator can be zero, we know that $x \neq -2$ and $x \neq 1$.

$$\dfrac{5}{x+2} = \dfrac{2}{x-1}$$

$$(x+2)(x-1)\left(\dfrac{5}{x+2}\right) = (x+2)(x-1)\left(\dfrac{2}{x-1}\right) \quad \text{Multiply both sides by } (x+2)(x-1), \text{ the LCD}$$

$$5(x-1) = 2(x+2)$$

$$5x - 5 = 2x + 4$$

$$3x = 9$$

$$x = 3$$

Because the only restrictions are $x \neq -2$ and $x \neq 1$, the solution set is $\{3\}$. (Check it!) ∎

EXAMPLE 4 Solve $\dfrac{2}{x-2} + 2 = \dfrac{x}{x-2}$.

Solution

No denominator can be zero, so $x \neq 2$.

$$\dfrac{2}{x-2} + 2 = \dfrac{x}{x-2}$$

$$(x-2)\left(\dfrac{2}{x-2} + 2\right) = (x-2)\left(\dfrac{x}{x-2}\right) \quad \text{Multiply by } x-2, \text{ the LCD}$$

$$2 + 2(x-2) = x$$

$$2 + 2x - 4 = x$$

$$2x - 2 = x$$

$$x = 2$$

Two cannot be a solution because it will produce a denominator of zero. There is no solution to the given equation; the solution set is $\varnothing$. ∎

Example 4 illustrates the importance of recognizing the restrictions that must be placed on possible values of a variable. We will indicate such restrictions at the beginning of our solution.

EXAMPLE 5 Solve $\dfrac{125 - n}{n} = 4 + \dfrac{10}{n}$.

Solution

$$\dfrac{125 - n}{n} = 4 + \dfrac{10}{n}, \quad n \neq 0 \quad \text{Note the necessary restriction}$$

$$n\left(\dfrac{125 - n}{n}\right) = n\left(4 + \dfrac{10}{n}\right) \quad \text{Multiply both sides by } n$$

$$125 - n = 4n + 10$$
$$115 = 5n$$
$$23 = n$$

The only restriction is $n \neq 0$, and the solution set is $\{23\}$.

3 Solve Word Problems

We are now ready to solve more problems, specifically those that translate into fractional equations.

EXAMPLE 6   **Apply Your Skill**

One number is 10 larger than another number. The indicated quotient of the smaller number divided by the larger number reduces to $\dfrac{3}{5}$. Find the numbers.

Solution

We let n represent the smaller number. Then $n + 10$ represents the larger number. The second sentence in the statement of the problem translates into the following equation:

$$\frac{n}{n + 10} = \frac{3}{5}, \quad n \neq -10$$
$$5n = 3(n + 10) \quad \text{Cross products are equal}$$
$$5n = 3n + 30$$
$$2n = 30$$
$$n = 15$$

If n is 15, then $n + 10$ is 25. Thus the numbers are 15 and 25. To check, consider the quotient of the smaller number divided by the larger number.

$$\frac{15}{25} = \frac{3 \cdot 5}{5 \cdot 5} = \frac{3}{5}$$

EXAMPLE 7   **Apply Your Skill**

One angle of a triangle has a measure of 40°, and the measures of the other two angles are in the ratio of 5 to 2. Find the measures of the other two angles.

Solution

The sum of the measures of the other two angles is $180° - 40° = 140°$. Let y represent the measure of one angle. Then $140 - y$ represents the measure of the other angle.

$$\frac{y}{140 - y} = \frac{5}{2}, \quad y \neq 140$$
$$2y = 5(140 - y) \quad \text{Cross products are equal}$$
$$2y = 700 - 5y$$
$$7y = 700$$
$$y = 100$$

If $y = 100$, then $140 - y = 40$. Therefore the measures of the other two angles of the triangle are 100° and 40°.

In Chapter 3, we solved some uniform motion problems in which the formula $d = rt$ played an important role. Let's consider another one of those problems; keep in mind that we can also write the formula $d = rt$ as $\dfrac{d}{r} = t$ or $\dfrac{d}{t} = r$.

EXAMPLE 8

Apply Your Skill

Wendy rides her bicycle 30 miles in the same time that it takes Kim to ride her bicycle 20 miles. If Wendy rides 5 miles per hour faster than Kim, find the rate of each.

Solution

Let r represent Kim's rate. Then $r + 5$ represents Wendy's rate. Let's record the information of this problem in a table.

	Distance	Rate	Time = $\dfrac{\text{Distance}}{\text{Rate}}$
Kim	20	r	$\dfrac{20}{r}$
Wendy	30	$r + 5$	$\dfrac{30}{r+5}$

We can use the fact that their times are equal as a guideline.

$$\underbrace{\dfrac{\text{Distance Kim rides}}{\text{Rate Kim rides}}}_{\text{Kim's Time}} = \underbrace{\dfrac{\text{Distance Wendy rides}}{\text{Rate Wendy rides}}}_{\text{Wendy's Time}}$$

$$\dfrac{20}{r} = \dfrac{30}{r+5}, \quad r \neq 0 \text{ and } r \neq -5$$

$$20(r + 5) = 30r$$
$$20r + 100 = 30r$$
$$100 = 10r$$
$$10 = r$$

Therefore, Kim rides at 10 miles per hour, and Wendy rides at $10 + 5 = 15$ miles per hour. ∎

CONCEPT QUIZ 6.5

For Problems 1–5, answer true or false.

1. The equation $\dfrac{5}{x-3} = \dfrac{2}{x+4}$ is a proportion that could be solved by using cross-multiplication.

2. The equation $\dfrac{x}{8} + 3x = \dfrac{x}{5}$ is a proportion that could be solved by using cross-multiplication.

3. The solution to the equation $\dfrac{4}{x-6} + 3 = \dfrac{x}{x-6}$ cannot be 6.

4. Any value of the variable that makes the numerator of an equation equal to zero cannot be a solution of the equation.

5. Zero cannot be a solution to a rational equation.

Section 6.5 Classroom Problem Set

Objective 1

1. Solve $\dfrac{x-3}{2} + \dfrac{2x-7}{3} = \dfrac{5}{6}$.

2. Solve $\dfrac{n+2}{7} + \dfrac{n}{3} = \dfrac{12}{7}$.

Objective 2

3. Solve $\dfrac{5}{x} + \dfrac{1}{4} = \dfrac{7}{x}$.

4. Solve $\dfrac{1}{3} + \dfrac{9}{n} = \dfrac{2}{n}$.

5. Solve $\dfrac{4}{x-2} = \dfrac{10}{x+4}$.

6. Solve $\dfrac{7}{a+3} = \dfrac{5}{a-9}$.

7. Solve $\dfrac{5}{x-5} + 3 = \dfrac{x}{x-5}$.

8. Solve $\dfrac{1}{x+2} = \dfrac{x}{x+2} + 3$.

9. Solve $\dfrac{70-x}{x} + 2 = \dfrac{84}{x}$.

10. Solve $\dfrac{3}{2} = \dfrac{x-1}{x} - 2$.

Objective 3

11. One number is 4 more than another number. The indicated quotient of the smaller number divided by the larger number reduces to $\dfrac{2}{3}$. Find the numbers.

12. What number must be subtracted from the numerator and denominator of $\dfrac{29}{31}$ to produce a fraction equivalent to $\dfrac{11}{12}$?

13. One angle of a triangle has a measure of 30°, and the measures of the other angles are in the ratio of 2 to 3. Find the measures of the other two angles.

14. The ratio of the measures of the complement of an angle to its supplement is 1 to 4. Find the measure of the angle.

15. Javier rides his bicycle 24 miles in the same time that it takes Aldrin to ride his bicycle 18 miles. If Javier rides 2 miles per hour faster than Aldrin, find the rate of each.

16. Kent drives his Mazda 270 miles in the same time that Dave drives his Toyota 250 miles. If Kent averages 4 miles per hour faster than Dave, find their respective rates.

THOUGHTS INTO WORDS

1. (a) Explain how to do the addition problem

    $\dfrac{3}{x+2} + \dfrac{5}{x-1}$.

    (b) Explain how to solve the equation

    $\dfrac{3}{x+2} + \dfrac{5}{x-1} = 0$.

2. How can you tell by inspection that $\dfrac{x}{x-4} = \dfrac{4}{x-4}$ has no solution?

3. How would you help someone solve the equation $\dfrac{1}{x} + \dfrac{2}{x} = \dfrac{3}{x}$?

Answers to the Concept Quiz

1. True   2. False   3. True   4. False   5. False

6.6 More Rational Equations and Problem Solving

OBJECTIVES

1. Solve Rational Equations
2. Solve Word Problems Including Rate-Time Problems

1 Solve Rational Equations

Let's begin this section by considering a few more rational equations. We will continue to solve them using the same basic techniques as in the preceding section.

EXAMPLE 1 Solve $\dfrac{10}{8x-2} - \dfrac{6}{4x-1} = \dfrac{1}{9}$.

Solution

$$\dfrac{10}{8x-2} - \dfrac{6}{4x-1} = \dfrac{1}{9}, \quad x \neq \dfrac{1}{4} \qquad \text{Do you see why } x \text{ cannot equal } \dfrac{1}{4}?$$

$$\dfrac{10}{2(4x-1)} - \dfrac{6}{4x-1} = \dfrac{1}{9} \qquad \text{Factor the first denominator}$$

$$18(4x-1)\left(\dfrac{10}{2(4x-1)} - \dfrac{6}{4x-1}\right) = 18(4x-1)\left(\dfrac{1}{9}\right) \qquad \text{Multiply both sides by } 18(4x-1), \text{ the LCD}$$

$$9(10) - 18(6) = 2(4x-1)$$
$$90 - 108 = 8x - 2$$
$$-18 = 8x - 2$$
$$-16 = 8x$$
$$-2 = x \qquad \text{Be sure that the solution } -2 \text{ checks!}$$

The solution set is $\{-2\}$. ∎

Remark: In the second step of the solution for Example 1, you may choose to reduce $\dfrac{10}{2(4x-1)}$ to $\dfrac{5}{4x-1}$. Then the left side, $\dfrac{5}{4x-1} - \dfrac{6}{4x-1}$, simplifies to $\dfrac{-1}{4x-1}$. This forms the proportion $\dfrac{-1}{4x-1} = \dfrac{1}{9}$, which can be solved easily using the *cross-multiplication method*.

EXAMPLE 2 Solve $\dfrac{2n}{n+3} + \dfrac{5n}{n^2-9} = 2$.

Solution

$$\dfrac{2n}{n+3} + \dfrac{5n}{n^2-9} = 2, \quad n \neq -3 \text{ and } n \neq 3$$

$$\dfrac{2n}{n+3} + \dfrac{5n}{(n+3)(n-3)} = 2$$

$$(n+3)(n-3)\left(\dfrac{2n}{n+3} + \dfrac{5n}{(n+3)(n-3)}\right) = (n+3)(n-3)(2)$$

$$2n(n-3) + 5n = 2(n^2 - 9)$$
$$2n^2 - 6n + 5n = 2n^2 - 18$$
$$-6n + 5n = -18 \quad \text{Add } -2n^2 \text{ to both sides}$$
$$-n = -18$$
$$n = 18$$

The solution set is {18}.

EXAMPLE 3 Solve $n + \dfrac{1}{n} = \dfrac{10}{3}$.

Solution

$$n + \frac{1}{n} = \frac{10}{3}, \quad n \neq 0$$
$$3n\left(n + \frac{1}{n}\right) = 3n\left(\frac{10}{3}\right)$$
$$3n^2 + 3 = 10n$$
$$3n^2 - 10n + 3 = 0$$
$$(3n - 1)(n - 3) = 0 \quad \text{Remember when we used the factoring techniques to help solve equations of this type in Chapter 5?}$$

$$3n - 1 = 0 \quad \text{or} \quad n - 3 = 0$$
$$3n = 1 \quad \text{or} \quad n = 3$$
$$n = \frac{1}{3}$$

The solution set is $\left\{\dfrac{1}{3}, 3\right\}$.

2 Solve Word Problems Including Rate-Time Problems

Recall that $\dfrac{2}{3}$ and $\dfrac{3}{2}$ are called multiplicative inverses, or *reciprocals*, of each other because their product is 1. In general, the reciprocal of any nonzero real number n is the number $\dfrac{1}{n}$. Let's use this idea to solve a problem.

EXAMPLE 4 **Apply Your Skill**

The sum of a number and its reciprocal is $\dfrac{26}{5}$. Find the number.

Solution

We let n represent the number. Then $\dfrac{1}{n}$ represents its reciprocal.

$$\text{Number} + \text{Its reciprocal} = \frac{26}{5}$$
$$\downarrow \qquad\qquad \downarrow \qquad\qquad \downarrow$$
$$n + \frac{1}{n} = \frac{26}{5}, \quad n \neq 0$$

6.6 More Rational Equations and Problem Solving

$$5n\left(n + \frac{1}{n}\right) = 5n\left(\frac{26}{5}\right) \quad \text{Multiply both sides by } 5n, \text{ the LCD}$$

$$5n^2 + 5 = 26n$$

$$5n^2 - 26n + 5 = 0$$

$$(5n - 1)(n - 5) = 0$$

$$5n - 1 = 0 \quad \text{or} \quad n - 5 = 0$$

$$5n = 1 \quad \text{or} \quad n = 5$$

$$n = \frac{1}{5}$$

If the number is $\frac{1}{5}$, its reciprocal is $\frac{1}{\frac{1}{5}} = 5$. If the number is 5, its reciprocal is $\frac{1}{5}$. ■

Now let's consider another uniform motion problem, which is a slight variation of those we studied in the previous section. Again, keep in mind that we always use the distance-rate-time relationships in these problems.

EXAMPLE 5   Apply Your Skill

To travel 60 miles, it takes Sue, riding a moped, 2 hours less than it takes LeAnn, riding a bicycle, to travel 50 miles (see Figure 6.1). Sue travels 10 miles per hour faster than LeAnn. Find the times and rates of both women.

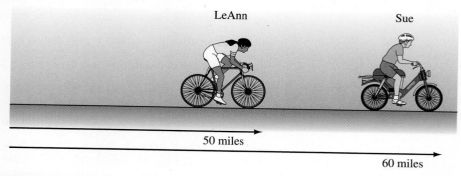

Figure 6.1

Solution

We let t represent LeAnn's time. Then $t - 2$ represents Sue's time. We can record the information from the problem in the table.

	Distance	Time	Rate $\left(r = \dfrac{d}{t}\right)$
LeAnn	50	t	$\dfrac{50}{t}$
Sue	60	$t-2$	$\dfrac{60}{t-2}$

We use the fact that Sue travels 10 miles per hour faster than LeAnn as a guideline to set up an equation.

$$\begin{array}{ccc}
\text{Sue's Rate} & = & \text{LeAnn's Rate} + 10 \\
\downarrow & \downarrow & \downarrow \\
\dfrac{60}{t-2} & = & \dfrac{50}{t} + 10, \quad t \neq 2 \text{ and } t \neq 0
\end{array}$$

Solving this equation yields

$$\frac{60}{t-2} = \frac{50}{t} + 10$$

$$t(t-2)\left(\frac{60}{t-2}\right) = t(t-2)\left(\frac{50}{t} + 10\right)$$

$$60t = 50(t-2) + 10t(t-2)$$
$$60t = 50t - 100 + 10t^2 - 20t$$
$$0 = 10t^2 - 30t - 100$$
$$0 = t^2 - 3t - 10$$
$$0 = (t-5)(t+2)$$

$$t - 5 = 0 \quad \text{or} \quad t + 2 = 0$$
$$t = 5 \quad \text{or} \quad t = -2$$

We must disregard the negative solution, so LeAnn's time is 5 hours, and Sue's time is $5 - 2 = 3$ hours. LeAnn's rate is $\dfrac{50}{5} = 10$ miles per hour, and Sue's rate is $\dfrac{60}{3} = 20$ miles per hour. (Be sure that all these results check back into the original problem!) ∎

There is another class of problems that we commonly refer to as work problems, or sometimes as *rate–time* problems. For example, if a certain machine produces 120 items in 10 minutes, then we say that it is producing at a rate of $\dfrac{120}{10} = 12$ items per minute. Likewise, if a person can do a certain job in 5 hours, then that person is working at a rate of $\dfrac{1}{5}$ of the job per hour. In general, if Q is the quantity of something done in t units of time, then the rate r is given by $r = \dfrac{Q}{t}$. The rate is stated in terms of *so much quantity per unit of time*. The uniform-motion problems we discussed earlier are a special kind of rate–time problem in which the *quantity* is distance. The use of tables to organize information, as we illustrated with the uniform-motion problems, is a convenient aid for some rate–time problems. Let's consider some examples.

EXAMPLE 6   Apply Your Skill

Printing press A can produce 35 flyers per minute, and press B can produce 50 flyers per minute. Printing press A is set up and starts a job, and then 15 minutes later printing press B is started, and both presses continue printing until 2225 flyers are produced. How long would printing press B be used?

Solution

We let m represent the number of minutes that printing press B is used. Then $m + 15$ represents the number of minutes that press A is used. The information in the problem can be organized in a table.

	Rate	Time	Quantity = Rate × Time
Press A	35	$m + 15$	$35(m + 15)$
Press B	50	m	$50m$

Since the total quantity (total number of flyers) is 2225 flyers, we can set up and solve the following equation:

$$35(m + 15) + 50m = 2225$$
$$35m + 525 + 50m = 2225$$
$$85m = 1700$$
$$m = 20$$

Therefore, printing press B must be used for 20 minutes. ∎

EXAMPLE 7 Apply Your Skill

Bill can mow a lawn in 45 minutes, and Jennifer can mow the same lawn in 30 minutes. How long would it take the two of them working together to mow the lawn? (See Figure 6.2.)

Figure 6.2

Remark: Before you look at the solution of this problem, *estimate* the answer. Remember that Jennifer can mow the lawn by herself in 30 minutes.

Solution

Bill's rate is $\dfrac{1}{45}$ of the lawn per minute, and Jennifer's rate is $\dfrac{1}{30}$ of the lawn per minute. If we let m represent the number of minutes that they work together, then $\dfrac{1}{m}$ represents the rate when working together. Therefore, since the sum of the individual rates must equal the rate working together, we can set up and solve the following equation:

$$\frac{1}{30} + \frac{1}{45} = \frac{1}{m}, \quad m \neq 0$$

$$90m\left(\frac{1}{30} + \frac{1}{45}\right) = 90m\left(\frac{1}{m}\right) \quad \text{Multiply both sides by } 90m, \text{ the LCD}$$

$$3m + 2m = 90$$
$$5m = 90$$
$$m = 18$$

It should take them 18 minutes to mow the lawn when working together. (How close was your estimate?) ∎

EXAMPLE 8

Apply Your Skill

It takes Amy twice as long to deliver papers as it does Nancy. How long would it take each girl by herself if they can deliver the papers together in 40 minutes?

Solution

We let m represent the number of minutes that it takes Nancy by herself. Then $2m$ represents Amy's time by herself. Therefore, Nancy's rate is $\frac{1}{m}$, and Amy's rate is $\frac{1}{2m}$. Since the combined rate is $\frac{1}{40}$, we can set up and solve the following equation:

$$\underset{\text{rate}}{\underset{\text{Nancy's}}{\frac{1}{m}}} + \underset{\text{rate}}{\underset{\text{Amy's}}{\frac{1}{2m}}} = \underset{\text{rate}}{\underset{\text{Combined}}{\frac{1}{40}}}, \quad m \neq 0$$

$$40m\left(\frac{1}{m} + \frac{1}{2m}\right) = 40m\left(\frac{1}{40}\right)$$

$$40 + 20 = m$$

$$60 = m$$

Therefore, Nancy can deliver the papers by herself in 60 minutes, and Amy can deliver them by herself in $2(60) = 120$ minutes. ∎

One final example of this section outlines another approach that some people find meaningful for work problems. This approach represents the fractional parts of a job. For example, if a person can do a certain job in 7 hours, then at the end of 3 hours, that person has finished $\frac{3}{7}$ of the job. (Again, we assume a constant rate of work.) At the end of 5 hours, $\frac{5}{7}$ of the job has been done—in general, at the end of h hours, $\frac{h}{7}$ of the job has been completed. Let's use this idea to solve a work problem.

EXAMPLE 9

Apply Your Skill

It takes Pat 12 hours to install a wood floor. After he had been working for 3 hours, he was joined by his brother Mike, and together they finished the floor in 5 hours. How long would it take Mike to install the floor by himself?

Solution

Let h represent the number of hours that it would take Mike to install the floor by himself.

Because it takes Pat 12 hours to do the entire floor, his working rate is $\frac{1}{12}$. The fractional part of the job that Pat does equals his working rate times his time. He works for 8 hours (3 hours before Mike and then 5 hours with Mike). Therefore, Pat's part of the job is $\frac{1}{12}(8) = \frac{8}{12}$. The fractional part of the job that Mike does equals his working rate times his time. Because h represents Mike's time to install the floor, his working rate is $\frac{1}{h}$. He works for 5 hours. Therefore, Mike's part of the job is $\frac{1}{h}(5) = \frac{5}{h}$. Adding the two fractional parts together results in one entire job being done. Let's also show this information in chart form to set up our guideline. Then we can write the equation and solve.

	Time to do entire job	Working rate	Time working	Fractional part of the job done
Pat	12	$\frac{1}{12}$	8	$\frac{8}{12}$
Mike	h	$\frac{1}{h}$	5	$\frac{5}{h}$

Fractional part of the job that Pat does

Fractional part of the job that Mike does

$$\frac{8}{12} + \frac{5}{h} = 1$$

$$12h\left(\frac{8}{12} + \frac{5}{h}\right) = 12h(1)$$

$$12h\left(\frac{8}{12}\right) + 12h\left(\frac{5}{h}\right) = 12h$$

$$8h + 60 = 12h$$

$$60 = 4h$$

$$15 = h$$

It would take Mike 15 hours to install the floor by himself. ■

We emphasize a point made earlier. Don't become discouraged if solving word problems is still giving you trouble. The development of problem-solving skills is a long-term objective. If you continue to work hard and give it your best shot, you will gradually become more and more confident in your approach to solving problems. Don't be afraid to try some different approaches on your own. Our problem-solving suggestions simply provide a framework for you to build on.

CONCEPT QUIZ 6.6

For Problems 1–5, answer true or false.

1. Multiplicative inverses are called reciprocals.
2. The product of a number and its reciprocal is 0.
3. The reciprocal of any nonzero number n is the number $\frac{1}{n}$.
4. If it takes a person x hours to do y tasks, then the person is working at a rate of $\frac{x}{y}$ tasks per hour.
5. If a person can do a certain job in 7 hours, then that person gets $\frac{3}{7}$ of the job done in 3 hours.

Section 6.6 Classroom Problem Set

Objective 1

1. Solve $\dfrac{7}{6x+2} - \dfrac{3}{3x+1} = \dfrac{1}{20}$.

2. Solve $\dfrac{2x}{x+1} - \dfrac{3}{x-1} = 2$.

3. Solve $\dfrac{x}{x+2} + \dfrac{x}{x^2-4} = 1$.

4. Solve $\dfrac{16}{t^2-16} + \dfrac{t}{2t-8} = \dfrac{1}{2}$.

5. Solve $y + \dfrac{1}{y} = \dfrac{5}{2}$.

6. Solve $\dfrac{17}{4} = n + \dfrac{1}{n}$.

Objective 2

7. The sum of a number and its reciprocal is $\dfrac{50}{7}$. Find the number.

8. The sum of a number and twice its reciprocal is $\dfrac{9}{2}$. Find the number.

9. To jog 16 miles, it takes Kim 1 hour less time than it takes Ryan to bicycle 36 miles. Ryan travels 4 miles per hour faster than Kim. Find the rates of both Kim and Ryan.

10. To travel 300 miles, it takes a freight train 2 hours longer than it takes an express train to travel 280 miles. The rate of the express train is 20 miles per hour faster than the rate of the freight train. Find the rates of both trains.

11. Derek, a sandwich maker, can make 3 sandwiches per minute. Marlene, another sandwich maker, can make 5 sandwiches per minute. Derek starts making sandwiches and 10 minutes later Marlene begins to make sandwiches. They both continue making sandwiches until the catering order for 110 sandwiches is completed. How long did Marlene work making the sandwiches?

12. Ann can write 9 math problems in an hour, whereas Laurie can write only 5 problems in an hour. Ann begins writing and 20 minutes, or $\dfrac{1}{3}$ hour, later Laurie begins writing. To finish 31 problems while writing together, how long did Laurie write problems?

13. Dean can wash and detail a car in 60 minutes, and Tasha can wash and detail the same car in 90 minutes. How long would it take the two of them working together to wash and detail the car?

14. Betty can do a job in 10 minutes. Doug can do the same job in 15 minutes. If they work together, how long will it take them to complete the job?

15. It takes Martin three times as long to clean an office as it does Julio. How long would it take each man to clean the office by himself if they can clean the office together in 30 minutes?

16. It takes Barry twice as long to deliver papers as it does Mike. How long would it take each of them if they deliver the paper together in 60 minutes?

17. It takes David 16 hours to build a deck. After he had been working for 7 hours, he was joined by his friend Carlos, and together they finished the deck in 5 hours. How long would it take Carlos to build the deck by himself?

18. Working together, Pam and Laura can complete a job in $1\dfrac{1}{2}$ hours. When working alone, it takes Laura 4 hours longer than Pam to do the job. How long does it take each of them working alone?

THOUGHTS INTO WORDS

1. Write a paragraph or two summarizing the new ideas about problem solving that you have acquired thus far in this course.

Answers to the Concept Quiz

1. True   2. False   3. True   4. False   5. True

Chapter 6   Review Problem Set

For Problems 1–8, simplify each algebraic fraction.

1. $\dfrac{56x^3y}{72xy^3}$

2. $\dfrac{x^2 - 9x}{x^2 - 6x - 27}$

3. $\dfrac{3n^2 - n - 10}{n^2 - 4}$

4. $\dfrac{16a^2 + 24a + 9}{20a^2 + 7a - 6}$

5. $\dfrac{x^2 - 16}{4 - x}$

6. $\dfrac{25 - y^2}{y + 5}$

7. $\dfrac{24 - 6a}{a - 4}$

8. $\dfrac{x^2 + 3x - 10}{2 + x - x^2}$

For Problems 9–12, multiply and express your answer in simplest form.

9. $\dfrac{7x^2y^2}{12y^3} \cdot \dfrac{18y}{28x}$

10. $\dfrac{x^2y}{x^2 + 2x} \cdot \dfrac{x^2 - x - 6}{y}$

11. $\dfrac{x^2 + 8x + 15}{x^2 + 4} \cdot \dfrac{3x^2 + 12}{x^2 + 6x + 5}$

12. $\dfrac{x^2 + 3xy + 2y^2}{x - y} \cdot \dfrac{x^2 - xy}{3x + 6y}$

For Problems 13–16, divide and express your answer in simplest form.

13. $\dfrac{n^2 - 2n - 24}{n^2 + 11n + 28} \div \dfrac{n^3 - 6n^2}{n^2 - 49}$

14. $\dfrac{4a^2 + 4a + 1}{(a + 6)^2} \div \dfrac{6a^2 - 5a - 4}{3a^2 + 14a - 24}$

15. $\dfrac{28a^2b^2}{-21a} \div \dfrac{-4a}{7b}$

16. $24x^3y^2 \div \dfrac{3x^2y + 6x^2}{y}$

For Problems 17–26, add or subtract as indicated, and express your answer in simplest form.

17. $\dfrac{12n + 8}{5} - \dfrac{2n - 7}{5}$

18. $\dfrac{3(x + 4)}{7x} + \dfrac{2(x - 6)}{7x}$

19. $\dfrac{3x + 4}{5} + \dfrac{2x - 7}{4}$

20. $\dfrac{7}{3x} + \dfrac{5}{4x} - \dfrac{2}{8x^2}$

21. $\dfrac{7}{n} + \dfrac{3}{n - 1}$

22. $\dfrac{2}{a - 4} - \dfrac{3}{a - 2}$

23. $\dfrac{2x}{x^2 - 3x} - \dfrac{3}{4x}$

24. $\dfrac{2}{x^2 + 7x + 10} + \dfrac{3}{x^2 - 25}$

25. $\dfrac{5x}{x^2 - 4x - 21} - \dfrac{3}{x - 7} + \dfrac{4}{x + 3}$

26. $\dfrac{3}{x - 7} + \dfrac{5}{x + 4} + \dfrac{4x}{x^2 - 3x - 28}$

For Problems 27–30, simplify each complex fraction.

27. $\dfrac{\dfrac{3}{x} - \dfrac{4}{y^2}}{\dfrac{4}{y} + \dfrac{5}{x}}$

28. $\dfrac{\dfrac{2}{x} - 1}{3 + \dfrac{5}{y}}$

29. $\dfrac{\dfrac{6}{mn} + \dfrac{3}{m}}{\dfrac{5}{m} + \dfrac{2}{n}}$

30. $\dfrac{\dfrac{x + 6}{4}}{\dfrac{3}{x} + \dfrac{1}{4}}$

For Problems 31–34, answer each question with a rational expression.

31. How long would it take to drive x miles when driving at a rate of y miles per hour?

32. If g gallons of heating oil cost d dollars, what is the price per gallon?

33. If Erick can assemble a computer in m minutes, what (fractional) part of the assembly has he done after 40 minutes?

34. Suppose the product of two numbers is p, and one of the factors is k. What is the other factor?

For Problems 35–48, solve each equation.

35. $\dfrac{2x - 1}{3} + \dfrac{3x - 2}{4} = \dfrac{5}{6}$

36. $\dfrac{6x - 4}{3} - \dfrac{2x - 5}{7} = -1$

37. $\dfrac{1}{2t} + \dfrac{3}{4} = \dfrac{5}{t}$

38. $\dfrac{5}{3x} - 2 = \dfrac{7}{2x} + \dfrac{1}{5x}$

39. $\dfrac{67 - x}{x} = 6 + \dfrac{4}{x}$

40. $\dfrac{5}{2n + 3} = \dfrac{6}{3n - 2}$

41. $\dfrac{x}{x - 3} + \dfrac{5}{x + 3} = 1$

42. $n + \dfrac{1}{n} = 2$

43. $\dfrac{n - 1}{n^2 + 8n - 9} - \dfrac{n}{n + 9} = 4$

44. $\dfrac{6}{7x} - \dfrac{1}{6} = \dfrac{5}{6x}$

45. $n + \dfrac{1}{n} = \dfrac{5}{2}$

46. $\dfrac{n}{5} = \dfrac{10}{n - 5}$

47. $\dfrac{-1}{2x - 5} + \dfrac{2x - 4}{4x^2 - 25} = \dfrac{5}{6x + 15}$

48. $1 + \dfrac{1}{n - 1} = \dfrac{1}{n^2 - n}$

For Problems 49–54, set up an equation and solve each problem.

49. It takes Nancy three times as long to complete a task as it does Becky. How long would it take each of them to complete the task if working together they can do it in 2 hours?

50. The sum of a number and twice its reciprocal is 3. Find the number.

51. The denominator of a fraction is twice the numerator. If 4 is added to the numerator and 18 to the denominator, a fraction that is equivalent to $\dfrac{4}{9}$ is produced. Find the original fraction.

52. Lanette can ride her moped 44 miles in the same time that Todd rides his bicycle 30 miles. If Lanette rides 7 miles per hour faster than Todd, find their rates.

53. Jim rode his bicycle 36 miles in 4 hours. For the first 20 miles, he rode at a constant rate, and then for the last 16 miles, he reduced his rate by 2 miles per hour. Find his rate for the last 16 miles.

54. An inlet pipe can fill a tank in 10 minutes. A drain can empty the tank in 12 minutes. If the tank is empty and both the inlet pipe and drain are open, how long will it be before the tank overflows?

Chapter 6 Practice Test

For Problems 1–4, simplify each algebraic fraction.

1. $\dfrac{72x^4y^5}{81x^2y^4}$

2. $\dfrac{x^2 + 6x}{x^2 - 36}$

3. $\dfrac{2n^2 - 7n - 4}{3n^2 - 8n - 16}$

4. $\dfrac{2x^3 + 7x^2 - 15x}{x^3 - 25x}$

For Problems 5–14, perform the indicated operations and express answers in simplest form.

5. $\left(\dfrac{8x^2y}{7x}\right)\left(\dfrac{21xy^3}{12y^2}\right)$

6. $\dfrac{x^2 - 49}{x^2 + 7x} \div \dfrac{x^2 - 4x - 21}{x^2 - 2x}$

7. $\dfrac{x^2 - 5x - 36}{x^2 - 15x + 54} \cdot \dfrac{x^2 - 2x - 24}{x^2 + 7x}$

8. $\dfrac{3x - 1}{6} - \dfrac{2x - 3}{8}$

9. $\dfrac{n + 2}{3} - \dfrac{n - 1}{5} + \dfrac{n - 6}{6}$

10. $\dfrac{3}{2x} - \dfrac{5}{6} + \dfrac{7}{9x}$

11. $\dfrac{6}{n} - \dfrac{4}{n - 1}$

12. $\dfrac{2x}{x^2 + 6x} - \dfrac{3}{4x}$

13. $\dfrac{9}{x^2 + 4x - 32} + \dfrac{5}{x + 8}$

14. $\dfrac{-3}{6x^2 - 7x - 20} - \dfrac{5}{3x^2 - 14x - 24}$

For Problems 15–22, solve each of the equations.

15. $\dfrac{x + 3}{5} - \dfrac{x - 2}{6} = \dfrac{23}{30}$

16. $\dfrac{5}{8x} - 2 = \dfrac{3}{x}$

17. $n + \dfrac{4}{n} = \dfrac{13}{3}$

18. $\dfrac{x}{8} = \dfrac{6}{x - 2}$

19. _____

20. _____

21. _____

22. _____

23. _____

24. _____

25. _____

19. $\dfrac{x}{x-1} + \dfrac{2}{x+1} = \dfrac{8}{3}$

20. $\dfrac{3}{2x+1} = \dfrac{5}{3x-6}$

21. $\dfrac{4}{n^2-n} - \dfrac{3}{n-1} = -1$

22. $\dfrac{3n-1}{3} + \dfrac{2n+5}{4} = \dfrac{4n-6}{9}$

For Problems 23–25, set up an equation and solve the problem.

23. The sum of a number and twice its reciprocal is $3\dfrac{2}{3}$. Find the number.

24. Wendy can ride her bicycle 42 miles in the same time that it takes Betty to ride her bicycle 36 miles. Wendy rides 2 miles per hour faster than Betty. Find Wendy's rate.

25. Garth can mow a lawn in 20 minutes, and Alex can mow the same lawn in 30 minutes. How long would it take the two of them working together to mow the lawn?

Chapters 1-6 Cumulative Review Problem Set

For Problems 1–8, evaluate each algebraic expression for the given values of the variables. First you may want to simplify the expression or change its form by factoring.

1. $3x - 2xy - 7x + 5xy$ for $x = \dfrac{1}{2}$ and $y = 3$

2. $7(a - b) - 3(a - b) - (a - b)$ for $a = -3$ and $b = -5$

3. $\dfrac{xy + yz}{y}$ for $x = \dfrac{2}{3}$, $y = \dfrac{5}{6}$, and $z = \dfrac{3}{4}$

4. $ab + b^2$ for $a = 0.4$ and $b = 0.6$

5. $x^2 - y^2$ for $x = -6$ and $y = 4$

6. $x^2 + 5x - 36$ for $x = -9$

7. $\dfrac{x^2 + 2x}{x^2 + 5x + 6}$ for $x = -6$

8. $\dfrac{x^2 + 3x - 10}{x^2 - 9x + 14}$ for $x = 4$

For Problems 9–16, evaluate each of the expressions.

9. 3^{-3}

10. $\left(\dfrac{2}{3}\right)^{-1}$

11. $\left(\dfrac{1}{2} + \dfrac{1}{3}\right)^0$

12. $\left(\dfrac{1}{3} + \dfrac{1}{4}\right)^{-1}$

13. -4^{-2}

14. $\left(\dfrac{2}{3}\right)^{-2}$

15. $\dfrac{1}{\left(\dfrac{2}{5}\right)^{-2}}$

16. $(-3)^{-3}$

For Problems 17–32, perform the indicated operations and express your answers in simplest form.

17. $\dfrac{7}{5x} + \dfrac{2}{x} - \dfrac{3}{2x}$

18. $\dfrac{4x}{5y} \div \dfrac{12x^2}{10y^2}$

19. $\dfrac{4}{x - 6} + \dfrac{3}{x + 4}$

20. $\dfrac{2}{x^2 - 4x} - \dfrac{3}{x^2}$

21. $\dfrac{x^2 - 8x}{x^2 - x - 56} \cdot \dfrac{x^2 - 49}{3xy}$

22. $\dfrac{5}{x^2 - x - 12} - \dfrac{3}{x - 4}$

23. $(-5x^2y)(7x^3y^4)$

24. $(9ab^3)^2$

25. $(-3n^2)(5n^2 + 6n - 2)$

26. $(5x - 1)(3x + 4)$

27. $(2x + 5)^2$

28. $(x + 2)(2x^2 - 3x - 1)$

29. $(x^2 - x - 1)(x^2 + 2x - 3)$

30. $(-2x - 1)(3x - 7)$

31. $\dfrac{24x^2y^3 - 48x^4y^5}{8xy^2}$

32. $(28x^2 - 19x - 20) \div (4x - 5)$

For Problems 33–42, factor each polynomial completely.

33. $3x^3 + 15x^2 + 27x$

34. $x^2 - 100$

35. $5x^2 - 22x + 8$

36. $8x^2 - 22x - 63$

37. $n^2 + 25n + 144$

38. $nx + ny - 2x - 2y$

39. $3x^3 - 3x$

40. $2x^3 - 6x^2 - 108x$

41. $36x^2 - 60x + 25$

42. $3x^2 - 5xy - 2y^2$

For Problems 43–57, solve each of the equations.

43. $3(x - 2) - 2(x + 6) = -2(x + 1)$

44. $x^2 = -11x$

45. $0.2x - 3(x - 0.4) = 1$

46. $\dfrac{3n - 1}{4} = \dfrac{5n + 2}{7}$

47. $5n^2 - 5 = 0$

48. $x^2 + 5x - 6 = 0$

49. $n + \dfrac{4}{n} = 4$

50. $\dfrac{2x + 1}{2} + \dfrac{3x - 4}{3} = 1$

51. $2(x - 1) - x(x - 1) = 0$

52. $\dfrac{3}{2x} - 1 = \dfrac{5}{3x} + 2$

53. $6t^2 + 19t - 7 = 0$

54. $(2x - 1)(x - 8) = 0$

55. $(x + 1)(x + 6) = 24$

56. $\dfrac{x}{x - 2} - \dfrac{7}{x + 1} = 1$

57. $\dfrac{1}{n} - \dfrac{2}{n - 1} = \dfrac{3}{n}$

For Problems 58–67, set up an equation or an inequality to help solve each problem.

58. One leg of a right triangle is 2 inches longer than the other leg. The hypotenuse is 4 inches longer than the shorter leg. Find the lengths of the three sides of the right triangle.

59. Twenty percent of what number is 15?

60. How many milliliters of a 65% solution of hydrochloric acid must be added to 40 milliliters of a 30% solution of hydrochloric acid to obtain a 55% hydrochloric acid solution?

61. The material for a landscaping border 28 feet long was bent into the shape of a rectangle. The length of the rectangle was 2 feet more than the width. Find the dimensions of the rectangle.

62. Two motorcyclists leave Daytona Beach at the same time and travel in opposite directions. If one travels at 55 miles per hour and the other travels at 65 miles per hour, how long will it take for them to be 300 miles apart?

63. Find the length of an altitude of a trapezoid with bases of 10 centimeters and 22 centimeters and an area of 120 square centimeters.

64. If a car uses 16 gallons of gasoline for a 352-mile trip, at the same rate of consumption, how many gallons will it use on a 594-mile trip?

65. Swati had scores of 89, 92, 87, and 90 on her first four history exams. What score must she get on the fifth exam to have an average of 90 or higher for the five exams?

66. Two less than three times a certain number is less than 10. Find all positive integers that satisfy this relationship.

67. One more than four times a certain number is greater than 15. Find all real numbers that satisfy this relationship.

Coordinate Geometry

Chapter 7 Warm-Up Problems

1. Plot these points on the number line: 1, 0, −2.

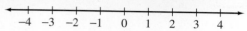

2. Evaluate the expressions for the given values of x.

   (a) $\frac{1}{2}x - 5$ if $x = 0, -2, 6$ (b) $x + 12$ if $x = -5, 1, \frac{5}{2}$

   (c) $-3x + 5$ if $x = -2, 0, 6$

3. Solve for the variable.

   (a) $3(0) - 2y = -6$ (b) $\frac{1}{2}x + 7(0) = -3$ (c) $4x - 3y = 15$ if $x = 0$

4. Solve for the variable.

   (a) $\frac{3}{5}x - 7 < 2$ (b) $6x - 1 > 2x - 3$ (c) $3x - 2(4x + 5) \leq 0$

5. Simplify.   (a) $\frac{7 - 3}{3 - 1}$ (b) $\frac{2 - (-1)}{0 - 3}$ (c) $\frac{-8 - 3}{5 - (-4)}$

6. Solve the proportions for the variable.

   (a) $\frac{30}{45} = \frac{15}{x}$ (b) $\frac{12}{y} = \frac{3}{4}$ (c) $\frac{4}{100} = \frac{m}{650}$

7. Solve for the indicated variable.   (a) $4x - 6y = 3$ for x

   (b) $5x - y = 6$ for y (c) $y - 3 = 2(x - 3)$ for y

8. Solve for x.   (a) $\frac{8 + 2}{3 - x} = 5$ (b) $\frac{-7 - x}{2 - 6} = 4$ (c) $\frac{12 + x}{7 + 9} = \frac{9}{8}$

René Descartes, a French mathematician of the 17th century, transformed geometric problems into an algebraic setting so that he could use the tools of algebra to solve those problems. This interface of algebraic and geometric ideas is the foundation of a branch of mathematics called analytic geometry, today more commonly called **coordinate geometry**.

We started to make this connection between algebra and geometry in Chapter 2 when graphing the solution sets of inequalities in one variable. For example, the solution set for $3x + 1 \geq 4$ is $\{x \mid x \geq 1\}$, and a geometric picture of this solution set is shown in Figure 7.1.

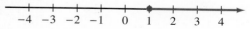

Figure 7.1

In this chapter we will associate pairs of real numbers with points in a geometric plane. This will provide the basis for obtaining pictures of algebraic equations and inequalities in two variables.

Video tutorials for all section learning objectives are available in a variety of delivery modes.

7.1 Cartesian Coordinate System
7.2 Graphing Linear Equations and Applications
7.3 Graphing Linear Inequalities
7.4 Slope of a Line
7.5 Writing Equations of Lines

INTERNET PROJECT

René Descartes devised the system of associating points on a plane with ordered pairs of real numbers. Do an Internet search to read about the legend associated with Descartes's inspiration for the rectangular coordinate system. The most commonly used coordinate system today is the latitude and longitude system. What are the reference planes used to define latitude and longitude?

7.1 Cartesian Coordinate System

OBJECTIVES

1. Plot Points on a Rectangular Coordinate System
2. Draw Graphs of Equations by Plotting Points
3. Solve Equations for the Specified Variable

1 Plot Points on a Rectangular Coordinate System

In Section 1.2 we introduced the real number line (Figure 7.2) as the result of setting up a one-to-one correspondence between the set of real numbers and the points on a line. Recall that the number associated with a point on the line is called the coordinate of that point.

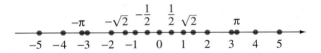

Figure 7.2

Now let's consider two lines (one vertical and one horizontal) that are perpendicular to each other at the point we associate with zero on both lines (Figure 7.3). We refer to these number lines as the horizontal and vertical **axes** or together as the **coordinate axes**; they partition the plane into four parts called **quadrants**. The quadrants are numbered counterclockwise from I to IV as indicated in Figure 7.3. The point of intersection of the two axes is called the **origin**.

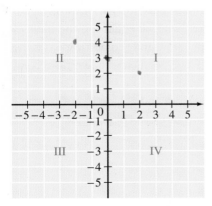

Figure 7.3

It is now possible to set up a one-to-one correspondence between ordered pairs of real numbers and the points in a plane. To each ordered pair of real numbers there corresponds a unique point in the plane, and to each point there corresponds a unique ordered pair of real numbers. We have indicated a part of this correspondence in

Figure 7.4. The ordered pair (3, 1) corresponds to point *A* and denotes that the point *A* is located 3 units to the right of and 1 unit up from the origin. (The ordered pair (0, 0) corresponds to the origin.) The ordered pair (−2, 4) corresponds to point *B* and denotes that point *B* is located 2 units to the left of and 4 units up from the origin. Make sure that you agree with all of the other points plotted in Figure 7.4.

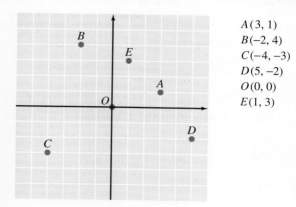

$A(3, 1)$
$B(-2, 4)$
$C(-4, -3)$
$D(5, -2)$
$O(0, 0)$
$E(1, 3)$

Figure 7.4

Remark: The notation (−2, 4) was used earlier in this text to indicate an interval of the real number line. Now we are using the same notation to indicate an ordered pair of real numbers. This double meaning should not be confusing because the context of the material will always indicate which meaning of the notation is being used. Throughout this chapter, we will be using the ordered-pair interpretation.

In general, we refer to the real numbers *a* and *b* in an ordered pair (*a*, *b*) associated with a point as the **coordinates of the point**. The first number, *a*, called the **abscissa**, is the directed distance of the point from the vertical axis measured parallel to the horizontal axis. The second number, *b*, called the **ordinate**, is the directed distance of the point from the horizontal axis measured parallel to the vertical axis [see Figure 7.5(a)]. Thus in the first quadrant, all points have a positive abscissa and a positive ordinate. In the second quadrant, all points have a negative abscissa and a positive ordinate. We have indicated the signs in all four quadrants in Figure 7.5(b). This system of associating points with ordered pairs of real numbers is called the **Cartesian coordinate system** or the **rectangular coordinate system**.

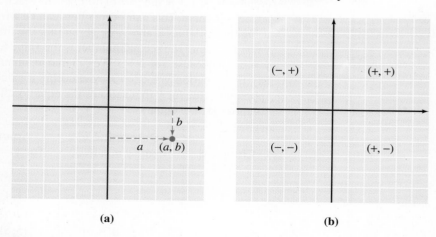

(a)   (b)

Figure 7.5

Plotting points on a rectangular coordinate system can be helpful when analyzing data to determine a trend or relationship. The following example shows the plot of some data.

EXAMPLE 1

The following chart shows the Friday and Saturday scores of golfers in terms of par. Plot the charted information on a rectangular coordinate system. For each golfer, let Friday's score be the first number in the ordered pair, and let Saturday's score be the second number in the ordered pair.

	Mark	Ty	Vinay	Bill	Herb	Rod
Friday's score	1	−2	−1	4	−3	0
Saturday's score	3	−2	0	7	−4	1

Solution

The ordered pairs are as follows:

Mark $(1, 3)$ Ty $(-2, -2)$
Vinay $(-1, 0)$ Bill $(4, 7)$
Herb $(-3, -4)$ Rod $(0, 1)$

The points are plotted on the rectangular coordinate system in Figure 7.6. In the study of statistics, this graph of the charted data would be called a scatterplot. For this plot, the points appear to approximate a straight-line path, which suggests that there is a linear correlation between Friday's score and Saturday's score.

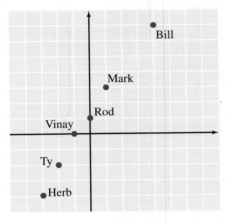

Figure 7.6

2 Draw Graphs of Equations by Plotting Points

In Section 2.6, the real number line was used to display the solution set of an inequality. Recall that the solution set of an inequality such as $x > 3$ has an infinite number of solutions. Hence the real number line is an effective way to display the solution set of an inequality.

Now we want to find the solution sets for equations with two variables. Let's begin by considering solutions for the equation $y = x + 3$. A solution of an equation in two variables is a pair of numbers that makes the equation a true statement. For the equation $y = x + 3$, the pair of numbers in which $x = 4$ and $y = 7$ makes the equation a true statement. This pair of numbers can be written as an ordered pair $(4, 7)$. When using the variables x and y, we agree that the first number of an ordered pair is the value for x, and the second number is the value for y. Likewise $(2, 5)$ is a solution for $y = x + 3$ because $5 = 2 + 3$. We can find an infinite number of ordered pairs that satisfy the equation $y = x + 3$ by choosing a value for x and determining the corresponding value for y that satisfies the equation. The following table lists some of the solutions for the equation $y = x + 3$.

Choose x	Determine y from $y = x + 3$	Solution for $y = x + 3$
0	3	$(0, 3)$
1	4	$(1, 4)$
3	6	$(3, 6)$
5	8	$(5, 8)$
−1	2	$(-1, 2)$
−3	0	$(-3, 0)$
−5	−2	$(-5, -2)$

Because the number of solutions for the equation $y = x + 3$ is infinite, we do not have a convenient way to list the solution set. This is similar to inequalities in which the solution set is infinite. For inequalities we used a real number line to display the solution set. Now we will use the rectangular coordinate system to display the solution set.

On a rectangular coordinate system where we label the horizontal axis as the x axis and the vertical axis as the y axis, we can locate the point associated with each ordered pair of numbers in the table. These points are shown in Figure 7.7(a). These points are only some of the infinite solutions of the equation $y = x + 3$. The straight line in Figure 7.7(b) that connects the points represents all the solutions of the equation and is called the graph of the equation $y = x + 3$.

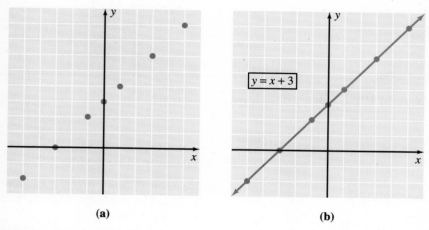

(a)   (b)

Figure 7.7

The next examples further illustrate the process of graphing equations.

EXAMPLE 2

Graph $y = 2x + 1$.

Solution

First we set up a table of some of the solutions for the equation $y = 2x + 1$. You can choose any value for the variable x and find the corresponding y value. The values we chose for x are in the table and include positive integers, zero, and negative integers.

x	y	Solutions (x, y)
0	1	(0, 1)
1	3	(1, 3)
2	5	(2, 5)
-1	-1	(-1, -1)
-2	-3	(-2, -3)
-3	-5	(-3, -5)

From the table we plot the points associated with the ordered pairs as shown in Figure 7.8(a). Connecting these points with a straight line produces the graph of the equation $y = 2x + 1$ as shown in Figure 7.8(b).

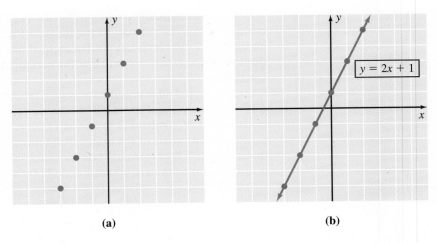

(a)   (b)

Figure 7.8

EXAMPLE 3

Graph $y = -\frac{1}{2}x + 3$.

Solution

First we set up a table of some of the solutions for the equation $y = -\frac{1}{2}x + 3$. You can choose any value for the variable x and find the corresponding y value; however, the values we chose for x are numbers that are divisible by 2. (This is not necessary but does produce integer values for y.)

x	y	Solutions (x, y)
0	3	(0, 3)
2	2	(2, 2)
4	1	(4, 1)
−2	4	(−2, 4)
−4	5	(−4, 5)

From the table we plot the points associated with the ordered pairs as shown in Figure 7.9(a). Connecting these points with a straight line produces the graph of the equation $y = -\frac{1}{2}x + 3$ as shown in Figure 7.9(b).

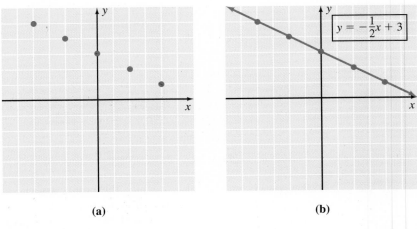

(a)   (b)

Figure 7.9

3 Solve Equations for the Specified Variable

Sometimes changing the form of an equation makes it easier to find solutions of the equation. In the next two examples, the equation will be solved for x or y.

EXAMPLE 4 Solve $4x + 9y = 12$ for y.

Solution

$$4x + 9y = 12$$
$$9y = -4x + 12 \quad \text{Subtracted } 4x \text{ from each side}$$
$$y = -\frac{4}{9}x + \frac{12}{9} \quad \text{Divided both sides by 9}$$
$$y = -\frac{4}{9}x + \frac{4}{3}$$

EXAMPLE 5 Solve $2x - 6y = 3$ for x.

Solution

$$2x - 6y = 3$$
$$2x = 6y + 3 \quad \text{Added } 6y \text{ to each side}$$
$$x = \frac{6}{2}y + \frac{3}{2} \quad \text{Divided both sides by 2}$$
$$x = 3y + \frac{3}{2}$$

To graph an equation in two variables, x and y, keep these steps in mind:

1. Solve the equation for y in terms of x or for x in terms of y, if it is not already in such a form.
2. Set up a table of ordered pairs that satisfies the equation.
3. Plot the points associated with the ordered pairs.
4. Connect the points.

We conclude this section with Example 6, which illustrates the steps.

EXAMPLE 6 Graph $6x + 3y = 9$.

Solution

First, let's change the form of the equation to make it easier to find solutions of the equation $6x + 3y = 9$. We can solve either for x in terms of y or for y in terms of x. Typically the equation is solved for y in terms of x, so that's what we show here.

$$6x + 3y = 9$$
$$3y = -6x + 9$$
$$y = \frac{-6x}{3} + \frac{9}{3}$$
$$y = -2x + 3$$

Now we can set up a table of values. Plotting these points and connecting them produces Figure 7.10.

x	y
0	3
1	1
2	−1
−1	5

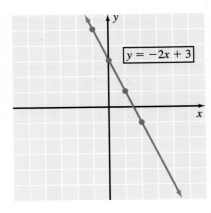

Figure 7.10

CONCEPT QUIZ 7.1

For Problems 1–5, answer true or false.

1. In a rectangular coordinate system, the coordinate axes partition the plane into four parts called quadrants.
2. Quadrants are named with Roman numerals and numbered clockwise.
3. The real numbers in an ordered pair are referred to as the coordinates of the point.
4. The equation $y = x + 3$ has an infinite number of ordered pairs that satisfy the equation.
5. The point of intersection of the coordinate axes is called the origin.

For Problems 6–10, match the points plotted in Figure 7.11 with their coordinates.

6. $(-3, 1)$
7. $(4, 0)$
8. $(3, -1)$
9. $(0, 4)$
10. $(-1, -3)$

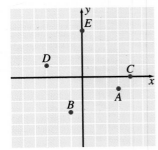

Figure 7.11

Section 7.1 Classroom Problem Set

Objective 1

1. The following chart shows for certain towns the amount of increase or decrease in the rainfall for the months of May and June compared to the average rainfall in inches. Plot the charted information on a rectangular coordinate system. For each town, let May's value be the first number in the ordered pair, and let June's value be the second number in the ordered pair.

	Tampa	Orlando	Miami	Naples	Tallahassee
May	−2	1	0	−1	3
June	3	3	−2	−4	−1

2. The following chart shows the amount of light and water above or below the normal amount given to a specific plant for five days. Plot the charted information on a rectangular coordinate system. Let the change in light be the first number in the ordered pair, and let the change in water be the second number in the ordered pair.

	Mon.	Tue.	Wed.	Thu.	Fri.
Change in light	1	−2	−1	4	−3
Change in water	−3	4	−1	0	−5

Objective 2

3. Graph $y = 3x - 2$.
4. Graph $y = -2x + 3$.
5. Graph $y = \frac{1}{3}x - 1$.
6. Graph $y = -\frac{3}{4}x + 2$.

Objective 3

7. Solve $2x - 3y = 9$ for y.
8. Solve $5x + 9y = 17$ for y.
9. Solve $3x + 5y = 7$ for x.
10. Solve $4x - 3y = 9$ for x.
11. Solve the equation for y and graph the equation.
    $$4x - 2y = 8$$
12. Solve the equation for y and graph the equation.
    $$3x + 6y = 12$$

THOUGHTS INTO WORDS

1. How would you convince someone that there are infinitely many ordered pairs of real numbers that satisfy the equation $x + y = 9$?

2. Explain why no points of the graph of the equation $y = x$ will be in the second quadrant.

Answers to the Concept Quiz
1. True 2. False 3. True 4. True 5. True 6. D 7. C 8. A 9. E 10. B

7.2 Graphing Linear Equations and Applications

OBJECTIVES

1. Find x and y Intercepts for Linear Equations
2. Graph Linear Equations
3. Apply Linear Equations

1 Find x and y Intercepts for Linear Equations

In the preceding section we graphed equations whose graphs were straight lines. In general any equation of the form $Ax + By = C$, where A, B, and C are constants (A and B not both zero) and x and y are variables, is a **linear equation in two variables**, and its graph is a straight line.

We should clarify two points about this description of a linear equation in two variables. First, the choice of x and y for variables is arbitrary. We could use any two letters to represent the variables. An equation such as $3m + 2n = 7$ can be considered a linear equation in two variables. So that we are not constantly changing the labeling of the coordinate axes when graphing equations, however, it is much easier to use the same two variables in all equations. Thus we will go along with convention and use x and y as our variables. Second, the statement "any equation of the form $Ax + By = C$" technically means "any equation of the form $Ax + By = C$ or *equivalent* to the form." For example, the equation $y = x + 3$, which has a straight line graph, is equivalent to $-x + y = 3$.

All of the following are examples of linear equations in two variables.

$$y = x + 3 \qquad y = -3x + 2 \qquad y = \frac{2}{5}x + 1 \qquad y = 2x$$

$$3x - 2y = 6 \qquad x - 4y = 5 \qquad 5x - y = 10 \qquad y = \frac{2x + 4}{3}$$

The knowledge that any equation of the form $Ax + By = C$ produces a straight line graph, along with the fact that two points determine a straight line, makes graphing linear equations in two variables a simple process. We merely find two solutions, plot the corresponding points, and connect the points with a straight line. It is probably wise to find a third point as a check point. Let's consider an example.

EXAMPLE 1

Graph $2x - 3y = 6$.

Solution

We recognize that equation $2x - 3y = 6$ is a linear equation in two variables, and therefore its graph will be a straight line. All that is necessary is to find two solutions and connect the points with a straight line. We will, however, also find a third solution to serve as a check point.

Let $x = 0$; then $2(0) - 3y = 6$
$$-3y = 6$$
$$y = -2 \qquad \text{Thus } (0, -2) \text{ is a solution}$$

Let $y = 0$; then $2x - 3(0) = 6$
$$2x = 6$$
$$x = 3 \qquad \text{Thus } (3, 0) \text{ is a solution}$$

Let $x = -3$; then $2(-3) - 3y = 6$
$$-6 - 3y = 6$$
$$-3y = 12$$
$$y = -4 \qquad \text{Thus } (-3, -4) \text{ is a solution}$$

We can plot the points associated with these three solutions and connect them with a straight line to produce the graph of $2x - 3y = 6$ in Figure 7.12.

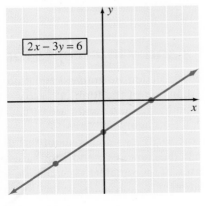

Figure 7.12

Let us review the approach to solving Example 1.

- We did not begin by solving the equation for x or y. We know the graph is a straight line, and there is no need for an extensive table of values.
- The first two solutions, $(0, -2)$ and $(3, 0)$, indicate where the line intersects the coordinate axes.
- The y coordinate of the point $(0, -2)$ is called the y intercept. The y intercept is -2.
- The x coordinate of the point $(3, 0)$ is called the x intercept. The x intercept is 3.
- To find the y intercept, let $x = 0$ and solve for y.
- To find the x intercept, let $y = 0$ and solve for x.
- The third solution, $(-3, -4)$, serves as a check. If $(-3, -4)$ had not been on the line we determined by the two intercepts, then we would have known that we made an error.

2 Graph Linear Equations

EXAMPLE 2   Graph $x + 2y = 4$ by plotting the intercepts and a check point.

Solution

Let's find the x and y intercepts, and then determine a third point as a check point.

To find the y intercept, let $x = 0$.

$$x + 2y = 4$$
$$0 + 2y = 4$$
$$2y = 4$$
$$y = 2$$

To find the x intercept, let $y = 0$.

$$x + 2y = 4$$
$$x + 2(0) = 4$$
$$x + 0 = 4$$
$$x = 4$$

Therefore the points $(0, 2)$ and $(4, 0)$ are on the graph.
To find a check point, let $x = 2$.

$$x + 2y = 4$$
$$2 + 2y = 4 \quad \text{when } x = 2$$
$$2y = 2$$
$$y = 1$$

Therefore, the point $(2, 1)$ is on the graph. We plot the points $(0, 2)$, $(4, 0)$, and $(2, 1)$ and connect them with a straight line to produce the graph in Figure 7.13.

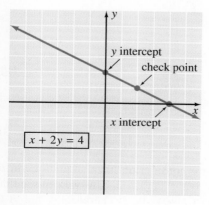

Figure 7.13

262 Chapter 7 Coordinate Geometry

EXAMPLE 3

Graph $2x + 3y = 7$.

Solution

The intercepts and a check point are given in the table. Finding intercepts may involve fractions, but the computation is usually easy. We plot the points from the table and show the graph of $2x + 3y = 7$ in Figure 7.14.

x	y	
0	$\frac{7}{3}$	Intercepts
$\frac{7}{2}$	0	
2	1	Check point

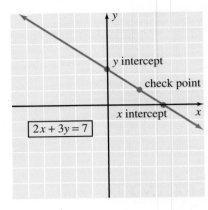

Figure 7.14

EXAMPLE 4

Graph $y = 2x$.

Solution

Notice that $(0, 0)$ is a solution; thus, this line intersects both axes at the origin. Since both the x intercept and the y intercept are determined by the origin, $(0, 0)$, we need another point to graph the line. Then a third point should be found to serve as a check point. These results are summarized in the table; the graph of $y = 2x$ is shown in Figure 7.15.

x	y	
0	0	Intercept
2	4	Additional point
−1	−2	Check point

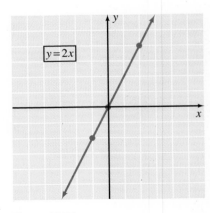

Figure 7.15

EXAMPLE 5

Graph $x = 3$.

Solution

Since we are considering linear equations in *two variables*, the equation $x = 3$ is equivalent to $x + 0(y) = 3$. Now we can see that any value of y can be used, but the x value must always be 3. Therefore, some of the solutions are $(3, 0)$, $(3, 1)$, $(3, 2)$,

(3, −1), and (3, −2). The graph of all the solutions is the vertical line indicated in Figure 7.16.

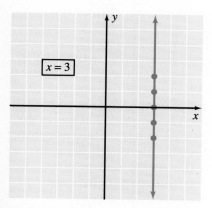

Figure 7.16

EXAMPLE 6 Graph $y = -3$.

Solution

Since we are considering equations in two variables, the equation $y = 4$ is equivalent to $0(x) + y = -3$. Now we can determine that for any value of x, the value of y will be -3. Therefore some of the solutions are $(-1, -3), (0, -3), (2, -3)$, and $(4, -3)$. The graph of all the solutions is a horizontal line indicated in Figure 7.17.

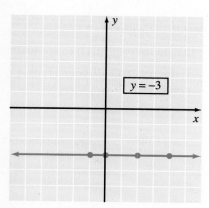

Figure 7.17

3 Apply Linear Equations

Linear equations in two variables can be used to model many different types of real-world problems. For example, suppose that a retailer wants to sell some items at a profit of 30% of the cost of each item. If we let s represent the selling price and c the cost of each item, then we can use the equation

$$s = c + 0.3c = 1.3c$$

to determine the selling price of each item on the basis of the cost of the item. For example, if the cost of an item is $4.50, then the retailer should sell it for $s = (1.3)(4.5) = \$5.85$.

By finding values that satisfy the equation $s = 1.3c$, we can create this table:

c	1	5	10	15	20
s	1.3	6.5	13	19.5	26

From the table we see that if the cost of an item is $15, then the retailer should sell it for $19.50 in order to make a profit of 30% of the cost.

Now let's get a picture (graph) of this linear relationship. We can label the horizontal axis c and the vertical axis s, and we can use the origin along with one ordered pair from the table to produce the straight line graph in Figure 7.18. (Because of the type of application, only nonnegative values for c and s are appropriate.)

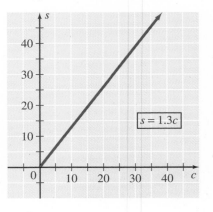

Figure 7.18

From the graph we can approximate s values on the basis of given c values. For example, if $c = 30$, then by reading up from 30 on the c axis to the line and then across to the s axis, we see that s is a little less than 40. (We get an exact s value of 39 by using the equation $s = 1.3c$.)

EXAMPLE 7

Apply Your Skill

Kristin has a cell phone plan that charges her two and one-half cents per minute of cell phone usage. The equation $C = 0.025m$, where m is the number of minutes used and C is the charge in dollars, describes this cell phone plan.

(a) Complete a table that shows the charges for 100 minutes, 300 minutes, 500 minutes, 800 minutes, and 1000 minutes.

(b) Graph the equation $C = 0.025m$.

(c) Use the graph from part b to approximate the charge when $m = 600$.

(d) Check the accuracy of your reading from the graph in part c by using the equation $C = 0.025m$.

Solution

(a) By finding values that satisfy the equation $C = 0.025m$, we can create this table:

m	100	300	500	800	1000
C	2.50	7.50	12.50	20.00	25.00

(b) Label the horizontal axis m and the vertical axis C and plot the points given in the table to create Figure 7.19. Connect the points with a straight line.

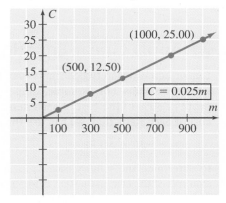

Figure 7.19

(c) On the horizontal axis, locate where $m = 600$. From that point read up until you intersect the graph and then go across to the vertical axis. We can see that when $m = 600$, C is approximately 15.00.

(d) To check the accuracy, use the equation $C = 0.025m$ and evaluate when $m = 600$.

$$C = 0.025m$$
$$C = 0.025(600) = 15 \quad \text{when } m = 600.$$

Many formulas that are used in various applications are linear equations in two variables. For example, the formula $C = \frac{5}{9}(F - 32)$, which converts temperatures from the Fahrenheit scale to the Celsius scale, is a linear relationship. The next example shows the graph of the formula $C = \frac{5}{9}(F - 32)$.

EXAMPLE 8

Apply Your Skill

The equation $C = \frac{5}{9}(F - 32)$, where F is the degrees Fahrenheit and C is the degrees Celsius, describes the relationship between the two temperature scales.

(a) Complete a table that shows the degrees Celsius for $-22°F$, $-13°F$, $5°F$, $32°F$, $50°F$, $68°F$, and $86°F$.

(b) Graph the equation $C = \frac{5}{9}(F - 32)$.

(c) Use the graph from part b to approximate the temperature in degrees Celsius when the temperature is $77°F$.

(d) Check the accuracy of your reading from the graph in part c by using the equation $C = \frac{5}{9}(F - 32)$.

Solution

(a) By finding values that satisfy the equation $C = \frac{5}{9}(F - 32)$, we can create this table:

F	-22	-13	5	32	50	68	86
C	-30	-25	-15	0	10	20	30

(b) To graph the equation $C = \frac{5}{9}(F - 32)$, we can label the horizontal axis F and the vertical axis C and plot two points that are given in the table. Figure 7.20 shows the graph of the equation.

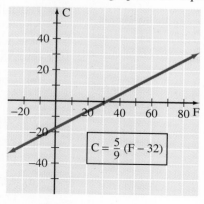

Figure 7.20

(c) From the graph we can approximate C values on the basis of given F values. For example, if F = 77°, then by reading up from 77 on the F axis to the line and then across to the C axis, we see that C is approximately 24°.

(d) To check the accuracy, use the equation $C = \frac{5}{9}(F - 32)$ and evaluate for F = 77°.

$$C = \frac{5}{9}(F - 32)$$

When F = 77,

$$C = \frac{5}{9}(77 - 32) = \frac{5}{9}(45) = 25$$ ∎

CONCEPT QUIZ 7.2

For Problems 1–7, answer true or false.

1. The equations $2x + y = 4$ and $y = -2x + 4$ are equivalent.
2. The y intercept of the graph of $3x + 4y = -12$ is -4.
3. The x intercept of the graph of $2x + 4y = -8$ is -4.
4. Determining just two points is sufficient to graph a straight line.
5. The graph of $y = 4$ is a vertical line.
6. The graph of $x = 3$ is a vertical line.
7. The graph of every linear equation has a y intercept.

Section 7.2 Classroom Problem Set

Objective 1

1. Graph $2x + 3y = 12$.
2. Graph $5x - 2y = -10$.

Objective 2

3. Graph $2x - y = 6$ by plotting the x and y intercepts and a check point.
4. Graph $3x + 5y = -15$ by plotting the x and y intercepts and a check point.
5. Graph $x - 2y = 5$ by plotting the x and y intercepts and a check point.
6. Graph $4x + 5y = 10$ by plotting the x and y intercepts and a check point.
7. Graph $y = -3x$.
8. Graph $x = 2y$.
9. Graph $x = -2$.
10. Graph $x = 5$.
11. Graph $y = 4$.
12. Graph $y = -3$.

Objective 3

13. Alex pays $5 a month and $1 per hour to use a video game program. The equation $C = m + 5$, where m is the number of hours played and C is the charge in dollars, describes the charges for this video game.

    (a) Complete a table that shows the charges for 1 hour, 3 hours, 5 hours, and 10 hours.

    (b) Graph the equation $C = m + 5$.

    (c) Use the graph from part b to approximate the charge when $m = 2.5$.

    (d) Check the accuracy of your reading from the graph in part c by using the equation $C = m + 5$.

14. The daily profit from an ice cream stand is given by the equation $p = 2n - 4$, where n represents the number of gallons of ice cream mix used in a day and p represents the number of dollars of profit.

    (a) Complete a table that shows the profits for 4 gallons, 6 gallons, 8 gallons, 10 gallons, and 12 gallons.

    (b) Graph the equation $p = 2n - 4$.

    (c) Use the graph from part b to approximate the profit when $n = 11$.

    (d) Check the accuracy of your reading from the graph in part c by using the equation $p = 2n - 4$.

15. Aidan is spending a semester studying in Ireland. While he is there he uses the formula $D = 1.5e$ to convert euros, e, to dollars, D.

    (a) Complete a table that shows the amount of dollars for 2 euros, 3 euros, 5 euros, and 10 euros.

    (b) Graph the equation $D = 1.5e$.

    (c) Use the graph from part b to approximate the dollars when $e = 6$.

    (d) Check the accuracy of your reading from the graph in part c by using the equation $D = 1.5e$.

16. The cost (c) of playing an online computer game for a time (t) is given by the equation $c = 3t + 5$. Label the horizontal axis t and the vertical axis c, and graph the equation for nonnegative values of t.

THOUGHTS INTO WORDS

1. Your friend is having trouble understanding why the graph of the equation $y = 3$ is a horizontal line containing the point $(0, 3)$. What might you do to help him?

2. How do we know that the graph of $y = -4x$ is a straight line that contains the origin?

3. Do all graphs of linear equations have x intercepts? Explain your answer.

4. How do we know that the graphs of $x - y = 4$ and $-x + y = -4$ are the same line?

Answers to the Concept Quiz
1. True   2. False   3. True   4. True   5. False   6. True   7. False

7.3 Graphing Linear Inequalities

OBJECTIVE

1. Graph Linear Inequalities

1 Graph Linear Inequalities

Linear inequalities in two variables are of the form $Ax + By > C$ or $Ax + By < C$, where A, B, and C are real numbers. (Combined linear equality and inequality statements are of the form $Ax + By \geq C$ or $Ax + By \leq C$.) Graphing linear inequalities is almost as easy as graphing linear equations. Our discussion leads to a simple step-by-step process. First we consider the following equation and related inequalities:

$$x - y = 2$$
$$x - y > 2$$
$$x - y < 2$$

The graph of $x - y = 2$ is shown in Figure 7.21. The line divides the plane into two half-planes, one above the line and one below the line. In Figure 7.22(a) we have indicated the coordinates for several points above the line. Note that for each point, the ordered pair of real numbers satisfies the inequality $x - y < 2$. This is true for *all points* in the half-plane above the line. Therefore, the graph of $x - y < 2$ is the half-plane above the line, indicated

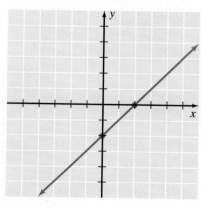

Figure 7.21

by the shaded region in Figure 7.22(b). We use a dashed line to indicate that points on the line do not satisfy $x - y < 2$.

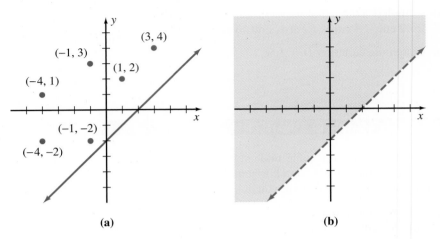

Figure 7.22

In Figure 7.23(a), we have indicated the coordinates of several points below the line $x - y = 2$. Note that for each point, the ordered pair of real numbers satisfies the inequality $x - y > 2$. This is true for *all points* in the half-plane below the line. Therefore, the graph of $x - y > 2$ is the half-plane below the line, indicated by the shaded region in Figure 7.23(b).

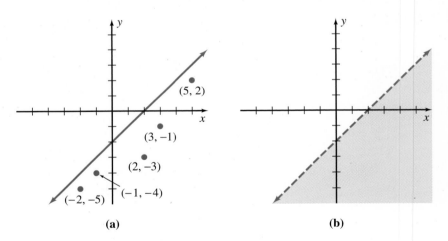

Figure 7.23

Based on this discussion, we suggest the following steps for graphing linear inequalities:

Step 1 Graph the corresponding equality. Use a solid line if the equality is included in the given statement and a dashed line if the equality is not included.

Step 2 Choose a *test point* not on the line and substitute its coordinates into the inequality statement. (The origin is a convenient point to use if it is not on the line.)

Step 3 Determine if the *test point* makes the inequality a true statement or a false statement. In other words, does the test point satisfy the inequality or not?

7.3 Graphing Linear Inequalities

Step 4 The graph of the inequality is

(a) the half-plane containing the test point if the test point makes the inequality a true statement, or

(b) the half-plane not containing the test point if the test point makes the inequality a false statement.

We can apply these steps to some examples.

EXAMPLE 1

Graph $2x + y > 4$.

Solution

Step 1 Graph $2x + y = 4$ as a dashed line, because equality is not included in the given statement $2x + y > 4$; see Figure 7.24(a).

Step 2 Choose the origin as a test point, and substitute its coordinates into the inequality.

$$2x + y > 4 \quad \text{becomes} \quad 2(0) + 0 > 4$$

Step 3 The inequality $2(0) + 0 > 4$ simplifies to $0 > 4$, which is a false statement.

Step 4 Since the test point does not satisfy the given inequality, the graph is the half-plane that does not contain the test point. Thus the graph of $2x + y > 4$ is the half-plane above the line, indicated in Figure 7.24(b).

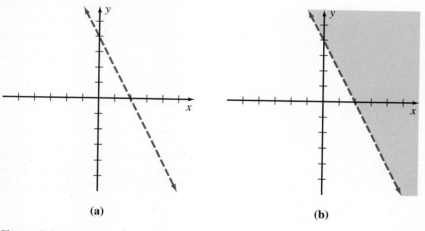

Figure 7.24

EXAMPLE 2

Graph $y \leq 2x$.

Solution

Step 1 Graph $y = 2x$ as a solid line, because equality is included in the given statement [see Figure 7.25(a)].

Step 2 Since the origin is on the line, we need to choose another point as a test point. Let's use $(3, 2)$.

$$y \leq 2x \quad \text{becomes} \quad 2 \leq 2(3)$$

Step 3 The inequality $2 \leq 2(3)$ simplifies to $2 \leq 6$, which is a true statement.

Step 4 Since the test point satisfies the given inequality, the graph is the half-plane that contains the test point. Thus the graph of $y \leq 2x$ is the line along with the half-plane below the line indicated in Figure 7.25(b).

270 Chapter 7 Coordinate Geometry

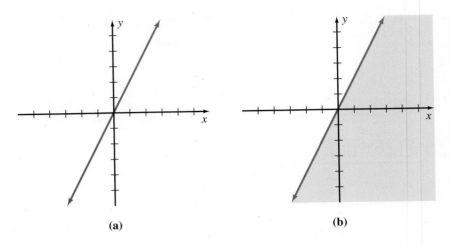

(a)　　　　　　　　　(b)

Figure 7.25

EXAMPLE 3   Graph $y < 4$.

Solution

Step 1   Graph $y = 4$ as a dashed line, because equality is not included in the given statement $y < 4$. Remember the equation $y = 4$ is equivalent to the equation $(0)x + y = 4$; hence for any value of x, the y value is always 4. Therefore the graph of $y = 4$ is a horizontal line as shown in Figure 7.26(a).

Step 2   Choose the origin as a test point, and substitute the coordinates into the inequality.

$$y < 4 \quad \text{becomes} \quad 0 < 4$$

Step 3   The inequality $0 < 4$ is a true statement.

Step 4   Because the test point satisfies the given inequality, the graph is the half-plane that contains the test point. Thus the graph of $y < 4$ is the half-plane below the line as indicated in Figure 7.26(b).

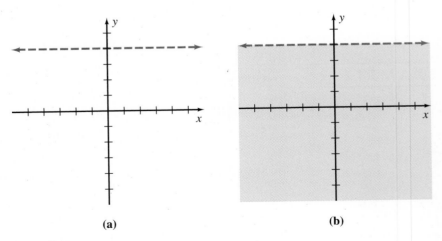

(a)　　　　　　　　　(b)

Figure 7.26

EXAMPLE 4

Graph $x \leq -2$.

Solution

Step 1 Graph $x = -2$ as a solid line, because equality is included in the given statement $x \leq -2$. Remember that the equation $x = -2$ is equivalent to the equation $x + (0)y = -2$; hence for any value of y, the x value is always -2. Therefore the graph of $x = -2$ is a vertical line as shown in Figure 7.27(a).

Step 2 Choose the origin as a test point, and substitute the coordinates into the inequality.

$$x \leq -2 \quad \text{becomes} \quad 0 \leq -2$$

Step 3 The inequality $0 \leq -2$ is a false statement.

Step 4 Because the test point does not satisfy the given inequality, the graph is the half-plane that does not contain the test point. Thus the graph of $x \leq -2$ is the line along with the half-plane to the left of the line as indicated in Figure 7.27(b).

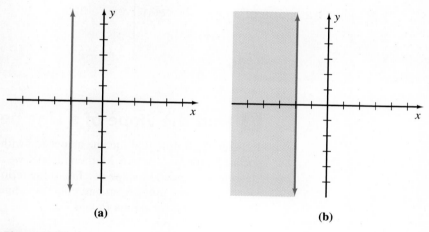

Figure 7.27

CONCEPT QUIZ 7.3

For Problems 1–5, answer true or false.

1. A dashed line on the graph of an inequality indicates that the points on the line do not satisfy the inequality.
2. The coordinates of any point can be used as a test point to determine the half-plane that is the solution of the inequality.
3. The ordered pair $(2, -3)$ satisfies the inequality $2x + y > 1$.
4. The solution set of $2x - y \geq 4$ would include the coordinates of the points on the line $2x - y = 4$.
5. The solution set of $x \geq 1$ would be the coordinates of the points on the line $x = 1$, because the graph of $x = 1$ is a vertical line.

Section 7.3 Classroom Problem Set

Objective 1

1. Graph $x + 2y > 6$.
2. Graph $4x - 3y < 12$.
3. Graph $y \geq -3x$.
4. Graph $y < x$.
5. Graph $y \geq -2$.
6. Graph $y < 3$.
7. Graph $x \geq 3$.
8. Graph $x > -2$.

THOUGHTS INTO WORDS

1. Explain how you would graph the inequality $-x - 2y > 4$.

2. Why is the point $(3, -2)$ not a good test point to use when graphing the inequality $3x - 2y \leq 13$?

Answers to the Concept Quiz
1. True 2. False 3. False 4. True 5. False

7.4 Slope of a Line

OBJECTIVES

1. Find the Slope of a Line between Two Points
2. Graph a Line Given a Point and the Slope
3. Apply the Concept of Slope to Solve Problems

1 Find the Slope of a Line between Two Points

In Figure 7.28, note that the line associated with $4x - y = 4$ is *steeper* than the line associated with $2x - 3y = 6$. Mathematically, we use the concept of *slope* to discuss the steepness of lines. The **slope** of a line is the ratio of the vertical change to the horizontal change as we move from one point on a line to another point. We indicate this in Figure 7.29 with the points P_1 and P_2.

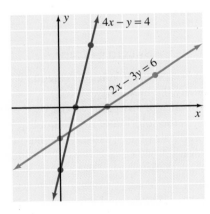

Figure 7.28

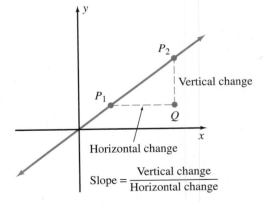

Figure 7.29

We can give a precise definition for slope by considering the coordinates of the points P_1, P_2, and Q in Figure 7.30. Since P_1 and P_2 represent any two points on the line, we assign the coordinates (x_1, y_1) to P_1 and (x_2, y_2) to P_2. The point Q is the same distance from the y axis as P_2 and the same distance from the x axis as P_1. Thus we assign the coordinates (x_2, y_1) to Q (see Figure 7.30). It should now be apparent that the vertical change is $y_2 - y_1$, and the horizontal change is $x_2 - x_1$. Thus we have the following definition for slope.

7.4 Slope of a Line 273

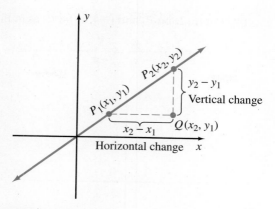

Figure 7.30

Definition 7.1

If points P_1 and P_2 with coordinates (x_1, y_1) and (x_2, y_2), respectively, are any two different points on a line, then the slope of the line (denoted by m) is

$$m = \frac{y_2 - y_1}{x_2 - x_1}, \qquad x_1 \neq x_2$$

Using Definition 7.1, we can easily determine the slope of a line if we know the coordinates of two points on the line.

EXAMPLE 1 Find the slope of the line determined by each of the following pairs of points.

(a) $(2, 1)$ and $(4, 6)$

(b) $(3, 2)$ and $(-4, 5)$

(c) $(-4, -3)$ and $(-1, -3)$

Solution

(a) Let $(2, 1)$ be P_1 and $(4, 6)$ be P_2 as in Figure 7.31; then we have

$$m = \frac{y_2 - y_1}{x_2 - x_1} = \frac{6 - 1}{4 - 2} = \frac{5}{2}$$

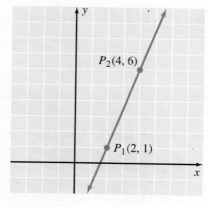

Figure 7.31

(b) Let $(3, 2)$ be P_1 and $(-4, 5)$ be P_2 as in Figure 7.32.

$$m = \frac{y_2 - y_1}{x_2 - x_1} = \frac{5 - 2}{-4 - 3} = \frac{3}{-7} = -\frac{3}{7}$$

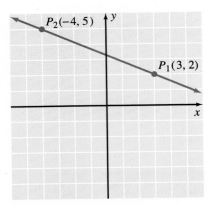

Figure 7.32

(c) Let $(-4, -3)$ be P_1 and $(-1, -3)$ be P_2 as in Figure 7.33.

$$m = \frac{y_2 - y_1}{x_2 - x_1} = \frac{-3 - (-3)}{-1 - (-4)} = \frac{0}{3} = 0$$

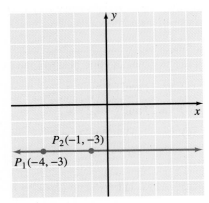

Figure 7.33 ■

The designation of P_1 and P_2 in such problems is arbitrary and does not affect the value of the slope. For example, in part a of Example 1, suppose that we let $(4, 6)$ be P_1 and $(2, 1)$ be P_2. Then we obtain

$$m = \frac{y_2 - y_1}{x_2 - x_1} = \frac{1 - 6}{2 - 4} = \frac{-5}{-2} = \frac{5}{2}$$

The parts of Example 1 illustrate the three basic possibilities for slope: that is, the slope of a line can be *positive*, *negative*, or *zero*.

- **Positive slope** A line that has a positive slope rises as we move from left to right, as in Example 1(a).
- **Negative slope** A line that has a negative slope falls as we move from left to right, as in Example 1(b).
- **Zero slope** A line that has a zero slope is a horizontal line, as in Example 1(c).

Finally, we need to realize that **the concept of slope is undefined for vertical lines**. This is because, for any vertical line, the change in x as we move from one point to another is zero. Thus the ratio $\dfrac{y_2 - y_1}{x_2 - x_1}$ will have a denominator of zero and be undefined. So in Definition 7.1, the restriction $x_1 \neq x_2$ is made.

Example 1 showed how to find the slope of a line when you know the coordinates of two points on the line. To find the slope when we are presented with the equation of the line, we need to first determine two points on the line. Example 2 provides the details on finding the slope when given the equation of a line.

EXAMPLE 2

Find the slope of the line determined by the equation $3x + 4y = 12$.

Solution

Since we can use any two points on the line to determine the slope of the line, let's find the intercepts.

If $x = 0$, then $3(0) + 4y = 12$

$$4y = 12$$
$$y = 3 \quad \text{Thus } (0, 3) \text{ is on the line}$$

If $y = 0$, then $3x + 4(0) = 12$

$$3x = 12$$
$$x = 4 \quad \text{Thus } (4, 0) \text{ is on the line}$$

Using $(0, 3)$ as P_1 and $(4, 0)$ as P_2, we have

$$m = \frac{y_2 - y_1}{x_2 - x_1} = \frac{0 - 3}{4 - 0} = \frac{-3}{4} = -\frac{3}{4}$$

■

2 Graph a Line Given a Point and the Slope

We need to emphasize one final idea pertaining to the concept of slope. The slope of a line is a **ratio** of vertical change to horizontal change. A slope of $\dfrac{3}{4}$ means that for every 3 units of vertical change, there is a corresponding 4 units of horizontal change. So starting at some point on the line, we could move to other points on the line as follows:

$\dfrac{3}{4} = \dfrac{6}{8}$ by moving 6 units *up* and 8 units to the *right*

$\dfrac{3}{4} = \dfrac{15}{20}$ by moving 15 units *up* and 20 units to the *right*

$\dfrac{3}{4} = \dfrac{\frac{3}{2}}{2}$ by moving $1\dfrac{1}{2}$ units *up* and 2 units to the *right*

$\dfrac{3}{4} = \dfrac{-3}{-4}$ by moving 3 units *down* and 4 units to the *left*

Likewise, a slope of $-\dfrac{5}{6}$ indicates that starting at some point on the line, we could move to other points on the line as follows:

$-\dfrac{5}{6} = \dfrac{-5}{6}$ by moving 5 units *down* and 6 units to the *right*

$-\dfrac{5}{6} = \dfrac{5}{-6}$ by moving 5 units *up* and 6 units to the *left*

$-\dfrac{5}{6} = \dfrac{-10}{12}$ by moving 10 units *down* and 12 units to the *right*

$-\dfrac{5}{6} = \dfrac{15}{-18}$ by moving 15 units *up* and 18 units to the *left*

EXAMPLE 3

Graph the line that passes through the point $(0, -2)$ and has a slope of $\dfrac{1}{3}$.

Solution

To begin, plot the point $(0, -2)$. And, because the slope $= \dfrac{\text{vertical change}}{\text{horizontal change}} = \dfrac{1}{3}$, we can locate another point on the line by starting from the point $(0, -2)$ and moving 1 unit up and 3 units to the right to obtain the point $(3, -1)$. Because two points determine a line, we can draw the line (Figure 7.34).

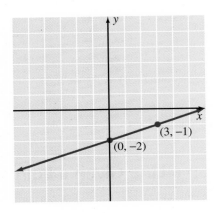

Figure 7.34

Remark: Because $m = \dfrac{1}{3} = \dfrac{-1}{-3}$, we can locate another point by moving 1 unit down and 3 units to the left from the point $(0, -2)$. ∎

EXAMPLE 4

Graph the line that passes through the point $(1, 3)$ and has a slope of -2.

Solution

To graph the line, plot the point $(1, 3)$. We know that $m = -2 = \dfrac{-2}{1}$. Furthermore, because the slope $= \dfrac{\text{vertical change}}{\text{horizontal change}} = \dfrac{-2}{1}$, we can locate another point on the line by starting from the point $(1, 3)$ and moving 2 units down and 1 unit to the right to obtain the point $(2, 1)$. Because two points determine a line, we can draw the line (Figure 7.35).

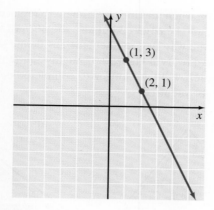

Figure 7.35

Remark: Because $m = -2 = \dfrac{-2}{1} = \dfrac{2}{-1}$, we can locate another point by moving 2 units up and 1 unit to the left from the point (1, 3). ∎

3 Apply the Concept of Slope to Solve Problems

The concept of slope has many real world applications even though the word "slope" is often not used. For example, the highway in Figure 7.36 is said to have a *grade* of 17%. This means that for every horizontal distance of 100 feet, the highway rises or drops 17 feet. In other words, the absolute value of the slope of the highway is $\dfrac{17}{100}$.

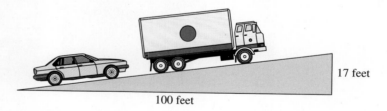

Figure 7.36

EXAMPLE 5

Apply Your Skill

A certain highway has a 3% grade. How many feet does it rise in a horizontal distance of 1 mile?

Solution

A 3% grade means a slope of $\dfrac{3}{100}$. Therefore, if we let y represent the unknown vertical distance and use the fact that 1 mile = 5280 feet, we can set up and solve the following proportion:

$$\dfrac{3}{100} = \dfrac{y}{5280}$$

$$100y = 3(5280) = 15{,}840$$

$$y = 158.4$$

The highway rises 158.4 feet in a horizontal distance of 1 mile. ∎

A roofer, when making an estimate to replace a roof, is concerned about not only the total area to be covered but also the *pitch* of the roof. (Contractors do not define pitch the same way that mathematicians define slope, but both terms refer to *steepness*.) The two roofs in Figure 7.37 might require the same number of shingles, but the roof on the left will take longer to complete because the pitch is so great that scaffolding will be required.

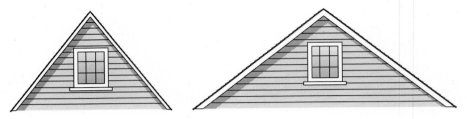

Figure 7.37

The concept of slope is also used in the construction of flights of stairs. The terms "rise" and "run" are commonly used, and the steepness (slope) of the stairs can be expressed as the ratio of rise to run. In Figure 7.38, the stairs on the left with the the ratio of $\frac{10}{11}$ are steeper than the stairs on the right, which have a ratio of $\frac{7}{11}$.

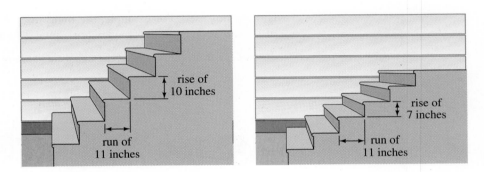

Figure 7.38

Technically, the concept of slope is involved in most situations involving an incline. Hospital beds are constructed so that both the head-end and the foot-end can be raised or lowered; that is, the slope of either end of the bed can be changed. Likewise, treadmills are designed so that the incline (slope) of the platform can be raised or lowered as desired. Perhaps you can think of other applications of the concept of slope.

CONCEPT QUIZ 7.4

For Problems 1–8, answer true or false.

1. The concept of slope of a line pertains to the steepness of the line.
2. The slope of a line is the ratio of the horizontal change to the vertical change moving from one point to another point on the line.
3. A line that has a negative slope falls as we move from left to right.
4. The slope of a vertical line is 0.
5. The slope of a horizontal line is 0.
6. A line cannot have a slope of 0.

7. A slope of $\dfrac{-5}{2}$ is the same as a slope of $-\dfrac{5}{-2}$.
8. A slope of 5 means that for every unit of horizontal change there is a corresponding 5 units of vertical change.

Section 7.4   Classroom Problem Set

Objective 1

1. Find the slope of the line determined by each of the following pairs of points.
   (a) $(1, 2)$ and $(3, 8)$
   (b) $(5, 3)$ and $(2, 4)$
   (c) $(-4, 2)$ and $(1, 2)$

2. Find the slope of the line determined by each of the following pairs of points.
   (a) $(-3, 4)$ and $(2, -6)$
   (b) $(9, 10)$ and $(6, 2)$
   (c) $(-2, 5)$ and $(1, -5)$

3. Find the slope of the line determined by the equation $2x - y = 4$.

4. Find the slope of the line determined by the equation $x - 4y = -6$.

Objective 2

5. Graph the line that passes through the point $(-2, -1)$ and has a slope of $\dfrac{2}{5}$.

6. Graph the line that passes through the point $(4, -5)$ and has a slope of -2.

7. Graph the line that passes through the point $(-3, 4)$ and has a slope of $-\dfrac{3}{2}$.

8. Graph the line that passes through the point $(-6, 2)$ and has a slope of $\dfrac{1}{4}$.

Objective 3

9. An exit ramp for an interstate highway has a grade of 2%. How many feet does it rise in a horizontal distance of 1250 feet?

10. If the ratio of rise to run is to be $\dfrac{3}{5}$ for some stairs, and the measure of the rise is 19 centimeters, find the measure of the run to the nearest centimeter.

THOUGHTS INTO WORDS

1. How would you explain the concept of slope to someone who was absent from class the day it was discussed?

2. If one line has a slope of $\dfrac{2}{3}$, and another line has a slope of 2, which line is steeper? Explain your answer.

3. Why do we say that the slope of a vertical line is undefined?

4. Suppose that a line has a slope of $\dfrac{3}{4}$ and contains the point $(5, 2)$. Are the points $(-3, -4)$ and $(14, 9)$ also on the line? Explain your answer.

Answers to the Concept Quiz

1. True 2. False 3. True 4. False 5. True 6. False 7. False 8. True

7.5 Writing Equations of Lines

OBJECTIVES

1. Find the Equation of a Line Given a Point and a Slope
2. Find the Equation of a Line Given Two Points
3. Find the Equation of a Line Given the Slope and y Intercept
4. Use the Point-Slope Form to Write Equations of Lines
5. Apply the Slope-Intercept Form of an Equation

1 Find the Equation of a Line Given a Point and a Slope

There are two basic types of problems in analytic or coordinate geometry:

1. Given an algebraic equation, find its geometric graph.
2. Given a set of conditions pertaining to a geometric figure, determine its algebraic equation.

We discussed problems of the first type in the first two sections of this chapter. Now we want to consider a few problems of the second type that deal with straight lines. In other words, given certain facts about a line, we need to be able to write its algebraic equation.

EXAMPLE 1 Find the equation of the line that has a slope of $\dfrac{3}{4}$ and contains the point $(1, 2)$.

Solution

First, let's graph the line as indicated in Figure 7.39. To draw the line, plot the point $(1, 2)$ and then because we know that the slope is $\dfrac{3}{4}$, we can find a second point by moving 3 units up and 4 units to the right. Now that we have a graph of the line, we can choose an arbitrary point (x, y) on the line that represents any point on the line other than the point $(1, 2)$. The slope between $(1, 2)$ and (x, y) is $\dfrac{3}{4}$; therefore we can write the following equation:

$$\frac{y - 2}{x - 1} = \frac{3}{4}$$

Now simplify this equation to write the equation of the line in standard form.

$$\frac{y - 2}{x - 1} = \frac{3}{4}$$

$3(x - 1) = 4(y - 2)$ Cross-multiply

$3x - 3 = 4y - 8$

$3x - 4y = -5$

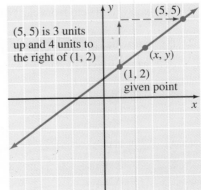

Figure 7.39

2 Find the Equation of a Line Given Two Points

EXAMPLE 2 Find the equation of the line that contains $(3, 4)$ and $(-2, 5)$.

Solution

First, we draw the line determined by the two given points in Figure 7.40. Since we know two points, we can find the slope.

$$m = \frac{y_2 - y_1}{x_2 - x_1} = \frac{5 - 4}{-2 - 3} = \frac{1}{-5} = -\frac{1}{5}$$

Now we can use the same approach as in Example 1. We form an equation using a variable point (x, y), one of the two given points (we choose P_1), and the slope of $-\frac{1}{5}$.

$$\frac{y - 4}{x - 3} = \frac{1}{-5} \qquad -\frac{1}{5} = \frac{1}{-5}$$

$$x - 3 = -5y + 20 \quad \text{Cross-multiply}$$

$$x + 5y = 23$$

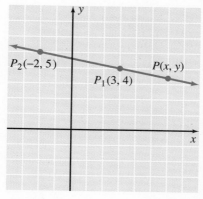

Figure 7.40

3 Find the Equation of a Line Given the Slope and y Intercept

EXAMPLE 3 Find the equation of the line that has a slope of $\frac{1}{4}$ and a y intercept of 2.

Solution

A y intercept of 2 means that the point $(0, 2)$ is on the line. Since the slope is $\frac{1}{4}$, we can find another point by moving 1 unit up and 4 units to the right of $(0, 2)$. The line is drawn in Figure 7.41. We choose a variable point (x, y) and proceed as in the preceding examples.

$$\frac{y - 2}{x - 0} = \frac{1}{4}$$

$$1(x - 0) = 4(y - 2)$$

$$x = 4y - 8$$

$$x - 4y = -8$$

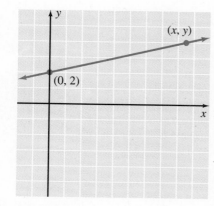

Figure 7.41

4 Use the Point-Slope Form to Write Equations of Lines

It may be helpful for you to pause for a moment and look back over Examples 1, 2, and 3. Notice that we used the same basic approach in all three examples; that is, we chose a variable point (x, y) and used it, along with another known point, to determine the equation of the line. You should also recognize that the approach we take in these examples can be generalized to produce some special forms of equations for straight lines.

To develop a general form, let us find the equation of the line that has a slope of m and contains the point (x_1, y_1). Then we can choose (x, y) to represent any other point on the line in Figure 7.42. The slope of the line is given by

$$m = \frac{y - y_1}{x - x_1}$$

Simplifying by cross-multiplication enables us to obtain

$$y - y_1 = m(x - x_1)$$

We refer to the equation

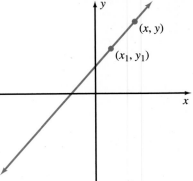

Figure 7.42

$$y - y_1 = m(x - x_1)$$

as the **point-slope form** of the equation of a straight line. Instead of the approach we used in Example 1, we could use the point-slope form to write the equation of a line with a given slope that contains a given point, as Example 4 illustrates.

EXAMPLE 4

Write the equation of the line that has a slope of $\frac{3}{5}$ and contains the point $(2, -4)$.

Solution

Substituting $\frac{3}{5}$ for m and $(2, -4)$ for (x_1, y_1) in the point-slope form, we obtain

$$y - y_1 = m(x - x_1)$$
$$y - (-4) = \frac{3}{5}(x - 2)$$
$$y + 4 = \frac{3}{5}(x - 2)$$
$$5(y + 4) = 3(x - 2) \quad \text{Multiply both sides by 5}$$
$$5y + 20 = 3x - 6$$
$$26 = 3x - 5y$$

∎

5 Apply the Slope-Intercept Form of an Equation

Now consider the equation of a line that has a slope of m and a y intercept of b (see Figure 7.43). A y intercept of b means that the line contains the point $(0, b)$; therefore, we can use the point-slope form.

$$y - y_1 = m(x - x_1)$$
$$y - b = m(x - 0) \quad y_1 = b \text{ and } x_1 = 0$$
$$y - b = mx$$
$$y = mx + b$$

7.5 Writing Equations of Lines

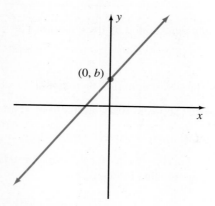

Figure 7.43

The equation

$$y = mx + b$$

is called the **slope-intercept form** of the equation of a straight line. We use it for three primary purposes, as the next three examples illustrate.

EXAMPLE 5 Find the equation of the line that has a slope of $\frac{1}{4}$ and a y intercept of 2.

Solution

This is a restatement of Example 3, but this time we will use the slope-intercept form of a line ($y = mx + b$) to write its equation. From the statement of the problem we know that $m = \frac{1}{4}$ and $b = 2$. Thus, substituting these values for m and b into $y = mx + b$, we obtain

$$y = mx + b$$
$$y = \frac{1}{4}x + 2$$
$$4y = x + 8$$
$$x - 4y = -8 \qquad \text{Same result as in Example 3} \qquad \blacksquare$$

Remark: It is acceptable to leave answers in slope-intercept form. We did not do that in Example 5 because we wanted to show that it was the same result as in Example 3.

EXAMPLE 6 Find the slope of the line with the equation $2x + 3y = 4$.

Solution

We can solve the equation for y in terms of x, and then compare it to the slope-intercept form to determine its slope.

$$2x + 3y = 4$$
$$3y = -2x + 4$$
$$y = -\frac{2}{3}x + \frac{4}{3}$$

Compare this result to $y = mx + b$, and you see that the slope of the line is $-\frac{2}{3}$. Furthermore, the y intercept is $\frac{4}{3}$. $\blacksquare$

EXAMPLE 7 Graph the line determined by the equation $y = \frac{2}{3}x - 1$.

Solution

Comparing the given equation to the general slope-intercept form, we see that the slope of the line is $\frac{2}{3}$, and the y intercept is -1. Because the y intercept is -1, we can plot the point $(0, -1)$. Then because the slope is $\frac{2}{3}$, let's move 3 units to the right and 2 units up from $(0, -1)$ to locate the point $(3, 1)$. The two points $(0, -1)$ and $(3, 1)$ determine the line in Figure 7.44.

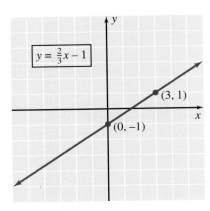

Figure 7.44

In general, *if the equation of a nonvertical line is written in slope-intercept form, the coefficient of x is the slope of the line, and the constant term is the y intercept.* (Remember that the concept of slope is not defined for a vertical line.) Let's consider a few more examples.

EXAMPLE 8 Find the slope and y intercept of each of the following lines and graph the lines.

(a) $5x - 4y = 12$ (b) $-y = 3x - 4$ (c) $y = 2$

Solution

(a) We change $5x - 4y = 12$ to slope-intercept form to get

$$5x - 4y = 12$$
$$-4y = -5x + 12$$
$$4y = 5x - 12$$
$$y = \frac{5}{4}x - 3$$

The slope of the line is $\frac{5}{4}$ (the coefficient of x), and the y intercept is -3 (the constant term). To graph the line, we plot the y intercept, -3. Then because the slope is $\frac{5}{4}$, we can determine a second point, $(4, 2)$, by moving 5 units up and 4 units to the right from the y intercept. The graph is shown in Figure 7.45.

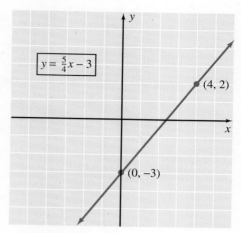

Figure 7.45

(b) We multiply both sides of the given equation by -1 to change it to slope-intercept form.

$$-y = 3x - 4$$
$$y = -3x + 4$$

The slope of the line is -3, and the y intercept is 4. To graph the line, we plot the y intercept, 4. Then because the slope is $-3 = \dfrac{-3}{1}$, we can find a second point, $(1, 1)$, by moving 3 units down and 1 unit to the right from the y intercept. The graph is shown in Figure 7.46.

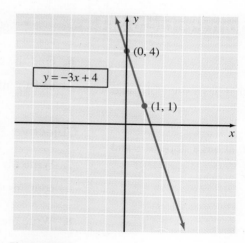

Figure 7.46

(c) We can write the equation $y = 2$ as

$$y = 0(x) + 2$$

The slope of the line is 0, and the y intercept is 2. To graph the line, we plot the y intercept, 2. Then because a line with a slope of 0 is horizontal, we draw a horizontal line through the y intercept. The graph is shown in Figure 7.47.

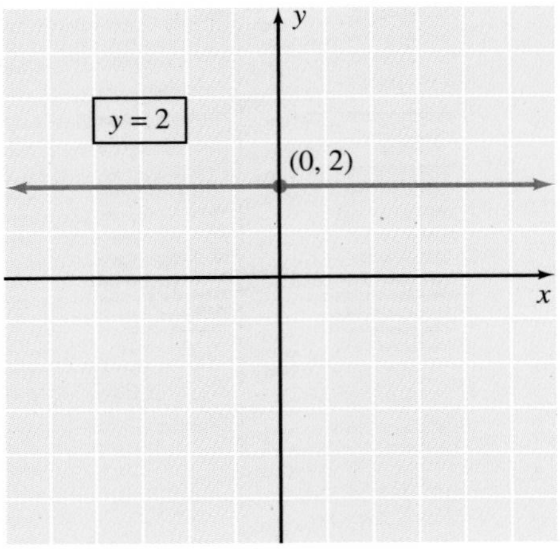

Figure 7.47

CONCEPT QUIZ 7.5

For Problems 1–5, answer true or false.

1. The equations $m = \dfrac{y - y_1}{x - x_1}$ and $y - y_1 = m(x - x_1)$ are equivalent.
2. In the slope-intercept form of an equation of a line $y = mx + b$, m is the slope.
3. The slope of a line determined by the equation $5x + 2y = 1$ is 5.
4. The slope of a line determined by the equation $3x - 2y = 4$ is $\dfrac{3}{2}$.
5. The concept of slope is not defined for the line $x = 3$.

Section 7.5 Classroom Problem Set

Objective 1

1. Find the equation of the line that has a slope of $\dfrac{2}{3}$ and contains the point $(4, 2)$.

2. Find the equation of the line that has a slope of $\dfrac{3}{5}$ and contains the point $(5, -6)$.

Objective 2

3. Find the equation of the line that contains $(-2, 3)$ and $(1, 5)$.

4. Find the equation of the line that contains $(-8, -7)$ and $(3, 4)$.

Objective 3

5. Find the equation of the line that has a slope of $\dfrac{1}{3}$ and a y intercept of -4.

6. Find the equation of the line that has a slope of $-\dfrac{5}{7}$ and a y intercept of -1.

Objective 4

7. Write the equation of a line that has a slope of $\dfrac{5}{2}$ and contains the point $(-1, 3)$.

8. Write the equation of a line that has a slope of $-\dfrac{3}{4}$ and contains the point $(-2, 1)$.

Objective 5

9. Find the equation of a line that has a slope of $-\dfrac{2}{3}$ and a y intercept of 5.

10. Find the equation of a line that has a slope of -1 and a y intercept of -7.

11. Find the slope of the line whose equation is $4x - 2y = 3$.

12. Find the slope of the line whose equation is $-5x + 7y = -14$.

13. Graph the line determined by the equation $y = \frac{3}{5}x - 2$.

14. Graph the line determined by the equation $y = \frac{2}{3}x + 1$.

15. Find the slope and y intercept of each of the following lines and graph the lines.

    (a) $3x + 2y = 6$

    (b) $-y = -x + 3$

    (c) $y = -4$

16. Find the slope and y intercept of each of the following lines and graph the lines.

    (a) $3x - 5y = 15$

    (b) $-2y = 4 + 3x$

    (c) $x = 3$

THOUGHTS INTO WORDS

1. Explain the importance of the slope-intercept form ($y = mx + b$) of the equation of a line.

2. How would you describe coordinate geometry to a group of elementary algebra students?

3. What does it mean to say that two points "determine" a line?

4. How can you tell by inspection that $y = 2x - 4$ and $y = -3x - 1$ are not parallel lines?

Answers to the Concept Quiz

1. True   2. True   3. False   4. True   5. True

Chapter 7   Review Problem Set

For Problems 1–6, find the x and y intercepts for the line with the given equation.

1. $x + 3y = -6$
2. $x - 2y = 4$
3. $4x - 5y = 20$
4. $2x - y = -1$
5. $y = 3x - 6$
6. $y = -2x + 8$

For Problems 7–20, graph each of the equations.

7. $2x - 5y = 10$
8. $y = -\frac{1}{3}x + 1$
9. $y = -2x$
10. $3x + 4y = 12$
11. $2x - 3y = 0$
12. $2x + y = 2$
13. $x - y = 4$
14. $x + 2y = -2$
15. $y = \frac{2}{3}x - 1$
16. $y = 3x$
17. $y = -3$
18. $y = 2$
19. $x = 4$
20. $x = -1$

21. The equation $C = 1.5e$ is used to convert euros, e, to Canadian dollars, C. Graph the equation and from the graph approximate the value of 9 euros in Canadian dollars.

22. A store that is going out of business is offering everything in the store at a sale price of 30% off the regular price. The equation $s = 0.7p$ gives the sale price, s, for any given price, p. Graph the equation and from the graph approximate the sale price of an item that was priced at $15.

23. A laundry service charges $4 for pickup and 50 cents per pound to wash clothes. The equation $C = 4 + 0.5p$, where C is the charge in dollars and p is the pounds of clothes, can be used to determine the charge. Graph the equation and approximate the charge for washing 18 pounds of clothes.

24. A u-pick strawberry patch charges $1.75 a pound for the strawberries you pick. The equation $C = \frac{7}{4}p$, where C is the cost in dollars and p is the pounds of strawberries, can be used to determine the cost. Graph the equation and approximate the cost for 7 pounds of strawberries.

For Problems 25–40, graph each of the inequalities.

25. $2x + y > -2$
26. $3x - y > 3$
27. $x - 3y < 3$
28. $x + 4y > 4$
29. $y \geq 2x - 1$
30. $y \leq -2x + 5$
31. $y < -x + 3$
32. $y \geq \frac{3}{2}x + 1$
33. $y > -3x$
34. $y > x$
35. $y < \frac{1}{2}x$
36. $y \geq -x$
37. $x \leq 4$
38. $x > -2$
39. $y > 2$
40. $y \geq 0$

For Problems 41–48, find the slope of the line determined by the two points.

41. $(3, -2)$ and $(1, 4)$
42. $(-1, 2)$ and $(0, 5)$
43. $(4, 1)$ and $(7, 0)$
44. $(2, 4)$ and $(4, 7)$
45. $(1, 5)$ and $(-3, 5)$
46. $(-2, 5)$ and $(-3, 6)$
47. $(3, 6)$ and $(3, 1)$
48. $(2, 12)$ and $(4, 7)$

49. The county building code stipulates that the rise to run for a wheelchair ramp has to have a ratio of 1 to 10. Find the length required for a ramp that has a rise of 5 feet.

50. If the ratio of rise to run is to be $\frac{7}{10}$ for some stairs and the measure of the rise is 30 centimeters, find the measure of the run to the nearest centimeter.

51. A ski slope must have a slope of less than $\frac{3}{20}$ to be considered a beginner's slope. One hill in consideration for the beginner's slope has a rise of 320 feet and a run of 1500 feet. Find the slope of the hill and decide if it can be classified as a beginner's slope.

52. The grade of a highway up a mountain is 8%. How much change in horizontal distance is there if the vertical height of the hill is 600 feet? Express the answer to the nearest foot.

53. Write the equation of a line that has a slope of $\frac{3}{4}$ and contains the point $(1, -2)$. Express the answer in standard form.

54. Write the equation of a line that has a slope of -2 and contains the point $(-5, 3)$. Express the answer in standard form.

55. Write the equation of a line that has a slope of -4 and contains the point $(1, -2)$. Express the answer in slope-intercept form.

56. Write the equation of a line that has a slope of $-\frac{3}{5}$ and contains the point $(-5, 3)$. Express the answer in slope-intercept form.

57. Write the equation of a line that contains the points $(2, -1)$ and $(0, 6)$. Express the answer in standard form.

58. Write the equation of a line that contains the points $(1, 6)$ and $(-1, 3)$. Express the answer in standard form.

59. Write the equation of a line that contains the points $(0, 0)$ and $(-4, 1)$. Express the answer in slope-intercept form.

60. Write the equation of a line that contains the points $(-1, 5)$ and $(2, 5)$. Express the answer in slope-intercept form.

For Problems 61–66, determine the slope and y intercept and graph the line.

61. $2x - 5y = 10$
62. $y = -\frac{1}{3}x + 1$
63. $x + 2y = 2$
64. $3x + y = -2$
65. $2x - y = 4$
66. $3x - 4y = 12$

Chapter 7 Practice Test

For Problems 1–4, determine the slope and y intercept for each equation.

1. $5x + 3y = 15$
2. $-2x + y = -4$
3. $y = -\dfrac{1}{2}x - 2$
4. $3x + y = 0$
5. Find the slope of the line determined by the points $(5, 1)$ and $(3, 4)$.
6. Find the slope of the line determined by the points $(3, -2)$ and $(5, -2)$.
7. Find x if the line through the points $(4, 7)$ and $(x, 13)$ has a slope of $\dfrac{3}{2}$.
8. Find y if the line through the points $(1, y)$ and $(6, 2)$ has a slope of $-\dfrac{3}{5}$.
9. If a line has a slope of $\dfrac{1}{4}$ and passes through the point $(3, 5)$, find the coordinates of two other points on the line.
10. If a line has a slope of -3 and passes through the point $(2, 1)$, find the coordinates of two other points on the line.
11. For the graphs given in Figure 7.48, match the lines with the following descriptions of their slopes.

    (a) Positive slope (b) Negative slope
    (c) Slope of 0   (d) Slope is undefined

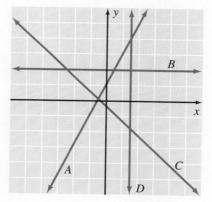

Figure 7.48

12. Suppose that a highway rises a distance of 85 feet over a horizontal distance of 1850 feet. Express the grade of the highway to the nearest tenth of a percent.
13. Find the x intercept of the graph of $y = 4x + 8$.
14. Find the y intercept of the graph of $2x - 3y = 12$.

For Problems 15–21, graph each equation or inequality.

15. $5x + 3y = 15$

16. $y = \dfrac{2}{3}x$

17. $y = 2x - 3$

18. $x = -2$

19. $y = 4$

20. $y \geq 2x - 4$

21. $x + 3y < -3$

22. An online tutoring service charges $4 for a question plus 25 cents per minute for the time spent. The equation $C = \dfrac{1}{4}t + 4$, where C is the cost in dollars and t is the time in minutes, describes the cost depending on the time used. Graph the equation and approximate the cost for 10 minutes of tutoring.

For Problems 23–25, express each equation in $Ax + By = C$ form, where A, B, and C are integers.

23. Determine the equation of the line that has a slope of $-\dfrac{3}{5}$ and a y intercept of 4.

24. Determine the equation of the line containing the point $(4, -2)$ and having a slope of $\dfrac{4}{9}$.

25. Determine the equation of the line that contains the points $(4, 6)$ and $(-2, -3)$.

Systems of Equations

Chapter 8   Warm-Up Problems

1. Is the ordered pair a solution for the equation $-3x + 2y = 10$?

   (a) $(0, 5)$ (b) $(6, 4)$ (c) $(-4, 11)$

2. Graph each equation.

   (a) $y = 2x - 3$ (b) $3x - y = 2$ (c) $4x + 2y = 1$

3. Find the x and y intercepts for each equation.

   (a) $x + y = 7$ (b) $3x - \frac{1}{2}y = 18$ (c) $y = -\frac{3}{4}x - \frac{1}{6}$

4. Translate the sentences into equations.

   (a) The sum of two numbers is 18.   (b) The difference of two numbers is 5.

   (c) Three times a number minus 6 is 24.

5. Use the multiplication property of equality.

   (a) Multiply $4x - 2y = 5$ by 3.   (b) Multiply $2x + y = 12$ by -1.

   (c) Multiply $12x - 4y = 24$ by $\frac{1}{4}$.

6. Solve for the variable in each equation.

   (a) $x + y + z = 1$ if $x = 5$ and $z = -4$ (b) $3y - z = 10$ if $z = 5$

   (c) $-2x + z = 5$ if $z = -7$

7. Solve for the variable in each equation.

   (a) $-3x + 2(x + 1) = 17$ (b) $6x - 4(-1 - x) = -36$

   (c) $2\left(\dfrac{9 - 6x}{3}\right) + 4x = 6$

8. How many gallons of water must be added to 10 gallons of a 20% salt solution to change it to a 15% salt solution?

8.1 Solving Linear Systems by Graphing

8.2 Elimination-by-Addition Method

8.3 Substitution Method

8.4 3 × 3 Systems of Equations

In Chapter 7 we studied linear equations in two variables. Now in Chapter 8 we will continue that work by considering two or more linear equations at the same time. The equations considered together form a *system of equations*. In this chapter we will discuss various techniques for solving systems of equations and also consider systems of linear inequalities. Systems of equations can readily be applied to solve many of the word problem types we've encountered in previous chapters. Many students find that using two variables and a system of equations to solve a word problem is easier than using only one variable. Although some of the word problems presented in this chapter could be solved without a system of equations, we encourage you to expand your problem-solving abilities by using a system of equations to solve them.

Video tutorials for all section learning objectives are available in a variety of delivery modes.

INTERNET PROJECT

In this chapter three methods for solving systems of equations are presented. Another method for solving systems of equations is Gaussian elimination and uses matrices. Do an Internet search on matrices in mathematics to find a definition of a matrix. What ancient civilizations were among the first to use matrices?

8.1 Solving Linear Systems by Graphing

OBJECTIVES

1. Determine If an Ordered Pair Is a Solution of a System of Equations
2. Solve Linear Systems of Equations by Graphing
3. Solve Linear Systems of Inequalities by Graphing

1 Determine If an Ordered Pair Is a Solution of a System of Equations

Suppose we graph $x - 2y = 4$ and $x + 2y = 8$ on the same set of axes, as shown in Figure 8.1. The ordered pair $(6, 1)$, which is associated with the point of intersection of the two lines, satisfies both equations. That is to say, $(6, 1)$ is the solution for $x - 2y = 4$ and $x + 2y = 8$. To check this, we can substitute 6 for x and 1 for y in both equations.

$x - 2y = 4$ becomes $6 - 2(1) = 4$

$x + 2y = 8$ becomes $6 + 2(1) = 8$

Thus we say that $\{(6, 1)\}$ is the solution set of the system

$$\begin{pmatrix} x - 2y = 4 \\ x + 2y = 8 \end{pmatrix}$$

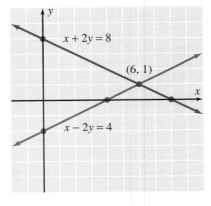

Figure 8.1

EXAMPLE 1

Determine if the ordered pair $(-2, 1)$ is a solution of the system of equations $\begin{pmatrix} 3x + y = -5 \\ x + 4y = 6 \end{pmatrix}$.

Solution

Substitute -2 for x and 1 for y in both equations.

$3x + y = -5$ becomes $3(-2) + 1 = -5$ A true statement

$x + 4y = 6$ becomes $-2 + 4(1) = 6$ A false statement

Because the ordered pair does not satisfy both equations, it is not a solution of the system of equations.

2 Solve Linear Systems of Equations by Graphing

Two or more linear equations in two variables considered together are called a **system of linear equations**. Here are three systems of linear equations:

$$\begin{pmatrix} x - 2y = 4 \\ x + 2y = 8 \end{pmatrix} \quad \begin{pmatrix} 5x - 3y = 9 \\ 3x + 7y = 12 \end{pmatrix} \quad \begin{pmatrix} 4x - y = 5 \\ 2x + y = 9 \\ 7x - 2y = 13 \end{pmatrix}$$

To **solve a system of linear equations** means to find all of the ordered pairs that are solutions of all of the equations in the system. There are several techniques for solving systems of linear equations. We will use three of them in this chapter—a graphing method in this section and two other methods in the following sections.

To solve a system of linear equations by **graphing**, we proceed as in the opening discussion of this section. We graph the equations on the same set of axes, and then the ordered pairs associated with any points of intersection are the solutions to the system. Let's consider another example.

EXAMPLE 2

Solve the system $\begin{pmatrix} x + y = 5 \\ x - 2y = -4 \end{pmatrix}$.

Solution

We can find the intercepts and a check point for each of the lines.

$x + y = 5$

x	y	
0	5	Intercepts
5	0	
2	3	Check point

$x - 2y = -4$

x	y	
0	5	Intercepts
-4	0	
-2	1	Check point

Figure 8.2 shows the graphs of the two equations. It appears that $(2, 3)$ is the solution of the system. To check it we can substitute 2 for x and 3 for y in both equations.

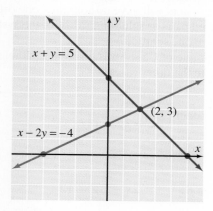

Figure 8.2

$x + y$ becomes $2 + 3 = 5$ A true statement

$x - 2y = -4$ becomes $2 - 2(3) = -4$ A true statement

Therefore, $\{(2, 3)\}$ is the solution set. ∎

It should be evident that solving systems of equations by graphing requires accurate graphs. In fact, unless the solutions are integers it is really quite difficult to obtain exact solutions from a graph. For this reason the systems in this section have integer solutions. Furthermore, checking a solution takes on additional significance when the graphing approach is used. By checking you can be absolutely sure that you are *reading* the correct solution from the graph.

Figure 8.3 shows the three possible cases for the graph of a system of two linear equations in two variables.

Case I The graphs of the two equations are two lines intersecting at one point. There is one solution, and we call the system a **consistent system**.

Case II The graphs of the two equations are parallel lines. There is no solution, and we call the system an **inconsistent system**.

Case III The graphs of the two equations are the same line. There are infinitely many solutions to the system. Any pair of real numbers that satisfies one of the equations also satisfies the other equation, and we say the **equations are dependent**.

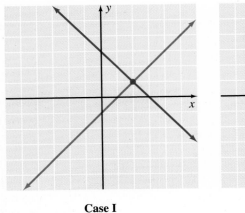
Case I

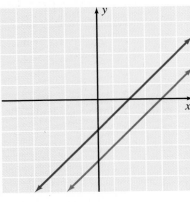
Case II

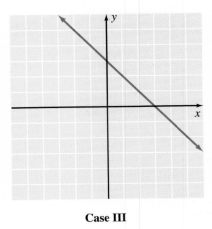
Case III

Figure 8.3

Thus as we solve a system of two linear equations in two variables, we know what to expect. The system will have no solutions, one ordered pair as a solution, or infinitely many ordered pairs as solutions. (Most of the systems that we will be working with in this text have one solution.)

An example of Case I was given in Example 2 (Figure 8.2). The next two examples illustrate the other cases.

EXAMPLE 3

Solve the system $\begin{pmatrix} 2x + 3y = 6 \\ 2x + 3y = 12 \end{pmatrix}$.

Solution

$2x + 3y = 6$

x	y
0	2
3	0
−3	4

$2x + 3y = 12$

x	y
0	4
6	0
3	2

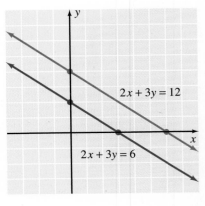

Figure 8.4 shows the graph of the system. Since the lines are parallel, there is no solution to the system. The solution set is $\emptyset$.

Figure 8.4

EXAMPLE 4

Solve the system $\begin{pmatrix} x + y = 3 \\ 2x + 2y = 6 \end{pmatrix}$.

Solution

x + y = 3	
x	y
0	3
3	0
1	2

2x + 2y = 6	
x	y
0	3
3	0
1	2

Figure 8.5 shows the graph of this system. Since the graphs of both equations are the same line, there are infinitely many solutions to the system. Any ordered pair of real numbers that satisfies one equation also satisfies the other equation.

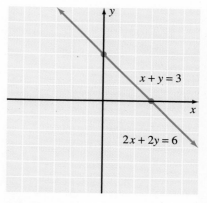

Figure 8.5

3 Solve Linear Systems of Inequalities by Graphing

You now have the skills to use a graphing approach to solve a system of linear inequalities. For example, the solution set of a system of linear inequalities, such as

$$\begin{pmatrix} x + y < 1 \\ x - y > 1 \end{pmatrix}$$

is the intersection of the solution sets of the individual inequalities. In Figure 8.6(a) we indicated the solution set for $x + y < 1$, and in Figure 8.6(b) we indicated the solution set for $x - y > 1$. Then in Figure 8.6(c) we shaded the region that represents the intersection of the two shaded regions in parts a and b, which is the solution of the given system. The shaded region in Figure 8.6(c) includes all points that are below the line $x + y = 1$ *and also* that are below the line $x - y = 1$.

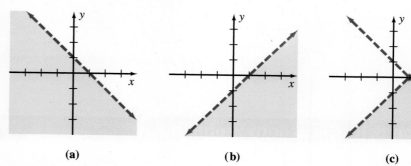

Figure 8.6

Let's solve another system of linear inequalities.

EXAMPLE 5

Solve the system $\begin{pmatrix} 2x + 3y \leq 6 \\ x - 4y < 4 \end{pmatrix}$.

Solution

Let's first graph the individual inequalities. The solution set for $2x + 3y \leq 6$ is shown in Figure 8.7(a), and the solution set for $x - 4y < 4$ is shown in Figure 8.7(b). (Note the solid line in part a and the dashed line in part b.) Then in Figure 8.7(c) we shaded the intersection of the graphs in parts a and b. Thus we represent the solution set for the given system by the shaded region in Figure 8.7(c). This region includes all points that are on or below the line $2x + 3y = 6$ *and also* that are above the line $x - 4y = 4$.

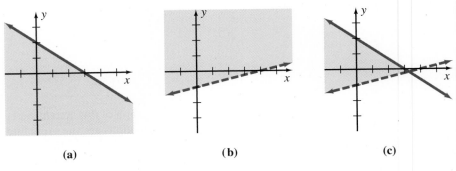

(a) (b) (c)

Figure 8.7

Remark: Remember that the shaded region in Figure 8.7(c) represents the solution set of the given system. Parts a and b were drawn only to help determine the final shaded region. With some practice, you will be able to go directly to part c without actually sketching the graphs of the individual inequalities.

CONCEPT QUIZ 8.1

For Problems 1–7, answer true or false.

1. To solve a system of equations means to find an ordered pair that satisfies any of the equations in the system of equations.
2. A consistent system of equations will have more than one solution.
3. If the graph of a system of two linear equations results in two distinct parallel lines, then the system has no solutions.
4. Every system of equations has a solution.
5. If the graphs of the two equations in a system are the same line, then the equations in the system are dependent.
6. To solve a system of two equations in variables x and y, it is sufficient to just find a value for x.
7. The ordered pair $(1, 4)$ satisfies the system of linear inequalities $\begin{pmatrix} x + y > 2 \\ 2x + y < 3 \end{pmatrix}$.

Section 8.1 Classroom Problem Set

Objective 1

1. Determine if the ordered pair $(3, -1)$ is a solution of the system of equations $\begin{pmatrix} 2x + y = 5 \\ 5x - 2y = 17 \end{pmatrix}$.

2. Determine if the ordered pair $(5, -4)$ is a solution of the system of equations $\begin{pmatrix} x - 3y = 17 \\ 5x - 2y = 17 \end{pmatrix}$.

Objective 2

3. Solve the system $\begin{pmatrix} x + y = 2 \\ x + 2y = 6 \end{pmatrix}$.

4. Solve the system $\begin{pmatrix} y = 2x + 5 \\ x + 3y = -6 \end{pmatrix}$.

5. Solve the system $\begin{pmatrix} 2x + 4y = 8 \\ x + 2y = 4 \end{pmatrix}$.

6. Solve the system $\begin{pmatrix} y = -2x + 3 \\ 6x + 3y = 9 \end{pmatrix}$.

7. Solve the system $\begin{pmatrix} 3x + y = 6 \\ 3x + y = 0 \end{pmatrix}$.

8. Solve the system $\begin{pmatrix} 3x - y = 3 \\ 3x - y = -3 \end{pmatrix}$.

Objective 3

9. Solve the system $\begin{pmatrix} x + y < 5 \\ 2x - 3y > 6 \end{pmatrix}$.

10. Solve the system $\begin{pmatrix} 4x + 3y \leq 12 \\ 4x - y \leq 4 \end{pmatrix}$.

THOUGHTS INTO WORDS

1. Discuss the strengths and weaknesses of solving a system of linear equations by graphing.

2. Determine a system of two linear equations for which the solution is (5, 7). Do any other systems have the same solution set? If so, find at least one more system.

3. Is it possible for a system of two linear equations to have exactly two solutions? Defend your answer.

Answers to the Concept Quiz
1. False 2. False 3. True 4. False 5. True 6. False 7. False

8.2 Elimination-by-Addition Method

OBJECTIVES

1. Use Elimination-by-Addition to Solve a System of Equations
2. Solve Word Problems Using a System of Equations

1 Use Elimination-by-Addition to Solve a System of Equations

We have used the addition property of equality (if $a = b$, then $a + c = b + c$) to help solve equations that contain one variable. An extension of the addition property forms the basis for another method of solving systems of linear equations. Property 8.1 states that when we add two equations, the resulting equation will be equivalent to the original ones. We can use this property to help solve a system of equations.

Property 8.1

For all real numbers a, b, c, and d, if $a = b$ and $c = d$, then

$$a + c = b + d$$

EXAMPLE 1

Solve $\begin{pmatrix} x + y = 12 \\ x - y = 2 \end{pmatrix}$.

Solution

$$x + y = 12$$
$$x - y = 2$$
$$2x = 14 \quad \text{This is the result of adding the two equations}$$

Solving this new equation in one variable, we obtain

$$2x = 14$$
$$\boxed{x = 7}$$

Now we can substitute the value of 7 for x in one of the original equations. Thus

$$x + y = 12$$
$$7 + y = 12$$
$$\boxed{y = 5}$$

To check, we see that

$$x + y = 12 \quad \text{and} \quad x - y = 2$$
$$7 + 5 = 12 \quad \text{and} \quad 7 - 5 = 2$$

Thus the solution set of the system is $\{(7, 5)\}$. ∎

Note in Example 1 that by adding the two original equations, we obtained a simple equation that contains only one variable. Adding equations to eliminate a variable is the key idea behind the **elimination-by-addition method** for solving systems of linear equations. The next example further illustrates this point.

EXAMPLE 2

Solve $\begin{pmatrix} 2x + 3y = -26 \\ 4x - 3y = 2 \end{pmatrix}$.

Solution

$$2x + 3y = -26$$
$$4x - 3y = 2$$
$$6x = -24 \quad \text{Added the two equations}$$
$$\boxed{x = -4}$$

We substitute -4 for x in one of the two original equations.

$$2x + 3y = -26$$
$$2(-4) + 3y = -26$$
$$-8 + 3y = -26$$
$$3y = -18$$
$$\boxed{y = -6}$$

The solution set is $\{(-4, -6)\}$. ∎

It may be necessary to change the form of one, or perhaps both, of the original equations before adding them. The next example demonstrates this idea.

EXAMPLE 3

Solve $\begin{pmatrix} y = x - 21 \\ x + y = -3 \end{pmatrix}$.

Solution

$$y = x - 21 \quad \underrightarrow{\text{Subtract } x \text{ from both sides}} \quad -x + y = -21$$
$$x + y = -3 \quad \underrightarrow{\text{Leave alone}} \quad x + y = -3$$
$$\phantom{x + y = -3 \quad \text{Leave alone}} \quad 2y = -24$$
$$\phantom{x + y = -3 \quad \text{Leave alone}} \quad \boxed{y = -12}$$

We substitute -12 for y in one of the original equations.

$$x + y = -3$$
$$x + (-12) = -3$$
$$\boxed{x = 9}$$

The solution set is $\{(9, -12)\}$.

Frequently, the multiplication property of equality needs to be applied first so that adding the equations will eliminate a variable, as the next example illustrates.

EXAMPLE 4

Solve $\begin{pmatrix} 2x + 5y = 29 \\ 3x - y = 1 \end{pmatrix}$.

Solution

Notice that adding the equations as they are would not eliminate a variable. However, we observe that multiplying the bottom equation by 5 and then adding this newly formed, but equivalent, equation to the top equation will eliminate the y variable terms.

$$2x + 5y = 29 \quad \underrightarrow{\text{Leave alone}} \quad 2x + 5y = 29$$
$$3x - y = 1 \quad \underrightarrow{\text{Multiply both sides by 5}} \quad 15x - 5y = 5$$
$$\phantom{3x - y = 1 \quad \text{Multiply both sides by 5}} \quad 17x = 34$$
$$\phantom{3x - y = 1 \quad \text{Multiply both sides by 5}} \quad x = 2$$

We substitute 2 for x in one of the original equations.

$$3x - y = 1$$
$$3(2) - y = 1$$
$$6 - y = 1$$
$$-y = -5$$
$$y = 5$$

The solution set is $\{(2, 5)\}$.

Notice in these problems that after finding the value of one of the variables, we substitute that number into one of the *original* equations to find the value of the other variable. It doesn't matter which of the two original equations you use, so pick the easiest one to solve.

Sometimes we need to apply the multiplication property to both equations. Let's look at an example of this type.

EXAMPLE 5

Solve $\begin{pmatrix} 2x + 3y = 4 \\ 9x - 2y = -13 \end{pmatrix}$.

Solution A

If we want to eliminate the x variable terms, then we want their coefficients to be equal in number (absolute value) and opposite in sign. To eliminate the x variable terms we can multiply the top equation by 9 and multiply the bottom equation by -2.

$2x + 3y = 4$ Multiply both sides by 9 → $18x + 27y = 36$
$9x - 2y = -13$ Multiply both sides by -2 → $-18x + 4y = 26$
$\overline{}$
$31y = 62$
$\boxed{y = 2}$

We substitute 2 for y in one of the original equations.

$2x + 3y = 4$
$2x + 3(2) = 4$
$2x + 6 = 4$
$2x = -2$
$\boxed{x = -1}$

The solution set is $\{(-1, 2)\}$.

Solution B

If we want to eliminate the y variable terms, then we want their coefficients to be equal in number (absolute value) and opposite in sign. To eliminate the y variable terms we can multiply the top equation by 2 and multiply the bottom equation by 3.

$2x + 3y = 4$ Multiply both sides by 2 → $4x + 6y = 8$
$9x - 2y = -13$ Multiply both sides by 3 → $27x - 6y = -39$
$\overline{}$
$31x = -31$
$\boxed{x = -1}$

We substitute -1 for x in one of the original equations.

$2x + 3y = 4$
$2(-1) + 3y = 4$
$-2 + 3y = 4$
$3y = 6$
$\boxed{y = 2}$

The solution set is $\{(-1, 2)\}$. ∎

Look carefully at Solutions A and B for Example 5. Especially notice the first steps, where we applied the multiplication property of equality to the two equations. In Solution A we multiplied by numbers so that adding the resulting equations eliminated the x variable. In Solution B we multiplied so that the y variable was eliminated when we added the resulting equations. Either approach will work; pick the one that involves the easiest computation.

EXAMPLE 6

Solve $\begin{pmatrix} 3x - 4y = 7 \\ 5x + 3y = 9 \end{pmatrix}$.

Solution

Let's eliminate the y variable terms by multiplying the top equation by 3 and the bottom equation by 4.

$3x - 4y = 7$ — Multiply both sides by 3 → $9x - 12y = 21$
$5x + 3y = 9$ — Multiply both sides by 4 → $20x + 12y = 36$

$$29x = 57$$
$$x = \frac{57}{29}$$

Since substituting $\frac{57}{29}$ for x in one of the original equations will produce some messy calculations, let's solve for y by eliminating the x variable terms.

$3x - 4y = 7$ — Multiply both sides by -5 → $-15x + 20y = -35$
$5x + 3y = 9$ — Multiply both sides by 3 → $15x + 9y = 27$

$$29y = -8$$
$$y = -\frac{8}{29}$$

The solution set is $\left\{ \left(\frac{57}{29}, -\frac{8}{29} \right) \right\}$. ■

2 Solve Word Problems Using a System of Equations

Many word problems that we solved earlier in this text using one equation in one variable can also be solved using a system of two linear equations in two variables. In fact, many times you may find that it seems natural to use two variables. It may also seem more meaningful at times to use variables other than x and y. Let's consider two examples.

EXAMPLE 7   Apply Your Skill

The difference of two numbers is 9. If four times the smaller number is subtracted from three times the larger number, the result is 21. Find the numbers.

Solution

We let x represent the larger number and y represent the smaller number. The problem translates into the following equations.

$x - y = 9$ The difference of two numbers is 9
$3x - 4y = 21$ Four times the smaller is subtracted from 3 times the larger

Solving this system by the elimination-by-addition method, we obtain

$x - y = 9$ — Multiply both sides by -3 → $-3x + 3y = -27$
$3x - 4y = 21$ — Leave alone → $3x - 4y = 21$

$$-y = -6$$
$$y = 6$$

Substitute 6 for y in one of the original equations.

$$x - y = 9$$
$$x - 6 = 9$$
$$x = 15$$

Therefore, the larger number is 15 and the smaller number is 6. ∎

EXAMPLE 8

Apply Your Skill

The cost of 3 tennis balls and 2 golf balls is $7. Furthermore, the cost of 6 tennis balls and 3 golf balls is $12. Find the cost of 1 tennis ball and the cost of 1 golf ball.

Solution

We can use t to represent the cost of one tennis ball and g the cost of one golf ball. The problem translates into the following system of equations.

$3t + 2g = 7$ The cost of 3 tennis balls and 2 golf balls is $7
$6t + 3g = 12$ The cost of 6 tennis balls and 3 golf balls is $12

Solving this system by the elimination-by-addition method, we obtain

$3t + 2g = 7$ Multiply both sides by -2 $-6t - 4g = -14$
$6t + 3g = 12$ Leave alone $6t + 3g = 12$
                                                $-g = -2$
                                                $g = 2$

We substitute 2 for g in one of the original equations.

$$3t + 2g = 7$$
$$3t + 2(2) = 7$$
$$3t + 4 = 7$$
$$3t = 3$$
$$t = 1$$

The cost of a tennis ball is $1 and the cost of a golf ball is $2. ∎

Before you tackle the word problems in this next problem set, it might be helpful to review the problem-solving suggestions we offered in Section 2.4. Those suggestions continue to apply here except that now you have the flexibility of using two equations and two unknowns. Don't forget that to check a word problem you need to see whether your answers satisfy the conditions stated in the original problem.

CONCEPT QUIZ 8.2

For Problems 1–6, answer true or false.

1. When two equations are added, the result is an equation that is equivalent to the original equations.
2. The objective of the elimination-by-addition method is to produce an equivalent equation in only one variable.
3. When solving a system of equations using the elimination-by-addition method, eliminate the x variable so that the x coefficients are equal in number (absolute value) and the same in sign.
4. For the system of equations $\begin{pmatrix} x + 2y = 5 \\ x - y = 3 \end{pmatrix}$, the y variable can be eliminated by just adding the equations.

5. When solving the system of equations $\begin{pmatrix} x + y = 5 \\ x - y = 3 \end{pmatrix}$, after determining that $x = 4$ you can find the value of y by substituting $x = 4$ into either equation.

6. When solving the system of equations $\begin{pmatrix} a + 2b = 5 \\ a - b = 3 \end{pmatrix}$, you must solve for the variable a first because it is the first variable in alphabetic order.

Section 8.2   Classroom Problem Set

Objective 1

1. Solve $\begin{pmatrix} 2x + y = 6 \\ x - y = 6 \end{pmatrix}$.

2. Solve $\begin{pmatrix} x - y = -14 \\ x + y = 6 \end{pmatrix}$.

3. Solve $\begin{pmatrix} x - 2y = 7 \\ 3x + 2y = 13 \end{pmatrix}$.

4. Solve $\begin{pmatrix} -3x + 2y = -21 \\ 3x - 7y = 36 \end{pmatrix}$.

5. Solve $\begin{pmatrix} y = x + 7 \\ x + 2y = 5 \end{pmatrix}$.

6. Solve $\begin{pmatrix} x + y = -10 \\ x = y + 6 \end{pmatrix}$.

7. Solve $\begin{pmatrix} 3x + 2y = -3 \\ 4x - y = 18 \end{pmatrix}$.

8. Solve $\begin{pmatrix} 4x - 5y = -36 \\ x + 2y = 30 \end{pmatrix}$.

9. Solve $\begin{pmatrix} 3x + 2y = 4 \\ 4x - 5y = 13 \end{pmatrix}$.

10. Solve $\begin{pmatrix} 7x + 5y = -6 \\ 4x + 3y = -4 \end{pmatrix}$.

11. Solve $\begin{pmatrix} 5x - 2y = 13 \\ 2x + 3y = 10 \end{pmatrix}$.

12. Solve $\begin{pmatrix} 6x - 2y = -3 \\ 5x - 9y = 1 \end{pmatrix}$.

Objective 2

13. The sum of two numbers is 31. If twice the smaller number is subtracted from the larger number, the result is 1. Find the numbers.

14. The difference of two numbers is 17. If the larger is increased by three times the smaller, the result is 37. Find the numbers.

15. The cost of 2 pizzas and 3 baskets of wings is $29. Also, the cost of 5 pizzas and 2 baskets of wings is $45. Find the cost of one pizza and the cost of a basket of wings.

16. A library buys a total of 35 books, which cost $546. The hardcover books cost $22 each and the softcover books cost $6 per book. How many books of each type did the library buy?

THOUGHTS INTO WORDS

1. Explain how you would solve the system $\begin{pmatrix} 2x - 3y = 5 \\ 4x + 7y = 9 \end{pmatrix}$ using the elimination-by-addition method.

2. Give a general description of how to apply the elimination-by-addition method.

Answers to the Concept Quiz
1. True   2. True   3. False   4. False   5. True   6. False

8.3 Substitution Method

OBJECTIVES

1. Use the Substitution Method to Solve a System of Equations
2. Determine Which Method to Use to Solve a System of Equations
3. Solve Mixture, Interest, and Geometry Word Problems Using a System of Equations

1 Use the Substitution Method to Solve a System of Equations

A third method of solving systems of equations is called the **substitution method**. Like the addition method, it produces exact solutions and can be used on any system of linear equations; however, some systems lend themselves more to the substitution method than others. We will consider a few examples to demonstrate the use of the substitution method.

EXAMPLE 1 Solve $\begin{pmatrix} y = x + 10 \\ x + y = 14 \end{pmatrix}$.

Solution

Because the first equation states that y equals $x + 10$, we can substitute $x + 10$ for y in the second equation.

$$x + y = 14 \quad \xrightarrow{\text{Substitute } x + 10 \text{ for } y} \quad x + (x + 10) = 14$$

Now we have an equation with one variable that can be solved in the usual way.

$$x + (x + 10) = 14$$
$$2x + 10 = 14$$
$$2x = 4$$
$$\boxed{x = 2}$$

Substituting 2 for x in one of the original equations, we can find the value of y.

$$y = x + 10$$
$$y = 2 + 10$$
$$\boxed{y = 12}$$

The solution set is $\{(2, 12)\}$. ∎

EXAMPLE 2 Solve $\begin{pmatrix} 3x + 5y = -7 \\ x = 2y + 5 \end{pmatrix}$.

Solution

Because the second equation states that x equals $2y + 5$, we can substitute $2y + 5$ for x in the first equation.

$$3x + 5y = -7 \quad \xrightarrow{\text{Substitute } 2y + 5 \text{ for } x} \quad 3(2y + 5) + 5y = -7$$

Solving this equation, we have

$$3(2y + 5) + 5y = -7$$
$$6y + 15 + 5y = -7$$

$$11y + 15 = -7$$
$$11y = -22$$
$$\boxed{y = -2}$$

Substituting -2 for y in one of the two original equations produces

$$x = 2y + 5$$
$$x = 2(-2) + 5$$
$$x = -4 + 5$$
$$\boxed{x = 1}$$

The solution set is $\{(1, -2)\}$. ∎

Note that the key idea behind the substitution method is the elimination of a variable, but the elimination is done by a substitution rather than by addition of the equations. The substitution method is especially convenient to use when at least one of the equations is of the form *y equals* or *x equals*. In Example 1 the first equation is of the form *y equals*, and in Example 2 the second equation is of the form *x equals*. Let's consider another example using the substitution method.

EXAMPLE 3

Solve $\begin{pmatrix} 2x + 3y = -30 \\ y = \frac{2}{3}x - 6 \end{pmatrix}$.

Solution

The second equation allows us to substitute $\frac{2}{3}x - 6$ for y in the first equation.

$$2x + 3y = -30 \quad \xrightarrow{\text{Substitute } \frac{2}{3}x - 6 \text{ for } y} \quad 2x + 3\left(\frac{2}{3}x - 6\right) = -30$$

Solving this equation produces

$$2x + 3\left(\frac{2}{3}x - 6\right) = -30$$
$$2x + 2x - 18 = -30$$
$$4x - 18 = -30$$
$$4x = -12$$
$$\boxed{x = -3}$$

Now we can substitute -3 for x in one of the original equations.

$$y = \frac{2}{3}x - 6$$
$$y = \frac{2}{3}(-3) - 6$$
$$y = -2 - 6$$
$$\boxed{y = -8}$$

The solution set is $\{(-3, -8)\}$. ∎

It may be necessary to change the form of one of the equations before we make a substitution. The following examples clarify this point.

EXAMPLE 4

Solve $\begin{pmatrix} 4x - 5y = 55 \\ x + y = -2 \end{pmatrix}$.

Solution

We can easily change the form of the second equation to make it ready for the substitution method.

$$x + y = -2$$
$$y = -2 - x \quad \text{Added } -x \text{ to both sides}$$

Now we can substitute $-2 - x$ for y in the first equation.

$$4x - 5y = 55 \quad \xrightarrow{\text{Substitute } -2 - x \text{ for } y} \quad 4x - 5(-2 - x) = 55$$

Solving this equation, we obtain

$$4x - 5(-2 - x) = 55$$
$$4x + 10 + 5x = 55$$
$$9x + 10 = 55$$
$$9x = 45$$
$$\boxed{x = 5}$$

Substituting 5 for x in one of the *original* equations produces

$$x + y = -2$$
$$5 + y = -2$$
$$\boxed{y = -7}$$

The solution set is $\{(5, -7)\}$. ∎

In Example 4, we could have started by changing the form of the first equation to make it ready for substitution. However, you should be able to look ahead and see that this would produce a fractional form to substitute. We were able to avoid any messy calculations with fractions by changing the form of the second equation instead of the first. Sometimes when using the substitution method, you cannot avoid fractional forms. The next example is a case in point.

EXAMPLE 5

Solve $\begin{pmatrix} 3x + 2y = 8 \\ 2x - 3y = -38 \end{pmatrix}$.

Solution

Looking ahead, we see that changing the form of either equation will produce a fractional form. Therefore, we will merely pick the first equation and solve for y.

$$3x + 2y = 8$$
$$2y = 8 - 3x \quad \text{Added } -3x \text{ to both sides}$$
$$y = \frac{8 - 3x}{2} \quad \text{Multiplied both sides by } \frac{1}{2}$$

Now we can substitute $\dfrac{8 - 3x}{2}$ for y in the second equation and determine the value of x.

$$2x - 3y = -38$$
$$2x - 3\left(\frac{8 - 3x}{2}\right) = -38$$

$$2x - \frac{24 - 9x}{2} = -38$$

$$4x - 24 + 9x = -76 \quad \text{Multiplied both sides by 2}$$

$$13x - 24 = -76$$

$$13x = -52$$

$$\boxed{x = -4}$$

Substituting -4 for x in one of the original equations, we have

$$3x + 2y = 8$$
$$3(-4) + 2y = 8$$
$$-12 + 2y = 8$$
$$2y = 20$$
$$\boxed{y = 10}$$

The solution set is $\{(-4, 10)\}$. ∎

2 Determine Which Method to Use to Solve a System of Equations

We have now studied three methods of solving systems of linear equations—the graphing method, the elimination-by-addition method, and the substitution method. As we indicated earlier, the graphing method is quite restrictive and works well only when the solutions are integers or when we need only approximate answers. Both the elimination-by-addition method and the substitution method can be used to obtain exact solutions for any system of linear equations in two variables. The method you choose may depend upon the original form of the equations. Next we consider two examples to illustrate this point.

EXAMPLE 6

Solve $\begin{pmatrix} 7x - 5y = -52 \\ y = 3x - 4 \end{pmatrix}$.

Solution

Because the second equation indicates that we can substitute $3x - 4$ for y, this system lends itself to the substitution method.

$$7x - 5y = -52 \quad \underrightarrow{\text{Substitute } 3x - 4 \text{ for } y} \quad 7x - 5(3x - 4) = -52$$

Solving this equation, we obtain

$$7x - 5(3x - 4) = -52$$
$$7x - 15x + 20 = -52$$
$$-8x + 20 = -52$$
$$-8x = -72$$
$$\boxed{x = 9}$$

Substituting 9 for x in one of the original equations produces

$$y = 3x - 4$$
$$y = 3(9) - 4$$
$$y = 27 - 4$$
$$\boxed{y = 23}$$

The solution set is $\{(9, 23)\}$. ∎

EXAMPLE 7 Solve $\begin{pmatrix} 10x + 7y = 19 \\ 2x - 6y = -11 \end{pmatrix}$.

Solution

Because changing the form of either of the two equations in preparation for the substitution method would produce a fractional form, we are probably better off using the elimination-by-addition method. Furthermore, notice that the coefficients of x lend themselves to this method.

$$10x + 7y = 19 \quad \text{Leave alone} \quad 10x + 7y = 19$$
$$2x - 6y = -11 \quad \text{Multiply by } -5 \quad -10x + 30y = 55$$
$$37y = 74$$
$$\boxed{y = 2}$$

Substituting 2 for y in the first equation of the given system produces

$$10x + 7y = 19$$
$$10x + 7(2) = 19$$
$$10x + 14 = 19$$
$$10x = 5$$
$$\boxed{x = \frac{5}{10} = \frac{1}{2}}$$

The solution set is $\left\{ \left(\frac{1}{2}, 2 \right) \right\}$. ∎

In Section 8.1, we explained that you can tell by graphing the equations whether the system has no solutions, one solution, or infinitely many solutions. That is, the two lines may be parallel (no solutions), they may intersect in one point (one solution), or they may coincide (infinitely many solutions). From a practical viewpoint, the systems that have one solution deserve most of our attention. However, we do need to be able to deal with the other situations as they arise. The next two examples demonstrate what occurs when we hit a "no solution" or "infinitely many solutions" situation when we are using either the elimination-by-addition or substitution method.

EXAMPLE 8 Solve the system $\begin{pmatrix} y = 2x - 1 \\ 6x - 3y = 7 \end{pmatrix}$.

Solution

Because the first equation indicates that we can substitute $2x - 1$ for y, this system lends itself to the substitution method.

$$6x - 3y = 7 \quad \text{Substitute } 2x - 1 \text{ for } y \quad 6x - 3(2x - 1) = 7$$

Now we solve this equation.

$$6x - 6x + 3 = 7$$
$$0 + 3 = 7$$
$$\boxed{3 = 7}$$

The false numerical statement, $3 = 7$, implies that the system has no solutions. Thus the solution set is $\emptyset$. (You may want to graph the two lines to verify this conclusion!) ∎

EXAMPLE 9

Solve the system $\begin{pmatrix} 2x - 3y = 4 \\ 10x - 15y = 20 \end{pmatrix}$.

Solution

We use the elimination-by-addition method.

$$\begin{array}{l} 2x - 3y = 4 \\ 10x - 15y = 20 \end{array} \xrightarrow[\text{Leave alone}]{\text{Multiply both sides by } -5} \begin{array}{l} -10x + 15y = -20 \\ 10x - 15y = 20 \\ \hline 0 + 0 = 0 \end{array}$$

The true numerical statement, $0 + 0 = 0$, implies that the system has infinitely many solutions. Any ordered pair that satisfies one of the equations also satisfies the other equation. ∎

3 Solve Mixture, Interest, and Geometry Word Problems Using a System of Equations

We will conclude this section with three word problems.

EXAMPLE 10 Apply Your Skill

Niki invested $5000, part of it at 4% interest and the rest at 6%. Her total interest earned for a year was $265. How much did she invest at each rate?

Solution

We let x represent the amount invested at 4%, and we let y represent the amount invested at 6%. The problem translates into the following system:

$x + y = 5000$ Niki invested $5000

$0.04x + 0.06y = 265$ Her total interest earned for a year was $265

Multiplying the second equation by 100 produces $4x + 6y = 26{,}500$. Then we have the following equivalent system to solve:

$$\begin{pmatrix} x + y = 5000 \\ 4x + 6y = 26{,}500 \end{pmatrix}$$

Using the elimination-by-addition method, we can proceed as follows:

$$\begin{array}{l} x + y = 5000 \\ 4x + 6y = 26{,}500 \end{array} \xrightarrow[\text{Leave alone}]{\text{Multiply both sides by } -4} \begin{array}{l} -4x - 4y = -20{,}000 \\ 4x + 6y = 26{,}500 \\ \hline 2y = 6500 \\ y = 3250 \end{array}$$

Substituting 3250 for y in $x + y = 5000$ yields

$x + y = 5000$

$x + 3250 = 5000$

$x = 1750$

Therefore, we know that Niki invested $1750 at 4% interest and $3250 at 6%. ∎

EXAMPLE 11 Apply Your Skill

A 25% chlorine solution is to be mixed with a 40% chlorine solution to produce 12 gallons of a 35% chlorine solution. How many gallons of each solution should be mixed?

Solution

Let x represent the gallons of 25% chlorine solution, and let y represent the gallons of 40% chlorine solution. Then one equation of the system will be $x + y = 12$. For the other equation we need to multiply the number of gallons of each solution by its percentage of chlorine, which gives us the equation $0.25x + 0.40y = 0.35(12)$. So we need to solve the following system:

$$\begin{pmatrix} x + y = 12 \\ 0.25x + 0.40y = 0.35(12) \end{pmatrix}$$

Let's simplify the second equation by finding the product of $0.35(12)$ and multiplying both sides of the equation by 100. Then the system becomes

$$\begin{pmatrix} x + y = 12 \\ 25x + 40y = 420 \end{pmatrix}$$

Using the elimination-by-addition method, we can proceed as follows:

$$\begin{array}{rl} x + y = 12 & \text{Multiply both sides by } -25 \\ 25x + 40y = 420 & \text{Leave alone} \end{array} \quad \begin{array}{r} -25x - 25y = -300 \\ 25x + 40y = 420 \\ \hline 15y = 120 \\ y = 8 \end{array}$$

Substituting 8 for y in $x + y = 12$ yields

$$x + y = 12$$
$$x + 8 = 12$$
$$x = 4$$

Therefore, we need 4 gallons of the 25% chlorine solution and 8 gallons of the 40% chlorine solution to make 12 gallons of 35% chlorine solution. ∎

EXAMPLE 12   **Apply Your Skill**

The length of a rectangle is 1 centimeter less than three times the width. The perimeter of the rectangle is 94 centimeters. Find the length and width of the rectangle.

Solution

We let w represent the width of the rectangle, and we let l represent the length of the rectangle (see Figure 8.8). The problem translates into the following system of equations:

$l = 3w - 1$ The length of the rectangle is 1 centimeter less than three times the width

$2l + 2w = 94$ The perimeter of the rectangle is 94 centimeters

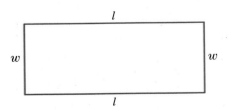

Figure 8.8

Multiplying both sides of the second equation by one-half produces the equivalent equation $l + w = 47$, so we have the following system to solve:

$$\begin{pmatrix} l = 3w - 1 \\ l + w = 47 \end{pmatrix}$$

The first equation indicates that we can substitute $3w - 1$ for l in the second equation.

$l + w = 47$ $\xrightarrow{\text{Substitute } 3w - 1 \text{ for } l}$ $3w - 1 + w = 47$

Solving this equation yields

$$3w - 1 + w = 47$$
$$4w - 1 = 47$$
$$4w = 48$$
$$\boxed{w = 12}$$

Substituting 12 for w in one of the original equations produces

$$l = 3w - 1$$
$$l = 3(12) - 1$$
$$l = 36 - 1$$
$$\boxed{l = 35}$$

The rectangle is 12 centimeters wide and 35 centimeters long. ∎

CONCEPT QUIZ 8.3

For Problems 1–7, answer true or false.

1. Every system of equations can be solved by the substitution method.
2. The substitution method eliminates a variable by substitution rather than by adding the equations.
3. To use the substitution method, one of the equations has to be solved for x.
4. The graphing method for solving a system of equations is practical for finding an exact solution.
5. Both the elimination-by-addition method and the substitution method will find exact solutions.
6. When solving a system of equations, obtaining a false numerical equation such as $1 = 5$ implies that the system has infinitely many solutions.
7. When solving a system of equations, obtaining a true numerical equation such as $3 = 3$ implies that any ordered pair that satisfies one equation also satisfies the other equation.

Section 8.3 Classroom Problem Set

Objective 1

1. Solve $\begin{pmatrix} x = y + 11 \\ x + 2y = -10 \end{pmatrix}$.

2. Solve $\begin{pmatrix} 4x - 3y = -6 \\ y = -3x - 2 \end{pmatrix}$.

3. Solve $\begin{pmatrix} 4x + 3y = -15 \\ y = 2x + 5 \end{pmatrix}$.

4. Solve. $\begin{pmatrix} 4y - 1 = x \\ 2x - 8y = 3 \end{pmatrix}$.

5. Solve $\begin{pmatrix} x + 4y = 22 \\ y = \dfrac{1}{2}x + 4 \end{pmatrix}$.

6. Solve $\begin{pmatrix} 4x - 5y = 6 \\ y = \dfrac{2}{3}x \end{pmatrix}$.

7. Solve $\begin{pmatrix} 2x + 3y = 4 \\ x + y = -1 \end{pmatrix}$.

8. Solve $\begin{pmatrix} 9x - 2y = -18 \\ 4x - y = -7 \end{pmatrix}$.

9. Solve $\begin{pmatrix} 2x + 3y = 8 \\ 3x - 2y = -14 \end{pmatrix}$.

10. Solve $\begin{pmatrix} 5x + 7y = 3 \\ 3x - 2y = 0 \end{pmatrix}$.

Objective 2

11. Solve $\begin{pmatrix} x = 2y - 6 \\ 4x - 3y = 16 \end{pmatrix}$.

12. Solve $\begin{pmatrix} y = 3x + 2 \\ 4x - 3y = -21 \end{pmatrix}$.

13. Solve $\begin{pmatrix} 9x + 5y = -15 \\ 3x - 2y = -27 \end{pmatrix}$.

14. Solve $\begin{pmatrix} 2x + 3y = 13 \\ 3x - 5y = -28 \end{pmatrix}$.

15. Solve $\begin{pmatrix} y = 3x + 2 \\ 3x - y = 4 \end{pmatrix}$.

16. Solve $\begin{pmatrix} x = 5y - 5 \\ 2x - 10y = 2 \end{pmatrix}$.

17. Solve $\begin{pmatrix} x - 2y = 4 \\ 3x - 6y = 12 \end{pmatrix}$.

18. Solve $\begin{pmatrix} 3x + 6y = 15 \\ x + 2y = 5 \end{pmatrix}$.

Objective 3

19. Simoni invested $7000 in two accounts. One account paid 5% interest and the other account paid 7% interest. Her total interest earned for the year was $440. How much did she invest at each rate?

20. Sydney invested $13,000, part of it at 5% and the rest at 6%. If her total yearly interest was $730, how much did she invest at each rate?

21. A 6% saline solution is to be mixed with a 12% saline solution to produce 30 liters of a 10% saline solution. How many gallons of each solution should be mixed?

22. A 10% salt solution is to be mixed with a 15% salt solution to produce 10 gallons of a 13% salt solution. How many gallons of the 10% solution and how many gallons of the 15% solution will be needed?

23. The length of a rectangle is 6 inches more than twice the width. The perimeter of the rectangle is 168 inches. Find the length and width of the rectangle.

24. Sam has three times as many nickels as pennies in his collection. Together his pennies and nickels have a value of $4.80. How many pennies and how many nickels does he have?

THOUGHTS INTO WORDS

1. Explain how you would solve the system $\begin{pmatrix} 5x - 4y = 10 \\ 3x - y = 6 \end{pmatrix}$ using the substitution method.

2. How do you decide whether to solve a system of linear equations by using the elimination-by-addition method or the substitution method?

3. What do you see as the strengths and weaknesses of the elimination-by-addition method and the substitution method?

Answers to the Concept Quiz

1. True **2.** True **3.** False **4.** False **5.** True **6.** False **7.** True

8.4 3 × 3 Systems of Equations

OBJECTIVES

1. Solve 3 × 3 Systems of Equations
2. Solve Word Problems Using 3 × 3 Systems of Equations

1 Solve 3 × 3 Systems of Equations

When we find the solution set of an equation in two variables, such as $2x + y = 9$, we are finding the ordered pairs that make the equation a true statement. Plotted in two dimensions, the graph of the solution set is a line.

Now consider an equation with three variables, such as $2x - y + 4z = 8$. A solution set of this equation is an ordered triple, (x, y, z), that makes the equation a true statement. For example, the ordered triple $(3, 2, 1)$ is a solution of $2x - y + 4z = 8$, because $2(3) - 2 + 4(1) = 8$. The graph of the solution set of an equation in three variables is a plane, not a line. In fact, graphing equations in three variables requires the use of a three-dimensional coordinate system.

A 3×3 (read "3 by 3") system of equations is a system of three linear equations in three variables. To solve a 3×3 system such as

$$\begin{pmatrix} 2x - y + 4z = 5 \\ 3x + 2y + 5z = 4 \\ 4x - 3y - z = 11 \end{pmatrix}$$

means to find all the ordered triples that satisfy all three equations. In other words, the solution set of the system is the intersection of the solution sets of all three equations in the system. Using a graphing approach to solve systems of three linear equations in three variables is not at all practical. However, a graphic analysis will provide insight into the types of possible solutions.

In general, each linear equation in three variables produces a plane. A system of three such equations produces three planes. There are various ways that the planes can intersect. In this course, however, you need to realize that a system of three linear equations in three variables produces one of the following possible solution sets.

1. There is *one ordered triple* that satisfies all three equations. The three planes have a common point of intersection as indicated in Figure 8.9.

Figure 8.9

2. There are *infinitely many* ordered triples in the solution set, all of which are coordinates of points on a line common to the planes. This can happen when three planes have a common line of intersection, as in Figure 8.10(a), or when two of the planes coincide and the third plane intersects them, as in Figure 8.10(b).

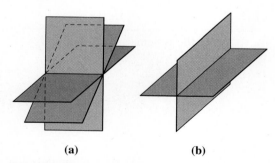

(a)   (b)

Figure 8.10

3. There are *infinitely many* ordered triples in the solution set, all of which are coordinates of points on a plane. This happens when the three planes coincide, as illustrated in Figure 8.11.

Figure 8.11

4. The solution set is *empty*; it is ∅. This can happen in various ways, as you can see in Figure 8.12. Notice that in each situation there are no points common to all three planes.

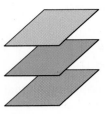
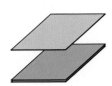

(a) Three parallel planes

(b) Two planes coincide and the third one is parallel to the coinciding planes.

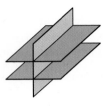
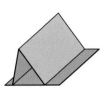

(c) Two planes are parallel and the third intersects them in parallel lines.

(d) No two planes are parallel, but two of them intersect in a line that is parallel to the third plane.

Figure 8.12

Now that you know what possibilities exist, we can consider finding the solution sets for some systems. Our approach will be the elimination-by-addition method, whereby systems are replaced with equivalent systems until we get a system that allows us to easily determine the solution set. We will start with an example that allows us to determine the solution set without changing to another, equivalent system.

EXAMPLE 1

Solve the system

$$\begin{pmatrix} 4x + 2y - z = -5 \\ 3y + z = -1 \\ 2z = 10 \end{pmatrix}$$ (1) (2) (3)

Solution

From equation (3), we can find the value of z.

$$2z = 10$$
$$z = 5$$

Now we can substitute 5 for z in equation (2).

$$3y + z = -1$$
$$3y + 5 = -1$$
$$3y = -6$$
$$y = -2$$

Finally, we can substitute -2 for y and 5 for z in equation (1).

$$4x + 2y - z = -5$$
$$4x + 2(-2) - 5 = -5$$
$$4x - 9 = -5$$
$$4x = 4$$
$$\boxed{x = 1}$$

The solution set is $\{(1, -2, 5)\}$. ∎

Notice the format of the equations in the system in Example 1. The first equation contains all three variables, the second equation has only two variables, and the third equation has only one variable. This allowed us to solve the third equation and then use "back substitution" to find the values of the other variables. Let's consider another example where we replace one equation to make an equivalent system.

EXAMPLE 2

Solve the system

$$\begin{pmatrix} 2x + 4y - 5z = -8 \\ y + 4z = 7 \\ 5y + 3z = 1 \end{pmatrix} \quad \begin{matrix} (1) \\ (2) \\ (3) \end{matrix}$$

Solution

In order to achieve the same format as in Example 1, we will need to eliminate the term with the y variable in equation (3). Using the concept of elimination, we can replace equation (3) with an equivalent equation we form by multiplying equation (2) by -5 and then adding that result to equation (3). The equivalent system is

$$\begin{pmatrix} 2x + 4y - 5z = -8 \\ y + 4z = 7 \\ -17z = -34 \end{pmatrix} \quad \begin{matrix} (4) \\ (5) \\ (6) \end{matrix}$$

From equation (6) we can find the value of z.

$$-17z = -34$$
$$\boxed{z = 2}$$

Now we can substitute 2 for z in equation (5).

$$y + 4z = 7$$
$$y + 4(2) = 7$$
$$\boxed{y = -1}$$

Finally, we can substitute -1 for y and 2 for z in equation (4).

$$2x + 4y - 5z = -8$$
$$2x + 4(-1) - 5(2) = -8$$
$$2x - 4 - 10 = -8$$
$$2x - 14 = -8$$
$$2x = 6$$
$$\boxed{x = 3}$$

The solution set is $\{(3, -1, 2)\}$. ∎

Now let's consider some examples where we replace more than one equation to make an equivalent system.

EXAMPLE 3

Solve the system

$$\begin{pmatrix} x + 2y - 3z = -1 \\ 3x - y + 2z = -13 \\ 2x + 3y - 5z = -4 \end{pmatrix} \begin{matrix} (1) \\ (2) \\ (3) \end{matrix}$$

Solution

We start by picking a pair of equations to form a new equation by eliminating a variable. We will use equations (1) and (2) to form a new equation while eliminating the x variable. We can replace equation (2) with an equation formed by multiplying equation (1) by -3 and adding the result to equation (2). The equivalent system is

$$\begin{pmatrix} x + 2y - 3z = -1 \\ -7y + 11z = -10 \\ 2x + 3y - 5z = -4 \end{pmatrix} \begin{matrix} (4) \\ (5) \\ (6) \end{matrix}$$

Now we take equation (4) and equation (6) and eliminate the same variable, x. We can replace equation (6) with a new equation formed by multiplying equation (4) by -2 and adding the result to equation (6). The equivalent system is

$$\begin{pmatrix} x + 2y - 3z = -1 \\ -7y + 11z = -10 \\ -y + z = -2 \end{pmatrix} \begin{matrix} (7) \\ (8) \\ (9) \end{matrix}$$

Now we take equations (8) and (9) and form a new equation by eliminating a variable. Either y or z can be eliminated. For this example we will eliminate y. We can replace equation (8) with a new equation formed by multiplying equation (9) by -7 and adding the result to equation (8). The equivalent system is

$$\begin{pmatrix} x + 2y - 3z = -1 \\ 4z = 4 \\ -y + z = -2 \end{pmatrix} \begin{matrix} (10) \\ (11) \\ (12) \end{matrix}$$

From equation (11), we can find the value of z.

$$4z = 4$$
$$z = 1$$

Now we substitute 1 for z in equation (12) and determine the value of y.

$$-y + z = -2$$
$$-y + 1 = -2$$
$$-y = -3$$
$$y = 3$$

Finally, we can substitute 3 for y and 1 for z in equation (10).

$$x + 2y - 3z = -1$$
$$x + 2(3) - 3(1) = -1$$
$$x + 6 - 3 = -1$$
$$x + 3 = -1$$
$$x = -4$$

The solution set is $\{(-4, 3, 1)\}$. ∎

EXAMPLE 4

Solve the system

$$\begin{pmatrix} 2x + 3y - z = 8 \\ 5x + 2y - 3z = 21 \\ 3x - 4y + 2z = 5 \end{pmatrix} \begin{matrix} (1) \\ (2) \\ (3) \end{matrix}$$

Solution

Studying the coefficients in the system indicates that eliminating the z terms from equations (2) and (3) would be easy to do. We can replace equation (2) with an equation formed by multiplying equation (1) by -3 and adding the result to equation (2). The equivalent system is

$$\begin{pmatrix} 2x + 3y - z = 8 \\ -x - 7y = -3 \\ 3x - 4y + 2z = 5 \end{pmatrix} \begin{matrix} (4) \\ (5) \\ (6) \end{matrix}$$

Now we replace equation (6) with an equation formed by multiplying equation (4) by 2 and adding the result to equation (6). The equivalent system is

$$\begin{pmatrix} 2x + 3y - z = 8 \\ -x - 7y = -3 \\ 7x + 2y = 21 \end{pmatrix} \begin{matrix} (7) \\ (8) \\ (9) \end{matrix}$$

Now we can eliminate the x term from equation (9). We replace equation (9) with an equation formed by multiplying equation (8) by 7 and adding the result to equation (9). The equivalent system is

$$\begin{pmatrix} 2x + 3y - z = 8 \\ -x - 7y = -3 \\ -47y = 0 \end{pmatrix} \begin{matrix} (10) \\ (11) \\ (12) \end{matrix}$$

From equation (12), we can determine the value of y.

$$-47y = 0$$
$$y = 0$$

Now we can substitute 0 for y in equation (11) and find the value of x.

$$-x - 7y = -3$$
$$-x - 7(0) = -3$$
$$-x = -3$$
$$x = 3$$

Finally, we can substitute 3 for x and 0 for y in equation (10).

$$2x + 3y - z = 8$$
$$2(3) + 3(0) - z = 8$$
$$6 - z = 8$$
$$-z = 2$$
$$z = -2$$

The solution set is $\{(3, 0, -2)\}$. ∎

EXAMPLE 5

Solve the system

$$\begin{pmatrix} x + 3y - 2z = 3 \\ 3x - 4y - z = 4 \\ 2x + 6y - 4z = 9 \end{pmatrix} \quad \begin{matrix}(1)\\(2)\\(3)\end{matrix}$$

Solution

Studying the coefficients indicates that it would be easy to eliminate the x terms from equations (2) and (3). We can replace equation (2) with an equation formed by multiplying equation (1) by -3 and adding the result to equation (2). Likewise, we can replace equation (3) with an equation formed by multiplying equation (1) by -2 and adding the result to equation (3). The equivalent system is

$$\begin{pmatrix} x + 3y - 2z = 3 \\ -13y + 5z = -5 \\ 0 + 0 + 0 = 3 \end{pmatrix} \quad \begin{matrix}(4)\\(5)\\(6)\end{matrix}$$

The false statement $0 = 3$ in equation (6) indicates that the system is inconsistent, and therefore the solution set is $\varnothing$. [If you were to graph this system, equations (1) and (3) would produce parallel planes, which is the situation depicted in Figure 8.12(c).] ∎

EXAMPLE 6

Solve the system

$$\begin{pmatrix} x + y + z = 6 \\ 3x + y - z = 2 \\ 5x + y - 3z = -2 \end{pmatrix} \quad \begin{matrix}(1)\\(2)\\(3)\end{matrix}$$

Solution

Studying the coefficients indicates that it would be easy to eliminate the y terms from equations (2) and (3). We can replace equation (2) with an equation formed by multiplying equation (1) by -1 and adding the result to equation (2). Likewise, we can replace equation (3) with an equation formed by multiplying equation (1) by -1 and adding the result to equation (3). The equivalent system is

$$\begin{pmatrix} x + y + z = 6 \\ 2x - 2z = -4 \\ 4x - 4z = -8 \end{pmatrix} \quad \begin{matrix}(4)\\(5)\\(6)\end{matrix}$$

Now we replace equation (6) with an equation formed by multiplying equation (5) by -2 and adding the result to equation (6). The equivalent system is

$$\begin{pmatrix} x + y + z = 6 \\ 2x - 2z = -4 \\ 0 + 0 = 0 \end{pmatrix} \quad \begin{matrix}(7)\\(8)\\(9)\end{matrix}$$

The true numerical statement $0 + 0 = 0$ in equation (9) indicates that the system has *infinitely many solutions*. ∎

2 Solve Word Problems Using 3 × 3 Systems of Equations

Now we will use the techniques we have presented to solve a geometric problem.

EXAMPLE 7   Apply Your Skill

In a certain triangle, the measure of $\angle A$ is 5° more than twice the measure of $\angle B$. The sum of the measures of $\angle B$ and $\angle C$ is 10° more than the measure of $\angle A$. Find the measures of all three angles.

Solution

We can solve this problem by setting up a system of three linear equations in three variables. We let

x = measure of $\angle A$

y = measure of $\angle B$

z = measure of $\angle C$

Knowing that the sum of the measures of the angles in a triangle is 180° gives us the equation $x + y + z = 180$. The information "the measure of $\angle A$ is 5° more than twice the measure of $\angle B$" gives us the equation $x = 2y + 5$ or an equivalent form $x - 2y = 5$. The information "the sum of the measures of $\angle B$ and $\angle C$ is 10° more than the measure of $\angle A$" gives us the equation $y + z = x + 10$ or an equivalent form $x - y - z = -10$. Putting the three equations together, we get the system of equations

$$\begin{pmatrix} x + y + z = 180 \\ x - 2y = 5 \\ x - y - z = -10 \end{pmatrix} \quad \begin{matrix}(1)\\(2)\\(3)\end{matrix}$$

To solve the system, we first replace equation (3) with an equation formed by adding equation (1) and equation (3). The equivalent system is

$$\begin{pmatrix} x + y + z = 180 \\ x - 2y = 5 \\ 2x = 170 \end{pmatrix} \quad \begin{matrix}(4)\\(5)\\(6)\end{matrix}$$

From equation (6), we can determine that $x = 85$.

Now we can substitute 85 for x in equation (5) and find the value of y.

$x - 2y = 5$

$85 - 2y = 5$

$-2y = -80$

$y = 40$

Finally, we can substitute 40 for y and 85 for x in equation (4) and find the value of z.

$x + y + z = 180$

$85 + 40 + z = 180$

$125 + z = 180$

$z = 55$

The measures of the angles are $\angle A = 85°$, $\angle B = 40°$, and $\angle C = 55°$. ∎

CONCEPT QUIZ 8.4

For Problems 1–8, answer true or false.

1. The graph of a linear equation in three variables is a line.
2. A system of three linear equations in three variables produces three planes when graphed.
3. Three planes can be related by intersecting in exactly two points.
4. One way three planes can be related is if two of the planes are parallel, and the third plane intersects them in parallel lines.

5. A system of three linear equations in three variables always has an infinite number of solutions.
6. A system of three linear equations in three variables can have one ordered triple as a solution.
7. The solution set of the system $\begin{pmatrix} 2x - y + 3z = 4 \\ y - z = 12 \\ 2z = 6 \end{pmatrix}$ is $\{(3, 15, 5)\}$.
8. The solution set of the system $\begin{pmatrix} x - y + z = 4 \\ x - y + z = 6 \\ 3y - 2z = 9 \end{pmatrix}$ is $\{(3, 1, 2)\}$.

Section 8.4 Classroom Problem Set

Objective 1

1. Solve the system $\begin{pmatrix} 3x - y + z = 15 \\ y + 2z = 6 \\ 3z = 12 \end{pmatrix}$.

2. Solve the system $\begin{pmatrix} 3x + y + 2z = 6 \\ 6y + 5z = -4 \\ -4z = 8 \end{pmatrix}$.

3. Solve the system $\begin{pmatrix} 3x + 4y - z = 14 \\ y + 4z = 17 \\ 2y + 3z = 19 \end{pmatrix}$.

4. Solve the system $\begin{pmatrix} 4x + 3y - 2z = 9 \\ 2x + y = 7 \\ 3x - 2y = 21 \end{pmatrix}$.

5. Solve the system $\begin{pmatrix} x + 3y - 2z = -1 \\ 2x - y + 3z = 5 \\ 3x + 2y - z = 8 \end{pmatrix}$.

6. Solve the system $\begin{pmatrix} 4x - 3y + z = 14 \\ 2x + y - 3z = 16 \\ 3x - 4y + 2z = 9 \end{pmatrix}$.

7. Solve the system $\begin{pmatrix} 2x + 3y + z = 9 \\ 3x - 2y - 2z = -16 \\ 5x + 4y + 3z = 9 \end{pmatrix}$.

8. Solve the system $\begin{pmatrix} x + 3y - 4z = 11 \\ 3x - y + 2z = 5 \\ 2x + 5y - z = 8 \end{pmatrix}$.

9. Solve the system $\begin{pmatrix} x + 2y - z = 4 \\ 3x - 4y + 2z = 5 \\ 2x + 4y - 2z = 10 \end{pmatrix}$.

10. Solve the system $\begin{pmatrix} 2x + 3y + z = 7 \\ x - 4y + 2z = -3 \\ 3x - y + 3z = 5 \end{pmatrix}$.

11. Solve the system $\begin{pmatrix} 2x + y + z = 8 \\ x + y - z = 5 \\ x - y + 5z = 1 \end{pmatrix}$.

12. Solve the system $\begin{pmatrix} x + 2y - z = 6 \\ 3x - y + 4z = -10 \\ 5x + 3y + 2z = 2 \end{pmatrix}$.

Objective 2

13. In a certain triangle, the measure of $\angle A$ is 15° less than $\angle B$. The sum of the measures of $\angle A$ and $\angle B$ is 10° more than the measure of $\angle C$. Find the measure of all three angles.

14. Two pounds of peaches, 1 pound of cherries, and 3 pounds of pears cost $5.64. One pound of peaches, 2 pounds of cherries, and 2 pounds of pears cost $4.65. Two pounds of peaches, 4 pounds of cherries, and 1 pound of pears cost $7.23. Find the price per pound for each item.

THOUGHTS INTO WORDS

1. Give a step-by-step description of how to solve this system of equations.
$$\begin{pmatrix} 2x + 3y - z = 3 \\ 4y + 3z = -2 \\ 6z = -12 \end{pmatrix}$$

2. Describe how you would solve this system of equations.
$$\begin{pmatrix} 2x + 3y - z = 7 \\ x + 2y = 4 \\ 3x - 4y = 2 \end{pmatrix}$$

Answers to the Concept Quiz
1. False **2.** True **3.** False **4.** True **5.** False **6.** True **7.** False **8.** False

Chapter 8 Review Problem Set

For Problems 1–4, decide whether the given ordered pair is a solution of the given system of equations.

1. $\begin{pmatrix} 3x - y = 0 \\ 6x + 2y = 12 \end{pmatrix}, (1, 3)$

2. $\begin{pmatrix} x + 3y = 10 \\ 2x - y = 10 \end{pmatrix}, (4, 2)$

3. $\begin{pmatrix} 2x - y = 13 \\ x + 3y = 10 \end{pmatrix}, (6, 21)$

4. $\begin{pmatrix} x = 5y + 2 \\ 4x - 3y = -26 \end{pmatrix}, (-8, -2)$

For Problems 5–10, use the graphing method to solve each system.

5. $\begin{pmatrix} 2x + y = 2 \\ x - y = -5 \end{pmatrix}$

6. $\begin{pmatrix} 2x + y = 8 \\ x - 2y = -6 \end{pmatrix}$

7. $\begin{pmatrix} 2x + 3y = 2 \\ y = -\dfrac{2}{3}x + 4 \end{pmatrix}$

8. $\begin{pmatrix} y = 2x - 2 \\ 2x - y = 4 \end{pmatrix}$

9. $\begin{pmatrix} 3x + y = 6 \\ y = -3x + 6 \end{pmatrix}$

10. $\begin{pmatrix} 2x + y = 4 \\ 4x + 2y = 8 \end{pmatrix}$

For Problems 11–16, indicate the solution set for each system of inequalities by shading the appropriate region on a graph.

11. $\begin{pmatrix} 4x + y < 4 \\ 2x - y > -4 \end{pmatrix}$

12. $\begin{pmatrix} 3x + y \geq -3 \\ x - y > -5 \end{pmatrix}$

13. $\begin{pmatrix} 3x + y \leq 3 \\ x - 2y > -4 \end{pmatrix}$

14. $\begin{pmatrix} x + 2y < 2 \\ 3x - 2y > 12 \end{pmatrix}$

15. $\begin{pmatrix} y > \dfrac{2}{5}x - 2 \\ y < 3x - 3 \end{pmatrix}$

16. $\begin{pmatrix} y < -x + 1 \\ y < 2x - 3 \end{pmatrix}$

For Problems 17–22, solve each system of equations using the elimination-by-addition method.

17. $\begin{pmatrix} 2x - y = 1 \\ 3x - 2y = -5 \end{pmatrix}$

18. $\begin{pmatrix} 9x + 2y = 140 \\ x + 5y = 135 \end{pmatrix}$

19. $\begin{pmatrix} 3x - 2y = -14 \\ 2x - y = -10 \end{pmatrix}$

20. $\begin{pmatrix} 4x + 3y = 8 \\ x + 2y = -3 \end{pmatrix}$

21. $\begin{pmatrix} 3x + 2y = -7 \\ 4x - 5y = -17 \end{pmatrix}$

22. $\begin{pmatrix} 4x - 3y = 6 \\ 3x - 5y = 10 \end{pmatrix}$

For Problems 23–28, solve each system of equations using the substitution method.

23. $\begin{pmatrix} 2x + 5y = 7 \\ x = -3y + 1 \end{pmatrix}$

24. $\begin{pmatrix} 5x - 7y = 9 \\ y = 3x - 2 \end{pmatrix}$

25. $\begin{pmatrix} y = -\dfrac{2}{3}x \\ \dfrac{1}{3}x - y = -9 \end{pmatrix}$

26. $\begin{pmatrix} y = 5x + 2 \\ 10x - 2y = 1 \end{pmatrix}$

27. $\begin{pmatrix} x - 2y = 9 \\ 3x + 5y = 16 \end{pmatrix}$

28. $\begin{pmatrix} 3x + 5y = 49 \\ 2x + 3y = 30 \end{pmatrix}$

For Problems 29–32, decide which method, elimination-by-addition or substitution, would be the most fitting to solve the given system of equations.

29. $\begin{pmatrix} 2x + 5y = 7 \\ 3x - 5y = 3 \end{pmatrix}$

30. $\begin{pmatrix} x + 4y = 8 \\ 2x - 4y = 1 \end{pmatrix}$

31. $\begin{pmatrix} y = \dfrac{1}{3}x \\ 2x - 3y = 9 \end{pmatrix}$

32. $\begin{pmatrix} 3x - 2y = 6 \\ x = y + 1 \end{pmatrix}$

For Problems 33–36, solve the system of equations.

33. $\begin{pmatrix} x + 3y - z = 1 \\ 2x - y + z = 3 \\ 3x + y + 2z = 12 \end{pmatrix}$

34. $\begin{pmatrix} 2x + 3y - z = 4 \\ x + 2y + z = 7 \\ 3x + y + 2z = 13 \end{pmatrix}$

35. $\begin{pmatrix} x - y - z = 4 \\ -3x + 2y + 5z = -21 \\ 5x - 3y - 7z = 30 \end{pmatrix}$

36. $\begin{pmatrix} 2x - y + z = -7 \\ -5x + 2y - 3z = 17 \\ 3x + y + 7z = -5 \end{pmatrix}$

For Problems 37–44, solve each problem by setting up and solving a system of two linear equations in two variables.

37. The sum of two numbers is 113. The larger number is 1 less than twice the smaller number. Find the numbers.

38. Last year Mark invested a certain amount of money at 6% annual interest and $50 more than that amount at 8%. He received $39.00 in interest. How much did he invest at each rate?

39. Cindy has 43 coins consisting of nickels and dimes. The total value of the coins is $3.40. How many coins of each kind does she have?

40. The length of a rectangle is 1 inch longer than three times the width. If the perimeter of the rectangle is 50 inches, find the length and width.

41. Two angles are complementary, and one of them is 6° less than twice the other one. Find the measure of each angle.

42. Two angles are supplementary, and the larger angle is 20° less than three times the smaller angle. Find the measure of each angle.

43. Four cheeseburgers and five milkshakes cost a total of $25.50. Two milkshakes cost $1.75 more than one cheeseburger. Find the cost of a cheeseburger, and also find the cost of a milkshake.

44. Three bottles of orange juice and two bottles of water cost $6.75. On the other hand, two bottles of juice and three bottles of water cost $6.15. Find the cost per bottle of each.

Chapter 8 Practice Test

For Problems 1–3, determine if the given ordered pair or ordered triple is a solution of the system of equations.

1. $\begin{pmatrix} 3x + y = -2 \\ x - 2y = -9 \end{pmatrix}, (-1, 1)$

2. $\begin{pmatrix} 5x + 2y = 17 \\ 3x - 2y = 7 \end{pmatrix}, (3, 1)$

3. $\begin{pmatrix} 2x + 3y - z = -8 \\ 4x - 2y + 3z = 4 \\ x + 5y - z = -6 \end{pmatrix}, (-2, 0, 4)$

For Problems 4–7, solve the system of equations by graphing.

4. $\begin{pmatrix} x + y = 5 \\ x - 2y = -4 \end{pmatrix}$

5. $\begin{pmatrix} 3x - 2y = 6 \\ 3x + y = -3 \end{pmatrix}$

6. $\begin{pmatrix} y = 2x - 4 \\ 2x - y = -3 \end{pmatrix}$

7. $\begin{pmatrix} y = -\frac{1}{2}x + 1 \\ x + 2y = 2 \end{pmatrix}$

For Problems 8–10, indicate the solution set for each system by shading the appropriate region.

8. $\begin{pmatrix} 3x + 2y < 6 \\ x - 2y > 2 \end{pmatrix}$

9. $\begin{pmatrix} x + 4y > -4 \\ 2x - 3y > -6 \end{pmatrix}$

10. $\begin{pmatrix} y \geq -\frac{1}{2}x + 3 \\ y \leq 2x + 2 \end{pmatrix}$

For Problems 11–13, solve the system of equations by the elimination-by-addition method.

11. $\begin{pmatrix} 2x + 3y = 26 \\ x - 3y = -14 \end{pmatrix}$

12. $\begin{pmatrix} x - 3y = -9 \\ 4x + 7y = 40 \end{pmatrix}$

13. $\begin{pmatrix} 4x - 3y = -11 \\ 5x + 2y = -8 \end{pmatrix}$

For Problems 14–17, solve the system of equations by the substitution method.

14. $\begin{pmatrix} y = -5x - 16 \\ 6x - 5y = 18 \end{pmatrix}$

15. $\begin{pmatrix} 2x + y = -2 \\ 4x - y = 5 \end{pmatrix}$

1. _____

2. _____

3. _____

4. _____

5. _____

6. _____

7. _____

8. _____

9. _____

10. _____

11. _____

12. _____

13. _____

14. _____

15. _____

16. $\begin{pmatrix} y = \dfrac{1}{3}x - 2 \\ x - 3y = 6 \end{pmatrix}$

17. $\begin{pmatrix} x = 3y - 4 \\ x - 3y = 2 \end{pmatrix}$

For Problems 18–20, solve the system of equations.

18. $\begin{pmatrix} 3x - 3y + 2z = -1 \\ 2y - z = 8 \\ 5z = -10 \end{pmatrix}$

19. $\begin{pmatrix} 2y + 3z = -9 \\ y - z = 8 \\ 3x + 4y + 2z = 5 \end{pmatrix}$

20. $\begin{pmatrix} x + y + 2z = 6 \\ -2x + y + 3z = 16 \\ 3x + 5y - z = -10 \end{pmatrix}$

For Problems 21–25, solve each problem by setting up and solving a system of equations.

21. The sum of two numbers is 96. The larger number is four less than three times the smaller number. Find the numbers.

22. Three reams of paper and 4 notebooks cost $19.63. Four reams of paper and 1 notebook cost $16.25. Find the cost of each item.

23. The length of a rectangle is 1 inch less than twice the width of the rectangle. If the perimeter of the rectangle is 40 inches, find the length of the rectangle.

24. One solution contains 30% alcohol and another solution contains 80% alcohol. Some of each of the two solutions are mixed to produce 5 liters of a 60% alcohol solution. How many liters of the 80% alcohol solution are used?

25. Last year Julio invested a certain amount of money at 5% (annual interest rate) and $500 more than that amount at 6%. He received $305 in interest. How much did he invest at each rate?

Roots and Radicals

Chapter 9 Warm-Up Problems

1. Multiply.

   (a) 15^2 (b) $\left(\dfrac{3}{7}\right)^2$ (c) $(0.6)^2$

2. Multiply.

   (a) $(-3)^3$ (b) $\left(\dfrac{1}{5}\right)^3$ (c) $-\left(-\dfrac{2}{3}\right)^3$

3. Simplify.

   (a) $13x^2 - 6x^2$ (b) $4m - 6m + 5m$ (c) $10x^2 - 3y^2$

4. Write as the product of prime factors.

   (a) 48   (b) 27   (c) 75

5. Multiply. Do not reduce to lowest terms.

   (a) $\dfrac{6}{13} \cdot \dfrac{13}{13}$ (b) $\dfrac{2}{5} \cdot \dfrac{5}{5}$ (c) $\dfrac{3}{2} \cdot \dfrac{2^2}{2^2}$

6. Simplify.

   (a) $(x^2)^3 \cdot x$ (b) $x^2 y^4 x y^3$ (c) $m^3 n^9 m^2 n$

7. Find the product.

   (a) $(x - 5)(x + 3)$ (b) $(m - 11)(m + 11)$ (c) $(3x - 2)(5x + 1)$

8. Solve for the variable.

   (a) $3x - 2 = 7$ (b) $7m + 3 = 4m + 2$ (c) $x^2 + 2x - 15 = 0$

9.1 Roots and Radicals
9.2 Simplifying Radicals
9.3 Simplifying Radicals of Quotients
9.4 Products and Quotients Involving Radicals
9.5 Solving Radical Equations

In Section 1.6 we used $\sqrt{2}$ and $\sqrt{3}$ as examples of irrational numbers. Irrational numbers in decimal form are nonrepeating decimals. For example, $\sqrt{2} = 1.414213562373\ldots$, where the three dots indicate that the decimal expansion continues indefinitely. In Chapter 1, we stated that we would return to the irrationals in Chapter 9. The time has come for us to expand our set of skills relative to the set of irrational numbers.

Video tutorials for all section learning objectives are available in a variety of delivery modes.

INTERNET PROJECT

Search the Internet for sites on square roots or cube roots. Find three sites with activities such as flash cards, matching, or timed drills to learn the roots of numbers. Write the URLs of these sites and a brief description of their activities to share with other students and the instructor.

9.1 Roots and Radicals

OBJECTIVES

1. Find the Square Root of a Perfect Square
2. Find the Cube Root of a Perfect Cube
3. Find an Approximation of a Square or Cube Root
4. Add and Subtract Radical Expressions
5. Solve Applications Using Formulas with Radicals

1 Find the Square Root of a Perfect Square

To **square a number** means to raise it to the second power—that is, to use the number as a factor twice.

$$4^2 = 4 \cdot 4 = 16 \quad \text{Read "four squared equals sixteen"}$$

$$10^2 = 10 \cdot 10 = 100$$

$$\left(\frac{1}{2}\right)^2 = \frac{1}{2} \cdot \frac{1}{2} = \frac{1}{4}$$

$$(-3)^2 = (-3)(-3) = 9$$

A **square root of a number** is one of its two equal factors. Thus 4 is a square root of 16 because $4 \cdot 4 = 16$. Likewise, -4 is also a square root of 16 because $(-4)(-4) = 16$. In general, a is a square root of b if $a^2 = b$. The following generalizations are a direct consequence of the previous statement.

1. Every positive real number has two square roots; one is positive and the other is negative. They are opposites of each other.
2. Negative real numbers have no real number square roots because any nonzero real number is positive when squared.
3. The square root of 0 is 0.

The symbol $\sqrt{}$, called a **radical sign**, is used to designate the nonnegative square root. The number under the radical sign is called the **radicand**. The entire expression (such as $\sqrt{16}$) is called a **radical**.

$\sqrt{16} = 4$ $\sqrt{16}$ indicates the nonnegative or principal square root of 16

$-\sqrt{16} = -4$ $-\sqrt{16}$ indicates the negative square root of 16

$\sqrt{0} = 0$ Zero has only one square root. Technically, we could write $-\sqrt{0} = -0 = 0$

$\sqrt{-4}$ is not a real number.

$-\sqrt{-4}$ is not a real number.

In general, the following definition is useful.

> **Definition 9.1**
>
> If $a \geq 0$ and $b \geq 0$, then $\sqrt{b} = a$ if and only if $a^2 = b$; a is called the **principal square root of b**.

If a is a number that is the square of an integer, then $\sqrt{a}$ and $-\sqrt{a}$ are rational numbers. For example, $\sqrt{1}$, $\sqrt{4}$, and $\sqrt{25}$ are the rational numbers 1, 2, and 5, respectively. The numbers 1, 4, and 25 are called **perfect squares** because each represents the square of some integer. The following chart contains the squares of the whole numbers from 1 through 20, inclusive. You should know these values so that you can immediately recognize such square roots as $\sqrt{81} = 9$, $\sqrt{144} = 12$, $\sqrt{289} = 17$, and so on from the list. Furthermore, perfect squares of multiples of 10 are easy to recognize. For example, because $30^2 = 900$, we know that $\sqrt{900} = 30$.

$1^2 = 1$	$8^2 = 64$	$15^2 = 225$
$2^2 = 4$	$9^2 = 81$	$16^2 = 256$
$3^2 = 9$	$10^2 = 100$	$17^2 = 289$
$4^2 = 16$	$11^2 = 121$	$18^2 = 324$
$5^2 = 25$	$12^2 = 144$	$19^2 = 361$
$6^2 = 36$	$13^2 = 169$	$20^2 = 400$
$7^2 = 49$	$14^2 = 196$	

Knowing this listing of perfect squares can also help you with square roots of some fractions. Consider the next example.

EXAMPLE 1

Evaluate the following square roots:

(a) $\sqrt{196}$ (b) $\sqrt{\dfrac{16}{25}}$ (c) $\sqrt{\dfrac{64}{49}}$ (d) $\sqrt{0.09}$

Solution

(a) $\sqrt{196} = 14$ because $14^2 = 196$ (b) $\sqrt{\dfrac{16}{25}} = \dfrac{4}{5}$ because $\left(\dfrac{4}{5}\right)^2 = \dfrac{16}{25}$

(c) $\sqrt{\dfrac{64}{49}} = \dfrac{8}{7}$ because $\left(\dfrac{8}{7}\right)^2 = \dfrac{64}{49}$ (d) $\sqrt{0.09} = 0.3$ because $(0.3)^2 = 0.09$ ∎

2 Find the Cube Root of a Perfect Cube

To **cube a number** means to raise it to the third power—that is, to use the number as a factor three times.

$2^3 = 2 \cdot 2 \cdot 2 = 8$ Read "two cubed equals eight"

$4^3 = 4 \cdot 4 \cdot 4 = 64$

$\left(\dfrac{2}{3}\right)^3 = \dfrac{2}{3} \cdot \dfrac{2}{3} \cdot \dfrac{2}{3} = \dfrac{8}{27}$

$(-2)^3 = (-2)(-2)(-2) = -8$

A cube root of a number is one of its three equal factors. Thus 2 is a cube root of 8 because $2 \cdot 2 \cdot 2 = 8$. (In fact, 2 is the only real number that is a cube root of 8.) Furthermore, -2 is a cube root of -8 because $(-2)(-2)(-2) = -8$. (In fact, -2 is the only real number that is a cube root of -8.)

In general, a is a cube root of b if $a^3 = b$. The following generalizations are a direct consequence of the previous statement.

1. Every positive real number has one positive real number cube root.
2. Every negative real number has one negative real number cube root.
3. The cube root of 0 is 0.

Remark: Technically, every nonzero real number has three cube roots, but only one of them is a real number. The other two roots are classified as complex numbers. We are restricting our work at this time to the set of real numbers.

The symbol $\sqrt[3]{}$ designates the cube root of a number. Thus we can write

$$\sqrt[3]{8} = 2 \qquad \sqrt[3]{\frac{1}{27}} = \frac{1}{3}$$

$$\sqrt[3]{-8} = -2 \qquad \sqrt[3]{-\frac{1}{27}} = -\frac{1}{3}$$

In general, the following definition is useful.

Definition 9.2
$\sqrt[3]{b} = a$ if and only if $a^3 = b$.

In Definition 9.2, if $b \geq 0$ then $a \geq 0$, whereas if $b < 0$ then $a < 0$. The number a is called the **principal cube root of b** or simply the **cube root of b**. In the radical $\sqrt[3]{b}$, the 3 is called the **index** of the radical. When working with square roots, we commonly omit writing the index; so we write $\sqrt{b}$ instead of $\sqrt[2]{b}$. The concept of root can be extended to fourth roots, fifth roots, sixth roots, and, in general, nth roots. However, in this text we will restrict our work to square roots and cube roots.

You should become familiar with the following perfect cubes so that you can recognize their roots without a calculator or a table.

$$2^3 = 8 \qquad 5^3 = 125 \qquad 8^3 = 512$$
$$3^3 = 27 \qquad 6^3 = 216 \qquad 9^3 = 729$$
$$4^3 = 64 \qquad 7^3 = 343 \qquad 10^3 = 1000$$

For example, you should recognize that $\sqrt[3]{343} = 7$.

Consider the following example on finding cube roots.

EXAMPLE 2   Evaluate the following cube roots:

(a) $\sqrt[3]{512}$ (b) $\sqrt[3]{\frac{8}{27}}$ (c) $\sqrt[3]{0.125}$

Solution

(a) $\sqrt[3]{512} = 8$ because $8^3 = 512$ (b) $\sqrt[3]{\frac{8}{27}} = \frac{2}{3}$ because $\left(\frac{2}{3}\right)^3 = \frac{8}{27}$

(c) $\sqrt[3]{0.125} = 0.5$ because $(0.5)^3 = 0.125$ ∎

3 Find an Approximation of a Square or Cube Root

So far we have found square and cube roots of perfect squares and perfect cubes. The square roots of a positive integer, which is *not* the square of an integer, are irrational numbers. Likewise the cube root of an integer, which is *not* the cube of an integer, is an irrational number. For example, $\sqrt{2}$, $-\sqrt{2}$, $\sqrt{23}$, $\sqrt[3]{10}$, $\sqrt[3]{-50}$, and $-\sqrt{75}$ are irrational numbers. Remember that irrational numbers have nonrepeating, nonterminating decimal representations. For example, $\sqrt{2} = 1.414213562373\ldots$, where the decimal never terminates and never repeats a block of digits.

For practical purposes, we often need to use a rational approximation of an irrational number. The calculator is a very useful tool for finding such approximations. Be sure you can use your calculator to approximate roots. Most calculators have a square root function key. To find a cube root, many calculators have a $\sqrt[n]{x}$ key, where you would enter 3 for *n* to find a cube root. Once you have found the root with your calculator, you will have to round off the result. In the following example, the roots have been rounded off to the nearest thousandth.

EXAMPLE 3

Use a calculator to find a rational approximation for each root. Express your answers to the nearest thousandth.

(a) $\sqrt{19}$ (b) $-\sqrt{38}$ (c) $\sqrt[3]{72}$ (d) $\sqrt[3]{-81}$

Solution

(a) $\sqrt{19} = 4.359$ (b) $-\sqrt{38} = -6.164$ (c) $\sqrt[3]{72} = 4.160$
(d) $\sqrt[3]{-81} = -4.327$ ■

4 Add and Subtract Radical Expressions

Recall our use of the distributive property as the basis for combining similar terms. Here are three examples:

$$3x + 2x = (3 + 2)x = 5x$$
$$7y - 4y = (7 - 4)y = 3y$$
$$9a^2 + 5a^2 = (9 + 5)a^2 = 14a^2$$

In a like manner, we can often simplify expressions that contain radicals by using the distributive property.

$$5\sqrt{2} + 7\sqrt{2} = (5 + 7)\sqrt{2} = 12\sqrt{2}$$
$$8\sqrt{5} - 2\sqrt{5} = (8 - 2)\sqrt{5} = 6\sqrt{5}$$
$$4\sqrt{7} + 6\sqrt{7} + 3\sqrt{11} - \sqrt{11} = (4 + 6)\sqrt{7} + (3 - 1)\sqrt{11} = 10\sqrt{7} + 2\sqrt{11}$$
$$6\sqrt[3]{7} + 4\sqrt[3]{7} - 2\sqrt[3]{7} = (6 + 4 - 2)\sqrt[3]{7} = 8\sqrt[3]{7}$$

Note that if we want *to simplify when adding or subtracting radical expressions, the radicals must have the same radicand and index.* Also note the form we use to indicate multiplication when a radical is involved. For example, $5 \cdot \sqrt{2}$ is written as $5\sqrt{2}$.

EXAMPLE 4 Simplify the following radical expressions:

(a) $6\sqrt{5} - 9\sqrt{5} + 7\sqrt{5}$ (b) $4\sqrt{7} + 6\sqrt{7} + 2\sqrt{3} + 7\sqrt{3}$

Solution

(a) $6\sqrt{5} - 9\sqrt{5} + 7\sqrt{5} = (6 - 9 + 7)\sqrt{5} = 4\sqrt{5}$

(b) $4\sqrt{7} + 6\sqrt{7} + 2\sqrt{3} + 7\sqrt{3} = (4 + 6)\sqrt{7} + (2 + 7)\sqrt{3}$
$= 10\sqrt{7} + 9\sqrt{3}$ ∎

Now suppose that we need to evaluate $5\sqrt{2} - \sqrt{2} + 4\sqrt{2} - 2\sqrt{2}$ to the nearest tenth. We can either evaluate the expression as it stands or first simplify it by combining radicals and then evaluate that result. Let's use the latter approach. (It would probably be a good idea for you to do it both ways for checking purposes.)

$5\sqrt{2} - \sqrt{2} + 4\sqrt{2} - 2\sqrt{2} = (5 - 1 + 4 - 2)\sqrt{2}$
$= 6\sqrt{2}$
$= 8.5$ to the nearest tenth

EXAMPLE 5 Find a rational approximation, to the nearest tenth, for

$7\sqrt{3} + 9\sqrt{5} + 2\sqrt{3} - 3\sqrt{5} + 13\sqrt{3}$

Solution

First, we simplify the given expression and then evaluate that result.

$7\sqrt{3} + 9\sqrt{5} + 2\sqrt{3} - 3\sqrt{5} + 13\sqrt{3} = (7 + 2 + 13)\sqrt{3} + (9 - 3)\sqrt{5}$
$= 22\sqrt{3} + 6\sqrt{5}$
$= 51.5$ to the nearest tenth ∎

5 Solve Applications Using Formulas with Radicals

Many real world applications involve radical expressions. For example, the *period* of a pendulum is the time it takes to swing from one side to the other and back (see Figure 9.1). A formula for the period is

$$T = 2\pi\sqrt{\frac{L}{32}}$$

where T is the period of the pendulum expressed in seconds, and L is the length of the pendulum in feet.

Figure 9.1

EXAMPLE 6 **Apply Your Skill**

Find the period, to the nearest tenth of a second, of a pendulum 2.5 feet long.

Solution

We use 3.14 as an approximation for π and substitute 2.5 for L in the formula.

$T = 2\pi\sqrt{\frac{L}{32}}$

$= 2(3.14)\sqrt{\frac{2.5}{32}}$

$= 1.8$ to the nearest tenth

The period is approximately 1.8 seconds. ∎

Police use the formula $S = \sqrt{30Df}$ to estimate a car's speed based on the length of skid marks (see Figure 9.2). In this formula, S represents the car's speed in miles per hour, D the length of skid marks measured in feet, and f a coefficient of friction. For a particular situation, the coefficient of friction is a constant that depends on the type and condition of the road surface.

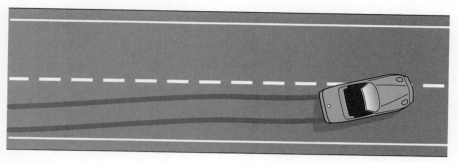

Figure 9.2

EXAMPLE 7   Apply Your Skill

Using 0.40 as a coefficient of friction, find how fast a car was moving if it skidded 225 feet. Express the answer to the nearest mile per hour.

Solution

We substitute 0.40 for f and 225 for D in the formula $S = \sqrt{30Df}$.

$$S = \sqrt{30(225)(0.40)} = 52 \quad \text{to the nearest whole number}$$

The car was traveling at approximately 52 miles per hour. ∎

CONCEPT QUIZ 9.1

For Problems 1–10, answer true or false.

1. Every positive real number has two square roots.
2. Negative real numbers have no real number square roots.
3. Numbers called "perfect squares" represent the square of some integer.
4. The square root of a real number is always a positive number.
5. The symbol $\sqrt{}$ is called a radical sign.
6. The $\sqrt{5}$ is a rational number.
7. The cube root of a number is one of its three equal factors.
8. Every negative real number has one negative cube root and two positive cube roots.
9. The symbol $\sqrt{}$ is used to designate all the square roots of a number.
10. The cube root of 0 is 0.

Section 9.1 Classroom Problem Set

Objective 1

1. Evaluate the following square roots.

   (a) $\sqrt{289}$ (b) $\sqrt{\dfrac{25}{81}}$ (c) $\sqrt{0.04}$

2. Evaluate the following square roots.

   (a) $\sqrt{2500}$ (b) $\sqrt{\dfrac{9}{16}}$ (c) $\sqrt{0.25}$

Objective 2

3. Evaluate the following cube roots.

   (a) $\sqrt[3]{64}$ (b) $\sqrt[3]{\dfrac{27}{1000}}$ (c) $\sqrt[3]{0.008}$

4. Evaluate the following cube roots.

   (a) $\sqrt[3]{-125}$ (b) $\sqrt[3]{-\dfrac{1}{64}}$ (c) $-\sqrt[3]{-1000}$

Objective 3

5. Use a calculator to find a rational approximation for each root. Express your answers to the nearest thousandth.

   (a) $\sqrt{425}$ (b) $-\sqrt{138}$
   (c) $\sqrt[3]{-450}$ (d) $\sqrt[3]{212}$

6. Use a calculator to find a rational approximation for each root. Express your answers to the nearest thousandth.

   (a) $\sqrt{250}$ (b) $-\sqrt{90}$
   (c) $-\sqrt[3]{100}$ (d) $-\sqrt[3]{-428}$

Objective 4

7. Simplify $6\sqrt{5} + 12\sqrt{3} - 2\sqrt{5} - 4\sqrt{3}$.
8. Simplify $8\sqrt{2} - 4\sqrt{3} - 9\sqrt{2} + 6\sqrt{3}$.
9. Find a rational solution, to the nearest tenth, for $6\sqrt{2} + 4\sqrt{7} + 5\sqrt{2} - 2\sqrt{7} - 3\sqrt{2}$.
10. Find a rational solution, to the nearest tenth, for $9\sqrt{6} - 3\sqrt{5} + 2\sqrt{6} - 7\sqrt{5} - \sqrt{6}$.

Objective 5

11. Find the period, to the nearest tenth, of a pendulum 3.2 feet long.
12. Find the period, to the nearest tenth, of a pendulum 4.0 feet long.
13. Using 0.35 as a coefficient of friction, find how fast a car was moving if it skidded 200 feet. Round the answer to the nearest mile per hour.
14. Using 0.35 as a coefficient of friction, find how fast a car was moving if it skidded 400 feet. Round the answer to the nearest mile per hour.

THOUGHTS INTO WORDS

1. Why is $\sqrt{-4}$ not a real number?
2. How could you find a whole number approximation for $\sqrt{1450}$ if you did not have a calculator available?

Answers to the Concept Quiz
1. True   2. True   3. True   4. False   5. True   6. False   7. True   8. False   9. False   10. True

9.2 Simplifying Radicals

OBJECTIVES

1. Change Radicals to Simplest Radical Form
2. Change Radicals That Contain Variables to Simplest Radical Form
3. Combine Radicals after Simplifying

1 Change Radicals to Simplest Radical Form

Note the following facts that pertain to square roots:

$$\sqrt{4 \cdot 9} = \sqrt{36} = 6$$
$$\sqrt{4}\sqrt{9} = 2 \cdot 3 = 6$$

Thus we observe that $\sqrt{4 \cdot 9} = \sqrt{4}\sqrt{9}$. This illustrates a general property.

9.2 Simplifying Radicals

Property 9.1

For any nonnegative real numbers a and b,
$$\sqrt{ab} = \sqrt{a}\sqrt{b}$$

In other words, we say that *the square root of a product is equal to the product of the square roots*.

Property 9.1 and the definition of square root provide the basis for expressing radical expressions in simplest radical form. For now, simplest radical form means that *the radicand contains no factors other than 1 that are perfect squares*. We present some examples to illustrate this meaning of simplest radical form.

EXAMPLE 1

Change each radical to simplest radical form.

(a) $\sqrt{8}$ (b) $\sqrt{45}$ (c) $\sqrt{48}$

Solution

(a) $\sqrt{8} = \sqrt{4 \cdot 2} = \sqrt{4}\sqrt{2} = 2\sqrt{2}$

  ↑ 4 is a perfect square   ↑ $\sqrt{4} = 2$

(b) $\sqrt{45} = \sqrt{9 \cdot 5} = \sqrt{9}\sqrt{5} = 3\sqrt{5}$

  ↑ 9 is a perfect square   ↑ $\sqrt{9} = 3$

(c) $\sqrt{48} = \sqrt{16 \cdot 3} = \sqrt{16}\sqrt{3} = 4\sqrt{3}$

  ↑ 16 is a perfect square   ↑ $\sqrt{16} = 4$

The first step in each example is to express the radicand of the given radical as the product of two factors, at least one of which is a perfect square other than 1. Observe the radicands of the final radicals. In each case, the radicand *cannot* be expressed as the product of two factors, at least one of which is a perfect square other than 1. We say that the final radicals, $2\sqrt{2}$, $3\sqrt{5}$, and $4\sqrt{3}$, are in simplest radical form.

You may vary the steps somewhat in changing to simplest radical form, but the final result should be the same. Consider another sequence of steps to change $\sqrt{48}$ to simplest form.

$$\sqrt{48} = \sqrt{4 \cdot 12} = \sqrt{4}\sqrt{12} = 2\sqrt{12} = 2\sqrt{4 \cdot 3} = 2\sqrt{4}\sqrt{3} = 2 \cdot 2\sqrt{3} = 4\sqrt{3}$$

↑ 4 is a perfect square   ↑ This is not in simplest form   ↑ 4 is a perfect square   ↑ Same result as in Example 3

Another variation of the technique for changing radicals to simplest form is to prime-factor the radicand and then to look for perfect squares in exponential form. We will redo the previous examples.

1. $\sqrt{8} = \sqrt{2 \cdot 2 \cdot 2} = \sqrt{2^2 \cdot 2} = \sqrt{2^2}\sqrt{2} = 2\sqrt{2}$

        ↑   ↑

    Prime 2^2 is a
    factors   perfect
    of 8   square

2. $\sqrt{45} = \sqrt{3 \cdot 3 \cdot 5} = \sqrt{3^2 \cdot 5} = \sqrt{3^2}\sqrt{5} = 3\sqrt{5}$

    Prime 3^2 is a
    factors   perfect
    of 45   square

3. $\sqrt{48} = \sqrt{2 \cdot 2 \cdot 2 \cdot 2 \cdot 3} = \sqrt{2^4 \cdot 3} = \sqrt{2^4}\sqrt{3} = 2^2\sqrt{3} = 4\sqrt{3}$

    Prime factors 2^4 is a $\sqrt{2^4} = 2^2$ because
    of 48   perfect square $2^2 \cdot 2^2 = 2^4$

The next example further illustrates the process of changing to simplest radical form. Only the major steps are shown, so be sure that you can fill in the details.

EXAMPLE 2   Change each radical to simplest radical form.

    (a) $\sqrt{56}$ **(b)** $\sqrt{75}$ **(c)** $\sqrt{108}$ **(d)** $5\sqrt{12}$

Solution

    (a) $\sqrt{56} = \sqrt{4}\sqrt{14} = 2\sqrt{14}$
    (b) $\sqrt{75} = \sqrt{25}\sqrt{3} = 5\sqrt{3}$
    (c) $\sqrt{108} = \sqrt{2 \cdot 2 \cdot 3 \cdot 3 \cdot 3} = \sqrt{2^2 \cdot 3^2}\sqrt{3} = 6\sqrt{3}$
    (d) $5\sqrt{12} = 5\sqrt{4}\sqrt{3} = 5 \cdot 2 \cdot \sqrt{3} = 10\sqrt{3}$ ∎

We can extend Property 9.1 to apply to cube roots.

> **Property 9.2**
>
> For any real numbers a and b,
> $$\sqrt[3]{ab} = \sqrt[3]{a}\sqrt[3]{b}$$

Now, using Property 9.2, we can simplify radicals involving cube roots. Here it is helpful to recognize the perfect cubes that we listed in the previous section.

EXAMPLE 3   Change each radical to simplest radical form.

    (a) $\sqrt[3]{24}$ **(b)** $\sqrt[3]{108}$ **(c)** $\sqrt[3]{375}$

Solution

    (a) $\sqrt[3]{24} = \sqrt[3]{8}\sqrt[3]{3} = 2\sqrt[3]{3}$

             ↑
       Perfect cube

(b) $\sqrt[3]{108} = \sqrt[3]{27}\sqrt[3]{4} = 3\sqrt[3]{4}$
↑
Perfect cube

(c) $\sqrt[3]{375} = \sqrt[3]{125}\sqrt[3]{3} = 5\sqrt[3]{3}$
↑
Perfect cube

2 Change Radicals That Contain Variables to Simplest Radical Form

Before we discuss the process of simplifying radicals that contain variables, there is one technicality that we should call to your attention. Let's look at some examples to clarify the point. Consider the radical $\sqrt{x^2}$.

Let $x = 3$; then $\sqrt{x^2} = \sqrt{3^2} = \sqrt{9} = 3$
Let $x = -3$; then $\sqrt{x^2} = \sqrt{(-3)^2} = \sqrt{9} = 3$

Thus if $x \geq 0$, then $\sqrt{x^2} = x$, *but* if $x < 0$, then $\sqrt{x^2} = -x$. Using the concept of absolute value, we can state that for all real numbers, $\sqrt{x^2} = |x|$.

Now consider the radical $\sqrt{x^3}$. Because x^3 is negative when x is negative, we need to restrict x to the nonnegative reals when working with $\sqrt{x^3}$. Therefore, we can write: If $x \geq 0$, then $\sqrt{x^3} = \sqrt{x^2}\sqrt{x} = x\sqrt{x}$, and no absolute value sign is necessary. Finally, let's consider the radical $\sqrt[3]{x^3}$.

Let $x = 2$; then $\sqrt[3]{x^3} = \sqrt[3]{2^3} = \sqrt[3]{8} = 2$
Let $x = -2$; then $\sqrt[3]{x^3} = \sqrt[3]{(-2)^3} = \sqrt[3]{-8} = -2$

Therefore, it is correct to write: $\sqrt[3]{x^3} = x$ for all real numbers, and again no absolute value sign is necessary.

The previous discussion indicates that technically every radical expression involving variables in the radicand needs to be analyzed individually as to the necessary restrictions imposed on the variables. However, to avoid considering such restrictions on a problem-to-problem basis, we shall merely assume that all variables represent positive real numbers.

EXAMPLE 4

Change each radical to simplest radical form.

(a) $\sqrt{x^2 y}$ **(b)** $\sqrt{4x^3}$ **(c)** $\sqrt{8xy^3}$ **(d)** $\sqrt{27x^5 y^3}$ **(e)** $\sqrt[3]{40x^4 y^5}$

Solution

(a) $\sqrt{x^2 y} = \sqrt{x^2}\sqrt{y} = x\sqrt{y}$

(b) $\sqrt{4x^3} = \sqrt{4x^2}\sqrt{x} = 2x\sqrt{x}$
↑
$4x^2$ is a perfect square because $(2x)(2x) = 4x^2$

(c) $\sqrt{8xy^3} = \sqrt{4y^2}\sqrt{2xy} = 2y\sqrt{2xy}$

(d) $\sqrt{27x^5y^3} = \sqrt{9x^4y^2}\sqrt{3xy} = 3x^2y\sqrt{3xy}$

$\sqrt{9x^4y^2} = 3x^2y$

(e) $\sqrt[3]{40x^4y^5} = \sqrt[3]{8x^3y^3}\sqrt[3]{5xy^2} = 2xy\sqrt[3]{5xy^2}$

Perfect cube

3 Combine Radicals after Simplifying

When simplifying expressions that contain radicals, we must often first change the radicals to simplest form, and then apply the distributive property.

EXAMPLE 5 Simplify $5\sqrt{8} + 3\sqrt{2}$.

Solution

$$5\sqrt{8} + 3\sqrt{2} = 5\sqrt{4}\sqrt{2} + 3\sqrt{2}$$
$$= 5 \cdot 2 \cdot \sqrt{2} + 3\sqrt{2}$$
$$= 10\sqrt{2} + 3\sqrt{2}$$
$$= (10 + 3)\sqrt{2}$$
$$= 13\sqrt{2}$$

EXAMPLE 6 Simplify $2\sqrt{27} - 5\sqrt{48} + 4\sqrt{3}$.

Solution

$$2\sqrt{27} - 5\sqrt{48} + 4\sqrt{3} = 2\sqrt{9}\sqrt{3} - 5\sqrt{16}\sqrt{3} + 4\sqrt{3}$$
$$= 2(3)\sqrt{3} - 5(4)\sqrt{3} + 4\sqrt{3}$$
$$= 6\sqrt{3} - 20\sqrt{3} + 4\sqrt{3}$$
$$= (6 - 20 + 4)\sqrt{3}$$
$$= -10\sqrt{3}$$

EXAMPLE 7 Simplify $\frac{1}{4}\sqrt{45} + \frac{1}{3}\sqrt{20}$.

Solution

$$\frac{1}{4}\sqrt{45} + \frac{1}{3}\sqrt{20} = \frac{1}{4}\sqrt{9}\sqrt{5} + \frac{1}{3}\sqrt{4}\sqrt{5}$$
$$= \frac{1}{4}(3)\sqrt{5} + \frac{1}{3}(2)\sqrt{5}$$
$$= \frac{3}{4}\sqrt{5} + \frac{2}{3}\sqrt{5}$$
$$= \left(\frac{3}{4} + \frac{2}{3}\right)\sqrt{5} = \left(\frac{9}{12} + \frac{8}{12}\right)\sqrt{5}$$
$$= \frac{17}{12}\sqrt{5}$$

EXAMPLE 8

Simplify $5\sqrt[3]{24} + 7\sqrt[3]{375}$.

Solution

$$5\sqrt[3]{24} + 7\sqrt[3]{375} = 5\sqrt[3]{8}\sqrt[3]{3} + 7\sqrt[3]{125}\sqrt[3]{3}$$
$$= 5(2)\sqrt[3]{3} + 7(5)\sqrt[3]{3}$$
$$= 10\sqrt[3]{3} + 35\sqrt[3]{3}$$
$$= 45\sqrt[3]{3}$$

CONCEPT QUIZ 9.2

For Problems 1–5, answer true or false.

1. The property $\sqrt{ab} = \sqrt{a}\sqrt{b}$ can be used to express the product of two radicals as one radical.
2. One condition for simplest radical form of a square root is that the radicand contains no factors other than 1 that are perfect squares.
3. For all real numbers, $\sqrt{x^2} = x$.
4. For all real numbers, $\sqrt[3]{x^3} = x$.
5. $4\sqrt{2} + 6\sqrt{3} = 10\sqrt{5}$

Section 9.2 Classroom Problem Set

Objective 1

1. Change each radical to simplest radical form.
   (a) $\sqrt{24}$ (b) $\sqrt{63}$ (c) $\sqrt{32}$

2. Change each radical to simplest radical form.
   (a) $\sqrt{54}$ (b) $\sqrt{80}$ (c) $\sqrt{44}$

3. Change each radical to simplest radical form.
   (a) $\sqrt{20}$ (b) $\sqrt{72}$ (c) $6\sqrt{50}$

4. Change each radical to simplest radical form.
   (a) $\sqrt{117}$ (b) $-6\sqrt{45}$ (c) $\frac{3}{2}\sqrt{8}$

5. Change each radical to simplest radical form.
   (a) $\sqrt[3]{16}$ (b) $\sqrt[3]{40}$ (c) $\sqrt[3]{81}$

6. Change each radical to simplest radical form.
   (a) $3\sqrt[3]{40}$ (b) $\sqrt[3]{375}$ (c) $\frac{2}{3}\sqrt[3]{81}$

Objective 2

7. Change each radical to simplest radical form.
   (a) $\sqrt{12x^2y}$ (b) $\sqrt{75a^3b^2}$ (c) $\sqrt[3]{16x^4y^3}$

8. Change each radical to simplest radical form.
   (a) $\sqrt{28x^3y}$ (b) $\sqrt{45a^2b^4}$ (c) $\sqrt[3]{56x^3y^5}$

Objective 3

9. Simplify $2\sqrt{45} + 4\sqrt{5}$.
10. Simplify $6\sqrt{48} + 5\sqrt{3}$.
11. Simplify $\sqrt{18} + 3\sqrt{50} - \sqrt{72}$.
12. Simplify $5\sqrt{12} + 3\sqrt{27} - 2\sqrt{75}$.
13. Simplify $\frac{1}{5}\sqrt{27} + \frac{3}{4}\sqrt{3}$.
14. Simplify $\frac{1}{3}\sqrt{12} - \frac{3}{2}\sqrt{48} + \frac{3}{4}\sqrt{108}$.
15. Simplify $7\sqrt[3]{24} + 4\sqrt[3]{81}$.
16. Simplify $9\sqrt[3]{16} - 2\sqrt[3]{54}$.

THOUGHTS INTO WORDS

1. Explain how you would help someone express $5\sqrt{72}$ in simplest radical form.

2. Explain your thought process when expressing $\sqrt{153}$ in simplest radical form.

Answers to the Concept Quiz
1. True   2. True   3. False   4. True   5. False

9.3 Simplifying Radicals of Quotients

OBJECTIVES

1. Simplify Roots of Quotients
2. Rationalize Denominators of Square Roots
3. Combine Radicals after Rationalizing the Denominators

1 Simplify Roots of Quotients

We now continue our work of simplifying radicals. We will now look at some examples that simplify the root of a quotient.

$$\sqrt{\frac{36}{9}} = \sqrt{4} = 2 \quad \text{and} \quad \frac{\sqrt{36}}{\sqrt{9}} = \frac{6}{3} = 2$$

Thus we see that $\sqrt{\frac{36}{9}} = \frac{\sqrt{36}}{\sqrt{9}}$.

$$\sqrt[3]{\frac{64}{8}} = \sqrt[3]{8} = 2 \quad \text{and} \quad \frac{\sqrt[3]{64}}{\sqrt[3]{8}} = \frac{4}{2} = 2$$

Thus we see that $\sqrt[3]{\frac{64}{8}} = \frac{\sqrt[3]{64}}{\sqrt[3]{8}}$.

We can state the following general property.

Property 9.3

For any nonnegative real numbers a and b, $b \neq 0$, $\sqrt{\frac{a}{b}} = \frac{\sqrt{a}}{\sqrt{b}}$.

For any real numbers a and b, $b \neq 0$, $\sqrt[3]{\frac{a}{b}} = \frac{\sqrt[3]{a}}{\sqrt[3]{b}}$.

To evaluate a radical such as $\sqrt{\frac{25}{4}}$, in which the numerator and denominator of the fractional radicand are perfect squares, you may use Property 9.3 or merely rely on the definition of square root.

$$\sqrt{\frac{25}{4}} = \frac{\sqrt{25}}{\sqrt{4}} = \frac{5}{2}$$

or

$$\sqrt{\frac{25}{4}} = \frac{5}{2} \quad \text{because} \quad \frac{5}{2} \cdot \frac{5}{2} = \frac{25}{4}$$

9.3 Simplifying Radicals of Quotients

Sometimes it is easier to do the indicated division first and then find the square root, as you can see in the next example.

$$\sqrt{\frac{324}{9}} = \sqrt{36} = 6$$

Now we can extend our concept of simplest radical form. An algebraic expression that contains a radical is said to be in **simplest radical form** if the following conditions are satisfied:

1. No fraction appears within a radical sign. $\left(\sqrt{\frac{2}{3}} \text{ violates this condition.}\right)$

2. No radical appears in the denominator. $\left(\frac{5}{\sqrt{8}} \text{ violates this condition.}\right)$

3. No radicand when expressed in prime factored form contains a factor raised to a power greater than or equal to the index. ($\sqrt{8} = \sqrt{2^3}$ violates this condition.)

The next examples show how to simplify expressions that do not meet these three stated conditions.

EXAMPLE 1

Simplify $\sqrt{\frac{13}{4}}$.

Solution

Apply Property 9.3 to write the expression as the quotient of two radicals.

$$\sqrt{\frac{13}{4}} = \frac{\sqrt{13}}{\sqrt{4}} = \frac{\sqrt{13}}{2}$$ ∎

EXAMPLE 2

Simplify $\sqrt[3]{\frac{16}{27}}$.

Solution

Apply Property 9.3 to write the expression as the quotient of two cube roots.

$$\sqrt[3]{\frac{16}{27}} = \frac{\sqrt[3]{16}}{\sqrt[3]{27}} = \frac{\sqrt[3]{16}}{3} = \frac{\sqrt[3]{8}\sqrt[3]{2}}{3} = \frac{2\sqrt[3]{2}}{3}$$

↑ Don't stop here. The radical in the numerator can be simplified. ∎

For Examples 3, 4, and 5, we show two methods of simplifying for each example. Both methods are correct but you may prefer one approach over the other.

EXAMPLE 3

Simplify $\frac{\sqrt{12}}{\sqrt{16}}$.

Solution A

$$\frac{\sqrt{12}}{\sqrt{16}} = \frac{\sqrt{12}}{4} = \frac{\sqrt{4}\sqrt{3}}{4} = \frac{2\sqrt{3}}{4} = \frac{\sqrt{3}}{2}$$

Solution B

$$\frac{\sqrt{12}}{\sqrt{16}} = \sqrt{\frac{12}{16}} = \sqrt{\frac{3}{4}} = \frac{\sqrt{3}}{\sqrt{4}} = \frac{\sqrt{3}}{2}$$

Reduce the fraction

The two approaches in this example show how important it is to think about the approach to a problem before you push the pencil.

Now let's consider an example where neither the numerator nor the denominator of the radicand is a perfect square. Keep in mind that an expression is not simplified if there is a radical in the denominator.

EXAMPLE 4   Simplify $\sqrt{\frac{2}{3}}$.

Solution A — Form of 1

$$\sqrt{\frac{2}{3}} = \frac{\sqrt{2}}{\sqrt{3}} = \frac{\sqrt{2}}{\sqrt{3}} \cdot \frac{\sqrt{3}}{\sqrt{3}} = \frac{\sqrt{6}}{3}$$

Solution B

$$\sqrt{\frac{2}{3}} = \sqrt{\frac{2}{3} \cdot \frac{3}{3}} = \sqrt{\frac{6}{9}} = \frac{\sqrt{6}}{\sqrt{9}} = \frac{\sqrt{6}}{3}$$

2 Rationalize Denominators of Square Roots

We refer to the process we used to simplify the radical in Example 4 as **rationalizing the denominator**. Notice that the denominator becomes a rational number. There is more than one way to rationalize the denominator, as the next example shows.

EXAMPLE 5   Simplify $\frac{\sqrt{5}}{\sqrt{8}}$.

Solution A

$$\frac{\sqrt{5}}{\sqrt{8}} = \frac{\sqrt{5}}{\sqrt{8}} \cdot \frac{\sqrt{8}}{\sqrt{8}} = \frac{\sqrt{40}}{8} = \frac{\sqrt{4}\sqrt{10}}{8} = \frac{2\sqrt{10}}{8} = \frac{\sqrt{10}}{4}$$

Solution B

$$\frac{\sqrt{5}}{\sqrt{8}} = \frac{\sqrt{5}}{\sqrt{8}} \cdot \frac{\sqrt{2}}{\sqrt{2}} = \frac{\sqrt{10}}{\sqrt{16}} = \frac{\sqrt{10}}{4}$$

Solution C

$$\frac{\sqrt{5}}{\sqrt{8}} = \frac{\sqrt{5}}{\sqrt{4}\sqrt{2}} = \frac{\sqrt{5}}{2\sqrt{2}} = \frac{\sqrt{5}}{2\sqrt{2}} \cdot \frac{\sqrt{2}}{\sqrt{2}} = \frac{\sqrt{10}}{4}$$

Study the following examples, and check that the answers are in simplest radical form according to the three conditions we listed on page 339.

EXAMPLE 6

Simplify each of these expressions.

(a) $\dfrac{3}{\sqrt{x}}$ (b) $\sqrt{\dfrac{2x}{3y}}$ (c) $\dfrac{3\sqrt{5}}{\sqrt{6}}$ (d) $\sqrt{\dfrac{4x^2}{9y}}$

Solution

(a) $\dfrac{3}{\sqrt{x}} = \dfrac{3}{\sqrt{x}} \cdot \dfrac{\sqrt{x}}{\sqrt{x}} = \dfrac{3\sqrt{x}}{x}$

(b) $\sqrt{\dfrac{2x}{3y}} = \dfrac{\sqrt{2x}}{\sqrt{3y}} = \dfrac{\sqrt{2x}}{\sqrt{3y}} \cdot \dfrac{\sqrt{3y}}{\sqrt{3y}} = \dfrac{\sqrt{6xy}}{3y}$

(c) $\dfrac{3\sqrt{5}}{\sqrt{6}} = \dfrac{3\sqrt{5}}{\sqrt{6}} \cdot \dfrac{\sqrt{6}}{\sqrt{6}} = \dfrac{3\sqrt{30}}{6} = \dfrac{\sqrt{30}}{2}$

(d) $\sqrt{\dfrac{4x^2}{9y}} = \dfrac{\sqrt{4x^2}}{\sqrt{9y}} = \dfrac{2x}{\sqrt{9}\sqrt{y}} = \dfrac{2x}{3\sqrt{y}} = \dfrac{2x}{3\sqrt{y}} \cdot \dfrac{\sqrt{y}}{\sqrt{y}} = \dfrac{2x\sqrt{y}}{3y}$ ∎

3 Combine Radicals after Rationalizing the Denominators

Let's return again to the idea of simplifying expressions that contain radicals. Sometimes it may appear as if no simplifying can be done; however, after the individual radicals have been changed to simplest form, the distributive property may apply.

EXAMPLE 7

Simplify $5\sqrt{2} + \dfrac{3}{\sqrt{2}}$.

Solution

$5\sqrt{2} + \dfrac{3}{\sqrt{2}} = 5\sqrt{2} + \dfrac{3}{\sqrt{2}} \cdot \dfrac{\sqrt{2}}{\sqrt{2}} = 5\sqrt{2} + \dfrac{3\sqrt{2}}{2}$

$= \left(5 + \dfrac{3}{2}\right)\sqrt{2} = \left(\dfrac{10}{2} + \dfrac{3}{2}\right)\sqrt{2}$

$= \dfrac{13}{2}\sqrt{2}$ or $\dfrac{13\sqrt{2}}{2}$ ∎

EXAMPLE 8

Simplify $\sqrt{\dfrac{3}{2}} + \sqrt{24}$.

Solution

$\sqrt{\dfrac{3}{2}} + \sqrt{24} = \dfrac{\sqrt{3}}{\sqrt{2}} + \sqrt{24} = \dfrac{\sqrt{3}}{\sqrt{2}} \cdot \dfrac{\sqrt{2}}{\sqrt{2}} + \sqrt{4}\sqrt{6}$

$= \dfrac{\sqrt{6}}{2} + 2\sqrt{6}$

$= \left(\dfrac{1}{2} + 2\right)\sqrt{6}$

$= \left(\dfrac{1}{2} + \dfrac{4}{2}\right)\sqrt{6}$

$= \dfrac{5}{2}\sqrt{6}$ ∎

CONCEPT QUIZ 9.3

For Problems 1–5, answer true or false.

1. A radical is not in simplest radical form if it has a fraction within the radical sign.
2. The radical $\dfrac{1}{\sqrt{5}}$ is in simplest radical form.
3. The process of *rationalizing the denominator* means the denominator has been changed from an irrational number to a rational number.
4. The expression $\dfrac{6}{\sqrt{12}}$ could have the denominator rationalized by multiplying the expression by $\dfrac{\sqrt{3}}{\sqrt{3}}$ or $\dfrac{\sqrt{12}}{\sqrt{12}}$.
5. The radical $\dfrac{\sqrt{8}}{5}$ is in simplest form because it does not have a radical in the denominator.

Section 9.3 Classroom Problem Set

Objective 1

1. Simplify $\sqrt{\dfrac{7}{9}}$.
2. Simplify $\sqrt{\dfrac{17}{4}}$.
3. Simplify $\sqrt[3]{\dfrac{81}{8}}$.
4. Simplify $\sqrt[3]{\dfrac{128}{27}}$.
5. Simplify $\sqrt{\dfrac{18}{81}}$.
6. Simplify $\sqrt{\dfrac{24}{25}}$.
7. Simplify $\sqrt{\dfrac{3}{5}}$.
8. Simplify $\sqrt{\dfrac{9}{48}}$.

Objective 2

9. Simplify $\sqrt{\dfrac{5}{12}}$.
10. Simplify $\dfrac{2\sqrt{5}}{7\sqrt{8}}$.

11. Simplify each of the expressions.
    (a) $\dfrac{4}{\sqrt{a}}$ (b) $\sqrt{\dfrac{5m}{2n}}$ (c) $\dfrac{5\sqrt{3}}{\sqrt{10}}$ (d) $\sqrt{\dfrac{9a^2}{16b}}$

12. Simplify each of the expressions.
    (a) $\sqrt{\dfrac{3}{x}}$ (b) $\sqrt{\dfrac{3y}{32x}}$ (c) $\dfrac{3\sqrt{7}}{4\sqrt{12}}$ (d) $\sqrt{\dfrac{25}{y^5}}$

Objective 3

13. Simplify $7\sqrt{3} + \dfrac{2}{\sqrt{3}}$.
14. Simplify $4\sqrt{\dfrac{1}{7}} + 6\sqrt{7}$.
15. Simplify $\sqrt{\dfrac{2}{5}} + \sqrt{90}$.
16. Simplify $-3\sqrt{2} + \sqrt{\dfrac{1}{2}}$.

THOUGHTS INTO WORDS

1. Your friend simplifies $\sqrt{\dfrac{6}{8}}$ as follows:

$$\sqrt{\dfrac{6}{8}} = \dfrac{\sqrt{6}}{\sqrt{8}} = \dfrac{\sqrt{6}}{\sqrt{8}} \cdot \dfrac{\sqrt{8}}{\sqrt{8}} = \dfrac{\sqrt{48}}{8} = \dfrac{\sqrt{16}\sqrt{3}}{8}$$

$$= \dfrac{4\sqrt{3}}{8} = \dfrac{\sqrt{3}}{2}$$

Could you show him a much shorter way to simplify this expression?

2. Is the expression $3\sqrt{2} + \sqrt{50}$ in simplest radical form? Why or why not?

Answers to the Concept Quiz

1. True   2. False   3. True   4. True   5. False

9.4 Products and Quotients Involving Radicals

OBJECTIVES

1. Multiply Radicals
2. Multiply Radicals Using the Distributive Property
3. Use Conjugates to Rationalize Denominators

1 Multiply Radicals

We use Property 9.1 ($\sqrt{ab} = \sqrt{a}\sqrt{b}$) and Property 9.2 ($\sqrt[3]{ab} = \sqrt[3]{a}\sqrt[3]{b}$) to multiply radical expressions, and in some cases, to simplify the resulting radical. The following examples illustrate several types of multiplication problems that involve radicals.

EXAMPLE 1

Multiply and simplify where possible.

(a) $\sqrt{3}\sqrt{12}$ (b) $\sqrt{3}\sqrt{15}$ (c) $\sqrt{7}\sqrt{8}$
(d) $\sqrt[3]{4}\sqrt[3]{6}$ (e) $(3\sqrt{2})(4\sqrt{3})$ (f) $(2\sqrt[3]{9})(4\sqrt[3]{6})$

Solution

(a) $\sqrt{3}\sqrt{12} = \sqrt{36} = 6$
(b) $\sqrt{3}\sqrt{15} = \sqrt{45} = \sqrt{9}\sqrt{5} = 3\sqrt{5}$
(c) $\sqrt{7}\sqrt{8} = \sqrt{56} = \sqrt{4}\sqrt{14} = 2\sqrt{14}$
(d) $\sqrt[3]{4}\sqrt[3]{6} = \sqrt[3]{24} = \sqrt[3]{8}\sqrt[3]{3} = 2\sqrt[3]{3}$
(e) $(3\sqrt{2})(4\sqrt{3}) = 3 \cdot 4 \cdot \sqrt{2} \cdot \sqrt{3} = 12\sqrt{6}$
(f) $(2\sqrt[3]{9})(4\sqrt[3]{6}) = 2 \cdot 4 \cdot \sqrt[3]{9} \cdot \sqrt[3]{6}$
$= 8\sqrt[3]{54}$
$= 8\sqrt[3]{27}\sqrt[3]{2}$
$= 8 \cdot 3 \cdot \sqrt[3]{2}$
$= 24\sqrt[3]{2}$

2 Multiply Radicals Using the Distributive Property

Recall how we use the distributive property when we find the product of a monomial and a polynomial. For example, $2x(3x + 4) = 2x(3x) + 2x(4) = 6x^2 + 8x$. Likewise, the distributive property and Properties 9.1 and 9.2 provide the basis for finding certain special products involving radicals. The next examples demonstrate this idea.

EXAMPLE 2

Multiply and simplify where possible.

(a) $\sqrt{2}(\sqrt{3} + \sqrt{5})$ (b) $\sqrt{3}(\sqrt{12} - \sqrt{6})$
(c) $\sqrt{8}(\sqrt{2} - 3)$ (d) $\sqrt{x}(\sqrt{x} + \sqrt{y})$
(e) $\sqrt[3]{2}(\sqrt[3]{4} + \sqrt[3]{10})$

Solution

(a) $\sqrt{2}(\sqrt{3} + \sqrt{5}) = \sqrt{2}\sqrt{3} + \sqrt{2}\sqrt{5} = \sqrt{6} + \sqrt{10}$

(b) $\sqrt{3}(\sqrt{12} - \sqrt{6}) = \sqrt{3}\sqrt{12} - \sqrt{3}\sqrt{6}$
$= \sqrt{36} - \sqrt{18}$
$= 6 - \sqrt{9}\sqrt{2}$
$= 6 - 3\sqrt{2}$

(c) $\sqrt{8}(\sqrt{2} - 3) = \sqrt{8}\sqrt{2} - (\sqrt{8})(3)$
$= \sqrt{16} - 3\sqrt{8}$
$= 4 - 3\sqrt{4}\sqrt{2}$
$= 4 - 6\sqrt{2}$

(d) $\sqrt{x}(\sqrt{x} + \sqrt{y}) = \sqrt{x}\sqrt{x} + \sqrt{x}\sqrt{y}$
$= \sqrt{x^2} + \sqrt{xy}$
$= x + \sqrt{xy}$

(e) $\sqrt[3]{2}(\sqrt[3]{4} + \sqrt[3]{10}) = \sqrt[3]{8} + \sqrt[3]{20} = 2 + \sqrt[3]{20}$ ∎

The distributive property plays a central role when we find the product of two binomials. For example, $(x + 2)(x + 3) = x(x + 3) + 2(x + 3) = x^2 + 3x + 2x + 6 = x^2 + 5x + 6$. We can find the product of two binomial expressions involving radicals in a similar fashion.

EXAMPLE 3

Multiply and simplify.

(a) $(\sqrt{3} + \sqrt{5})(\sqrt{2} + \sqrt{6})$ (b) $(\sqrt{7} - 3)(\sqrt{7} + 6)$

Solution

(a) $(\sqrt{3} + \sqrt{5})(\sqrt{2} + \sqrt{6}) = \sqrt{3}(\sqrt{2} + \sqrt{6}) + \sqrt{5}(\sqrt{2} + \sqrt{6})$
$= \sqrt{3}\sqrt{2} + \sqrt{3}\sqrt{6} + \sqrt{5}\sqrt{2} + \sqrt{5}\sqrt{6}$
$= \sqrt{6} + \sqrt{18} + \sqrt{10} + \sqrt{30}$
$= \sqrt{6} + 3\sqrt{2} + \sqrt{10} + \sqrt{30}$

(b) $(\sqrt{7} - 3)(\sqrt{7} + 6) = \sqrt{7}(\sqrt{7} + 6) - 3(\sqrt{7} + 6)$
$= \sqrt{7}\sqrt{7} + 6\sqrt{7} - 3\sqrt{7} - 18$
$= 7 + 6\sqrt{7} - 3\sqrt{7} - 18$
$= -11 + 3\sqrt{7}$ ∎

If the binomials are of the form $(a + b)(a - b)$, then we can use the multiplication pattern $(a + b)(a - b) = a^2 - b^2$.

EXAMPLE 4

Multiply and simplify.

(a) $(\sqrt{6} + 2)(\sqrt{6} - 2)$ (b) $(3 - \sqrt{5})(3 + \sqrt{5})$
(c) $(\sqrt{8} + \sqrt{5})(\sqrt{8} - \sqrt{5})$

Solution

(a) $(\sqrt{6} + 2)(\sqrt{6} - 2) = (\sqrt{6})^2 - 2^2 = 6 - 4 = 2$
(b) $(3 - \sqrt{5})(3 + \sqrt{5}) = 3^2 - (\sqrt{5})^2 = 9 - 5 = 4$
(c) $(\sqrt{8} + \sqrt{5})(\sqrt{8} - \sqrt{5}) = (\sqrt{8})^2 - (\sqrt{5})^2 = 8 - 5 = 3$ ∎

Note that in each part of Example 4, the final product contains no radicals. This happens whenever we multiply expressions such as $\sqrt{a} + \sqrt{b}$ and $\sqrt{a} - \sqrt{b}$, where a and b are rational numbers.

$$(\sqrt{a} + \sqrt{b})(\sqrt{a} - \sqrt{b}) = (\sqrt{a})^2 - (\sqrt{b})^2 = a - b$$

Expressions such as $\sqrt{8} + \sqrt{5}$ and $\sqrt{8} - \sqrt{5}$ are called **conjugates** of each other. Likewise, $\sqrt{6} + 2$ and $\sqrt{6} - 2$ are conjugates, as are $3 - \sqrt{5}$ and $3 + \sqrt{5}$. Now let's see how we can use conjugates to rationalize denominators.

3 Use Conjugates to Rationalize Denominators

EXAMPLE 5   Simplify $\dfrac{4}{\sqrt{5} + \sqrt{2}}$.

Solution

To simplify the expression, the denominator needs to be a rational number. Let's multiply the numerator and denominator by $\sqrt{5} - \sqrt{2}$, which is the conjugate of the denominator.

$$\frac{4}{\sqrt{5} + \sqrt{2}} = \frac{4}{\sqrt{5} + \sqrt{2}} \cdot \frac{\sqrt{5} - \sqrt{2}}{\sqrt{5} - \sqrt{2}} \qquad \frac{\sqrt{5} - \sqrt{2}}{\sqrt{5} - \sqrt{2}} \text{ is merely a form of 1}$$

$$= \frac{4(\sqrt{5} - \sqrt{2})}{(\sqrt{5} + \sqrt{2})(\sqrt{5} - \sqrt{2})}$$

$$= \frac{4(\sqrt{5} - \sqrt{2})}{5 - 2}$$

$$= \frac{4(\sqrt{5} - \sqrt{2})}{3} \quad \text{or} \quad \frac{4\sqrt{5} - 4\sqrt{2}}{3}$$

Either answer is acceptable

The next four examples illustrate further the process of rationalizing and simplifying expressions that contain binomial denominators.

EXAMPLE 6   Rationalize the denominator and simplify $\dfrac{\sqrt{3}}{\sqrt{6} - 2}$.

Solution

To simplify, we want to multiply the numerator and denominator by the conjugate of the denominator, which is $\sqrt{6} + 2$.

$$\frac{\sqrt{3}}{\sqrt{6} - 2} = \frac{\sqrt{3}}{\sqrt{6} - 2} \cdot \frac{\sqrt{6} + 2}{\sqrt{6} + 2}$$

$$= \frac{\sqrt{3}(\sqrt{6} + 2)}{(\sqrt{6} - 2)(\sqrt{6} + 2)}$$

$$= \frac{\sqrt{18} + 2\sqrt{3}}{6 - 4}$$

$$= \frac{3\sqrt{2} + 2\sqrt{3}}{2} \qquad \sqrt{18} = \sqrt{9}\sqrt{2} = 3\sqrt{2}$$

EXAMPLE 7

Rationalize the denominator and simplify $\dfrac{\sqrt{x} + 2}{\sqrt{x} - 3}$.

Solution

To rationalize the denominator, we need to multiply the numerator and denominator by $\sqrt{x} + 3$, which is the conjugate of the denominator.

$$\dfrac{\sqrt{x} + 2}{\sqrt{x} - 3} = \dfrac{\sqrt{x} + 2}{\sqrt{x} - 3} \cdot \dfrac{\sqrt{x} + 3}{\sqrt{x} + 3}$$

$$= \dfrac{(\sqrt{x} + 2)(\sqrt{x} + 3)}{(\sqrt{x} - 3)(\sqrt{x} + 3)}$$

$$= \dfrac{x + 2\sqrt{x} + 3\sqrt{x} + 6}{x - 9}$$

$$= \dfrac{x + 5\sqrt{x} + 6}{x - 9}$$ ∎

EXAMPLE 8

Rationalize the denominator and simplify $\dfrac{3 + \sqrt{2}}{\sqrt{2} - 6}$.

Solution

To change the denominator to a rational number, we can multiply the numerator and denominator by $\sqrt{2} + 6$, which is the conjugate of the denominator.

$$\dfrac{3 + \sqrt{2}}{\sqrt{2} - 6} = \dfrac{3 + \sqrt{2}}{\sqrt{2} - 6} \cdot \dfrac{\sqrt{2} + 6}{\sqrt{2} + 6}$$

$$= \dfrac{(3 + \sqrt{2})(\sqrt{2} + 6)}{(\sqrt{2} - 6)(\sqrt{2} + 6)}$$

$$= \dfrac{3\sqrt{2} + 18 + 2 + 6\sqrt{2}}{2 - 36}$$

$$= \dfrac{9\sqrt{2} + 20}{-34}$$

$$= -\dfrac{9\sqrt{2} + 20}{34} \qquad \dfrac{a}{-b} = -\dfrac{a}{b}$$ ∎

CONCEPT QUIZ 9.4

For Problems 1–5, answer true or false.

1. The product of two radicals always results in an expression that has a radical, even after simplifying.
2. Just like multiplying binomials, the product of $(\sqrt{3} + 4)(\sqrt{5} - 6)$ can be found by applying the distributive property.
3. The conjugate of $(4 + \sqrt{3})$ is $(-4 - \sqrt{3})$.
4. The product of a binomial radical expression and its conjugate is a rational number.
5. To rationalize the denominator for the expression $\dfrac{5\sqrt{3}}{6 - \sqrt{3}}$, we would multiply the expression by $\dfrac{\sqrt{3}}{\sqrt{3}}$.

Section 9.4   Classroom Problem Set

Objective 1

1. For each of the following, multiply and simplify where possible.
   (a) $\sqrt{2}\sqrt{14}$ (b) $(5\sqrt{3})(2\sqrt{7})$
   (c) $\sqrt[3]{10}\sqrt[3]{4}$ (d) $(2\sqrt[3]{5})(3\sqrt[3]{50})$

2. For each of the following, multiply and simplify where possible.
   (a) $\sqrt{3}\sqrt{10}$ (b) $(5\sqrt{2})(4\sqrt{12})$
   (c) $\sqrt[3]{12}\sqrt[3]{2}$ (d) $(5\sqrt[3]{10})(7\sqrt[3]{3})$

Objective 2

3. For each of the following, multiply and simplify where possible.
   (a) $\sqrt{3}(\sqrt{7} - \sqrt{5})$ (b) $\sqrt{2}(\sqrt{18} + 5)$
   (c) $\sqrt{a}(\sqrt{a} - \sqrt{b})$ (d) $\sqrt[3]{3}(\sqrt[3]{3} + \sqrt[3]{18})$

4. For each of the following, multiply and simplify where possible.
   (a) $\sqrt{8}(\sqrt{3} - \sqrt{2})$ (b) $\sqrt{12}(\sqrt{6} - \sqrt{8})$
   (c) $4\sqrt{a}(\sqrt{a^3} - \sqrt{b})$ (d) $\sqrt[3]{2}(\sqrt[3]{27} - \sqrt[3]{24})$

5. For each of the following, multiply and simplify.
   (a) $(\sqrt{6} - \sqrt{2})(\sqrt{3} + \sqrt{7})$
   (b) $(\sqrt{5} + 4)(\sqrt{5} + 2)$

6. For each of the following, multiply and simplify.
   (a) $(\sqrt{5} + \sqrt{6})(\sqrt{8} - \sqrt{3})$
   (b) $(\sqrt{7} - 6)(\sqrt{7} + 5)$

7. For each of the following, multiply and simplify.
   (a) $(\sqrt{5} - 6)(\sqrt{5} + 6)$ (b) $(4 - \sqrt{3})(4 + \sqrt{3})$
   (c) $(\sqrt{6} + \sqrt{2})(\sqrt{6} - \sqrt{2})$

8. For each of the following, multiply and simplify.
   (a) $(\sqrt{3} - 6)(\sqrt{3} + 1)$
   (b) $(5 + \sqrt{10})(5 - \sqrt{10})$
   (c) $(5\sqrt{6} - \sqrt{2})(5\sqrt{6} + \sqrt{2})$

Objective 3

9. Simplify $\dfrac{3}{\sqrt{7} + \sqrt{2}}$.

10. Simplify $\dfrac{2}{\sqrt{6} + \sqrt{5}}$.

11. Simplify $\dfrac{\sqrt{2}}{\sqrt{10} + 3}$.

12. Simplify $\dfrac{\sqrt{6}}{\sqrt{3} - 1}$.

13. Simplify $\dfrac{\sqrt{a} + 3}{\sqrt{a} - 4}$.

14. Simplify $\dfrac{\sqrt{x} - 5}{\sqrt{x} + 2}$.

15. Simplify $\dfrac{5 - \sqrt{6}}{\sqrt{6} - 2}$.

16. Simplify $\dfrac{1 - \sqrt{2}}{\sqrt{2} - 8}$.

THOUGHTS INTO WORDS

1. Explain how the distributive property has been used in this chapter.

2. How would you help someone rationalize the denominator and simplify the expression $\dfrac{\sqrt{4}}{\sqrt{12} + \sqrt{8}}$?

Answers to the Concept Quiz

1. False   2. True   3. False   4. True   5. False

9.5 Solving Radical Equations

OBJECTIVES

1. Solve Radical Equations of the Form $\sqrt{ax + b} = c$
2. Solve Radical Equations of the Form $\sqrt{ax + b} = \sqrt{cx + d}$
3. Solve Radical Equations of the Form $\sqrt{ax + b} = cx + d$
4. Solve and Apply Radical Formulas

1 Solve Radical Equations of the Form $\sqrt{ax + b} = c$

Equations that contain radicals with variables in the radicand are called **radical equations**. Here are some examples of radical equations that involve one variable.

$$\sqrt{x} = 3 \qquad \sqrt{2x + 1} = 5 \qquad \sqrt{3x + 4} = -4$$

$$\sqrt{5s - 2} = \sqrt{2s + 19} \qquad \sqrt{2y - 4} = y - 2 \qquad \sqrt{x + 6} = x$$

In order to solve such equations we need the following property of equality.

Property 9.4

For real numbers a and b, if $a = b$, then

$$a^2 = b^2$$

Property 9.4 states that we can **square both sides of an equation**. However, squaring both sides of an equation sometimes produces results that do not satisfy the original equation. Let's consider two examples to illustrate the point.

EXAMPLE 1 Solve $\sqrt{x} = 3$.

Solution

$$\sqrt{x} = 3$$
$$(\sqrt{x})^2 = 3^2 \quad \text{Square both sides}$$
$$x = 9$$

Since $\sqrt{9} = 3$, the solution set is $\{9\}$.   ∎

EXAMPLE 2 Solve $\sqrt{x} = -3$.

Solution

$$\sqrt{x} = -3$$
$$(\sqrt{x})^2 = (-3)^2 \quad \text{Square both sides}$$
$$x = 9$$

Because $\sqrt{9} \neq -3$, 9 is not a solution and the solution set is $\emptyset$.   ∎

In general, squaring both sides of an equation produces an equation that has all of the solutions of the original equation, but it may also have some extra solutions that do not satisfy the original equation. (Such extra solutions are called **extraneous solutions** or **roots**.) Therefore, when using the "squaring" property (Property 9.4), you *must* check each potential solution in the original equation.

We now consider some examples to demonstrate different situations that arise when solving radical equations.

EXAMPLE 3

Solve $\sqrt{2x + 1} = 5$.

Solution

$$\sqrt{2x + 1} = 5$$
$$(\sqrt{2x + 1})^2 = 5^2 \quad \text{Square both sides}$$
$$2x + 1 = 25$$
$$2x = 24$$
$$x = 12$$

✔ Check

$$\sqrt{2x + 1} = 5$$
$$\sqrt{2(12) + 1} \stackrel{?}{=} 5$$
$$\sqrt{24 + 1} \stackrel{?}{=} 5$$
$$\sqrt{25} \stackrel{?}{=} 5$$
$$5 = 5$$

The solution set is $\{12\}$. ∎

EXAMPLE 4

Solve $\sqrt{3x + 4} = -4$.

Solution

$$\sqrt{3x + 4} = -4$$
$$(\sqrt{3x + 4})^2 = (-4)^2 \quad \text{Square both sides}$$
$$3x + 4 = 16$$
$$3x = 12$$
$$x = 4$$

✔ Check

$$\sqrt{3x + 4} = -4$$
$$\sqrt{3(4) + 4} \stackrel{?}{=} -4$$
$$\sqrt{16} \stackrel{?}{=} -4$$
$$4 \neq -4$$

Since 4 does not check (4 is an extraneous root), the equation $\sqrt{3x + 4} = -4$ has no real number solutions. The solution set is $\emptyset$. ∎

EXAMPLE 5

Solve $3\sqrt{2y + 1} = 5$.

Solution

$$3\sqrt{2y + 1} = 5$$

$$\sqrt{2y + 1} = \frac{5}{3} \qquad \text{Divided both sides by 3}$$

$$(\sqrt{2y + 1})^2 = \left(\frac{5}{3}\right)^2$$

$$2y + 1 = \frac{25}{9}$$

$$2y = \frac{25}{9} - 1$$

$$2y = \frac{25}{9} - \frac{9}{9}$$

$$2y = \frac{16}{9}$$

$$\frac{1}{2}(2y) = \frac{1}{2}\left(\frac{16}{9}\right) \qquad \text{Multiplied both sides by } \frac{1}{2}$$

$$y = \frac{8}{9}$$

✔ **Check**

$$3\sqrt{2y + 1} = 5$$

$$3\sqrt{2\left(\frac{8}{9}\right) + 1} \stackrel{?}{=} 5$$

$$3\sqrt{\frac{16}{9} + \frac{9}{9}} \stackrel{?}{=} 5$$

$$3\sqrt{\frac{25}{9}} \stackrel{?}{=} 5$$

$$3\left(\frac{5}{3}\right) \stackrel{?}{=} 5$$

$$5 = 5$$

The solution set is $\left\{\frac{8}{9}\right\}$. ■

2 Solve Radical Equations of the Form $\sqrt{ax + b} = \sqrt{cx + d}$

EXAMPLE 6

Solve $\sqrt{5s - 2} = \sqrt{2s + 19}$.

Solution

$$\sqrt{5s - 2} = \sqrt{2s + 19}$$

$$(\sqrt{5s - 2})^2 = (\sqrt{2s + 19})^2 \qquad \text{Square both sides}$$

$$5s - 2 = 2s + 19$$
$$3s = 21$$
$$s = 7$$

✔ **Check**

$$\sqrt{5s - 2} = \sqrt{2s + 19}$$
$$\sqrt{5(7) - 2} \stackrel{?}{=} \sqrt{2(7) + 19}$$
$$\sqrt{33} = \sqrt{33}$$

The solution set is {7}.

3 Solve Radical Equations of the Form $\sqrt{ax + b} = cx + d$

EXAMPLE 7 Solve $\sqrt{2y - 4} = y - 2$.

Solution

$$\sqrt{2y - 4} = y - 2$$
$$(\sqrt{2y - 4})^2 = (y - 2)^2 \quad \text{Square both sides}$$
$$2y - 4 = y^2 - 4y + 4$$
$$0 = y^2 - 6y + 8$$
$$0 = (y - 4)(y - 2) \quad \text{Factor the right side}$$

$y - 4 = 0$ or $y - 2 = 0$ Remember the property: $ab = 0$
$y = 4$ or $y = 2$ if and only if $a = 0$ or $b = 0$

✔ **Check**

$$\sqrt{2y - 4} = y - 2 \qquad \sqrt{2y - 4} = y - 2$$
$$\sqrt{2(4) - 4} \stackrel{?}{=} 4 - 2 \qquad \sqrt{2(2) - 4} \stackrel{?}{=} 2 - 2$$
$$\sqrt{4} \stackrel{?}{=} 2 \quad \text{or} \quad \sqrt{0} \stackrel{?}{=} 0$$
$$2 = 2 \quad \text{or} \quad 0 = 0$$

The solution set is {2, 4}.

EXAMPLE 8 Solve $\sqrt{x} + 6 = x$.

Solution

$$\sqrt{x} + 6 = x$$
$$\sqrt{x} = x - 6 \quad \text{We added } -6 \text{ to both sides so that the term with}$$
$$(\sqrt{x})^2 = (x - 6)^2 \quad \text{the radical is alone on one side of the equation}$$
$$x = x^2 - 12x + 36$$
$$0 = x^2 - 13x + 36$$
$$0 = (x - 4)(x - 9)$$

$x - 4 = 0$ or $x - 9 = 0$ Apply $ab = 0$ if and only if $a = 0$ or $b = 0$
$x = 4$ or $x = 9$

✓ Check

$$\sqrt{x} + 6 = x \qquad\qquad \sqrt{x} + 6 = x$$
$$\sqrt{4} + 6 \stackrel{?}{=} 4 \qquad\qquad \sqrt{9} + 6 \stackrel{?}{=} 9$$
$$2 + 6 \stackrel{?}{=} 4 \quad \text{or} \quad 3 + 6 \stackrel{?}{=} 9$$
$$8 \neq 4 \qquad\qquad 9 = 9$$

The solution set is $\{9\}$. ∎

Note in Example 8 that we changed the form of the original equation, $\sqrt{x} + 6 = x$, to $\sqrt{x} = x - 6$ before we squared both sides. Squaring both sides of $\sqrt{x} + 6 = x$ produces $x + 12\sqrt{x} + 36 = x^2$, a more complex equation that still contains a radical. So again, it pays to think ahead a few steps before carrying out the details of the problem.

4 Solve and Apply Radical Formulas

In Section 9.1 we used the formula $S = \sqrt{30Df}$ to approximate how fast a car was traveling based on the length of skid marks. (Remember that S represents the speed of the car in miles per hour, D the length of skid marks measured in feet, and f a coefficient of friction.) This same formula can be used to estimate the lengths of skid marks that are produced by cars traveling at different rates on various types of road surfaces. To use the formula for this purpose, we change the form of the equation by solving for D.

$$\sqrt{30Df} = S$$
$$30Df = S^2 \qquad \text{The result of squaring both sides of the original equation}$$
$$D = \frac{S^2}{30f} \qquad \text{D, S, and f are positive numbers, so this final equation and the original one are equivalent}$$

EXAMPLE 9   Apply Your Skill

Suppose that for a particular road surface, the coefficient of friction is 0.35. How far will a car traveling at 60 miles per hour skid when the brakes are applied?

Solution

We substitute 0.35 for f and 60 for S in the formula $D = \dfrac{S^2}{30f}$.

$$D = \frac{60^2}{30(0.35)} = 343 \quad \text{to the nearest whole number}$$

The car will skid approximately 343 feet. ∎

Remark: Pause for a moment and think about the result in Example 9. The coefficient of friction of 0.35 applies to a wet concrete road surface. Note that a car traveling at 60 miles per hour will skid farther than the length of a football field.

CONCEPT QUIZ 9.5

For Problems 1–5, answer true or false.

1. Equations that contain radicals with variables in the radicand are called radical equations.
2. Squaring both sides of a radical equation produces an equivalent equation.
3. Extra solutions that do not satisfy the original equation are called simultaneous solutions.

Section 9.5 Classroom Problem Set

Objective 1

1. Solve $\sqrt{y} = 9$.
2. Solve $\sqrt{2x} = 6$.
3. Solve $\sqrt{a} = -12$.
4. Solve $3\sqrt{x} = -18$.
5. Solve $\sqrt{3y+1} = 4$.
6. Solve $\sqrt{2n-3} = 13$.
7. Solve $\sqrt{2y+8} = -10$.
8. Solve $\sqrt{5a+2} = -1$.
9. Solve $2\sqrt{3x-6} = 24$.
10. Solve $7\sqrt{a+2} = 14$.

Objective 2

11. Solve $\sqrt{2y+5} = \sqrt{4y-7}$.
12. Solve $\sqrt{8x-6} = \sqrt{4x+11}$.

Objective 3

13. Solve $\sqrt{2y-8} = y - 4$.
14. Solve $\sqrt{x+7} = x + 7$.
15. Solve $2 + \sqrt{y} = y$.
16. Solve $\sqrt{x^2+2x+3} = x + 2$.

Objective 4

17. Suppose that for a particular road surface the coefficient of friction is 0.40. How far will a car traveling at 55 miles per hour skid when the brakes are applied?
18. Suppose that for a particular road surface the coefficient of friction is 0.40. How far will a car traveling at 90 miles per hour skid when the brakes are applied?

THOUGHTS INTO WORDS

1. Explain in your own words why possible solutions for radical equations *must* be checked.
2. Your friend attempts to solve the equation

    $$3 + 2\sqrt{x} = x$$

    as follows:

    $$(3 + 2\sqrt{x})^2 = x^2$$
    $$9 + 12\sqrt{x} + 4x = x^2$$

    At this step, she stops and doesn't know how to proceed. What help can you give her?

Answers to the Concept Quiz

1. True 2. False 3. False 4. True 5. True

Chapter 9 Review Problem Set

For Problems 1–10, evaluate each expression without the use of a calculator.

1. $\sqrt{64}$
2. $-\sqrt{49}$
3. $\sqrt{1600}$
4. $\sqrt{\dfrac{81}{25}}$
5. $-\sqrt{\dfrac{4}{9}}$
6. $\sqrt{\dfrac{49}{36}}$
7. $\sqrt[3]{27}$
8. $\sqrt[3]{-125}$
9. $-\sqrt[3]{-64}$
10. $\sqrt[3]{\dfrac{8}{125}}$

For Problems 11–16, use a calculator to find a rational approximation of each root. Round each answer to the nearest hundredth.

11. $\sqrt{158}$
12. $\sqrt{1250}$
13. $-\sqrt{860}$
14. $\sqrt{250}$
15. $\sqrt[3]{612}$
16. $\sqrt[3]{-789}$

For Problems 17–22, change each radical to simplest radical form.

17. $\sqrt{20}$
18. $\sqrt{32}$
19. $5\sqrt{8}$
20. $\sqrt{80}$
21. $\sqrt[3]{24}$
22. $\sqrt[3]{250}$

For Problems 23–28, change each radical to simplest radical form. All variables represent positive real numbers.

23. $\sqrt{12x^2}$
24. $\sqrt{50x^2y}$
25. $3\sqrt{20a^3b^2}$
26. $\sqrt{48a^4}$
27. $\sqrt[3]{8x^2y^3}$
28. $\sqrt[3]{-27x^4}$

For Problems 29–34, simplify each radical expression. All variables represent positive real numbers.

29. $2\sqrt{50} + 3\sqrt{72} - 5\sqrt{8}$
30. $\sqrt{48} + 2\sqrt{75} - \sqrt{12}$
31. $\sqrt{8x} - 3\sqrt{18x}$
32. $\sqrt{20y} + \sqrt{45y} - 3\sqrt{5y}$
33. $\sqrt[3]{-27x^4} + x\sqrt[3]{8x}$
34. $\sqrt[3]{-27x^2} + 3\sqrt[3]{-8x^2}$

For Problems 35–38, change each radical to simplest radical form. All variables represent positive real numbers.

35. $\dfrac{\sqrt[3]{40}}{\sqrt[3]{5}}$
36. $\dfrac{\sqrt[3]{40}}{\sqrt[3]{8}}$
37. $\sqrt{\dfrac{2x^3}{9}}$
38. $\dfrac{\sqrt{72a}}{\sqrt{16b^2}}$

For Problems 39–50, rationalize the denominator and simplify. All variables represent positive real numbers.

39. $\dfrac{\sqrt{36}}{\sqrt{7}}$
40. $\dfrac{3\sqrt{2}}{\sqrt{5}}$
41. $\sqrt{\dfrac{7}{8}}$
42. $\sqrt{\dfrac{8}{24}}$
43. $\sqrt{\dfrac{4}{x}}$
44. $\dfrac{\sqrt{2x}}{\sqrt{5y}}$
45. $\dfrac{4\sqrt{3}}{\sqrt{12}}$
46. $\dfrac{5\sqrt{2}}{2\sqrt{3}}$
47. $\dfrac{-3\sqrt{2}}{\sqrt{27}}$
48. $\dfrac{4\sqrt{6}}{3\sqrt{12}}$
49. $\dfrac{3\sqrt{x}}{4\sqrt{y^3}}$
50. $\dfrac{-2\sqrt{x^2y}}{5\sqrt{xy}}$

For Problems 51–54, simplify each expression.

51. $\sqrt{\dfrac{2}{3}} - 2\sqrt{54}$
52. $3\sqrt{10} + \sqrt{\dfrac{2}{5}}$
53. $4\sqrt{20} - \dfrac{3}{\sqrt{5}} + \sqrt{45}$
54. $\sqrt{\dfrac{3}{5}} + 3\sqrt{60}$

For Problems 55–64, find the products and express your answers in simplest radical form.

55. $(\sqrt{6})(\sqrt{12})$
56. $(2\sqrt{3})(3\sqrt{6})$
57. $(-5\sqrt{8})(2\sqrt{2})$
58. $(2\sqrt[3]{7})(5\sqrt[3]{4})$
59. $\sqrt[3]{2}(\sqrt[3]{3} + \sqrt[3]{4})$
60. $3\sqrt{5}(\sqrt{8} - 2\sqrt{12})$
61. $(\sqrt{3} + \sqrt{5})(\sqrt{3} + \sqrt{7})$
62. $(2\sqrt{3} + 3\sqrt{2})(\sqrt{3} - 5\sqrt{2})$
63. $(\sqrt{6} + 2\sqrt{7})(3\sqrt{6} - \sqrt{7})$
64. $(3 + 2\sqrt{5})(4 - 3\sqrt{5})$

For Problems 65–68, rationalize the denominator and simplify.

65. $\dfrac{5}{\sqrt{7} - \sqrt{5}}$
66. $\dfrac{\sqrt{6}}{\sqrt{3} - \sqrt{2}}$
67. $\dfrac{2}{3\sqrt{2} - \sqrt{6}}$
68. $\dfrac{\sqrt{6}}{3\sqrt{7} + 2\sqrt{10}}$

For Problems 69–80, solve each of the equations.

69. $\sqrt{5x + 6} = 6$
70. $\sqrt{2x - 1} = 13$
71. $\sqrt{8 - x} = 6$
72. $\sqrt{-2x} = 8$
73. $\sqrt{2x + 3} = -2$
74. $\sqrt{3x + 1} = -5$
75. $\sqrt{6x + 1} = \sqrt{3x + 13}$
76. $\sqrt{3x - 4} = \sqrt{x + 7}$
77. $\sqrt{y + 5} = y + 5$
78. $\sqrt{2x - 4} = x - 6$
79. $\sqrt{-3a + 10} = a - 2$
80. $3 - \sqrt{2x - 1} = 2$

81. The time T, measured in seconds, that it takes an object to fall d feet (neglecting air resistance) is given by the formula $T = \sqrt{\dfrac{d}{16}}$. Find the times that it takes objects to fall 100 feet, 350 feet, and 500 feet, respectively. Express the answers to the nearest tenth of a second.

82. Solve the formula $\sqrt{22ab} = c$ for b.

Chapter 9 Practice Test

1. Evaluate $-\sqrt{\dfrac{64}{49}}$.

2. Evaluate $\sqrt{0.0025}$.

For Problems 3–5, use a calculater to evaluate each of the following to the nearest tenth.

3. $\sqrt{8}$

4. $-\sqrt{32}$

5. $\dfrac{3}{\sqrt{2}}$

For Problems 6–14, change each radical expression to simplest radical form. All variables represent positive real numbers.

6. $\sqrt{45}$

7. $-4\sqrt[3]{54}$

8. $\dfrac{2\sqrt{3}}{3\sqrt{6}}$

9. $\sqrt{\dfrac{25}{2}}$

10. $\dfrac{\sqrt{24}}{\sqrt{36}}$

11. $\sqrt{\dfrac{5}{8}}$

12. $\sqrt[3]{-250x^4y^3}$

13. $\dfrac{\sqrt{3x}}{\sqrt{5y}}$

14. $\dfrac{3}{4}\sqrt{48x^3y^2}$

For Problems 15–18, find the indicated products and express the answers in simplest radical form.

15. $(\sqrt{8})(\sqrt{12})$

16. $(6\sqrt[3]{5})(4\sqrt[3]{2})$

17. $\sqrt{6}(2\sqrt{12} - 3\sqrt{8})$

18. $(2\sqrt{5} + \sqrt{3})(\sqrt{5} - 3\sqrt{3})$

19. Rationalize the denominator and simplify:

$\dfrac{\sqrt{6}}{\sqrt{12} + \sqrt{2}}$

20. Simplify $2\sqrt{24} - 4\sqrt{54} + 3\sqrt{96}$.

1. _____
2. _____
3. _____
4. _____
5. _____
6. _____
7. _____
8. _____
9. _____
10. _____
11. _____
12. _____
13. _____
14. _____
15. _____
16. _____
17. _____
18. _____
19. _____
20. _____

For Problems 21–24, solve each equation.

21. $\sqrt{3x + 1} = 4$

22. $\sqrt{2x - 5} = -4$

23. $\sqrt{n - 3} = 3 - n$

24. $\sqrt{3x + 6} = x + 2$

25. Use the formula $T = 2\pi\sqrt{\dfrac{L}{32}}$, in which T is the period of a pendulum expressed in seconds, and L is the length of the pendulum in feet, to find the period of a pendulum 3.5 feet long. Express your answer to the nearest tenth.

Chapters 1–9 Cumulative Review Problem Set

For Problems 1–6, evaluate each of the numerical expressions.

1. -2^6
2. $\left(\dfrac{1}{4}\right)^{-3}$
3. $\left(\dfrac{1}{3} - \dfrac{1}{4}\right)^{-2}$
4. $-\sqrt{64}$
5. $\sqrt{\dfrac{4}{9}}$
6. $3^0 + 3^{-1} + 3^{-2}$

For Problems 7–10, evaluate each algebraic expression for the given values of the variables.

7. $3(2x - 1) - 4(2x + 3) - (x + 6)$ for $x = -4$
8. $(3x^2 - 4x - 6) - (3x^2 + 3x + 1)$ for $x = 6$
9. $2(a - b) - 3(2a + b) + 2(a - 3b)$ for $a = -2$ and $b = 3$
10. $x^2 - 2xy + y^2$ for $x = 5$ and $y = -2$

For Problems 11–25, perform the indicated operations, and express your answers in simplest form using positive exponents only.

11. $\dfrac{3}{4x} + \dfrac{5}{2x} - \dfrac{7}{x}$
12. $\dfrac{3}{x - 2} - \dfrac{4}{x + 3}$
13. $\dfrac{3x}{7y} \div \dfrac{6x}{35y^2}$
14. $\dfrac{x - 2}{x^2 + x - 6} \cdot \dfrac{x^2 + 6x + 9}{x^2 - x - 12}$
15. $\dfrac{7}{x^2 + 3x - 18} - \dfrac{8}{x - 3}$
16. $(-3xy)(-4y^2)(5x^3y)$
17. $(-4x^{-5})(2x^3)$
18. $\dfrac{-12a^{-2}b^3}{4a^{-5}b^4}$
19. $(3n^4)^{-1}$
20. $(9x - 2)(3x + 4)$
21. $(-x - 1)(5x + 7)$
22. $(3x + 1)(2x^2 - x - 4)$
23. $\dfrac{15x^6y^8 - 20x^3y^5}{5x^3y^2}$
24. $(10x^3 - 8x^2 - 17x - 3) \div (5x + 1)$
25. $\dfrac{\dfrac{1}{x} - \dfrac{1}{y}}{\dfrac{1}{xy}}$

26. If 2 gallons of paint will cover 1500 square feet of walls, how many gallons are needed for 3500 square feet?

27. 18 is what percent of 72?

28. Solve $V = \dfrac{1}{3}Bh$ for B if $V = 432$ and $h = 12$.

29. How many feet of fencing are needed to enclose a rectangular garden that measures 25 feet by 40 feet?

30. Find the total surface area of a sphere that has a radius 5 inches long. Use 3.14 as an approximation for π.

31. Write each number in scientific notation.
    (a) 85,000
    (b) 0.0009
    (c) 0.00000104
    (d) 53,000,000

For Problems 32–37, factor each expression completely.

32. $12x^3 + 14x^2 - 40x$
33. $12x^2 - 27$
34. $xy + 3x - 2y - 6$
35. $30 + 19x - 5x^2$
36. $4x^4 - 4$
37. $21x^2 + 22x - 8$

For Problems 38–43, change each radical expression to simplest radical form.

38. $4\sqrt{28}$
39. $-\sqrt{45}$
40. $\sqrt{\dfrac{36}{5}}$
41. $\dfrac{5\sqrt{8}}{6\sqrt{12}}$
42. $\sqrt{72xy^5}$
43. $\dfrac{-2\sqrt{ab^2}}{5\sqrt{b}}$

For Problems 44–46, find each product, and express your answer in simplest radical form.

44. $(3\sqrt{8})(4\sqrt{2})$
45. $6\sqrt{2}(9\sqrt{8} - 3\sqrt{12})$
46. $(3\sqrt{2} - \sqrt{7})(3\sqrt{2} + \sqrt{7})$

For Problems 47 and 48, rationalize the denominator and simplify.

47. $\dfrac{4}{\sqrt{3} + \sqrt{2}}$
48. $\dfrac{-6}{3\sqrt{5} - \sqrt{6}}$

For Problems 49 and 50, simplify each of the radical expressions.

49. $3\sqrt{50} - 7\sqrt{72} + 4\sqrt{98}$
50. $\dfrac{2}{3}\sqrt{20} - \dfrac{3}{4}\sqrt{45} + \sqrt{80}$

For Problems 51–55, graph each of the equations.

51. $3x - 6y = -6$
52. $y = \dfrac{1}{3}x + 4$
53. $y = -\dfrac{2}{5}x + 3$
54. $y - 2x = 0$
55. $y = -x$

For Problems 56–58, graph each linear inequality.

56. $y \geq 2x - 6$

57. $3x - 2y < -6$

58. $22x - 4y > 8$

For Problems 59–64, solve each of the problems.

59. Find the slope of the line determined by the points $(-3, 6)$ and $(2, -4)$.

60. Find the slope of the line determined by the equation $4x - 7y = 12$.

61. Write the equation of the line that has a slope of $\dfrac{2}{3}$ and contains the point $(7, 2)$.

62. Write the equation of the line that contains the points $(-4, 1)$ and $(-1, -3)$.

63. Write the equation of the line that has a slope of $-\dfrac{1}{4}$ and a y intercept of -3.

64. Find the slope of a line whose equation is $3x - 2y = 12$.

For Problems 65–68, solve each of the systems by using either the substitution method or the addition method.

65. $\begin{pmatrix} y = 3x - 5 \\ 3x + 4y = -5 \end{pmatrix}$

66. $\begin{pmatrix} 4x - 3y = -20 \\ 3x + 5y = 14 \end{pmatrix}$

67. $\begin{pmatrix} \dfrac{1}{2}x - \dfrac{2}{3}y = -11 \\ \dfrac{1}{3}x + \dfrac{5}{6}y = 8 \end{pmatrix}$

68. $\begin{pmatrix} 2x + 7y = 22 \\ 4x - 5y = -13 \end{pmatrix}$

For Problems 69–80, solve each equation.

69. $-2(n - 1) + 4(2n - 3) = 4(n + 6)$

70. $\dfrac{4}{x - 1} = \dfrac{-1}{x + 6}$

71. $\dfrac{t - 1}{3} - \dfrac{t + 2}{4} = -\dfrac{5}{12}$

72. $-7 - 2n - 6n = 7n - 5n + 12$

73. $\dfrac{n - 5}{2} = 3 - \dfrac{n + 4}{5}$

74. $0.11x + 0.14(x + 400) = 181$

75. $\dfrac{x}{60 - x} = 7 + \dfrac{4}{60 - x}$

76. $1 + \dfrac{x + 1}{2x} = \dfrac{3}{4}$

77. $x^2 + 4x - 12 = 0$

78. $2x^2 - 8 = 0$

79. $\sqrt{3x - 6} = 9$

80. $\sqrt{3n - 2} = 7$

For Problems 81–86, solve each of the inequalities.

81. $-3n - 4 \leq 11$

82. $-5 > 3n - 4 - 7n$

83. $2(x - 2) + 3(x + 4) > 6$

84. $\dfrac{1}{2}n - \dfrac{2}{3}n < -1$

85. $\dfrac{x + 1}{2} + \dfrac{x - 2}{6} < \dfrac{3}{8}$

86. $\dfrac{x - 3}{7} - \dfrac{x - 2}{4} \leq \dfrac{9}{14}$

For Problems 87–97, set up an equation, an inequality, or a system of equations to help solve each problem.

87. If two angles are supplementary and the larger angle is 15° less than twice the smaller angle, find the measure of each angle.

88. The sum of two numbers is 50. If the larger number is 2 less than three times the smaller number, find the numbers.

89. The sum of the squares of two consecutive odd whole numbers is 130. Find the numbers.

90. Suppose that Nick has 47 coins consisting of nickels, dimes, and quarters. The number of dimes is 1 more than twice the number of nickels, and the number of quarters is 4 more than three times the number of nickels. Find the number of coins of each denomination.

91. A home valued at \$140,000 is assessed \$2940 in real estate taxes; at the same rate, what would be the taxes on a home assessed at \$180,000?

92. A retailer has some skirts that cost her \$30 each. She wants to sell them at a profit of 60% of the cost. What price should she charge for the skirts?

93. Rosa leaves a town traveling in her car at a rate of 45 miles per hour. One hour later Polly leaves the same town traveling the same route at a rate of 55 miles per hour. How long will it take Polly to overtake Rosa?

94. How many milliliters of pure acid must be added to 100 milliliters of a 10% acid solution to obtain a 20% solution?

95. Suppose that Andy has scores of 85, 90, and 86 on his first three algebra tests. What score must he get on the fourth algebra test to have an average of 88 or higher for the four tests?

96. The Cubs have won 70 games and lost 72 games. They have 20 more games to play. To win more than 50% of all their games, how many of the 20 games remaining must they win?

97. Seth can do a job in 20 minutes. Butch can do the same job in 30 minutes. If they work together, how long will it take them to complete the job?

Quadratic Equations

Chapter 10 Warm-Up Problems

1. Solve by factoring.

   (a) $x^2 - 6x = 0$ (b) $9m^2 - 18m + 1 = 0$ (c) $n^2 - 64 = 0$

2. Simplify the roots.

   (a) $\sqrt{169}$ (b) $\sqrt{\dfrac{25}{4}}$ (c) $\sqrt{75}$

3. If the legs of a right triangle are 3 and 4, find the length of the hypotenuse. (*Hint:* Use the Pythagorean theorem.)

4. Factor the perfect square trinomials.

   (a) $9x^2 - 30x + 25$ (b) $m^2 - 12m + 36$ (c) $4x^2 + 2x + 1$

5. Fill in the blanks.

   (a) $\dfrac{1}{2}(6) = $ _____ ; $3^2 = $ _____

   (b) $\dfrac{1}{2}(3) = $ _____ ; $\left(\dfrac{3}{2}\right)^2 = $ _____

   (c) $\dfrac{1}{2}\left(-\dfrac{8}{3}\right) = $ _____ ; $\left(\dfrac{4}{3}\right)^2 = $ _____

6. Simplify.

   (a) $8^2 - 4(3)(2)$ (b) $(-6)^2 - 4(1)(7)$ (c) $5^2 - 4(-2)(2)$

7. Simplify.

   (a) $\sqrt{7^2 - 4(1)(-18)}$ (b) $\sqrt{(-1)^2 - 4(1)(-12)}$

   (c) $\sqrt{(6)^2 - 4(-8)(2)}$

8. Simplify.

   (a) $\dfrac{-3 + \sqrt{3^2 - 4(2)(-27)}}{2(2)}$ (b) $\dfrac{-(-6) - \sqrt{(-6)^2 - 4(1)(9)}}{2(1)}$

   (c) $\dfrac{-13 + \sqrt{13^2 - 4(6)(-5)}}{2(6)}$

10.1 Quadratic Equations

10.2 Completing the Square

10.3 Quadratic Formula

10.4 Solving Quadratic Equations—Which Method?

10.5 Solving Problems Using Quadratic Equations

Solving equations has been a central theme of this text. We now pause for a moment and reflect on the different types of equations that we have solved.

Video tutorials for all section learning objectives are available in a variety of delivery modes.

INTERNET PROJECT

Quadratic equations have two solutions. The derivation of the word *quadratic* is from Latin. In the English language, the prefix *quad* means "four." Do an Internet search to discover the origins of the name quadratic for these equations that have two answers.

Type of equation	Examples
First-degree equations of one variable	$4x + 3 = 7x + 1; 3(x - 6) = 9$
Second-degree equations of one variable that are factorable	$x^2 + 3x = 0; x^2 + 5x + 6 = 0;$ $x^2 - 4 = 0; x^2 + 10x + 25 = 0$
Rational equations	$\dfrac{3}{x} + \dfrac{2}{x} = 4; \dfrac{5}{a-2} = \dfrac{6}{a+3};$ $\dfrac{2}{x^2 - 4} + \dfrac{5}{x+2} = \dfrac{6}{x-2}$
Radical equations	$\sqrt{x} = 4; \sqrt{y+2} = 3;$ $\sqrt{a+1} = \sqrt{2a-7}$
Systems of equations	$\begin{pmatrix} 2x + 3y = 4 \\ 5x - y = 7 \end{pmatrix};$ $\begin{pmatrix} 3a + 5b = 9 \\ 7a - 9b = 12 \end{pmatrix}$

As indicated in the chart, we have learned how to solve some second-degree equations, but only those for which the quadratic polynomial is factorable. In this chapter, we will expand our work to include more general types of second-degree equations in one variable and thus broaden our problem-solving capabilities.

10.1 Quadratic Equations

OBJECTIVES

1. Solve Quadratic Equations by Factoring
2. Solve Quadratic Equations by Applying the Square-Root Property
3. Solve Word Problems Involving the Pythagorean Theorem

1 Solve Quadratic Equations by Factoring

A second-degree equation in one variable contains the variable with an exponent of 2, but no higher power. Such equations are also called **quadratic equations**. Here are some examples of quadratic equations.

$$x^2 = 25 \qquad y^2 + 6y = 0 \qquad x^2 + 7x - 4 = 0$$
$$4y^2 + 2y - 1 = 0 \qquad 5x^2 + 2x - 1 = 2x^2 + 6x - 5$$

We can also define a quadratic equation in the variable x as any equation that can be written in the form $ax^2 + bx + c = 0$, where a, b, and c are real numbers and $a \neq 0$. We refer to the form $ax^2 + bx + c = 0$ as the **standard form** of a quadratic equation.

In Chapter 5, you solved quadratic equations (we didn't use the term "quadratic" at that time) by factoring and applying the property, $ab = 0$ if and only if $a = 0$ or $b = 0$. Let's review a few examples of that type.

EXAMPLE 1

Solve $x^2 - 13x = 0$.

Solution

$$x^2 - 13x = 0$$
$$x(x - 13) = 0 \quad \text{Factor left side of equation}$$
$$x = 0 \quad \text{or} \quad x - 13 = 0 \quad \text{Apply } ab = 0 \text{ if and only if } a = 0 \text{ or } b = 0$$
$$x = 0 \quad \text{or} \quad x = 13$$

The solution set is $\{0, 13\}$. Don't forget to check these solutions!

EXAMPLE 2

Solve $n^2 + 2n - 24 = 0$.

Solution

$$n^2 + 2n - 24 = 0$$
$$(n + 6)(n - 4) = 0 \quad \text{Factor left side}$$
$$n + 6 = 0 \quad \text{or} \quad n - 4 = 0 \quad \text{Apply } ab = 0 \text{ if and only if } a = 0 \text{ or } b = 0$$
$$n = -6 \quad \text{or} \quad n = 4$$

The solution set is $\{-6, 4\}$.

EXAMPLE 3

Solve $x^2 + 6x + 9 = 0$.

Solution

$$x^2 + 6x + 9 = 0$$
$$(x + 3)(x + 3) = 0 \quad \text{Factor left side}$$
$$x + 3 = 0 \quad \text{or} \quad x + 3 = 0 \quad \text{Apply } ab = 0 \text{ if and only if } a = 0 \text{ or } b = 0$$
$$x = -3 \quad \text{or} \quad x = -3$$

The solution set is $\{-3\}$.

EXAMPLE 4

Solve $y^2 = 49$.

Solution

$$y^2 = 49$$
$$y^2 - 49 = 0$$
$$(y + 7)(y - 7) = 0 \quad \text{Factor left side}$$
$$y + 7 = 0 \quad \text{or} \quad y - 7 = 0 \quad \text{Apply } ab = 0 \text{ if and only if } a = 0 \text{ or } b = 0$$
$$y = -7 \quad \text{or} \quad y = 7$$

The solution set is $\{-7, 7\}$.

Note the type of equation that we solved in Example 4. We can generalize from that example and consider the equation $x^2 = a$, where a is any nonnegative real number. We can solve this equation as follows:

$$x^2 = a$$
$$x^2 = (\sqrt{a})^2 \qquad (\sqrt{a})^2 = a$$
$$x^2 - (\sqrt{a})^2 = 0$$
$$(x - \sqrt{a})(x + \sqrt{a}) = 0 \qquad \text{Factor left side}$$
$$x - \sqrt{a} = 0 \quad \text{or} \quad x + \sqrt{a} = 0 \qquad \text{Apply } ab = 0 \text{ if and only if}$$
$$x = \sqrt{a} \quad \text{or} \quad x = -\sqrt{a} \qquad a = 0 \text{ or } b = 0$$

The solutions are $\sqrt{a}$ and $-\sqrt{a}$. The process we used to solve this equation is a general property (stated formally below), which can be applied to solve certain types of quadratic equations.

2 Solve Quadratic Equations by Applying the Square-Root Property

> **Property 10.1**
>
> For any nonnegative real number a,
>
> $x^2 = a$ if and only if $x = \sqrt{a}$ or $x = -\sqrt{a}$
>
> (The statement "$x = \sqrt{a}$ or $x = -\sqrt{a}$" can be written as $x = \pm\sqrt{a}$.)

Property 10.1 is sometimes referred to as the **square-root property**. This property, along with our knowledge of square roots, makes it very easy to solve quadratic equations of the form $x^2 = a$.

EXAMPLE 5   Solve $x^2 = 81$.

Solution

$$x^2 = 81$$
$$x = \pm\sqrt{81} \qquad \text{Apply Property 10.1}$$
$$x = \pm 9$$

The solution set is $\{-9, 9\}$.

EXAMPLE 6   Solve $x^2 = 8$.

Solution

$$x^2 = 8$$
$$x = \pm\sqrt{8}$$
$$x = \pm 2\sqrt{2} \qquad \sqrt{8} = \sqrt{4}\sqrt{2} = 2\sqrt{2}$$

The solution set is $\{-2\sqrt{2}, 2\sqrt{2}\}$.

EXAMPLE 7   Solve $5n^2 = 12$.

Solution

$$5n^2 = 12$$
$$n^2 = \frac{12}{5} \qquad \text{Divided both sides by 5}$$

$$n = \pm\sqrt{\frac{12}{5}}$$

$$n = \pm\frac{2\sqrt{15}}{5} \qquad \sqrt{\frac{12}{5}} = \frac{\sqrt{12}}{\sqrt{5}} = \frac{\sqrt{12}}{\sqrt{5}} \cdot \frac{\sqrt{5}}{\sqrt{5}} = \frac{\sqrt{60}}{5} = \frac{2\sqrt{15}}{5}$$

The solution set is $\left\{-\dfrac{2\sqrt{15}}{5}, \dfrac{2\sqrt{15}}{5}\right\}$. ∎

EXAMPLE 8

Solve $(x - 2)^2 = 16$.

Solution

$$(x - 2)^2 = 16$$
$$x - 2 = \pm 4$$
$$x - 2 = 4 \quad \text{or} \quad x - 2 = -4$$
$$x = 6 \quad \text{or} \quad x = -2$$

The solution set is $\{-2, 6\}$. ∎

EXAMPLE 9

Solve $(x + 5)^2 = 27$.

Solution

$$(x + 5)^2 = 27$$
$$x + 5 = \pm\sqrt{27}$$
$$x + 5 = \pm 3\sqrt{3} \qquad \sqrt{27} = \sqrt{9}\sqrt{3} = 3\sqrt{3}$$
$$x + 5 = 3\sqrt{3} \quad \text{or} \quad x + 5 = -3\sqrt{3}$$
$$x = -5 + 3\sqrt{3} \quad \text{or} \quad x = -5 - 3\sqrt{3}$$

The solution set is $\{-5 - 3\sqrt{3}, -5 + 3\sqrt{3}\}$. ∎

It may be necessary to change the form before we can apply Property 10.1. The next example illustrates this procedure.

EXAMPLE 10

Solve $3(2x - 3)^2 + 8 = 44$.

Solution

$$3(2x - 3)^2 + 8 = 44$$
$$3(2x - 3)^2 = 36$$
$$(2x - 3)^2 = 12$$
$$2x - 3 = \pm\sqrt{12} \qquad \text{Apply Property 10.1}$$
$$2x - 3 = \sqrt{12} \quad \text{or} \quad 2x - 3 = -\sqrt{12}$$
$$2x = 3 + \sqrt{12} \quad \text{or} \quad 2x = 3 - \sqrt{12}$$
$$x = \frac{3 + \sqrt{12}}{2} \quad \text{or} \quad x = \frac{3 - \sqrt{12}}{2}$$
$$x = \frac{3 + 2\sqrt{3}}{2} \quad \text{or} \quad x = \frac{3 - 2\sqrt{3}}{2} \qquad \sqrt{12} = \sqrt{4}\sqrt{3} = 2\sqrt{3}$$

The solution set is $\left\{\dfrac{3 - 2\sqrt{3}}{2}, \dfrac{3 + 2\sqrt{3}}{2}\right\}$. ∎

Note that quadratic equations of the form $x^2 = a$, where a is a *negative* number, have no real number solutions. For example, $x^2 = -4$ has no real number solutions, because any real number squared is nonnegative. In a like manner, an equation such as $(x + 3)^2 = -14$ has no real number solutions.

3 Solve Word Problems Involving the Pythagorean Theorem

Our work with radicals, Property 10.1, and the Pythagorean theorem merges to form the basis for solving a variety of problems that pertain to right triangles. First, let's restate the Pythagorean theorem.

Pythagorean Theorem

If for a right triangle, a and b are the measures of the legs, and c is the measure of the hypotenuse, then

$$a^2 + b^2 = c^2$$

(The *hypotenuse* is the side opposite the right angle, and the *legs* are the other two sides, as shown in Figure 10.1.)

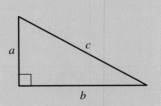

Figure 10.1

EXAMPLE 11   Apply Your Skill

Find c in Figure 10.2.

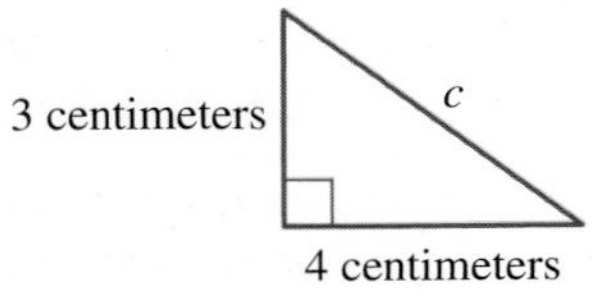

Figure 10.2

Solution

Applying the Pythagorean theorem, we have

$c^2 = a^2 + b^2$

$c^2 = 3^2 + 4^2$

$c^2 = 9 + 16$

$c^2 = 25$

$c = 5$

The length of c is 5 centimeters.   ∎

Remark: Don't forget that the equation $c^2 = 25$ does have two solutions, 5 and -5. However, because we are finding the lengths of line segments, we can disregard the negative solutions.

EXAMPLE 12   Apply Your Skill

A 50-foot rope hangs from the top of a flagpole. When pulled taut to its full length, the rope reaches a point on the ground 18 feet from the base of the pole. Find the height of the pole to the nearest tenth of a foot.

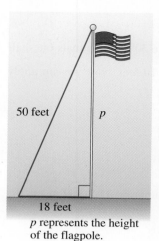

Figure 10.3
p represents the height of the flagpole.

Solution

We can sketch Figure 10.3 and record the given information. Using the Pythagorean theorem, we solve for p as follows:

$$p^2 + 18^2 = 50^2$$
$$p^2 + 324 = 2500$$
$$p^2 = 2176$$
$$p = \sqrt{2176} = 46.6 \quad \text{To the nearest tenth}$$

The height of the flagpole is approximately 46.6 feet. ∎

An isosceles triangle has two sides of the same length. Thus an **isosceles right triangle** is a right triangle that has both legs of the same length. The next example presents a problem involving an isosceles right triangle.

EXAMPLE 13   Apply Your Skill

Find the length of each leg of an isosceles right triangle if the hypotenuse is 8 meters long.

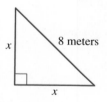

Figure 10.4

Solution

We sketch an isosceles right triangle in Figure 10.4 and let x represent the length of each leg. Then we can determine x by applying the Pythagorean theorem.

$$x^2 + x^2 = 8^2$$
$$2x^2 = 64$$
$$x^2 = 32$$
$$x = \sqrt{32} = \sqrt{16}\sqrt{2} = 4\sqrt{2}$$

Each leg is $4\sqrt{2}$ meters long. ∎

Remark: In Example 12, we made no attempt to express $\sqrt{2176}$ in simplest radical form because the answer was to be given as a rational approximation to the nearest tenth. However, in Example 13 we left the final answer in radical form and expressed it in simplest radical form.

Another special kind of right triangle is one that contains acute angles of 30° and 60°. In such a right triangle, often referred to as a 30°–60° right triangle, *the side opposite the 30° angle is equal in length to one-half the length of the hypotenuse.* This relationship, along with the Pythagorean theorem, provides us with another problem-solving technique.

EXAMPLE 14   Apply Your Skill

Suppose that a 20-foot ladder is leaning against a building and makes an angle of 60° with the ground. How far up the building does the top of the ladder reach? Express your answer to the nearest tenth of a foot.

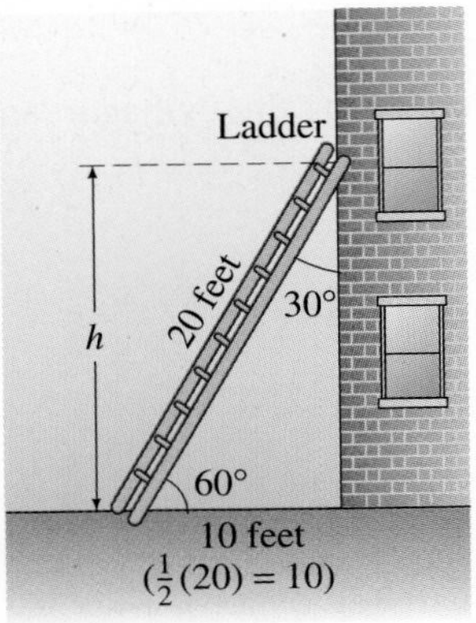

Solution

Figure 10.5 illustrates this problem. The side opposite the 30° angle equals one-half of the hypotenuse, so the length of that side is $\dfrac{1}{2}(20) = 10$ feet. Now we can apply the Pythagorean theorem.

$$h^2 + 10^2 = 20^2$$
$$h^2 + 100 = 400$$
$$h^2 = 300$$
$$h = \sqrt{300} = 17.3 \quad \text{to the nearest tenth}$$

The top of the ladder touches the building at approximately 17.3 feet from the ground. ∎

Figure 10.5

CONCEPT QUIZ 10.1

For Problems 1–10, answer true or false.

1. If an equation has any variable raised to the second power, it is a quadratic equation.
2. Any quadratic equation in the variable x can be written in the form $ax^2 + bx + c = 0$, where $a, b,$ and c are real numbers and $a \neq 0$.
3. We refer to the form $ax^2 + bx + c = 0$ as the normal form of a quadratic equation.
4. The quadratic equation $-2x^2 + 6x - 9 = 0$ is in standard form.
5. The statement $x = \pm\sqrt{2}$ means that $x = -\sqrt{2}$ or $x = \sqrt{2}$.
6. The Pythagorean theorem applies to any triangle.
7. In a right triangle, the hypotenuse is the side opposite the right angle.
8. If for a right triangle, a and b are the measures of the legs, and c is the measure of the hypotenuse, then $a + b = c$.
9. An isosceles right triangle is a right triangle in which two legs are of equal length.
10. In a 30°–60° right triangle, the side opposite the 30° angle is equal in length to one-third the length of the hypotenuse.

Section 10.1 Classroom Problem Set

Objective 1

1. Solve $y^2 + 7y = 0$.
2. Solve $8n^2 = -56n$.
3. Solve $x^2 - 4x - 21 = 0$.
4. Solve $4y^2 - 21y - 18 = 0$.
5. Solve $a^2 - 10a + 25 = 0$.
6. Solve $9x^2 + 12x + 4 = 0$.

Objective 2

7. Solve $n^2 = 144$.
8. Solve $a^2 = \dfrac{4}{81}$.
9. Solve $y^2 = 64$.
10. Solve $4a^2 = 36$.
11. Solve $n^2 = 12$.
12. Solve $x^2 = 24$.
13. Solve $3y^2 = 8$.
14. Solve $12a^2 = 49$.
15. Solve $(a + 5)^2 = 9$.
16. Solve $(4x + 3)^2 = 1$.
17. Solve $(y - 7)^2 = 20$.
18. Solve $(3n - 2)^2 - 28 = 0$.
19. Solve $3(5n + 1)^2 + 4 = 58$.
20. Solve $2(7x - 1)^2 + 5 = 37$.

Objective 3

21. Find c in Figure 10.6.

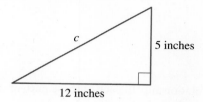

Figure 10.6

22. Let a and b represent the lengths of the legs of a right triangle, and c represent the length of the hypotenuse. Find a if $b = 6$ meters and $c = 8$ meters. Write the answer in simplest radical form.

23. A guy wire that is 30 feet in length hangs from the top of a tower. When pulled taut to its full length, the wire reaches a point 12 feet from the base of the tower. Find the height of the tower to the nearest tenth of a foot.

24. A 42-foot guy wire makes an angle of 60° with the ground. The wire is attached to the top of a telephone pole. Find the distance from the base of the pole to the point on the pole where the wire is attached. Find the answer to the nearest tenth of a foot.

25. Find the length of each leg of an isosceles right triangle if the hypotenuse is 12 feet long.

26. Let a and b represent the lengths of the legs of an isosceles right triangle, and c represent the length of the hypotenuse of the same right triangle. Find b and c if $a = 4$.

27. Suppose that a 16-foot ladder is leaning against a building and makes an angle of 60° with the ground. How far up on the building does the top of the ladder reach? Express your answer to the nearest tenth of a foot.

28. The diagonal of a square parking lot is 50 meters. Find to the nearest meter the length of a side of the lot.

THOUGHTS INTO WORDS

1. Explain why the equation $(x - 4)^2 + 14 = 2$ has no real number solutions.

2. Suppose that your friend solved the equation $(x + 3)^2 = 25$ as follows:

$$(x + 3)^2 = 25$$
$$x^2 + 6x + 9 = 25$$
$$x^2 + 6x - 16 = 0$$
$$(x + 8)(x - 2) = 0$$
$$x + 8 = 0 \quad \text{or} \quad x - 2 = 0$$
$$x = -8 \quad \text{or} \quad x = 2$$

Is this a correct approach to the problem? Can you suggest an easier approach to the problem?

Answers to the Concept Quiz

1. False **2.** True **3.** False **4.** True **5.** True **6.** False **7.** True **8.** False **9.** True **10.** False

10.2 Completing the Square

OBJECTIVE

1 Solve Quadratic Equations by Completing the Square

1 Solve Quadratic Equations by Completing the Square

Thus far we have solved quadratic equations by factoring or by applying Property 10.1 (if $x^2 = a$, then $x = \sqrt{a}$ or $x = -\sqrt{a}$). In this section, we will consider another method called *completing the square*, which will give us the power to solve *any* quadratic equation.

We studied a factoring technique in Chapter 5 that was based on recognizing **perfect square trinomials**. In each of the following equations, the trinomial on the right side, which is the result of squaring a binomial on the left side, is a perfect square trinomial.

$$(x + 5)^2 = x^2 + 10x + 25$$
$$(x + 7)^2 = x^2 + 14x + 49$$
$$(x - 3)^2 = x^2 - 6x + 9$$
$$(x - 6)^2 = x^2 - 12x + 36$$

We need to pay attention to the following special relationship. In each of these perfect square trinomials, *the constant term is equal to the square of one-half of the coefficient of the x term*. For example,

$$x^2 + 10x + 25 \qquad \frac{1}{2}(10) = 5 \quad \text{and} \quad 5^2 = 25$$

$$x^2 - 12x + 36 \qquad \frac{1}{2}(12) = 6 \quad \text{and} \quad 6^2 = 36$$

This relationship allows us to form a perfect square trinomial by adding the proper constant term. For example, suppose that we want to form a perfect square trinomial from $x^2 + 8x$. Because $\frac{1}{2}(8) = 4$ and $4^2 = 16$, we can form the perfect square trinomial $x^2 + 8x + 16$.

Now we can use the preceding ideas to help solve some quadratic equations.

EXAMPLE 1

Solve $x^2 + 8x - 1 = 0$ by the method of completing the square.

Solution

$$x^2 + 8x - 1 = 0$$
$$x^2 + 8x = 1 \qquad \text{Isolated the } x^2 \text{ and } x \text{ terms}$$
$$\frac{1}{2}(8) = 4 \quad \text{and} \quad 4^2 = 16 \qquad \text{Took } \frac{1}{2} \text{ of the coefficient of the } x \text{ term and then squared the result}$$
$$x^2 + 8x + 16 = 1 + 16 \qquad \text{Added 16 to both sides of the equation}$$
$$(x + 4)^2 = 17 \qquad \text{Factored the perfect square trinomial}$$

Now we can proceed as we did with similar equations in the last section.

$$x + 4 = \pm\sqrt{17}$$
$$x + 4 = \sqrt{17} \qquad \text{or} \qquad x + 4 = -\sqrt{17}$$
$$x = -4 + \sqrt{17} \qquad \text{or} \qquad x = -4 - \sqrt{17}$$

The solution set is $\{-4 - \sqrt{17}, -4 + \sqrt{17}\}$. ∎

Observe that the method of completing the square to solve a quadratic equation is just what the name implies. We form a perfect square trinomial; then we change the equation to the necessary form for using the property, if $x^2 = a$, then $x = \sqrt{a}$ or $x = -\sqrt{a}$. Let's consider another example.

EXAMPLE 2

Solve $x^2 - 2x - 11 = 0$ by the method of completing the square.

Solution

$$x^2 - 2x - 11 = 0$$
$$x^2 - 2x = 11 \qquad \text{Isolated the } x^2 \text{ and } x \text{ terms}$$

$$\frac{1}{2}(2) = 1 \quad \text{and} \quad 1^2 = 1 \qquad \text{Took } \frac{1}{2} \text{ of the coefficient of the } x \text{ term and then squared the result}$$

$$x^2 - 2x + 1 = 11 + 1 \qquad \text{Added 1 to both sides of the equation}$$

$$(x - 1)^2 = 12 \qquad \text{Factored the perfect square trinomial}$$

$$x - 1 = \pm\sqrt{12}$$

$$x - 1 = \pm 2\sqrt{3}$$

$$x - 1 = 2\sqrt{3} \quad \text{or} \quad x - 1 = -2\sqrt{3}$$

$$x = 1 + 2\sqrt{3} \quad \text{or} \quad x = 1 - 2\sqrt{3}$$

The solution set is $\{1 - 2\sqrt{3}, 1 + 2\sqrt{3}\}$. ∎

In the next example, the coefficient of the x term is odd, which means that taking one-half of it puts us into the realm of fractions. The use of common fractions rather than decimals makes our previous work with radicals relevant.

EXAMPLE 3

Solve $x^2 - 3x + 1 = 0$ by the method of completing the square.

Solution

$$x^2 - 3x + 1 = 0$$

$$x^2 - 3x = -1$$

$$\frac{1}{2}(3) = \frac{3}{2} \quad \text{and} \quad \left(\frac{3}{2}\right)^2 = \frac{9}{4} \qquad \text{Took } \frac{1}{2} \text{ of the coefficient of the } x \text{ term and then squared the result}$$

$$x^2 - 3x + \frac{9}{4} = -1 + \frac{9}{4} \qquad \text{Added } \frac{9}{4} \text{ to both sides of the equation}$$

$$\left(x - \frac{3}{2}\right)^2 = \frac{5}{4} \qquad \text{Factored the perfect square trinomial}$$

$$x - \frac{3}{2} = \pm\sqrt{\frac{5}{4}}$$

$$x - \frac{3}{2} = \pm\frac{\sqrt{5}}{2}$$

$$x - \frac{3}{2} = \frac{\sqrt{5}}{2} \quad \text{or} \quad x - \frac{3}{2} = -\frac{\sqrt{5}}{2}$$

$$x = \frac{3}{2} + \frac{\sqrt{5}}{2} \quad \text{or} \quad x = \frac{3}{2} - \frac{\sqrt{5}}{2}$$

$$x = \frac{3 + \sqrt{5}}{2} \quad \text{or} \quad x = \frac{3 - \sqrt{5}}{2}$$

The solution set is $\left\{\dfrac{3 - \sqrt{5}}{2}, \dfrac{3 + \sqrt{5}}{2}\right\}$. ∎

The relationship for a perfect square trinomial that states *the constant term is equal to the square of one-half of the coefficient of the x term* holds only if the coefficient of x^2 is 1. Thus we need to make a slight adjustment when solving quadratic equations that have a coefficient of x^2 other than 1. The next example shows how to make this adjustment.

EXAMPLE 4

Solve $2x^2 + 12x - 3 = 0$ by the method of completing the square.

Solution

$$2x^2 + 12x - 3 = 0$$

$$2x^2 + 12x = 3$$

$$x^2 + 6x = \frac{3}{2} \quad \text{Multiply both sides by } \frac{1}{2}$$

$$x^2 + 6x + 9 = \frac{3}{2} + 9 \quad \left[\frac{1}{2}(6)\right]^2 = 3^2 = 9; \text{ add 9 to both sides of the equation}$$

$$(x + 3)^2 = \frac{21}{2}$$

$$x + 3 = \pm\sqrt{\frac{21}{2}}$$

$$x + 3 = \pm\frac{\sqrt{42}}{2} \quad \sqrt{\frac{21}{2}} = \frac{\sqrt{21}}{\sqrt{2}} = \frac{\sqrt{21}}{\sqrt{2}} \cdot \frac{\sqrt{2}}{\sqrt{2}} = \frac{\sqrt{42}}{2}$$

$$x + 3 = \frac{\sqrt{42}}{2} \quad \text{or} \quad x + 3 = -\frac{\sqrt{42}}{2}$$

$$x = -3 + \frac{\sqrt{42}}{2} \quad \text{or} \quad x = -3 - \frac{\sqrt{42}}{2}$$

$$x = \frac{-6 + \sqrt{42}}{2} \quad \text{or} \quad x = \frac{-6 - \sqrt{42}}{2}$$

The solution set is $\left\{\dfrac{-6 - \sqrt{42}}{2}, \dfrac{-6 + \sqrt{42}}{2}\right\}$. ∎

As we mentioned earlier, we can use the method of completing the square to solve *any* quadratic equation. To illustrate this point, we will use this method to solve an equation that we could also solve by factoring.

EXAMPLE 5 Solve $x^2 + 2x - 8 = 0$ by the method of completing the square and by factoring.

Solution A

By completing the square,

$$x^2 + 2x - 8 = 0$$

$$x^2 + 2x = 8$$

$$x^2 + 2x + 1 = 8 + 1 \quad \left[\frac{1}{2}(2)\right]^2 = 1^2 = 1; \text{ add 1 to both sides of the equation}$$

$$(x + 1)^2 = 9$$

$$x + 1 = \pm 3$$

$$x + 1 = 3 \quad \text{or} \quad x + 1 = -3$$

$$x = 2 \quad \text{or} \quad x = -4$$

The solution set is $\{-4, 2\}$.

Solution B

By factoring,

$$x^2 + 2x - 8 = 0$$

$$(x + 4)(x - 2) = 0$$

$$x + 4 = 0 \quad \text{or} \quad x - 2 = 0$$

$$x = -4 \quad \text{or} \quad x = 2$$

The solution set is $\{-4, 2\}$. ∎

We don't claim that using the method of completing the square with an equation such as the one in Example 5 is easier than the factoring technique. However, it is important for you to recognize that the method of completing the square will work with any quadratic equation.

Our final example of this section demonstrates that the method of completing the square will identify those quadratic equations that have no real number solutions.

EXAMPLE 6

Solve $x^2 + 10x + 30 = 0$ by the method of completing the square.

Solution

$$x^2 + 10x + 30 = 0$$
$$x^2 + 10x = -30$$
$$x^2 + 10x + 25 = -30 + 25$$
$$(x + 5)^2 = -5$$

We can stop here and reason as follows: Any value of x will yield a nonnegative value for $(x + 5)^2$; thus, it cannot equal -5. The original equation, $x^2 + 10x + 30 = 0$, has no solutions in the set of real numbers. ∎

CONCEPT QUIZ 10.2

For Problems 1–7, answer true or false.

1. In a perfect square trinomial, the constant term is equal to one-half the coefficient of the x term.
2. The method of completing the square can be used to solve any quadratic equation.
3. Every quadratic equation solved by completing the square will have real number solutions.
4. The completing-the-square method cannot be used if the equation could be solved by factoring.
5. To use the completing-the-square method for solving the equation $3x^2 + 2x = 5$, we would first divide both sides of the equation by 3.
6. The equation $x^2 + 2x = 0$ cannot be solved by using the completing-the-square method because there is no constant term.
7. To solve the equation $x^2 - 5x = 1$ by completing the square, we would start by adding $\dfrac{25}{4}$ to both sides of the equation.

Section 10.2 Classroom Problem Set

Objective 1

1. Solve $y^2 + 10y - 6 = 0$ by the method of completing the square.
2. Solve $x^2 + 8x + 3 = 0$ by the method of completing the square.
3. Solve $y^2 - 6y + 2 = 0$ by the method of completing the square.
4. Solve $a^2 - 4a = 1$ by the method of completing the square.
5. Solve $a^2 + 5a - 3 = 0$ by the method of completing the square.
6. Solve $a^2 + 5a = 2$ by the method of completing the square.
7. Solve $3y^2 + 6y - 2 = 0$ by the method of completing the square.
8. Solve $3x^2 + 2x - 2 = 0$ by the method of completing the square.
9. Solve $n^2 + n - 12 = 0$ by the method of completing the square and by factoring.
10. Solve $6a^2 = 11a + 10$ by the method of completing the square and by factoring.
11. Solve $y^2 + 6y + 22 = 0$ by the method of completing the square.
12. Solve $n^2 + 12n + 40 = 0$ by the method of completing the square.

THOUGHTS INTO WORDS

1. Give a step-by-step description of how to solve the equation $3x^2 + 10x - 8 = 0$ by completing the square.

2. An error has been made in the following solution. Find it and explain how to correct it.

$$4x^2 - 4x + 1 = 0$$
$$4x^2 - 4x = -1$$
$$4x^2 - 4x + 4 = -1 + 4$$
$$(2x - 2)^2 = 3$$
$$2x - 2 = \pm\sqrt{3}$$
$$2x - 2 = \sqrt{3} \quad \text{or} \quad 2x - 2 = -\sqrt{3}$$
$$2x = 2 + \sqrt{3} \quad \text{or} \quad 2x = 2 - \sqrt{3}$$
$$x = \frac{2 + \sqrt{3}}{2} \quad \text{or} \quad x = \frac{2 - \sqrt{3}}{2}$$

The solution set is $\left\{\dfrac{2 + \sqrt{3}}{2}, \dfrac{2 - \sqrt{3}}{2}\right\}$.

Answers to the Concept Quiz
1. False 2. True 3. False 4. False 5. True 6. False 7. True

10.3 Quadratic Formula

OBJECTIVE

1 Solve Quadratic Equations by Using the Quadratic Formula

1 Solve Quadratic Equations by Using the Quadratic Formula

We can use the method of completing the square to solve any quadratic equation. The equation $ax^2 + bx + c = 0$, where a, b, and c are real numbers with $a \neq 0$, can represent *any* quadratic equation. These two ideas merge to produce the *quadratic formula*, a formula that we can use to solve any quadratic equation. The merger is formed when using the method of completing the square to solve the equation $ax^2 + bx + c = 0$ as follows:

$$ax^2 + bx + c = 0$$

$$ax^2 + bx = -c$$

$$x^2 + \frac{b}{a}x = -\frac{c}{a} \qquad \text{Multiply both sides by } \frac{1}{a}$$

$$x^2 + \frac{b}{a}x + \frac{b^2}{4a^2} = -\frac{c}{a} + \frac{b^2}{4a^2} \qquad \text{Complete the square by adding } \frac{b^2}{4a^2} \text{ to both sides}$$

$$\left(x + \frac{b}{2a}\right)^2 = \frac{b^2 - 4ac}{4a^2} \qquad \text{The right side is combined into a single term with LCD } 4a^2$$

$$x + \frac{b}{2a} = \pm\sqrt{\frac{b^2 - 4ac}{4a^2}}$$

$$x + \frac{b}{2a} = \pm\frac{\sqrt{b^2 - 4ac}}{\sqrt{4a^2}}$$

$$x + \frac{b}{2a} = \pm\frac{\sqrt{b^2 - 4ac}}{2a} \qquad \sqrt{4a^2} = |2a| \text{ but } 2a \text{ can be used because of } \pm$$

$$x = -\frac{b}{2a} \pm \frac{\sqrt{b^2 - 4ac}}{2a}$$

$$x = \frac{-b \pm \sqrt{b^2 - 4ac}}{2a}$$

The solutions are $\dfrac{-b + \sqrt{b^2 - 4ac}}{2a}$ and $\dfrac{-b - \sqrt{b^2 - 4ac}}{2a}$.

We usually state the **quadratic formula** as follows:

$$x = \frac{-b \pm \sqrt{b^2 - 4ac}}{2a}$$

We can use it to solve any quadratic equation by expressing the equation in standard form and by substituting the values for a, b, and c into the formula. Consider the following examples.

EXAMPLE 1

Solve $x^2 + 7x + 10 = 0$ by using the quadratic formula.

Solution

The given equation is in standard form with $a = 1$, $b = 7$, and $c = 10$. Let's substitute these values into the quadratic formula and simplify.

$$x = \frac{-b \pm \sqrt{b^2 - 4ac}}{2a}$$

$$x = \frac{-7 \pm \sqrt{7^2 - 4(1)(10)}}{2(1)}$$

$$x = \frac{-7 \pm \sqrt{9}}{2}$$

$$x = \frac{-7 \pm 3}{2}$$

$$x = \frac{-7 + 3}{2} \quad \text{or} \quad x = \frac{-7 - 3}{2}$$

$$x = -2 \quad \text{or} \quad x = -5$$

The solution set is $\{-5, -2\}$. ∎

EXAMPLE 2

Solve $x^2 - 3x = 1$ by using the quadratic formula.

Solution

First we need to change the equation to the standard form of $ax^2 + bx + c = 0$.

$$x^2 - 3x = 1$$
$$x^2 - 3x - 1 = 0$$

We need to think of $x^2 - 3x - 1 = 0$ as $x^2 + (-3)x + (-1) = 0$ to determine the values $a = 1$, $b = -3$, and $c = -1$. Let's substitute these values into the quadratic formula and simplify.

$$x = \frac{-(-3) \pm \sqrt{(-3)^2 - 4(1)(-1)}}{2(1)}$$

$$x = \frac{3 \pm \sqrt{9 + 4}}{2}$$

$$x = \frac{3 \pm \sqrt{13}}{2}$$

The solution set is $\left\{\dfrac{3 - \sqrt{13}}{2}, \dfrac{3 + \sqrt{13}}{2}\right\}$. ∎

EXAMPLE 3 Solve $15n^2 - n - 2 = 0$ by using the quadratic formula.

Solution

Remember that although we commonly use the variable x in the statement of the quadratic formula, any variable could be used. Writing the equation as $15n^2 + (-1)n + (-2) = 0$ gives us the standard form of $an^2 + bn + c = 0$ with $a = 15$, $b = -1$, and $c = -2$. Now we can solve the equation by substituting into the quadratic formula and simplifying.

$$n = \frac{-(-1) \pm \sqrt{(-1)^2 - 4(15)(-2)}}{2(15)}$$

$$n = \frac{1 \pm \sqrt{1 + 120}}{30}$$

$$n = \frac{1 \pm \sqrt{121}}{30}$$

$$n = \frac{1 \pm 11}{30}$$

$$n = \frac{1 + 11}{30} \quad \text{or} \quad n = \frac{1 - 11}{30}$$

$$n = \frac{12}{30} \quad \text{or} \quad n = \frac{-10}{30}$$

$$n = \frac{2}{5} \quad \text{or} \quad n = -\frac{1}{3}$$

The solution set is $\left\{-\dfrac{1}{3}, \dfrac{2}{5}\right\}$. ∎

EXAMPLE 4 Solve $t^2 - 2t - 4 = 0$ by using the quadratic formula.

Solution

Writing the equation as $t^2 + (-2)t + (-4) = 0$ gives us the standard form of $at^2 + bt + c = 0$, with $a = 1$, $b = -2$, and $c = -4$. Now we can solve by substituting into the quadratic formula and simplifying.

$$t = \frac{-(-2) \pm \sqrt{(-2)^2 - 4(1)(-4)}}{2(1)}$$

$$t = \frac{2 \pm \sqrt{4 + 16}}{2}$$

$$t = \frac{2 \pm \sqrt{20}}{2} = \frac{2 \pm \sqrt{4}\sqrt{5}}{2} = \frac{2 \pm 2\sqrt{5}}{2}$$

Now the fraction can be simplified. One way to simplify is to factor the common factor of 2 out of the numerator and then reduce the fraction.

$$t = \frac{2 \pm 2\sqrt{5}}{2} = \frac{2(1 \pm \sqrt{5})}{2} = 1 \pm \sqrt{5}$$

The solution set is $\{1 - \sqrt{5}, 1 + \sqrt{5}\}$. ∎

We can easily identify quadratic equations that have no real number solutions when we use the quadratic formula. The final example of this section illustrates this point.

EXAMPLE 5

Solve $x^2 - 2x + 8 = 0$ by using the quadratic formula.

Solution

$$x = \frac{-(-2) \pm \sqrt{(-2)^2 - 4(1)(8)}}{2(1)}$$

$$x = \frac{2 \pm \sqrt{4 - 32}}{2}$$

$$x = \frac{2 \pm \sqrt{-28}}{2}$$

Since $\sqrt{-28}$ is not a real number, we conclude that the given equation has no real number solutions. ∎

CONCEPT QUIZ 10.3

For Problems 1–5, answer true or false.

1. The quadratic formula can be used to solve any quadratic equation.
2. When solving quadratic equations using the quadratic formula, there are always two distinct, different answers.
3. The quadratic formula cannot be used if the quadratic equation can be solved by factoring.
4. To solve $3x^2 + 5x - 6 = 0$ by the quadratic formula, we would substitute 3 for a, 5 for b, and -6 for c in the quadratic formula.
5. To solve $2x^2 + 7x = 5$ by the quadratic formula, we would substitute 2 for a, 7 for b, and 5 for c in the quadratic formula.

Section 10.3 Classroom Problem Set

Objective 1

1. Solve $x^2 + 3x - 18 = 0$ by using the quadratic formula.
2. Solve $a^2 = 8a - 12$ by using the quadratic formula.
3. Solve $x^2 + 5x = 3$ by using the quadratic formula.
4. Solve $2y^2 - y - 4 = 0$ by using the quadratic formula.
5. Solve $6y^2 + 5y - 4 = 0$ by using the quadratic formula.
6. Solve $12x^2 + 19x = -5$ by using the quadratic formula.
7. Solve $x^2 + 4x - 6 = 0$ by using the quadratic formula.
8. Solve $3x^2 + 4x - 1 = 0$ by using the quadratic formula.
9. Solve $y^2 + 3y + 11 = 0$.
10. Solve $3t^2 + 6t = 5$.

THOUGHTS INTO WORDS

1. Explain how to use the quadratic formula to solve the equation $x^2 = 2x + 6$.
2. Your friend states that the equation $-x^2 - 6x + 16 = 0$ must be changed to $x^2 + 6x - 16 = 0$ (by multiplying both sides by -1) before the quadratic formula can be applied. Is he right about this, and if not, how would you convince him that he is wrong?
3. Another of your friends claims that the quadratic formula can be used to solve the equation $x^2 - 4 = 0$. How would you react to this claim?

Answers to the Concept Quiz
1. True **2.** False **3.** False **4.** True **5.** False

10.4 Solving Quadratic Equations—Which Method?

OBJECTIVE

1 Solve Quadratic Equations by Any of the Three Basic Methods

1 Solve Quadratic Equations by Any of the Three Basic Methods

We now summarize the three basic methods of solving quadratic equations by solving a specific quadratic equation using each technique. Consider the equation $x^2 + 4x - 12 = 0$.

Factoring Method

$x^2 + 4x - 12 = 0$
$(x + 6)(x - 2) = 0$
$x + 6 = 0$ or $x - 2 = 0$
$x = -6$ or $x = 2$

The solution set is $\{-6, 2\}$.

Completing the Square Method

$x^2 + 4x - 12 = 0$
$x^2 + 4x = 12$
$x^2 + 4x + 4 = 12 + 4$
$(x + 2)^2 = 16$
$x + 2 = \pm\sqrt{16}$
$x + 2 = 4$ or $x + 2 = -4$
$x = 2$ or $x = -6$

The solution set is $\{-6, 2\}$.

Quadratic Formula Method

$x^2 + 4x - 12 = 0$

$$x = \frac{-4 \pm \sqrt{4^2 - 4(1)(-12)}}{2(1)}$$

$$x = \frac{-4 \pm \sqrt{64}}{2}$$

$$x = \frac{-4 \pm 8}{2}$$

$x = \dfrac{-4 + 8}{2}$ or $x = \dfrac{-4 - 8}{2}$

$x = 2$ or $x = -6$

The solution set is $\{-6, 2\}$.

We have also discussed the use of Property 10.1 ($x^2 = a$ if and only if $x = \pm\sqrt{a}$) for certain types of quadratic equations. For example, we can solve $x^2 = 4$ easily by applying Property 10.1 and obtaining $x = \sqrt{4}$ or $x = -\sqrt{4}$; thus, the solutions are 2 and -2.

Which method should you use to solve a particular quadratic equation? Let's consider some examples in which the different techniques are used. Keep in mind that this is a decision you must make as the need arises. So become as familiar as you can with the strengths and weaknesses of each method.

EXAMPLE 1

Solve $2x^2 + 12x - 54 = 0$.

Solution

First, it is very helpful to recognize a factor of 2 in each of the terms on the left side.

$$2x^2 + 12x - 54 = 0$$
$$x^2 + 6x - 27 = 0 \quad \text{Multiply both sides by } \frac{1}{2}$$

Now you should recognize that the left side can be factored. Thus we can proceed as follows:

$$(x + 9)(x - 3) = 0$$
$$x + 9 = 0 \quad \text{or} \quad x - 3 = 0$$
$$x = -9 \quad \text{or} \quad x = 3$$

The solution set is $\{-9, 3\}$. ∎

EXAMPLE 2

Solve $(4x + 3)^2 = 16$.

Solution

The form of this equation lends itself to the use of Property 10.1 ($x^2 = a$ if and only if $x = \pm\sqrt{a}$).

$$(4x + 3)^2 = 16$$
$$4x + 3 = \pm\sqrt{16}$$
$$4x + 3 = 4 \quad \text{or} \quad 4x + 3 = -4$$
$$4x = 1 \quad \text{or} \quad 4x = -7$$
$$x = \frac{1}{4} \quad \text{or} \quad x = -\frac{7}{4}$$

The solution set is $\left\{-\frac{7}{4}, \frac{1}{4}\right\}$. ∎

EXAMPLE 3

Solve $n + \dfrac{1}{n} = 5$.

Solution

First, we need to *clear the equation of fractions* by multiplying both sides by n.

$$n + \frac{1}{n} = 5, \quad n \neq 0$$
$$n\left(n + \frac{1}{n}\right) = 5(n)$$
$$n^2 + 1 = 5n$$

Now we can change the equation to standard form.

$$n^2 - 5n + 1 = 0$$

Because the left side cannot be factored using integers, we must solve the equation by using either the method of completing the square or the quadratic formula. Using the formula, we obtain

$$n = \frac{-(-5) \pm \sqrt{(-5)^2 - 4(1)(1)}}{2(1)}$$

$$n = \frac{5 \pm \sqrt{21}}{2}$$

The solution set is $\left\{\frac{5 - \sqrt{21}}{2}, \frac{5 + \sqrt{21}}{2}\right\}$. ∎

EXAMPLE 4

Solve $t^2 = \sqrt{2}t$.

Solution

A quadratic equation without a constant term can be solved easily by the factoring method.

$$t^2 = \sqrt{2}t$$
$$t^2 - \sqrt{2}t = 0$$
$$t(t - \sqrt{2}) = 0$$
$$t = 0 \quad \text{or} \quad t - \sqrt{2} = 0$$
$$t = 0 \quad \text{or} \quad t = \sqrt{2}$$

The solution set is $\{0, \sqrt{2}\}$. (Check each of these solutions in the given equation.) ∎

EXAMPLE 5

Solve $x^2 - 28x + 192 = 0$.

Solution

Determining whether or not the left side is factorable presents a bit of a problem because of the size of the constant term. Therefore, let's not concern ourselves with trying to factor; instead we will use the quadratic formula.

$$x^2 - 28x + 192 = 0$$

$$x = \frac{-(-28) \pm \sqrt{(-28)^2 - 4(1)(192)}}{2(1)}$$

$$x = \frac{28 \pm \sqrt{784 - 768}}{2}$$

$$x = \frac{28 \pm \sqrt{16}}{2}$$

$$x = \frac{28 + 4}{2} \quad \text{or} \quad x = \frac{28 - 4}{2}$$

$$x = 16 \quad \text{or} \quad x = 12$$

The solution set is $\{12, 16\}$. ∎

EXAMPLE 6

Solve $x^2 + 12x = 17$.

Solution

The form of this equation, and the fact that the coefficient of x is even, makes the method of completing the square a reasonable approach.

$$x^2 + 12x = 17$$
$$x^2 + 12x + 36 = 17 + 36$$
$$(x + 6)^2 = 53$$
$$x + 6 = \pm\sqrt{53}$$
$$x = -6 \pm \sqrt{53}$$

The solution set is $\{-6 - \sqrt{53}, -6 + \sqrt{53}\}$. ∎

CONCEPT QUIZ 10.4

Match the equation with the method most appropriate for solving the equation.

1. $x^2 - 5x - 3 = 0$
2. $y^2 + 6y = 14$
3. $(2x - 1)^2 = 25$
4. $m^2 + 2m - 15 = 0$

A. Factoring
B. Square root property
C. Completing the square
D. Quadratic formula

Section 10.4 Classroom Problem Set

Objective 1

1. Solve $3x^2 - 3x - 6 = 0$.
2. Solve $2a^2 + 10a - 28 = 0$.
3. Solve $(5x - 1)^2 = 49$.
4. Solve $(3n - 1)^2 = 25$.
5. Solve $y - \dfrac{2}{y} = -3$.
6. Solve $1 = \dfrac{3}{n} - n$.
7. Solve $x^2 = 10x$.
8. Solve $3n = 2n^2$.
9. Solve $t^2 + 30t + 216 = 0$.
10. Solve $x^2 - 28x + 187 = 0$.
11. Solve $y^2 + 10y = 15$.
12. Solve $\dfrac{2}{3n - 1} = \dfrac{n + 2}{6}$.

THOUGHTS INTO WORDS

1. Which method would you use to solve the equation $x^2 + 30x = -216$? Explain your reasons for making this choice.

2. Explain how you would solve the equation $0 = -x^2 - x + 6$.

3. How can you tell by inspection that the equation $x^2 + x + 4 = 0$ has no real number solutions?

Answers to the Concept Quiz

1. D **2.** C **3.** B **4.** A

10.5 Solving Problems Using Quadratic Equations

OBJECTIVE

1 Solve Word Problems Involving Quadratic Equations

1 Solve Word Problems Involving Quadratic Equations

The following diagram indicates our approach in this text.

Develop skills $\longrightarrow$ Use skills to solve equations $\longrightarrow$ Use equations to solve word problems

Now you should be ready to use your skills relative to solving systems of equations (Chapter 8) and quadratic equations to solve additional types of word problems. Before you consider such problems, let's review and update the problem-solving suggestions we offered in Chapter 2.

Suggestions for Solving Word Problems

1. Read the problem carefully and make certain that you understand the meanings of all the words. Be especially alert for any technical terms used in the statement of the problem.
2. Read the problem a second time (perhaps even a third time) to get an overview of the situation being described and to determine the known facts as well as what is to be found.
3. Sketch any figure, diagram, or chart that might be helpful in analyzing the problem.
*4. Choose *meaningful* variables to represent the unknown quantities. Use one or two variables, whichever seems easiest. The term "meaningful" refers to the choice of letters to use as variables. Choose letters that have some significance for the problem under consideration. For example, if the problem deals with the length and width of a rectangle, then *l* and *w* are natural choices for the variables.
*5. Look for *guidelines* that you can use to help set up equations. A guideline might be a formula such as *area of a rectangular region equals length times width*, or a statement of a relationship such as *the product of the two numbers is 98*.
*6. (a) Form an equation that contains the variable and translates the conditions of the guideline from English into algebra; or
    (b) form two equations that contain the two variables and translate the guidelines from English into algebra.
*7. Solve the equation (system of equations) and use the solution (solutions) to determine all facts requested in the problem.
8. **Check all answers back in the original statement of the problem.**

The asterisks indicate those suggestions that have been revised to include using systems of equations to solve problems. Keep these suggestions in mind as you study the examples and work the problems in this section.

EXAMPLE 1

Apply Your Skill

The length of a rectangular region is 2 centimeters more than its width. The area of the region is 35 square centimeters. Find the length and width of the rectangle.

Solution

We let l represent the length, and we let w represent the width (see Figure 10.7). We can use the area formula for a rectangle, $A = lw$, and the statement "the length of a rectangular region is 2 centimeters greater than its width" as guidelines to form a system of equations.

$$\begin{pmatrix} lw = 35 \\ l = w + 2 \end{pmatrix}$$

The second equation indicates that we can substitute $w + 2$ for l. Making this substitution in the first equation yields

$$(w + 2)(w) = 35$$

Solving this quadratic equation by factoring, we get

$$w^2 + 2w = 35$$
$$w^2 + 2w - 35 = 0$$
$$(w + 7)(w - 5) = 0$$
$$w + 7 = 0 \quad \text{or} \quad w - 5 = 0$$
$$w = -7 \quad \text{or} \quad w = 5$$

The width of a rectangle cannot be a negative number, so we discard the solution -7. Thus the width of the rectangle is 5 centimeters, and the length $(w + 2)$ is 7 centimeters. ∎

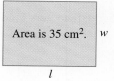

Figure 10.7

EXAMPLE 2

Apply Your Skill

Find two consecutive whole numbers whose product is 506.

Solution

We let n represent the smaller whole number. Then $n + 1$ represents the next larger whole number. The phrase "whose product is 506" translates into the equation

$$n(n + 1) = 506$$

Changing this quadratic equation into standard form produces

$$n^2 + n = 506$$
$$n^2 + n - 506 = 0$$

Because of the size of the constant term, let's not try to factor; instead, we use the quadratic formula.

$$n = \frac{-1 \pm \sqrt{1^2 - 4(1)(-506)}}{2(1)}$$

$$n = \frac{-1 \pm \sqrt{2025}}{2}$$

$$n = \frac{-1 \pm 45}{2} \qquad \sqrt{2025} = 45$$

$$n = \frac{-1 + 45}{2} \quad \text{or} \quad n = \frac{-1 - 45}{2}$$

$$n = 22 \quad \text{or} \quad n = -23$$

Since we are looking for whole numbers, we discard the solution -23. Therefore, the whole numbers are 22 and 23. ∎

EXAMPLE 3   Apply Your Skill

The perimeter of a rectangular lot is 100 meters, and its area is 616 square meters. Find the length and width of the lot.

Solution

We let l represent the length and w represent the width (see Figure 10.8).

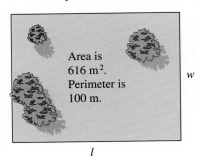

Figure 10.8

Then

$$\begin{pmatrix} lw = 616 \\ 2l + 2w = 100 \end{pmatrix} \begin{array}{l} \leftarrow \text{Area is 616 m}^2 \\ \leftarrow \text{Perimeter is 100 m} \end{array}$$

Multiplying the second equation by $\dfrac{1}{2}$ produces $l + w = 50$, which can be changed to $l = 50 - w$. Substituting $50 - w$ for l in the first equation produces the quadratic equation

$$(50 - w)(w) = 616$$
$$50w - w^2 = 616$$
$$w^2 - 50w = -616$$

Using the method of completing the square, we have

$$w^2 - 50w + 625 = -616 + 625$$
$$(w - 25)^2 = 9$$
$$w - 25 = \pm 3$$
$$w - 25 = 3 \quad \text{or} \quad w - 25 = -3$$
$$w = 28 \quad \text{or} \quad w = 22$$

If $w = 28$, then $l = 50 - w = 22$. If $w = 22$, then $l = 50 - w = 28$. The rectangle is 28 meters by 22 meters or 22 meters by 28 meters. ∎

EXAMPLE 4   Apply Your Skill

Find two numbers such that their sum is 2 and their product is -1.

Solution

We let n represent one of the numbers and m represent the other number.

$$\begin{pmatrix} n + m = 2 \\ nm = -1 \end{pmatrix} \begin{array}{l} \leftarrow \text{Their sum is 2} \\ \leftarrow \text{Their product is } -1 \end{array}$$

We can change the first equation to $m = 2 - n$; then we can substitute $2 - n$ for m in the second equation.

$$n(2 - n) = -1$$
$$2n - n^2 = -1$$
$$-n^2 + 2n + 1 = 0$$
$$n^2 - 2n - 1 = 0 \quad \text{Multiply both sides by } -1$$
$$n = \frac{-(-2) \pm \sqrt{(-2)^2 - 4(1)(-1)}}{2(1)}$$
$$n = \frac{2 \pm \sqrt{8}}{2} = \frac{2 \pm 2\sqrt{2}}{2} = 1 \pm \sqrt{2}$$

If $n = 1 + \sqrt{2}$, then $m = 2 - (1 + \sqrt{2})$
$$= 2 - 1 - \sqrt{2}$$
$$= 1 - \sqrt{2}$$

If $n = 1 - \sqrt{2}$, then $m = 2 - (1 - \sqrt{2})$
$$= 2 - 1 + \sqrt{2}$$
$$= 1 + \sqrt{2}$$

The numbers are $1 + \sqrt{2}$ and $1 - \sqrt{2}$. Perhaps you should check these numbers in the original statement of the problem! ∎

Finally, let's consider a uniform motion problem similar to those we solved in Chapter 7. Now we have the flexibility of using two equations in two variables.

EXAMPLE 5

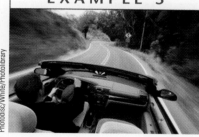

Apply Your Skill

Larry drove 156 miles in 1 hour more than it took Mike to drive 108 miles. Mike drove at an average rate of 2 miles per hour faster than Larry. How fast did each one travel?

Solution

We can represent the unknown rates and times like this:

Let r represent Larry's rate.

Let t represent Larry's time.

Then $r + 2$ represents Mike's rate, and $t - 1$ represents Mike's time.

Because *distance equals rate times time*, we can set up the following system:

$$\begin{pmatrix} rt = 156 \\ (r + 2)(t - 1) = 108 \end{pmatrix}$$

Solving the first equation for r produces $r = \dfrac{156}{t}$. Substituting $\dfrac{156}{t}$ for r in the second equation and simplifying, we obtain

$$\left(\frac{156}{t} + 2\right)(t - 1) = 108$$
$$156 - \frac{156}{t} + 2t - 2 = 108$$
$$2t - \frac{156}{t} + 154 = 108$$
$$2t - \frac{156}{t} + 46 = 0$$

$$2t^2 - 156 + 46t = 0 \quad \text{Multiply both sides by } t, t \neq 0$$
$$2t^2 + 46t - 156 = 0$$
$$t^2 + 23t - 78 = 0 \quad \text{Divide both sides by 2}$$

We can solve this quadratic equation by factoring.

$$(t + 26)(t - 3) = 0$$
$$t + 26 = 0 \quad \text{or} \quad t - 3 = 0$$
$$t = -26 \quad \text{or} \quad t = 3$$

We must disregard the negative solution. So Mike's time is $3 - 1 = 2$ hours. Larry's rate is $\frac{156}{3} = 52$ miles per hour, and Mike's rate is $52 + 2 = 54$ miles per hour. ■

CONCEPT QUIZ 10.5

For Problems 1–6, answer true or false.

1. When solving a word problem, a guideline helps to determine the equation for solving the problem.
2. When solving a word problem, you must choose x as the variable to represent the unknown quantities.
3. When reading a word problem, you should take note of any formulas that can be applied in solving the problem.
4. Some word problems could be solved using only one variable or solved using two variables and a system of equations.
5. When solving a word problem, a sketch, diagram, or chart is useless.
6. When solving a word problem, it is only necessary to check the answers in the equation determined to solve the problem.

Section 10.5 Classroom Problem Set

Objective 1

1. The length of a rectangular piece of sheet metal is 3 inches more than the width. The area of the sheet metal is 88 square inches. Find the length and width of the rectangle.

2. The length of a rectangular region is 4 centimeters greater than the width. The area of the region is 45 square centimeters. Find the length and width of the rectangle.

3. Find two consecutive whole numbers whose product is 342.

4. Each of three consecutive whole numbers is squared. The three results are added, and the sum is 245. Find the three whole numbers.

5. The perimeter of a rectangular garden is 84 feet, and its area is 360 square feet. Find the length and width of the garden.

6. The perimeter of a rectangle is 132 yards and its area is 1080 square yards. Find the length and width of the rectangle.

7. Find two numbers such that their sum is 10 and their product is 6.

8. Find two numbers such that their sum is 4 and their product is 1.

9. In an experimental solar car race, John's car went 90 miles in 1 hour less time than Mark's car went 80 miles. John's car went at an average rate of 10 miles per hour faster than Mark's car. How fast did each car travel?

10. By increasing the speed of a car by 10 miles per hour, it is possible to make a trip of 200 miles in 1 hour less time. What was the original speed for the trip?

THOUGHTS INTO WORDS

1. Return to Example 1 of this section and explain how the problem could be solved using one variable and one equation.

2. Write a page or two on the topic "using algebra to solve problems."

Answers to the Concept Quiz
1. True 2. False 3. True 4. True 5. False 6. False

Chapter 10 Review Problem Set

For Problems 1–4, solve each equation by factoring.

1. $x^2 + 2x - 24 = 0$
2. $x^2 - 13x + 40 = 0$
3. $x^2 + 5x = 14$
4. $2x^2 + 8x = 0$

For Problems 5–8, solve each equation by applying the square-root property.

5. $(x - 5)^2 = 64$
6. $(x + 4)^2 = 64$
7. $5x^2 = 80$
8. $(3x + 1)^2 = 40$

For Problems 9–12, solve each equation by completing the square.

9. $y^2 + 14y = 20$
10. $x^2 + 8x = 24$
11. $w^2 - 2w - 7 = 0$
12. $n^2 - 10n - 4 = 0$

For Problems 13–16, solve by using the quadratic formula.

13. $x^2 + 3x - 5 = 0$
14. $x^2 + 7x - 6 = 0$
15. $3y^2 - 2y - 4 = 0$
16. $2x^2 + 4x - 3 = 0$

For Problems 17–38, solve each quadratic equation.

17. $(2x + 7)^2 = 25$
18. $x^2 + 8x = -3$
19. $21x^2 - 13x + 2 = 0$
20. $x^2 = 17x$
21. $n - \dfrac{4}{n} = -3$
22. $n^2 - 26n + 165 = 0$
23. $3a^2 + 7a - 1 = 0$
24. $4x^2 - 4x + 1 = 0$
25. $5x^2 + 6x + 7 = 0$
26. $3x^2 + 18x + 15 = 0$
27. $3(x - 2)^2 - 2 = 4$
28. $x^2 + 4x - 14 = 0$
29. $y^2 = 45$
30. $x(x - 6) = 27$
31. $x^2 = x$
32. $n^2 - 4n - 3 = 6$
33. $n^2 - 44n + 480 = 0$
34. $\dfrac{x^2}{4} = x + 1$
35. $\dfrac{5x - 2}{3} = \dfrac{2}{x + 1}$
36. $\dfrac{-1}{3x - 1} = \dfrac{2x + 1}{-2}$
37. $\dfrac{5}{x - 3} + \dfrac{4}{x} = 6$
38. $\dfrac{1}{x + 2} - \dfrac{2}{x} = 3$

For Problems 39–48, set up an equation or a system of equations to help solve each problem.

39. Standing at a point 16 yards from the base of a tower, it is obvious that the distance to the top of the tower is 4 yards more than the height of the tower. Find the height of the tower.

40. The difference in the lengths of the two legs of a right triangle is 2 yards. If the length of the hypotenuse is $2\sqrt{13}$ yards, find the length of each leg.

41. The length of the hypotenuse of an isosceles right triangle is 12 inches. Find the length of each leg.

42. In a 30°–60° right triangle, the side opposite the 60° angle is 8 centimeters long. Find the length of the hypotenuse.

43. The perimeter of a rectangle is 42 inches, and its area is 108 square inches. Find the width and length of the rectangle.

44. Find two consecutive whole numbers whose product is 342.

45. Each of three consecutive odd whole numbers is squared. The three results are added and the sum is 251. Find the numbers.

46. The combined area of two squares is 50 square meters. Each side of the larger square is three times as long as a side of the smaller square. Find the lengths of the sides of each square.

47. A company has a rectangular parking lot 40 meters wide and 60 meters long. They plan to increase the area of the lot by 1100 square meters by adding a strip of equal width to one side and one end. Find the width of the strip to be added.

48. Jay traveled 225 miles in 2 hours less time than it took Jean to travel 336 miles. If Jay's rate was 3 miles per hour slower than Jean's rate, find each rate.

Chapter 10 Practice Test

1. The two legs of a right triangle are 4 inches and 6 inches long. Find the length of the hypotenuse. Express your answer in simplest radical form.

2. A diagonal of a rectangular plot of ground measures 14 meters. If the width of the rectangle is 5 meters, find the length to the nearest meter.

3. A diagonal of a square piece of paper measures 10 inches. Find, to the nearest inch, the length of a side of the square.

4. In a 30°–60° right triangle, the side opposite the 30° angle is 4 centimeters long. Find the length of the side opposite the 60° angle. Express your answer in simplest radical form.

For Problems 5–20, solve each equation.

5. $(3x + 2)^2 = 49$
6. $4x^2 = 64$
7. $8x^2 - 10x + 3 = 0$
8. $x^2 - 3x - 5 = 0$
9. $n^2 + 2n = 9$
10. $(2x - 1)^2 = -16$
11. $y^2 + 10y = 24$
12. $2x^2 - 3x - 4 = 0$
13. $\dfrac{x-2}{3} = \dfrac{4}{x+1}$
14. $\dfrac{2}{x-1} + \dfrac{1}{x} = \dfrac{5}{2}$
15. $n(n - 28) = -195$
16. $n + \dfrac{3}{n} = \dfrac{19}{4}$
17. $(2x + 1)(3x - 2) = -2$
18. $(7x + 2)^2 - 4 = 21$
19. $(4x - 1)^2 = 27$
20. $n^2 - 5n + 7 = 0$

For Problems 21–25, set up an equation or a system of equations to help solve each problem.

21. A room contains 120 seats. The number of seats per row is 1 less than twice the number of rows. Find the number of seats per row.

22. Abu rode his bicycle 56 miles in 2 hours less time than it took Stan to ride his bicycle 72 miles. If Abu's rate was 2 miles per hour faster than Stan's rate, find Abu's rate.

23. Find two consecutive odd whole numbers whose product is 255.

24. The combined area of two squares is 97 square feet. Each side of the larger square is 1 foot more than twice the length of a side of the smaller square. Find the length of a side of the larger square.

25. The perimeter of a rectangle is 56 inches and its area is 180 square inches. Find the length and width of the rectangle.

Solutions to Warm-Up Problems

CHAPTER 1

1. (a)

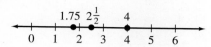

(b) The value of x is $\frac{1}{2}$.

2. (a) tens place (b) hundredths place
   (c) thousandths

3. (a) $12 + 7 = 19$
   (b) $8 + 14 + 6 = 22 + 6$
   $= 28$
   (c) $23 + 0 + 18 = 23 + 18$
   $= 41$

4. (a) $12 - 12 = 0$
   (b) $37 - 11 = 26$
   (c) $123 - 15 = 108$

5. (a) $(32)(4) = 128$
   (b) $(14)(5) = 70$
   (c) $(42)(0) = 0$ Any number multiplied by 0 is equal to 0.

6. (a) $18 \div 3 = 6$
   (b) $2 \div 3 = 0.\overline{6} \approx 0.67$ This is a repeating decimal number.
   (c) $9 \div 0$ is undefined.   Division by 0 is not allowed.

7. (a) $2 \cdot 2 \cdot 5 \cdot 7 = 4 \cdot 5 \cdot 7$
   $= 20 \cdot 7$
   $= 140$
   (b) $3 \cdot 5 \cdot 5 \cdot 11 = 15 \cdot 5 \cdot 11$
   $= 75 \cdot 11$
   $= 825$
   (c) $2 \cdot 2 \cdot 2 \cdot 3 \cdot 3 \cdot 3 = 4 \cdot 2 \cdot 3 \cdot 3 \cdot 3$
   $= 8 \cdot 3 \cdot 3 \cdot 3$
   $= 24 \cdot 3 \cdot 3$
   $= 72 \cdot 3$
   $= 216$

8. (a)
$$\begin{array}{r} 16 \\ 6\overline{)96} \\ \underline{6} \\ 36 \\ \underline{36} \\ 0 \end{array}$$

(b)
$$\begin{array}{r} 17 \\ 14\overline{)238} \\ \underline{14} \\ 98 \\ \underline{98} \\ 0 \end{array}$$

(c)
$$\begin{array}{r} 48.5 \\ 4\overline{)194} \\ \underline{16} \\ 34 \\ \underline{32} \\ 20 \\ \underline{20} \\ 0 \end{array}$$

9. (a) $3\frac{2}{3} = \frac{3 \times 3 + 2}{3} = \frac{11}{3}$
   (b) $6\frac{5}{8} = \frac{8 \times 6 + 5}{8} = \frac{53}{8}$
   (c) $9\frac{1}{2} = \frac{2 \times 9 + 1}{2} = \frac{19}{2}$

10. (a) $\frac{12}{20} = \frac{4 \cdot 3}{4 \cdot 5} = \frac{\cancel{4} \cdot 3}{\cancel{4} \cdot 5} = \frac{3}{5}$
    (b) $\frac{6}{16} = \frac{2 \cdot 3}{2 \cdot 8} = \frac{\cancel{2} \cdot 3}{\cancel{2} \cdot 8} = \frac{3}{8}$
    (c) $\frac{18}{36} = \frac{2 \cdot 9}{4 \cdot 9} = \frac{2 \cdot \cancel{9}}{4 \cdot \cancel{9}} = \frac{\cancel{2} \cdot 1}{\cancel{2} \cdot 2} = \frac{1}{2}$

CHAPTER 2

1. (a) i. 13,456 is thirteen thousand, four hundred fifty-six.

    ii. 4.091 is four and ninety-one thousandths.

    iii. $\dfrac{5}{4}$ is five-fourths.

    (b) i. 1,020,164 ii. $\dfrac{2}{3}$ iii. 83.205

2. (a) $0.57 - 0.29 = 0.28$

    (b) $1.73 + 0.097$ Must align decimal points: $\begin{array}{r}1.73\\+0.097\\\hline 1.827\end{array}$

    (c) $0.62 - 0.62 = 0$

3. (a) Change subtraction to the addition of the opposite. $-5 - 5 = -5 + (-5) = -10$

    (b) When signs are different, subtract and retain the sign of the number with the largest absolute value. $-15 + 7 = -|15 - 7| = -8$ The sign is negative because $|-15| = 15$, which is larger than $|7| = 7$.

    (c) Change subtraction to the addition of the opposite. $-6 - (-6) = -6 + 6 = 0$ Note that opposites, additive inverses, always sum to 0.

4. (a) $6 = 2 \cdot 3$
    $3 = 3$
    $8 = 2 \cdot 2 \cdot 2$
    LCD $= 2 \cdot 2 \cdot 2 \cdot 3 = 24$

    (b) $20 = 2 \cdot 2 \cdot 5$
    $12 = 2 \cdot 2 \cdot 3$
    LCD $= 2 \cdot 2 \cdot 3 \cdot 5 = 60$

    (c) $9 = 3 \cdot 3$
    $15 = 3 \cdot 5$
    $21 = 3 \cdot 7$
    LCD $= 3 \cdot 3 \cdot 5 \cdot 7 = 315$

5. (a) Find the LCD, make equivalent fractions, then add.
    $4 = 2 \cdot 2$
    $3 = 3$
    LCD $= 2 \cdot 2 \cdot 3 = 12$
    $\dfrac{1}{4} = \dfrac{1 \cdot 3}{4 \cdot 3} = \dfrac{3}{12}$
    $\dfrac{2}{3} = \dfrac{2 \cdot 4}{3 \cdot 4} = \dfrac{8}{12}$
    $\dfrac{3}{12} + \dfrac{8}{12} = \dfrac{3+8}{12} = \dfrac{11}{12}$

    (b) The denominators are the same, so add the numerators.
    $\dfrac{6}{11} + \dfrac{3}{11} = \dfrac{6+3}{11} = \dfrac{9}{11}$

    (c) $8 = 2 \cdot 2 \cdot 2$
    $14 = 2 \cdot 7$
    LCD $= 2 \cdot 2 \cdot 2 \cdot 7 = 56$
    $\dfrac{3}{8} = \dfrac{3 \cdot 7}{8 \cdot 7} = \dfrac{21}{56}$
    $\dfrac{5}{14} = \dfrac{5 \cdot 4}{14 \cdot 4} = \dfrac{20}{56}$
    $\dfrac{21}{56} + \dfrac{20}{56} = \dfrac{21+20}{56} = \dfrac{41}{56}$

6. The product of reciprocals is always 1.

    (a) $\left(\dfrac{6}{7}\right)\left(\dfrac{7}{6}\right) = \dfrac{6 \cdot 7}{7 \cdot 6} = \dfrac{42}{42} = 1$

    (b) $(13)\left(\dfrac{1}{13}\right) = \dfrac{13 \cdot 1}{13} = \dfrac{13}{13} = 1$

    (c) A negative number times a positive number results in a negative number.
    $\left(-\dfrac{3}{10}\right)\left(\dfrac{10}{3}\right) = -\dfrac{3 \cdot 10}{10 \cdot 3} = -\dfrac{30}{30} = -1$

7. (a) $18\left(\dfrac{2x}{3}\right) = \dfrac{18}{1} \cdot \dfrac{2x}{3} = \dfrac{3 \cdot 6 \cdot 2 \cdot x}{1 \cdot 3}$
    $= \dfrac{\cancel{3} \cdot 6 \cdot 2 \cdot x}{1 \cdot \cancel{3}} = \dfrac{12x}{1} = 12x$

    (b) $\dfrac{3}{2}(24y) = \dfrac{3}{2}\left(\dfrac{24y}{1}\right) = \dfrac{3 \cdot 2 \cdot 12 \cdot y}{2 \cdot 1}$
    $= \dfrac{3 \cdot \cancel{2} \cdot 12 \cdot y}{\cancel{2} \cdot 1} = \dfrac{3 \cdot 12 \cdot y}{1} = 36y$

    (c) Use the distributive property.
    $9\left(\dfrac{5n}{6} - \dfrac{2n}{3}\right) = 9\left(\dfrac{5n}{6}\right) - 9\left(\dfrac{2n}{3}\right)$
    $= \dfrac{\cancel{9}^3}{1}\left(\dfrac{5n}{\cancel{6}_2}\right) - \dfrac{\cancel{9}^3}{1}\left(\dfrac{2n}{\cancel{3}_1}\right) = \dfrac{15n}{2} - 6n$
    $= \dfrac{15n}{2} - \dfrac{12n}{2} = \dfrac{15n - 12n}{2} = \dfrac{3n}{2} = \dfrac{3}{2}n$

8. (a) $5(2x + 7) = 5(2x) + 5(7) = 10x + 35$

    (b) $-4(7x + 10) = -4(7x) - 4(10) = -28x - 40$

    (c) $-1(3n - 2) = -1(3n) - 1(-2) = -3n + 2$

9. Combine like terms.

    (a) $3x + 2y - x = 3x + 2y + (-x)$
    $= 3x + (-x) + 2y = 2x + 2y$

    (b) $16m - 9m = (16 - 9)m = 7m$

    (c) Apply the distributive property, then add like terms.
    $2(x + 7) + 3x = 2x + 14 + 3x = 2x + 3x + 14$
    $= (2 + 3)x + 14 = 5x + 14$

10. To find an average, add the terms and then divide by the number of terms.

(a) $\dfrac{8 + 21 + 16}{3} = \dfrac{45}{3} = 15$ Divided by 3 since there are three terms.

(b) $\dfrac{92 + 83 + 79 + 94}{4} = \dfrac{348}{4} = 87$ Divided by 4 since there are four terms.

(c) $\dfrac{138 + 174 + x}{3} = \dfrac{312 + x}{3}$

CHAPTER 3

1. (a) Divide 5 by 40. $\dfrac{5}{40} = 0.125$

   (b) Divide 400 by 0.25. Move the decimal point 2 places to the right in the numerator and denominator, then divide.
   $\dfrac{400}{0.25} = \dfrac{40000}{25} = 1600$

   (c) Multiply 0.36 and 50 by multiplying 36 and 50. Then move the decimal point 2 places to the left. $(0.36)(50) = 18$

2. (a) Multiply 3 times 12, count the number of places to the right of the decimal point in both numbers, then move the decimal point 2 places to the left. $0.3(1.2) = 0.36$

   (b) $0.27\left(\dfrac{1}{9}\right) = \dfrac{0.27}{9} = 0.03$

   (c) Change the divisor to an integer by multiplying the numerator and denominator by 10.
   $\dfrac{4.4}{-0.2} = \dfrac{4.4}{-0.2} \cdot \dfrac{10}{10} = \dfrac{44}{-2} = -22$

3. Follow the order of operations.
   (a) $15 + 0.4(120) = 15 + 48 = 63$
   (b) $256 + 0.25(256) = 256 + 64 = 320$
   (c) $0.15(2.3) - 0.78 = 0.345 - 0.78 = -0.435$

4. When multiplying by a power of 10, move the decimal point to the right.
   (a) $10(0.1n + 2) = 1n + 20$ Moved the decimal 1 place to the right.
   (b) $100(0.75x - 0.5y) = 75x - 50y$ Moved the decimal 2 places to the right.
   (c) $0.04(2000) + 0.025(1500) = 80 + 37.5 = 117.5$

5. Follow the order of operations.
   (a) $\dfrac{1}{2}(12)(9) = 6(9) = 54$
   (b) $2(7)^2 + 2(7)(5) = 2 \cdot 49 + 2(7)(5)$
   $= 98 + 2(7)(5)$
   $= 98 + 14(5)$
   $= 98 + 70 = 168$

   (c) Use the commutative property for multiplication.
   $\dfrac{1}{2}(15)(8 + 12) = \dfrac{1}{2}(15)(20)$
   $= \dfrac{1}{2}(20)(15)$
   $= 10(15) = 150$

6. Use the distributive property.
   (a) $-3(4x - 9) = (-3)(4x) + (-3)(-9)$
   $= -12x + 27$
   (b) $6\left(\dfrac{1}{2}n - 5\right) = 6\left(\dfrac{1}{2}n\right) + 6(-5) = 3n - 30$
   (c) $\dfrac{2}{3}(63 - p) = \dfrac{2}{3}(63) - \dfrac{2}{3}(p)$
   $= 2(21) - \dfrac{2}{3}p = 42 - \dfrac{2}{3}p$

7. Combine like terms.
   (a) $x + \dfrac{3}{4}x - 2 = \left(1 + \dfrac{3}{4}\right)x - 2 = \left(\dfrac{4}{4} + \dfrac{3}{4}\right)x - 2$
   $= \dfrac{7}{4}x - 2$

   (b) Use the distributive property. Reduce the fraction to lowest terms.
   $\dfrac{3}{8}n + \dfrac{5}{8}(n + 2) = \dfrac{3}{8}n + \dfrac{5}{8}n + \dfrac{5}{8}(2)$
   $= \left(\dfrac{3}{8} + \dfrac{5}{8}\right)n + \dfrac{10}{8}$
   $= \dfrac{8}{8}x + \dfrac{10}{8} = x + \dfrac{5}{4}$

   (c) $x + (x + 2) + (x + 4)$
   $= x + x + x + 2 + 4 = 3x + 6$

8. Remember, $a = 1 \cdot a$.
   (a) $x - 0.7x = 1x - 0.7x = (1 - 0.7)x = 0.3x$
   (b) $n + 0.6n = 1n + 0.6n = (1 + 0.6)n = 1.6n$
   (c) $0.03(400 + p) = 0.03(400) + 0.03p = 12 + 0.03p$

9. (a) $3\dfrac{2}{3} = \dfrac{3 \times 3 + 2}{3} = \dfrac{11}{3}$

   (b) $6\dfrac{5}{8} = \dfrac{8 \times 6 + 5}{8} = \dfrac{53}{8}$

   (c) $9\dfrac{1}{2} = \dfrac{2 \times 9 + 1}{2} = \dfrac{19}{2}$

10. Reduce to lowest terms.

    (a) $\dfrac{12}{20} = \dfrac{4 \cdot 3}{4 \cdot 5} = \dfrac{\cancel{4} \cdot 3}{\cancel{4} \cdot 5} = \dfrac{3}{5}$

    (b) $\dfrac{6}{16} = \dfrac{2 \cdot 3}{2 \cdot 8} = \dfrac{\cancel{2} \cdot 3}{\cancel{2} \cdot 8} = \dfrac{3}{8}$

    (c) $\dfrac{18}{36} = \dfrac{2 \cdot 9}{4 \cdot 9} = \dfrac{2 \cdot \cancel{9}}{4 \cdot \cancel{9}} = \dfrac{\cancel{2} \cdot 1}{\cancel{2} \cdot 2} = \dfrac{1}{2}$

CHAPTER 4

1. (a) $6 - (-7) = 6 + 7 = 13$
   (b) $-4 - 3 = -4 + (-3) = -7$
   (c) $-1 - (-5) = -1 + 5 = 4$

2. (a) $\dfrac{60}{4} = 15$ (b) $\dfrac{48}{-6} = -8$

   (c) $\dfrac{-24}{-4} = 6$

3. (a) $\left(\dfrac{3}{4}\right)\left(\dfrac{3}{4}\right)\left(\dfrac{3}{4}\right) = \dfrac{3 \cdot 3 \cdot 3}{4 \cdot 4 \cdot 4} = \dfrac{27}{64}$

   (b) $(0.4)^2 = (0.4)(0.4) = 0.16$

   (c) $\left(-\dfrac{2}{5}\right)^3 = \left(-\dfrac{2}{5}\right)\left(-\dfrac{2}{5}\right)\left(-\dfrac{2}{5}\right)$
   $= -\dfrac{2 \cdot 2 \cdot 2}{5 \cdot 5 \cdot 5} = -\dfrac{8}{125}$

4. (a) $\dfrac{1.21}{1.1} = 1.1$

   (b) $\dfrac{-5.6}{0.28} = -20$

   (c) $6(3.5) = 21$

5. (a) $12\overline{)108}$ with quotient 9, 108, remainder 0

   (b) $9\overline{)39}$ with quotient 4, 36, remainder 3; $4\dfrac{3}{9} = 4\dfrac{1}{3}$

   (c) $16\overline{)456}$ with quotient 28, 32, 136, 128, remainder 8; $16\dfrac{8}{16} = 16\dfrac{1}{2}$

6. (a) $3(4x^2 + 6x - 1) = 3(4x^2) + 3(6x) + 3(-1)$
   $= 12x^2 + 18x - 3$
   (b) $-(y^3 - 2) = -y^3 + 2$
   (c) $-(-3n^2 + 8n - 2) = 3n^2 - 8n + 2$

7. (a) $2x + 3 - 7x + 4 = 2x + (-7x) + 3 + 4$
   $= -5x + 7$
   (b) $8x^2 - 3x^2 + 4y + y = (8 - 3)x^2 + (4 + 1)y$
   $= 5x^2 + 5y$
   (c) $6m^2 - (-3m^2) = 6m^2 + 3m^2$
   $= (6 + 3)m^2 = 9m^2$

8. (a) $x^2 + 6x + 4x + 24 = x^2 + (6 + 4)x + 24$
   $= x^2 + 10x + 24$
   (b) $m^2 - 3m + 5m - 15 = m^2 + (-3 + 5)m - 15$
   $= m^2 + 2m - 15$
   (c) $12n^2 - 18n - 8n + 12$
   $= 12n^2 + (-18 + (-8))n + 12$
   $= 12n^2 - 26n + 12$
   (d) $x^2 - 4x - 4x + 16 = x^2 + [-4 + (-4)]x + 16$
   $= x^2 - 8x + 16$

9. $A = \text{length} \cdot \text{width}$
   (a) $A = 17 \cdot 12 = 204$
   (b) $A = x \cdot 15 = 15x$

CHAPTER 5

1. (a) $8 = 2 \cdot 2 \cdot 2$
   $18 = 2 \cdot 3 \cdot 3$
   GCF $= 2$
   (b) $36 = 2 \cdot 2 \cdot 3 \cdot 3$
   $42 = 2 \cdot 3 \cdot 7$
   GCF $= 2 \cdot 3 = 6$
   (c) $16 = 2 \cdot 2 \cdot 2 \cdot 2$
   $24 = 2 \cdot 2 \cdot 2 \cdot 3$
   $60 = 2 \cdot 2 \cdot 3 \cdot 5$
   GCF $= 2 \cdot 2 = 4$

2. (a) $3(8) = 24; 3 + 8 = 11$
   (b) $-11(6) = -66; -11 + 6 = -5$
   (c) $(-2)(-18) = 36; -2 + (-18) = -20$

3. (a) $(-2)^2 - 7(-2) = 4 + 14 = 18$
   (b) $(-10)(-10 + 8) = (-10)(-2) = 20$
   (c) $(12 + 3)(12 - 12) = 15(0) = 0$

4. (a) $x^2 + (x + 2)^2 = x^2 + x^2 + 4x + 4$
   $= 2x^2 + 4x + 4$
   (b) $5m(2n - 1) - 3m(2n - 1)$
   $= 5m(2n) + (5m)(-1) - 3m(2n) - 3m(-1)$
   $= 10mn - 5m - 6mn + 3m$
   $= 10mn - 6mn - 5m + 3m = 4mn - 2m$
   (c) $-5xy(x^2 - 1) = -5xy(x^2) - 5xy(-1)$
   $= -5x^3y + 5xy$

5. (a) $x^2 + 7x - 8x - 56 = x^2 - x - 56$
   (b) $m^2 - 9m + 9m - 81 = m^2 - 81$ This is the difference of two squares.
   (c) $4y^2 + 14y + 14y + 49 = 4y^2 + 28y + 49$ This is a perfect square trinomial.

6. (a) $(6x - 7)(3x + 1) = (6x)(3x) + 6x(1)$
   $+ (-7)(3x) + (-7)(1)$
   $= 18x^2 + 6x - 21x - 7$
   $= 18x^2 - 15x - 7$
   (b) $(2m - 3)^2 = (2m - 3)(2m - 3)$
   $= 4m^2 - 6m - 6m + 9$
   $= 4m^2 - 12m + 9$
   (c) $(x - 4y)(x + 4y) = x^2 + 4xy - 4xy - 16y^2$
   $= x^2 - 16y^2$

7. (a) $m - 8 = 0$
   $m - 8 + 8 = 0 + 8$
   $m = 8$
   (b) $4x - 12 = 0$
   $4x - 12 + 12 = 0 + 12$
   $4x = 12$
   $\left(\dfrac{1}{4}\right)4x = \left(\dfrac{1}{4}\right)12$
   $x = 3$
   (c) $3 - 6n = 0$
   $3 - 3 - 6n = 0 - 3$
   $-6n = -3$
   $\left(-\dfrac{1}{6}\right)(-6n) = \left(-\dfrac{1}{6}\right)(-3)$
   $n = \dfrac{1}{2}$

8. The area of a triangle is $A = \dfrac{1}{2}bh$.
   (a) $A = \dfrac{1}{2}(16)(5) = 8(5) = 40$
   (b) $A = \dfrac{1}{2}(3x)(9) = \dfrac{27}{2}x$
   (c) $A = \dfrac{1}{2}m(m + 7) = \dfrac{m(m + 7)}{2}$

CHAPTER 6

1. (a) $\dfrac{5}{9} + \dfrac{6}{9} = \dfrac{5 + 6}{9} = \dfrac{11}{9}$
   (b) $\dfrac{3x}{11} - \left(-\dfrac{8x}{11}\right) = \dfrac{3x - (-8x)}{11}$
   $= \dfrac{3x + 8x}{11} = \dfrac{11x}{11} = 1x = x$
   (c) $\dfrac{9}{14} - \dfrac{8}{21} = \dfrac{9}{14}\left(\dfrac{3}{3}\right) - \dfrac{8}{21}\left(\dfrac{2}{2}\right) = \dfrac{27}{42} - \dfrac{16}{42} = \dfrac{11}{42}$

2. (a) $32 = 2 \cdot 2 \cdot 2 \cdot 2 \cdot 2$
   $18 = 2 \cdot 3 \cdot 3$
   LCD $= 2 \cdot 2 \cdot 2 \cdot 2 \cdot 2 \cdot 3 \cdot 3 = 2^5 3^2 = 288$
   (b) $6 = 2 \cdot 3$
   $21 = 3 \cdot 7$
   LCD $= 2 \cdot 3 \cdot 7 = 42$
   (c) $4x^2y = 2 \cdot 2 \cdot x \cdot x \cdot y$
   $6xy = 2 \cdot 3 \cdot x \cdot y$
   LCD $= 2 \cdot 2 \cdot 3 \cdot x \cdot x \cdot y = 12x^2y$

3. (a) $\dfrac{8x}{26x} = \dfrac{2 \cdot 4 \cdot x}{2 \cdot 13 \cdot x} = \dfrac{\cancel{2} \cdot 4 \cdot \cancel{x}}{\cancel{2} \cdot 13 \cdot \cancel{x}} = \dfrac{4}{13}$
   (b) $\dfrac{-56m}{77mn} = -\dfrac{7 \cdot 8 \cdot m}{7 \cdot 11 \cdot m \cdot n}$
   $= -\dfrac{\cancel{7} \cdot 8 \cdot \cancel{m}}{\cancel{7} \cdot 11 \cdot \cancel{m} \cdot n} = -\dfrac{8}{11n}$

(c) $\dfrac{-22x^3y^4}{-46xy^5} = \dfrac{2 \cdot 11 \cdot x^3y^4}{2 \cdot 23 \cdot xy^5}$

$= \dfrac{2 \cdot 11 \cdot \cancel{x^3}x^2 \cancel{y^4}}{\cancel{2} \cdot 23 \cdot \cancel{x}\cancel{y^5}y}$

4. (a) $\dfrac{3}{4} \div \dfrac{5}{12} = \dfrac{3}{4} \cdot \dfrac{12}{5} = \dfrac{3 \cdot 12}{4 \cdot 5} = \dfrac{3 \cdot \cancel{4} \cdot 3}{\cancel{4} \cdot 5} = \dfrac{9}{5}$

(b) $\dfrac{4}{x} \div \dfrac{6}{x^2} = \dfrac{4}{x} \cdot \dfrac{x^2}{6} = \dfrac{2 \cdot 2 \cdot \cancel{x} \cdot x}{2 \cdot 3 \cdot \cancel{x}} = \dfrac{2x}{3} = \dfrac{2}{3}x$

(c) $\dfrac{-14x^2y}{21xy} = -\dfrac{2 \cdot 7 \cdot \cancel{x} \cdot x \cdot \cancel{y}}{3 \cdot 7 \cdot \cancel{x} \cdot \cancel{y}} = -\dfrac{2x}{3} = -\dfrac{2}{3}x$

5. (a) $20\left(\dfrac{1}{4}x + \dfrac{3}{5}\right) = 20\left(\dfrac{1}{4}x\right) + 20\left(\dfrac{3}{5}\right)$

$= 5x + 4(3) = 5x + 12$

(b) $18\left(\dfrac{3}{2} - \dfrac{1}{3}\right) = 18\left(\dfrac{3}{2}\right) + 18\left(-\dfrac{1}{3}\right)$

$= \overset{9}{\cancel{18}}\left(\dfrac{3}{\cancel{2}}\right) + \overset{6}{\cancel{18}}\left(-\dfrac{1}{\cancel{3}}\right)$

$= 27 - 6 = 21$

(c) $4x\left(\dfrac{1}{2x} - \dfrac{9}{4x}\right) = 4x\left(\dfrac{1}{2x}\right) + 4x\left(-\dfrac{9}{4x}\right)$

$= \overset{2}{\cancel{4x}}\left(\dfrac{1}{\cancel{2x}}\right) + \cancel{4x}\left(-\dfrac{9}{\cancel{4x}}\right)$

$= 2 - 9 = -7$

6. (a) $5x + 3 - 4x + 1 = 5x - 4x + 3 + 1 = x + 4$

(b) $(6x + 1) - (2x + 3) = 6x + 1 - 2x - 3$

$= 6x - 2x + 1 - 3$

$= 4x - 2$

(c) $\dfrac{3(m - 1) - (m - 1)}{2}$

$= \dfrac{3m - 3 - m + 1}{2} = \dfrac{2m - 2}{2}$

$= \dfrac{2(m - 1)}{2} = m - 1$

7. (a) $x^2 + 7x - 30 = x^2 + (10 - 3)x + (10)(-3)$

$= x^2 + 10x - 3x - 30$

$= x(x + 10) - 3(x + 10)$

$= (x + 10)(x - 3)$

(b) $12x^2 - 5x - 3 = 12x^2 + 4x - 9x - 3$

$= 4x(3x + 1) - 3(3x + 1)$

$= (3x + 1)(4x - 3)$

(c) $12x^2y + 8xy^2 = 4xy(3x) + 4xy(2y)$

$= 4xy(3x + 2y)$

8. (a) The reciprocal of $\dfrac{57}{20}$ is $\dfrac{20}{57}$. Note that $\dfrac{57}{20} \cdot \dfrac{20}{57}$

$= \dfrac{57 \cdot 20}{20 \cdot 57} = 1.$

(b) Solve the distance formula, $d = rt$, for t: $t = \dfrac{d}{r}$.

Substitute 260 for d and 65 for r.

$t = \dfrac{260}{65} = 4$ hours

(c) Solve the distance formula, $d = rt$, for r: $r = \dfrac{d}{t}$.

Substitute 570 for d and x for t.

$r = \dfrac{570}{x}$ miles per hour

CHAPTER 7

1.

number line showing points at -2, 0, 1, 2 on a scale from -3 to 4

2. (a) if $x = 0$: $\dfrac{1}{2}(0) - 5 = -5$

if $x = -2$: $\dfrac{1}{2}(-2) - 5 = -1 - 5 = -6$

if $x = 6$: $\dfrac{1}{2}(6) - 5 = 3 - 5 = -2$

(b) if $x = -5$: $-5 + 12 = 7$

if $x = 1$: $1 + 12 = 13$

if $x = \dfrac{5}{2}$: $\dfrac{5}{2} + 12 = \dfrac{5}{2} + \dfrac{24}{2} = \dfrac{29}{2}$

(c) if $x = -2$: $-3(-2) + 5 = 6 + 5 = 11$

if $x = 0$: $-3(0) + 5 = 0 + 5 = 5$

if $x = 6$: $-3(6) + 5 = -18 + 5 = -13$

3. (a) $3(0) - 2y = -6$

$0 - 2y = -6$

$\left(-\dfrac{1}{2}\right)(-2y) = \left(-\dfrac{1}{2}\right)(-6)$

$y = 3$

(b) $\dfrac{1}{2}x + 7(0) = -3$

$\dfrac{1}{2}x + 0 = -3$

$$2 \cdot \frac{1}{2}x = -3 \cdot 2$$
$$x = -6$$

(c) $\quad 4x - 3y = 15 \quad \text{if } x = 0$
$$4(0) - 3y = 15$$
$$-3y = 15$$
$$\left(-\frac{1}{3}\right)(-3y) = \left(-\frac{1}{3}\right)15$$
$$y = -5$$

4. (a) $\quad \frac{3}{5}x - 7 < 2$
$$\frac{3}{5}x - 7 + 7 < 2 + 7$$
$$\frac{3}{5}x < 9$$
$$\left(\frac{5}{3}\right)\left(\frac{3}{5}x\right) < \left(\frac{5}{3}\right)9$$
$$x < \frac{45}{3} = 15$$

(b) $\quad 6x - 1 > 2x - 3$
$$6x - 1 + 1 > 2x - 3 + 1$$
$$6x > 2x - 2$$
$$6x - 2x > 2x - 2x - 2$$
$$4x > -2$$
$$\left(\frac{1}{4}\right)4x > \left(\frac{1}{4}\right)(-2)$$
$$x > -\frac{1}{2}$$

(c) $3x - 2(4x + 5) \leq 0$
$$3x - 8x - 10 \leq 0$$
$$-5x - 10 \leq 0$$
$$-5x - 10 + 10 \leq 0 + 10$$
$$-5x \leq 10$$
$$\left(-\frac{1}{5}\right)(-5x) \geq \left(-\frac{1}{5}\right)10 \quad \text{Reverse the inequality sign.}$$
$$x \geq -2$$

5. (a) $\quad \dfrac{7-3}{3-1} = \dfrac{4}{2} = \dfrac{2}{1} = 2$

(b) $\quad \dfrac{2-(-1)}{0-3} = \dfrac{3}{-3} = -\dfrac{1}{1} = -1$

(c) $\quad \dfrac{-8-3}{5-(-4)} = \dfrac{-11}{9} = -\dfrac{11}{9}$

6. (a) $\quad \dfrac{30}{45} = \dfrac{15}{x}$
$$30x = 45 \cdot 15$$
$$30x = 675$$
$$x = \frac{675}{30} = 22.5$$

(b) $\quad \dfrac{12}{y} = \dfrac{3}{4}$
$$12 \cdot 4 = 3y$$
$$48 = 3y$$
$$16 = y$$

(c) $\quad \dfrac{4}{100} = \dfrac{m}{650}$
$$100m = 4 \cdot 650$$
$$100m = 2600$$
$$m = \frac{2600}{100} = 26$$

7. (a) $\quad 4x - 6y = 3$
$$4x - 6y + 6y = 3 + 6y$$
$$4x = 3 + 6y$$
$$\frac{4x}{4} = \frac{3 + 6y}{4}$$
$$x = \frac{3}{4} + \frac{6}{4}y$$
$$x = \frac{3}{4} + \frac{3}{2}y$$

(b) $\quad 5x - y = 6$
$$5x - 5x - y = 6 - 5x$$
$$-y = 6 - 5x$$
$$y = -6 + 5x$$
$$y = 5x - 6$$

(c) $\quad y - 3 = 2(x - 3)$
$$y - 3 = 2x - 6$$
$$y - 3 + 3 = 2x - 6 + 3$$
$$y = 2x - 3$$

8. (a) $\quad \dfrac{8+2}{3-x} = 5$
$$8 + 2 = 5(3 - x)$$
$$10 = 15 - 5x$$
$$10 - 15 = 15 - 15 - 5x$$
$$-5 = -5x$$
$$\frac{-5}{-5} = \frac{-5x}{-5}$$
$$1 = x$$

(b)
$$\frac{-7-x}{2-6} = 4$$
$$\frac{-7-x}{-4} = 4$$
$$-7-x = 4(-4)$$
$$-7-x = -16$$
$$-7+7-x = -16+7$$
$$-x = -9$$
$$x = 9$$

(c)
$$\frac{12+x}{7+9} = \frac{9}{8}$$
$$\frac{12+x}{16} = \frac{9}{8}$$
$$8(12+x) = 9 \cdot 16$$
$$96+8x = 144$$
$$96-96+8x = 144-96$$
$$8x = 48$$
$$x = \frac{48}{8} = 6$$

CHAPTER 8

1. (a) $(0, 5)$
$$-3x + 2y = 10$$
$$-3(0) + 2(5) = 10$$
$$0 + 10 = 10$$
$$10 = 10$$
Solution

(b) $(6, 4)$
$$-3x + 2y = 10$$
$$-3(6) + 2(4) = 10$$
$$-18 + 8 = 10$$
$$-10 = 10$$
Not a solution

(c) $(-4, 11)$
$$-3x + 2y = 10$$
$$-3(-4) + 2(11) = 10$$
$$12 + 22 = 10$$
$$32 = 10$$
Not a solution

2. (a) $y = 2x - 3$

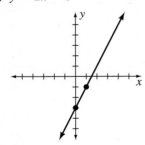

(b) $3x - y = 2$
$$-y = -3x + 2$$
$$y = 3x - 2$$

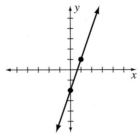

(c) $4x + 2y = 1$
$$2y = -4x + 1$$
$$y = -\frac{4}{2}x + \frac{1}{2}$$
$$y = -2x + \frac{1}{2}$$

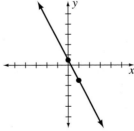

3. (a) Let $x = 0$
$$x + y = 7$$
$$0 + y = 7$$
$$y = 7$$
y intercept is $(0, 7)$.

Let $y = 0$
$$x + y = 7$$
$$x + 0 = 7$$
$$x = 7$$
x intercept is $(7, 0)$.

(b) Let $x = 0$

$$3x - \frac{1}{2}y = 18$$

$$3(0) - \frac{1}{2}y = 18$$

$$-\frac{1}{2}y = 18$$

$$-2\left(-\frac{1}{2}y\right) = (-2)18$$

$$y = -36$$

y intercept is $(0, -36)$.

Let $y = 0$

$$3x - \frac{1}{2}y = 18$$

$$3x - \frac{1}{2}(0) = 18$$

$$3x = 18$$

$$\frac{3x}{3} = \frac{18}{3}$$

$$x = 6$$

x intercept is $(6, 0)$.

(c) Let $x = 0$

$$y = -\frac{3}{4}x - \frac{1}{6}$$

$$y = -\frac{3}{4}(0) - \frac{1}{6}$$

$$y = -\frac{1}{6}$$

y intercept is $\left(0, -\frac{1}{6}\right)$.

Let $y = 0$

$$y = -\frac{3}{4}x - \frac{1}{6}$$

$$0 = -\frac{3}{4}x - \frac{1}{6}$$

$$\frac{1}{6} = -\frac{3}{4}x$$

$$\left(-\frac{4}{3}\right)\left(\frac{1}{6}\right) = \left(-\frac{4}{3}\right)\left(-\frac{3}{4}x\right)$$

$$-\frac{2}{9} = x$$

x intercept is $\left(-\frac{2}{9}, 0\right)$.

4. (a) $x + y = 18$
(b) $x - y = 5$
(c) $3x - 6 = 24$

5. (a)
$$3(4x - 2y) = 3(5)$$
$$3(4x) + 3(-2y) = 15$$
$$12x - 6y = 15$$

(b)
$$(-1)(2x + y) = (-1)(12)$$
$$(-1)(2x) + (-1)(y) = -12$$
$$-2x - y = -12$$

(c)
$$\left(\frac{1}{4}\right)(12x) - \left(\frac{1}{4}\right)(4y) = \left(\frac{1}{4}\right)24$$
$$3x - y = 6$$

6. (a)
$$5 + y + (-4) = 1$$
$$1 + y = 1$$
$$1 - 1 + y = 1 - 1$$
$$y = 0$$

(b)
$$3y - 5 = 10$$
$$3y - 5 + 5 = 10 + 5$$
$$3y = 15$$
$$\frac{3y}{3} = \frac{15}{3}$$
$$y = 5$$

(c)
$$-2x + (-7) = 5$$
$$-2x + (-7) + 7 = 5 + 7$$
$$-2x = 12$$
$$\frac{-2x}{-2} = \frac{12}{-2}$$
$$x = -6$$

7. (a)
$$-3x + 2(x + 1) = 17$$
$$-3x + 2x + 2 = 17$$
$$-x + 2 = 17$$
$$-x + 2 - 2 = 17 - 2$$
$$-x = 15$$
$$x = -15$$

(b)
$$6x - 4(-1 - x) = -36$$
$$6x + 4 + 4x = -36$$
$$10x + 4 = -36$$
$$10x + 4 - 4 = -36 - 4$$
$$10x = -40$$
$$\frac{10x}{10} = \frac{-40}{10}$$
$$x = -4$$

(c) $3\left[2\left(\dfrac{9-6x}{3}\right)+4x\right]=3(6)$

$3\left[2\left(\dfrac{9-6x}{3}\right)\right]+3(4x)=3(6)$

$2(9-6x)+12x=18$

$18-12x+12x=18$

$18=18$

Infinitely many solutions

8. $10(0.2)+x(0)=0.15(x+10)$

$2=0.15x+1.5$

$2-1.5=0.15x+1.5-1.5$

$0.5=0.15x$

$\dfrac{0.5}{0.15}=\dfrac{0.15x}{0.15}$

$\dfrac{10}{3}=x$

$3\dfrac{1}{3}$ gallons of water

CHAPTER 9

1. (a) $15^2=15\cdot15=225$
   (b) $\left(\dfrac{3}{7}\right)^2=\left(\dfrac{3}{7}\right)\left(\dfrac{3}{7}\right)=\dfrac{3\cdot3}{7\cdot7}=\dfrac{9}{49}$
   (c) $(0.6)^2=(0.6)(0.6)=0.36$

2. (a) $(-3)^3=(-3)(-3)(-3)=-27$
   (b) $\left(\dfrac{1}{5}\right)^3=\left(\dfrac{1}{5}\right)\left(\dfrac{1}{5}\right)\left(\dfrac{1}{5}\right)=\dfrac{1\cdot1\cdot1}{5\cdot5\cdot5}=\dfrac{1}{125}$
   (c) $-\left(-\dfrac{2}{3}\right)^3=-\left(-\dfrac{2}{3}\right)\left(-\dfrac{2}{3}\right)\left(-\dfrac{2}{3}\right)$
   $=-\left(\dfrac{-2\cdot-2\cdot-2}{3\cdot3\cdot3}\right)=-\left(-\dfrac{8}{27}\right)=\dfrac{8}{27}$

3. (a) $13x^2-6x^2=(13-6)x^2=7x^2$
   (b) $4m-6m+5m=(4-6+5)m=3m$
   (c) $10x^2-3y^2$ Already simplified.

4. (a) $48=2\cdot2\cdot2\cdot2\cdot3=2^4\cdot3$
   (b) $27=3\cdot3\cdot3=3^3$
   (c) $75=3\cdot25=3\cdot5^2$

5. (a) $\dfrac{6}{13}\cdot\dfrac{13}{13}=\dfrac{6\cdot13}{13\cdot13}=\dfrac{78}{169}$
   (b) $\dfrac{2}{5}\cdot\dfrac{5}{5}=\dfrac{2\cdot5}{5\cdot5}=\dfrac{10}{25}$
   (c) $\dfrac{3}{2}\cdot\dfrac{2^2}{2^2}=\dfrac{3\cdot2^2}{2\cdot2^2}=\dfrac{3\cdot4}{2^3}=\dfrac{12}{8}$

6. (a) $(x^2)^3\cdot x=x^{2\cdot3}\cdot x=x^6\cdot x=x^{6+1}=x^7$
   (b) $x^2y^4xy^3=x^2\cdot x\cdot y^4\cdot y^3=x^{2+1}y^{4+3}=x^3y^7$
   (c) $m^3n^9m^2n=m^3\cdot m^2\cdot n^9\cdot n=m^{3+2}n^{9+1}=m^5n^{10}$

7. (a) $(x-5)(x+3)=x^2+3x-5x-15$
   $=x^2-2x-15$
   (b) $(m-11)(m+11)=m^2-11^2=m^2-121$
   (c) $(3x-2)(5x+1)=15x^2+3x-10x-2$
   $=15x^2-7x-2$

8. (a) $3x-2=7$
   $3x-2+2=7+2$
   $3x=9$
   $\dfrac{3x}{3}=\dfrac{9}{3}$
   $x=3$

   (b) $7m+3=4m+2$
   $7m-4m+3=4m-4m+2$
   $3m+3=2$
   $3m+3-3=2-3$
   $3m=-1$
   $\dfrac{3m}{3}=\dfrac{-1}{3}$
   $m=-\dfrac{1}{3}$

   (c) $x^2+2x-15=0$
   $(x+5)(x-3)=0$
   $x+5=0$ or $x-3=0$
   $x=-5$ or $x=3$

CHAPTER 10

1. (a) $x^2 - 6x = 0$

$x(x - 6) = 0$

$x = 0$ or $x - 6 = 0$

$x = 0$ or $x = 6$

(b) $9m^2 - 18m + 1 = 0$

$(9m - 1)(9m - 1) = 0$

$9m - 1 = 0$ or $9m - 1 = 0$

$9m = 1$ or $9m = 1$

$\dfrac{9m}{9} = \dfrac{1}{9}$ or $\dfrac{9m}{9} = \dfrac{1}{9}$

$m = \dfrac{1}{9}$ or $m = \dfrac{1}{9}$

(c) $n^2 - 64 = 0$

$(n - 8)(n + 8) = 0$

$n - 8 = 0$ or $n + 8 = 0$

$n = 8$ or $n = -8$

2. (a) $\sqrt{169} = 13$ since $13^2 = 169$

(b) $\sqrt{\dfrac{25}{4}} = \dfrac{\sqrt{25}}{\sqrt{4}} = \dfrac{5}{2}$ since $\left(\dfrac{5}{2}\right)^2 = \dfrac{25}{4}$

(c) $\sqrt{75} = \sqrt{5^2 \cdot 3} = \sqrt{5^2} \cdot \sqrt{3} = 5\sqrt{3}$

since $(5\sqrt{3})^2 = 25 \cdot 3 = 75$

3. Use the Pythagorean theorem: $a^2 + b^2 = c^2$

Let $a = 3$ and $b = 4$.

$3^2 + 4^2 = c^2$

$9 + 16 = c^2$

$25 = c^2$

$c = 5$

4. (a) $9x^2 - 30x + 25 = (3x)^2 - 2(3x)(5) + 5^2$

$= (3x - 5)(3x - 5) = (3x - 5)^2$

(b) $m^2 - 12m + 36 = m^2 - 2(6)m + 6^2$

$= (m - 6)(m - 6) = (m - 6)^2$

(c) $4x^2 + 2x + 1$ Not a perfect square trinomial

5. (a) $\dfrac{1}{2}(6) = 3; \; 3^2 = 9$

(b) $\dfrac{1}{2}(3) = \dfrac{3}{2}; \; \left(\dfrac{3}{2}\right)^2 = \dfrac{9}{4}$

(c) $\dfrac{1}{2}\left(-\dfrac{8}{3}\right) = -\dfrac{4}{3}; \; \left(-\dfrac{4}{3}\right)^2 = \dfrac{16}{9}$

6. Use the order of operations.

(a) $8^2 - 4(3)(2) = 64 - 4(3)(2) = 64 - 12(2)$

$= 64 - 24 = 40$

(b) $(-6)^2 - 4(1)(7) = 36 - 4(1)(7) = 36 - 4(7)$

$= 36 - 28 = 8$

(c) $5^2 - 4(-2)(2) = 25 - 4(-2)(2) = 25 + 8(2)$

$= 25 + 16 = 41$

7. (a) $\sqrt{7^2 - 4(1)(-18)} = \sqrt{49 - 4(1)(-18)}$

$= \sqrt{49 - (-72)}$

$= \sqrt{49 + 72} = \sqrt{121} = 11$

(b) $\sqrt{(-1)^2 - 4(1)(-12)} = \sqrt{1 - (-48)}$

$= \sqrt{1 + 48} = \sqrt{49} = 7$

(c) $\sqrt{(6)^2 - 4(-8)(2)} = \sqrt{36 - (-64)}$

$= \sqrt{36 + 64} = \sqrt{100} = 10$

8. (a) $\dfrac{-3 + \sqrt{3^2 - 4(2)(-27)}}{2(2)} = \dfrac{-3 + \sqrt{9 - (-216)}}{4}$

$= \dfrac{-3 + \sqrt{225}}{4}$

$= \dfrac{-3 + 15}{4} = \dfrac{12}{4} = 3$

(b) $\dfrac{-(-6) - \sqrt{(-6)^2 - 4(1)(9)}}{2(1)} = \dfrac{6 - \sqrt{36 - 36}}{2}$

$= \dfrac{6 - \sqrt{0}}{2}$

$= \dfrac{6}{2} = 3$

(c) $\dfrac{-13 + \sqrt{13^2 - 4(6)(-5)}}{2(6)}$

$= \dfrac{-13 + \sqrt{169 - (-120)}}{12} = \dfrac{-13 + \sqrt{289}}{12}$

$= \dfrac{-13 + 17}{12} = \dfrac{4}{12} = \dfrac{1}{3}$

Solutions to Odd-Numbered Classroom Problem Sets

CHAPTER 1

Section 1.1

1. $5 - 3 + 12 - 2 - 1 = 2 + 12 - 2 - 1$
$= 14 - 2 - 1$
$= 12 - 1$
$= 11$

3. $12(6 - 4) = 12(2)$
$= 24$

5. $(9 - 1) \div (9 - 5) = (8) \div (4)$
$= 2$

7. $\dfrac{(8-5)(1+3)}{6} + \dfrac{12}{11-7} = \dfrac{(3)(4)}{6} + \dfrac{12}{4}$
$= \dfrac{12}{6} + 3$
$= 2 + 3$
$= 5$

9. $16 \cdot 2 + 6 = 32 + 6$
$= 38$

11. $3 + 5 \cdot 2 - 10 \div 2 = 3 + 10 - 5$
$= 13 - 5$
$= 8$

13. $24 \div 6 \cdot 2 + 4 \cdot 8 \div 2 = 4 \cdot 2 + 32 \div 2$
$= 8 + 16$
$= 24$

15. $12 + 4[5(2 + 8)] = 12 + 4[5(10)]$
$= 12 + 4[50]$
$= 12 + 200$
$= 212$

17. $3 \cdot 5^2 - 7 \cdot 2^3 = 3 \cdot (5 \cdot 5) - 7 \cdot (2 \cdot 2 \cdot 2)$
$= 3 \cdot 25 - 7 \cdot 8$
$= 75 - 56$
$= 19$

19. $4x + 3y = 4(5) + 3(2)$ when $x = 5, y = 2$
$= 20 + 6$
$= 26$

21. $10a - 5b = 10(8) - 5(3)$ when $a = 8, b = 3$
$= 80 - 15$
$= 65$

23. $3xy - 2x + 4xyz = 3(6)(2) - 2(6) + 4(6)(2)(4)$
when $x = 6, y = 2, z = 4$
$= 18(2) - 12 + 24(2)(4)$
$= 36 - 12 + 48(4)$
$= 36 - 12 + 192$
$= 24 + 192$
$= 216$

25. $\dfrac{7c + d}{2c - d} = \dfrac{7(4) + 2}{2(4) - 2}$ when $c = 4, d = 2$
$= \dfrac{28 + 2}{8 - 2}$
$= \dfrac{30}{6}$
$= 5$

27. $(x + 3y)(4x - 2y) = (8 + 3 \cdot 3)(4 \cdot 8 - 2 \cdot 3)$
when $x = 8, y = 3$
$= (8 + 9)(32 - 6)$
$= (17)(26)$
$= 442$

29. $a^2 + 2ab + b^2 = (7)^2 + 2(7)(4) + (4)^2$
when $a = 7, b = 4$
$= 49 + 14(4) + 16$
$= 49 + 56 + 16$
$= 105 + 16$
$= 121$

31. $A = \dfrac{bh}{2} = \dfrac{(15)(4)}{2}$ when $b = 15, h = 4$
$= \dfrac{60}{2}$
$= 30$

Section 1.2

1. (a) $-24 + (-32) = -(|-24| + |-32|)$
$= -(24 + 32)$
$= -56$

(b) $-40 + 75 = |75| - |-40|$
$= 75 - 40$
$= 35$

(c) $86 + (-31) = |86| - |-31|$
$= 86 - 31$
$= 55$

(d) $15 + (-15) = |15| - |-15|$
$= 15 - 15$
$= 0$

3. (a) $-20 - (-38) = -20 + 38$
$= 18$

(b) $35 - 75 = 35 + (-75)$
$= -40$

(c) $81 - (-31) = 81 + 31$
$= 112$

(d) $25 - (-25) = 25 + 25$
$= 50$

5. (a) $(6)(-2) = -(|6| \cdot |-2|)$
$= -(6 \cdot 2)$
$= -12$

(b) $(-1)(-5) = |-1| \cdot |-5|$
$= 1 \cdot 5$
$= 5$

(c) $(0)(-8) = 0$

(d) $(-4)(6) = -(|-4| \cdot |6|)$
$= -(4 \cdot 6)$
$= -24$

7. (a) $\dfrac{-18}{6} = -\left(\dfrac{|-18|}{|6|}\right)$
$= -\left(\dfrac{18}{6}\right)$
$= -3$

(b) $\dfrac{0}{-8} = 0$

(c) $\dfrac{-50}{-2} = \dfrac{|-50|}{|-2|}$
$= \dfrac{50}{2}$
$= 25$

(d) $\dfrac{80}{-4} = -\left(\dfrac{|80|}{|-4|}\right)$
$= -\left(\dfrac{80}{4}\right)$
$= -20$

9. $5(-3) - 4(-2) - 7(1) = -15 - (-8) - 7$
$= -15 + 8 - 7$
$= -7 - 7$
$= -14$

11. $\dfrac{20 + 4(-2)}{-3} = \dfrac{20 + (-8)}{-3}$
$= \dfrac{12}{-3}$
$= -4$

13. $4x - 5y = 4(-2) - 5(-6)$ when $x = -2, y = -6$
$= -8 - (-30)$
$= -8 + 30$
$= 22$

15. $-3a - 4b = -3(-1) - 4(5)$ when $a = -1, b = 5$
$= 3 - 20$
$= -17$

17. $\dfrac{5(F - 32)}{9} = \dfrac{5(-4 - 32)}{9}$ when $F = -4$
$= \dfrac{5(-36)}{9}$
$= \dfrac{-180}{9}$
$= -20$

19. $-835{,}000 + 320{,}000 + 410{,}000 + (-120{,}000)$
$= -515{,}000 + 410{,}000 + (-120{,}000)$
$= -105{,}000 + (-120{,}000)$
$= -225{,}000$

Over four years the boat manufacturer lost $225,000.

21. $4(-2) + 3(-1) + 2(3) = -8 + (-3) + 6$
$= -11 + 6$
$= -5$

Kay was 5 under par.

Section 1.3

1. $-37 + [37 + (-8)] = [-37 + 37] + (-8)$
$= 0 + (-8)$
$= -8$

3. $50[(-2)(-37)] = [(50)(-2)](-37)$
$= (-100)(-37)$
$= 3700$

5. $-18 + 24 + (-8) + (-12) + 32$

24	-18	56
32	-8	-38
56	-12	18
	-38	

The sum is 18.

7. $(-15)(-2 + 10) = (-15)(-2) + (-15)(10)$
$= 30 + (-150)$
$= -120$

9. $-19(-34 + 34) = -19(0)$
$= 0$

11. $27(12) + 27(-2) = 27[12 + (-2)]$
$= 27(10)$
$= 270$

13. (a) $6a - 3a + 7 - 5a + 1 = 6a - 3a - 5a + 7 + 1$
$= (6 - 3 - 5)a + (7 + 1)$
$= -2a + 8$

(b) $2y - 3x + 5x - 6y = -3x + 5x + 2y - 6y$
$= (-3 + 5)x + (2 - 6)y$
$= 2x - 4y$

15. (a) $2(a - 2) - 5(a + 6) = 2(a) - 2(2) - 5(a) - 5(6)$
$= 2a - 4 - 5a - 30$
$= 2a - 5a - 4 - 30$
$= -3a - 34$

(b) $7(4 - 2b) - 4(5 - b)$
$= 7(4) - 7(2b) - 4(5) - 4(-b)$
$= 28 - 14b - 20 + 4b$
$= -14b + 4b + 28 - 20$
$= -10b + 8$

17. $5a + 3b - 2a - 7b = 5a - 2a + 3b - 7b$
$= 3a - 4b \quad \text{for } a = 6, b = -5$
$= 3(6) - 4(-5)$
$= 18 - (-20)$
$= 18 + 20 = 38$

19. $2xy + 3x - 2y - 4xy$
$= 2xy - 4xy + 3x - 2y$
$= -2xy + 3x - 2y \quad \text{for } x = -3, y = -1$
$= -2(-3)(-1) + 3(-3) - 2(-1)$
$= 6(-1) + (-9) + 2$
$= -6 + (-9) + 2$
$= -15 + 2 = -13$

21. $6(a - 3) - 2(a + 4) = 6a - 18 - 2a - 8$
$= 6a - 2a - 18 - 8$
$= 4a - 26 \quad \text{when } a = 7$
$= 4(7) - 26$
$= 28 - 26$
$= 2$

Section 1.4

1. (a) $36 = 2 \cdot 18$
$= 2 \cdot 2 \cdot 9$
$= 2 \cdot 2 \cdot 3 \cdot 3$
$= 2^2 \cdot 3^2$

(b) $42 = 2 \cdot 21$
$= 2 \cdot 3 \cdot 7$

3. $42 = 2 \cdot 3 \cdot 7$
$60 = 2 \cdot 2 \cdot 3 \cdot 5 = 2^2 \cdot 3 \cdot 5$
The GCF $= 2 \cdot 3 = 6$.

5. $63 = 3 \cdot 3 \cdot 7 = 3^2 \cdot 7$
$54 = 2 \cdot 3 \cdot 3 \cdot 3 = 2 \cdot 3^3$
The GCF $= 3^2 = 9$.

7. $49 = 7 \cdot 7 = 7^2$
$80 = 2 \cdot 2 \cdot 2 \cdot 2 \cdot 5 = 2^4 \cdot 5$
The GCF $= 1$.

9. $36 = 2 \cdot 2 \cdot 3 \cdot 3 = 2^2 \cdot 3^2$
$72 = 2 \cdot 2 \cdot 2 \cdot 3 \cdot 3 = 2^3 \cdot 3^2$
$90 = 2 \cdot 3 \cdot 3 \cdot 5 = 2 \cdot 3^2 \cdot 5$
The GCF $= 2 \cdot 3^2 = 18$.

11. $12 = 2 \cdot 2 \cdot 3 = 2^2 \cdot 3$
$30 = 2 \cdot 3 \cdot 5 = 2 \cdot 3 \cdot 5$
The LCM $= 2^2 \cdot 3 \cdot 5 = 4 \cdot 3 \cdot 5 = 60$.

13. $20 = 2 \cdot 2 \cdot 5 = 2^2 \cdot 5$
$75 = 3 \cdot 5 \cdot 5 = 3 \cdot 5^2$
The LCM $= 2^2 \cdot 3 \cdot 5^2 = 4 \cdot 3 \cdot 25 = 300$.

15. $12 = 2 \cdot 2 \cdot 3 = 2^2 \cdot 3$
$18 = 2 \cdot 3 \cdot 3 = 2 \cdot 3^2$
$30 = 2 \cdot 3 \cdot 5 = 2 \cdot 3 \cdot 5$
The LCM $= 2^2 \cdot 3^2 \cdot 5 = 4 \cdot 9 \cdot 5 = 180$.

17. $15 = 3 \cdot 5 = 3 \cdot 5$
$8 = 2 \cdot 2 \cdot 2 = 2^3$
The LCM $= 2^3 \cdot 3 \cdot 5 = 8 \cdot 3 \cdot 5 = 120$.

Section 1.5

1. $\dfrac{28}{32} = \dfrac{4 \cdot 7}{4 \cdot 8} = \dfrac{4}{4} \cdot \dfrac{7}{8} = 1 \cdot \dfrac{7}{8} = \dfrac{7}{8}$

3. $\dfrac{25}{35} = \dfrac{\cancel{5} \cdot 5}{\cancel{5} \cdot 7} = \dfrac{5}{7}$

5. $-\dfrac{60}{144} = -\dfrac{2 \cdot 2 \cdot \cancel{3} \cdot 5}{2 \cdot 2 \cdot 2 \cdot 2 \cdot \cancel{3} \cdot 3} = -\dfrac{5}{12}$

7. $\dfrac{5}{7} \cdot \dfrac{3}{10} = \dfrac{5 \cdot 3}{7 \cdot 10} = \dfrac{\cancel{5} \cdot 3}{7 \cdot 2 \cdot \cancel{5}} = \dfrac{3}{14}$

9. $\dfrac{5}{6} \cdot \dfrac{12}{25} = \dfrac{\overset{1}{\cancel{5}} \cdot \overset{2}{\cancel{12}}}{\underset{1}{\cancel{6}} \cdot \underset{5}{\cancel{25}}} = \dfrac{2}{5}$

11. $\left(\dfrac{6}{18}\right)\left(-\dfrac{10}{14}\right) = -\dfrac{\overset{1}{\cancel{6}} \cdot \overset{5}{\cancel{10}}}{\underset{3}{\cancel{18}} \cdot \underset{7}{\cancel{14}}} = -\dfrac{5}{21}$

13. $\left(-\dfrac{10}{3}\right)\left(-\dfrac{42}{18}\right) = \dfrac{2 \cdot 5 \cdot 2 \cdot \cancel{3} \cdot 7}{\cancel{3} \cdot 2 \cdot 3 \cdot 3} = \dfrac{70}{9}$

15. (a) $\dfrac{1}{5} \div \dfrac{2}{15} = \dfrac{1}{5} \cdot \dfrac{15}{2} = \dfrac{\cancel{15}^{3}}{\cancel{5} \cdot 2} = \dfrac{3}{2}$

   (b) $-\dfrac{3}{8} \div 4 = -\dfrac{3}{8} \cdot \dfrac{1}{4} = -\dfrac{3}{32}$

17. $\dfrac{1}{7} + \dfrac{2}{3} = \dfrac{1}{7} \cdot \dfrac{3}{3} + \dfrac{2}{3} \cdot \dfrac{7}{7}$
   $= \dfrac{3}{21} + \dfrac{14}{21}$
   $= \dfrac{3 + 14}{21}$
   $= \dfrac{17}{21}$

19. $\dfrac{2}{5} + \dfrac{1}{3} = \dfrac{2}{5} \cdot \dfrac{3}{3} + \dfrac{1}{3} \cdot \dfrac{5}{5}$
   $= \dfrac{6}{15} + \dfrac{5}{15}$
   $= \dfrac{6 + 5}{15}$
   $= \dfrac{11}{15}$

21. $\dfrac{13}{60} + \dfrac{5}{18}$
   $60 = 2 \cdot 2 \cdot 3 \cdot 5 = 2^2 \cdot 3 \cdot 5$
   $18 = 2 \cdot 3 \cdot 3 = 2 \cdot 3^2$
   The LCD $= 2^2 \cdot 3^2 \cdot 5 = 4 \cdot 9 \cdot 5 = 180$.
   $\dfrac{13}{60} + \dfrac{5}{18} = \dfrac{13}{60} \cdot \dfrac{3}{3} + \dfrac{5}{18} \cdot \dfrac{10}{10}$
   $= \dfrac{39}{180} + \dfrac{50}{180}$
   $= \dfrac{39 + 50}{180}$
   $= \dfrac{89}{180}$

23. $\dfrac{8}{15} - \dfrac{3}{10}$
   $15 = 3 \cdot 5$
   $10 = 2 \cdot 5$
   The LCD $= 2 \cdot 3 \cdot 5 = 30$.
   $\dfrac{8}{15} - \dfrac{3}{10} = \dfrac{8}{15} \cdot \dfrac{2}{2} - \dfrac{3}{10} \cdot \dfrac{3}{3}$
   $= \dfrac{16}{30} - \dfrac{9}{30}$
   $= \dfrac{16 - 9}{30}$
   $= \dfrac{7}{30}$

25. $\dfrac{3}{10} + \left(\dfrac{-7}{12}\right)$
   $10 = 2 \cdot 5$
   $12 = 2 \cdot 2 \cdot 3 = 2^2 \cdot 3$
   The LCD $= 2^2 \cdot 3 \cdot 5 = 4 \cdot 3 \cdot 5 = 60$.
   $\dfrac{3}{10} + \left(\dfrac{-7}{12}\right) = \dfrac{3}{10} \cdot \dfrac{6}{6} + \left(\dfrac{-7}{12} \cdot \dfrac{5}{5}\right)$
   $= \dfrac{18}{60} + \left(\dfrac{-35}{60}\right)$
   $= \dfrac{18 + (-35)}{60}$
   $= \dfrac{-17}{60} = -\dfrac{17}{60}$

27. $\dfrac{5}{8} + (-2)$ LCD $= 8$
   $\dfrac{5}{8} + (-2) = \dfrac{5}{8} + \left(-\dfrac{2}{1} \cdot \dfrac{8}{8}\right)$
   $= \dfrac{5}{8} + \left(-\dfrac{16}{8}\right)$
   $= \dfrac{5 + (-16)}{8}$
   $= \dfrac{-11}{8} = -\dfrac{11}{8}$

29. $\dfrac{3}{14} + \dfrac{1}{4} \cdot \dfrac{4}{7} - \dfrac{1}{2} \cdot \dfrac{3}{7} = \dfrac{3}{14} + \dfrac{1}{7} - \dfrac{3}{14}$
   $= \dfrac{3}{14} - \dfrac{3}{14} + \dfrac{1}{7}$
   $= 0 + \dfrac{1}{7}$
   $= \dfrac{1}{7}$

31. $\dfrac{3}{11} \div \dfrac{2}{11} + \left(\dfrac{-1}{3}\right)\left(\dfrac{1}{4}\right) + \dfrac{5}{6}$
   $= \dfrac{3}{11} \cdot \dfrac{11}{2} + \left(\dfrac{-1}{3}\right)\left(\dfrac{1}{4}\right) + \dfrac{5}{6}$
   $= \dfrac{3}{2} + \left(\dfrac{-1}{3}\right)\left(\dfrac{1}{4}\right) + \dfrac{5}{6}$
   $= \dfrac{3}{2} + \left(\dfrac{-1}{12}\right) + \dfrac{5}{6}$ LCD $= 12$
   $= \dfrac{3}{2} \cdot \dfrac{6}{6} + \left(\dfrac{-1}{12}\right) + \dfrac{5}{6} \cdot \dfrac{2}{2}$

$$= \frac{18}{12} + \left(\frac{-1}{12}\right) + \frac{10}{12}$$

$$= \frac{18 + (-1) + 10}{12}$$

$$= \frac{27}{12} = \frac{3 \cdot 3 \cdot 3}{2 \cdot 2 \cdot 3} = \frac{9}{4}$$

33. $20\left(\frac{1}{2} + \frac{1}{5}\right) = 20\left(\frac{1}{2}\right) + 20\left(\frac{1}{5}\right)$

$$= 10 + 4$$
$$= 14$$

35. $\frac{3}{8}\left(\frac{1}{3} + \frac{1}{5}\right) = \frac{3}{8}\left(\frac{1}{3}\right) + \frac{3}{8}\left(\frac{1}{5}\right)$

$$= \frac{1}{8} + \frac{3}{40}$$

$$= \frac{1}{8} \cdot \frac{5}{5} + \frac{3}{40}$$

$$= \frac{5}{40} + \frac{3}{40}$$

$$= \frac{5 + 3}{40}$$

$$= \frac{8}{40} = \frac{\cancel{8}^{1}}{\cancel{8} \cdot 5} = \frac{1}{5}$$

37. Divide 120 by $\frac{3}{4}$ to find the number of installations.

$$120 \div \frac{3}{4} = \frac{120}{1} \cdot \frac{4}{3}$$

$$= \frac{\cancel{120}^{40}}{1} \cdot \frac{4}{\cancel{3}_{1}}$$

$$= \frac{40 \cdot 4}{1 \cdot 1}$$

$$= 160$$

There will be 160 installations.

39. First add the lengths of the pieces needed to be cut.

$$1\frac{3}{8} + 6\frac{1}{4} = \frac{11}{8} + \frac{25}{4} = \frac{11}{8} + \frac{50}{8} = \frac{61}{8}$$

To find the remaining piece, subtract $\frac{61}{8}$ from 12.

$$12 - \frac{61}{8} = \frac{96}{8} - \frac{61}{8} = \frac{35}{8} = 4\frac{3}{8}$$

The length of the remaining piece is $4\frac{3}{8}$ feet.

Section 1.6

1. (a) $17.3 + 21.05 + 0.4$

   17.3
   21.05
   +0.4
   ─────
   38.75

(b) $12.45 + 1.8 + 0.04 + 1.261$

   12.45
   1.8
   0.04
   +1.261
   ──────
   15.551

3. (a) $13.6 - 10.35$

   13.6
   −10.35
   ──────
   3.25

(b) $8.427 - 0.32$

   8.427
   −0.32
   ─────
   8.107

5. (a) $28.4(1.32)$

   28.4   one digit
   ×1.32   two digits
   ──────
   568
   8520
   28400
   ──────
   37.488   three digits

(b) $0.017(0.02)$

   0.017   three digits
   ×0.02   two digits
   ───────
   0.00034   five digits

7. (a) $29.16 \div 1.2 \rightarrow 291.6 \div 12$

$$\begin{array}{r} 24.3 \\ 12\overline{)291.6} \\ \underline{24} \\ 51 \\ \underline{48} \\ 36 \\ \underline{36} \\ 0 \end{array}$$

(b) $20.528 \div 0.04 \rightarrow 2052.8 \div 4$

$$\begin{array}{r} 513.2 \\ 4\overline{)2052.8} \\ \underline{20} \\ 5 \\ \underline{4} \\ 12 \\ \underline{12} \\ 8 \\ \underline{8} \\ 0 \end{array}$$

9. $5.3a - 3.06b - 0.08b - 0.01a$

$$= 5.3a - 0.01a - 3.06b - 0.08b$$
$$= (5.3 - 0.01)a + (-3.06 - 0.08)b$$
$$= 5.29a + (-3.14b)$$
$$= 5.29a - 3.14b$$

11. $\frac{2}{3}x - \frac{1}{4}y = \frac{2}{3}\left(\frac{3}{7}\right) - \frac{1}{4}(-2)$ when $x = \frac{3}{7}, y = -2$

$= \frac{2}{7} + \frac{2}{4}$

$= \frac{2}{7} + \frac{1}{2}$

$= \frac{2}{7} \cdot \frac{2}{2} + \frac{1}{2} \cdot \frac{7}{7}$

$= \frac{4}{14} + \frac{7}{14}$

$= \frac{11}{14}$

13. $\frac{1}{4}x + \frac{3}{5}x - \frac{1}{3}x = \left(\frac{1}{4} + \frac{3}{5} - \frac{1}{3}\right)x$

$= \left(\frac{1}{4} \cdot \frac{15}{15} + \frac{3}{5} \cdot \frac{12}{12} - \frac{1}{3} \cdot \frac{20}{20}\right)x$

$= \left(\frac{15}{60} + \frac{36}{60} - \frac{20}{60}\right)x$

$= \frac{31}{60}x$ when $x = -\frac{6}{7}$

$= \frac{31}{60}\left(-\frac{6}{7}\right)$

$= -\frac{31 \cdot \overset{1}{\cancel{6}}}{\underset{10}{\cancel{60}} \cdot 7}$

$= -\frac{31}{70}$

15. $-2x + 4y$ when $x = 2.1, y = 0.07$
$-2(2.1) + 4(0.07) = -4.2 + 0.28$
$= -3.92$

17. $5.6x + 0.8x - 1.8x - 0.4$
$= (5.6 + 0.8 - 1.8)x - 0.4$
$= 4.6x - 0.4$ when $x = -0.3$
$= 4.6(-0.3) - 0.4$
$= -1.38 - 0.4$
$= -1.78$

19. Add the number of ounces.
$4.6 + 1.25 + 8 + 0.02 = 5.85 + 8 + 0.02$
$= 13.85 + 0.02$
$= 13.87$
The pharmacist prepared 13.87 ounces.

CHAPTER 2

Section 2.1

1. $x - 11 = -4$
$x - 11 + 11 = -4 + 11$
$x = 7$
The solution set is $\{7\}$.

3. $y + 6 = 2$
$y + 6 - 6 = 2 - 6$
$y = -4$
The solution set is $\{-4\}$.

5. $x - \frac{1}{5} = \frac{1}{2}$
$x - \frac{1}{5} + \frac{1}{5} = \frac{1}{2} + \frac{1}{5}$
$x = \frac{5}{10} + \frac{2}{10}$
$x = \frac{7}{10}$
The solution set is $\left\{\frac{7}{10}\right\}$.

7. $0.08 = x + 0.45$
$0.08 - 0.45 = x + 0.45 - 0.45$
$-0.37 = x$
The solution set is $\{-0.37\}$.

9. $\frac{2}{3}x = 12$
$\frac{3}{2}\left(\frac{2}{3}x\right) = \frac{3}{2}(12)$
$x = \frac{36}{2} = 18$
The solution set is $\{18\}$.

11. $3x = 49$
$\frac{3x}{3} = \frac{49}{3}$
$x = \frac{49}{3}$
The solution set is $\left\{\frac{49}{3}\right\}$.

13. $\frac{3}{5}y = -\frac{1}{4}$
$\frac{5}{3}\left(\frac{3}{5}y\right) = \frac{5}{3}\left(-\frac{1}{4}\right)$
$y = -\frac{5}{12}$
The solution set is $\left\{-\frac{5}{12}\right\}$.

15. $\dfrac{x}{4} = -\dfrac{3}{2}$

$4\left(\dfrac{x}{4}\right) = 4\left(-\dfrac{3}{2}\right)$

$y = -\dfrac{12}{2} = -6$

The solution set is $\{-6\}$.

17. $0.3y = 18.6$

$\dfrac{0.3y}{0.3} = \dfrac{18.6}{0.3}$

$y = 62$

The solution set is $\{62\}$.

Section 2.2

1. The sum of two numbers is 56.
One of the numbers is y.
The other number is $56 - y$.

3. The difference of two numbers is 20.
The smaller number is x.
The larger number is $20 + x$.

5. The product of two numbers is 45.
One of the numbers is x.
The other number is $\dfrac{45}{x}$.

7. In 10 minutes the fax transmits $12(10) = 120$ pages.
In 20 minutes the fax transmits $12(20) = 240$ pages.
In y minutes the fax transmits $12(y) = (12y)$ pages.

9. Four quarters and six dimes would be $4(25) + 6(10) = 160$ cents. So q quarters and d dimes would be $(25q + 10d)$ cents.

11. Suppose an electric car travels at 30 miles per hour for 2 hours. Using Distance = Rate · Time,

$d = (30)(2)$

$d = 60$

The car travels 60 miles. Then an electric car traveling p miles per hour for 4 hours would travel $d = p(4) = (4p)$ miles.

13. The cost of the trip is e. The cost per day is $\dfrac{e}{8}$ euros.

15. Because 1 centimeter = 10 millimeters, multiply x by 10. The bolts are $(10x)$ millimeters apart.

17. To change from inches to feet, divide p inches by 12.
So the height of the plasma TV is $\left(\dfrac{p}{12}\right)$ feet.

19. Let x represent the width of a rectangle. Then $2x + 3$ represents the length of the rectangle.
Use $P = 2l + 2w$.
$P = 2(2x + 3) + 2(x)$
$P = 4x + 6 + 2x$
$P = (6x + 6)$ inches

21. Let s represent the length of the side of a square in yards. Because 1 yard = 3 feet, the side of the square is $(3s)$ feet. To find the area, use $A = s^2$.
$A = (3s)^2$
$A = (9s^2)$ square feet

Section 2.3

1. $4x + 3 = -17$

$4x + 3 - 3 = -17 - 3$

$4x = -20$

$\dfrac{4x}{4} = \dfrac{-20}{4}$

$x = -5$

The solution set is $\{-5\}$.

3. $3x - 2 = 13$

$3x - 2 + 2 = 13 + 2$

$3x = 15$

$\dfrac{3x}{3} = \dfrac{15}{3}$

$x = 5$

The solution set is $\{5\}$.

5. $1 - 5y = 36$

$1 - 5y - 1 = 36 - 1$

$-5y = 35$

$\dfrac{-5y}{-5} = \dfrac{35}{-5}$

$y = -7$

The solution set is $\{-7\}$.

7. $24 = 4a - 3$

$24 + 3 = 4a - 3 + 3$

$27 = 4a$

$\dfrac{27}{4} = \dfrac{4a}{4}$

$\dfrac{27}{4} = a$

The solution set is $\left\{\dfrac{27}{4}\right\}$.

9. Let n represent the number to be found.

$n + 21 = 78$

$n + 21 - 21 = 78 - 21$

$n = 57$

The number is 57.

11. Let a represent Dahlia's age now. Then $a - 5$ represents her age 5 years ago.
$$a - 5 = 19$$
$$a - 5 + 5 = 19 + 5$$
$$a = 24$$
Dahlia is 24 years old now.

13. Let x represent the price per minute.
$$\begin{pmatrix}\text{Number of}\\\text{minutes}\end{pmatrix} \cdot \begin{pmatrix}\text{Price per}\\\text{minute}\end{pmatrix} = \begin{pmatrix}\text{Total}\\\text{price}\end{pmatrix}$$
$$34 \cdot x = 2.72$$
$$34x = 2.72$$
$$\frac{34x}{34} = \frac{2.72}{34}$$
$$x = 0.08$$
Damien pays $0.08 per minute.

15. Let n represent the number to be found.
$$4n - 6 = 30$$
$$4n - 6 + 6 = 30 + 6$$
$$4n = 36$$
$$\frac{4n}{4} = \frac{36}{4}$$
$$n = 9$$
The number is 9.

17. Let x represent the cost of each DVD movie.
$$\begin{pmatrix}\text{Cost of}\\\text{DVD player}\end{pmatrix} + 5 \cdot \begin{pmatrix}\text{Cost of}\\\text{DVD movie}\end{pmatrix} = 439.00$$
$$349 + 5x = 439$$
$$349 + 5x - 349 = 439 - 349$$
$$5x = 90$$
$$\frac{5x}{5} = \frac{90}{5}$$
$$x = 18$$
Each DVD movie costs $18.

Section 2.4

1.
$$2x - 7 - 5x = 14$$
$$-3x - 7 = 14$$
$$-3x - 7 + 7 = 14 + 7$$
$$-3x = 21$$
$$\frac{-3x}{-3} = \frac{21}{-3}$$
$$x = -7$$
The solution set is $\{-7\}$.

3.
$$x - 5 = 6x - 45$$
$$x - 5 - 6x = 6x - 45 - 6x$$
$$-5x - 5 = -45$$
$$-5x - 5 + 5 = -45 + 5$$
$$-5x = -40$$
$$\frac{-5x}{-5} = \frac{-40}{-5}$$
$$x = 8$$
The solution set is $\{8\}$.

5.
$$2a + 24 = 6a + 44$$
$$2a + 24 - 6a = 6a + 44 - 6a$$
$$-4a + 24 = 44$$
$$-4a + 24 - 24 = 44 - 24$$
$$-4a = 20$$
$$\frac{-4a}{-4} = \frac{20}{-4}$$
$$a = -5$$
The solution set is $\{-5\}$.

7. Let n represent the first even number. Then $n + 2$ represents the next even number.
$$n + (n + 2) = 126$$
$$2n + 2 = 126$$
$$2n + 2 - 2 = 126 - 2$$
$$2n = 124$$
$$\frac{2n}{2} = \frac{124}{2}$$
$$n = 62$$
If $n = 62$, then $n + 2 = 64$. So the numbers are 62 and 64.

9. Let x represent the number of deliveries.
$$50 + 0.75x = 77$$
$$50 + 0.75x - 50 = 77 - 50$$
$$0.75x = 27$$
$$\frac{0.75x}{0.75} = \frac{27}{0.75}$$
$$x = 36$$
The driver must make 36 deliveries.

11. Let a represent the measure of the larger angle. Then $a - 56$ represents the measure of the smaller angle. The sum of the measures of supplementary angles is 180°.
$$a + (a - 56) = 180$$
$$2a - 56 = 180$$
$$2a = 236$$
$$\frac{2a}{2} = \frac{236}{2}$$
$$a = 118$$
If $a = 118$, then $a - 56 = 62$. The two supplementary angles have measures of 118° and 62°.

13. Let a represent the measure of the first angle. Then $3a$ represents the measure of the third angle, and $5 + 3a$ represents the measure of the second angle. The sum of the measures of the angles of a triangle is 180°.

$$a + (5 + 3a) + 3a = 180$$
$$7a + 5 = 180$$
$$7a + 5 - 5 = 180 - 5$$
$$7a = 175$$
$$\frac{7a}{7} = \frac{175}{7}$$
$$a = 25$$

If $a = 25$, then the first angle measures 25°, the second angle measures $5 + 3a = 80°$, and the third angle measures $3a = 75°$.

Section 2.5

1. $3(x - 4) = -5(x + 2)$
$$3x - 12 = -5x - 10$$
$$8x - 12 = -10$$
$$8x = 2$$
$$\frac{8x}{8} = \frac{2}{8}$$
$$x = \frac{1}{4}$$

The solution set is $\left\{\frac{1}{4}\right\}$.

3. $2(x - 4) - 7(x + 2) = 3$
$$2x - 8 - 7x - 14 = 3$$
$$-5x - 22 = 3$$
$$-5x = 25$$
$$\frac{-5x}{-5} = \frac{25}{-5}$$
$$x = -5$$

The solution set is $\{-5\}$.

5. $\frac{4}{5}x + \frac{2}{3} = \frac{7}{15}$

The LCD of 3, 5, and 15 is 15

$$15\left(\frac{4}{5}x + \frac{2}{3}\right) = 15\left(\frac{7}{15}\right)$$
$$15\left(\frac{4}{5}x\right) + 15\left(\frac{2}{3}\right) = 1(7)$$
$$3(4x) + 5(2) = 7$$
$$12x + 10 = 7$$
$$12x = -3$$
$$\frac{12x}{12} = \frac{-3}{12}$$
$$x = -\frac{1}{4}$$

The solution set is $\left\{-\frac{1}{4}\right\}$.

7. $\frac{7y}{12} - \frac{5}{6} = \frac{3}{8}$

The LCD of 6, 8, and 12 is 24

$$24\left(\frac{7y}{12} - \frac{5}{6}\right) = 24\left(\frac{3}{8}\right)$$
$$24\left(\frac{7y}{12}\right) - 24\left(\frac{5}{6}\right) = 3(3)$$
$$2(7y) - 4(5) = 9$$
$$14y - 20 = 9$$
$$14y = 29$$
$$\frac{14y}{14} = \frac{29}{14}$$
$$y = \frac{29}{14}$$

The solution set is $\left\{\frac{29}{14}\right\}$.

9. $\frac{x - 4}{3} + \frac{x + 2}{5} = \frac{7}{10}$

The LCD of 3, 5, and 10 is 30

$$30\left(\frac{x - 4}{3} + \frac{x + 2}{5}\right) = 30\left(\frac{7}{10}\right)$$
$$30\left(\frac{x - 4}{3}\right) + 30\left(\frac{x + 2}{5}\right) = 3(7)$$
$$10(x - 4) + 6(x + 2) = 21$$
$$10x - 40 + 6x + 12 = 21$$
$$16x - 28 = 21$$
$$16x = 49$$
$$\frac{16x}{16} = \frac{49}{16}$$
$$x = \frac{49}{16}$$

The solution set is $\left\{\frac{49}{16}\right\}$.

11. $\frac{x - 1}{8} - \frac{x - 4}{3} = \frac{5}{4}$

The LCD of 3, 4, and 8 is 24

$$24\left(\frac{x - 1}{8} - \frac{x - 4}{3}\right) = 24\left(\frac{5}{4}\right)$$
$$24\left(\frac{x - 1}{8}\right) - 24\left(\frac{x - 4}{3}\right) = 6(5)$$

$$3(x - 1) - 8(x - 4) = 30$$
$$3x - 3 - 8x + 32 = 30$$
$$-5x + 29 = 30$$
$$-5x = 1$$
$$\frac{-5x}{-5} = \frac{1}{-5}$$
$$x = -\frac{1}{5}$$

The solution set is $\left\{-\frac{1}{5}\right\}$.

13. $$6x + 1 = 2(3x + 2)$$
$$6x + 1 = 6x + 4$$
$$6x + 1 - 6x = 6x + 4 - 6x$$
$$1 = 4$$

The result is a false statement. Therefore, the equation is a contradiction, and the solution set is ∅.

15. $$3(x - 1) + 2x + 7 = 5x + 4$$
$$3x - 3 + 2x + 7 = 5x + 4$$
$$5x + 4 = 5x + 4$$
$$5x + 4 - 5x = 5x + 4 - 5x$$
$$4 = 4$$

The result is a true statement. Therefore, the equation is an identity, and any real number is a solution. The solution set is {all reals}.

17. Let q represent the number of quarters. Then $20 - q$ represents the number of dimes.

$$\binom{\text{Value of quarters}}{\text{in cents}} + \binom{\text{Value of dimes}}{\text{in cents}} = 395$$
$$25q + 10(20 - q) = 395$$
$$25q + 200 - 10q = 395$$
$$15q + 200 = 395$$
$$15q = 195$$
$$\frac{15q}{15} = \frac{195}{15}$$
$$q = 13$$

If $q = 13$, then $20 - q = 7$, so Michaela has 13 quarters and 7 dimes.

19. Let n represent the number to be found.

$$\frac{3}{4}n - 12 = \frac{1}{2}n \quad \text{The LCD of 2 and 4 is 4}$$
$$4\left(\frac{3}{4}n - 12\right) = 4\left(\frac{1}{2}n\right)$$
$$4\left(\frac{3}{4}n\right) - 4(12) = 2(n)$$
$$3n - 48 = 2n$$

$$n - 48 = 0$$
$$n = 48$$

The number is 48.

21. Let x represent the hourly rate. Then $\frac{3}{2}x$ represents $1\frac{1}{2}$ times his hourly rate.

$$\binom{\text{Wages for}}{\text{first 40 hours}} + \binom{\text{Wages for}}{\text{overtime hours}} = 696$$
$$40x + 12\left(\frac{3}{2}x\right) = 696$$
$$40x + 6(3x) = 696$$
$$40x + 18x = 696$$
$$58x = 696$$
$$\frac{58x}{58} = \frac{696}{58}$$
$$x = 12$$

His hourly rate is $12.00.

23. Let n represent the first whole number. Then $n + 1$ represents the second whole number, and $n + 2$ represents the third whole number.

$$3n + 2(n + 1) + (n + 2) = 76$$
$$3n + 2n + 2 + n + 2 = 76$$
$$6n + 4 = 76$$
$$6n = 72$$
$$\frac{6n}{6} = \frac{72}{6}$$
$$n = 12$$

If $n = 12$, then $n + 1 = 13$ and $n + 2 = 14$. So the numbers are 12, 13, and 14.

Section 2.6

1. $$x + 5 > 2$$
$$x + 5 - 5 > 2 - 5$$
$$x > -3$$

The solution set is $\{x | x > -3\}$.

```
—+—○—+—+—+—+—+—+—
-4 -3 -2 -1 0  1  2  3  4
```

3. $$x - 3 \leq -1$$
$$x - 3 + 3 \leq -1 + 3$$
$$x \leq 2$$

The solution set is $\{x | x \leq 2\}$.

```
—+—+—+—+—+—●—+—+—
-3 -2 -1 0  1  2  3  4
```

5. $$8 > 5 + x$$
$$8 - 5 > 5 + x - 5$$
$$3 > x \text{ means } x < 3.$$

The solution set is $\{x|x < 3\}$.

7. $7x > 21$

$\dfrac{7x}{7} > \dfrac{21}{7}$

$x > 3$

The solution set is $\{x|x > 3\}$.

9. $\dfrac{2}{3}x \le \dfrac{4}{7}$

$\dfrac{3}{2}\left(\dfrac{2}{3}x\right) \le \dfrac{3}{2}\left(\dfrac{4}{7}\right)$

$x \le \dfrac{6}{7}$

The solution set is $\left\{x \mid x \le \dfrac{6}{7}\right\}$.

11. $-5x > 10$

$\dfrac{-5x}{-5} < \dfrac{10}{-5}$

$x < -2$

The solution set is $\{x|x < -2\}$.

13. $2x - 1 > 7$

$2x > 8$

$\dfrac{2x}{2} > \dfrac{8}{2}$

$x > 4$

The solution set is $\{x|x > 4\}$.

15. $-4y + 5 < 17$

$-4y < 12$

$\dfrac{-4y}{-4} > \dfrac{12}{-4}$

$y > -3$

The solution set is $\{y|y > -3\}$.

Section 2.7

1. $7x + 8 \le 3x + 12$

$4x + 8 \le 12$

$4x \le 4$

$\dfrac{4x}{4} \le \dfrac{4}{4}$

$x \le 1$

The solution set is $\{x|x \le 1\}$.

3. $2(x + 1) + 4(x - 2) \ge 3(x + 6)$

$2x + 2 + 4x - 8 \ge 3x + 18$

$6x - 6 \ge 3x + 18$

$3x - 6 \ge 18$

$3x \ge 24$

$\dfrac{3x}{3} \ge \dfrac{24}{3}$

$x \ge 8$

The solution set is $\{x|x \ge 8\}$.

5. $-\dfrac{4}{3}y + \dfrac{5}{6}y < \dfrac{7}{6}$ The LCD of 3 and 6 is 6

$6\left(-\dfrac{4}{3}y + \dfrac{5}{6}y\right) < 6\left(\dfrac{7}{6}\right)$

$6\left(-\dfrac{4}{3}y\right) + 6\left(\dfrac{5}{6}y\right) < 7$

$2(-4y) + 5y < 7$

$-8y + 5y < 7$

$-3y < 7$

$\dfrac{-3y}{-3} > \dfrac{7}{-3}$

$y > -\dfrac{7}{3}$

The solution set is $\left\{y \mid y > -\dfrac{7}{3}\right\}$.

7. $\dfrac{x+4}{3} - \dfrac{x+2}{4} \le \dfrac{5}{12}$

The LCD of 3, 4, and 12 is 12

$12\left(\dfrac{x+4}{3} - \dfrac{x+2}{4}\right) \le 12\left(\dfrac{5}{12}\right)$

$12\left(\dfrac{x+4}{3}\right) - 12\left(\dfrac{x+2}{4}\right) \le 5$

$4(x + 4) - 3(x + 2) \le 5$

$4x + 16 - 3x - 6 \le 5$

$x + 10 \le 5$

$x \le -5$

The solution set is $\{x|x \le -5\}$.

9. $x > 0$ and $x < 5$

$x > 0$

$x < 5$

$x > 0$ and $x < 5$

The solution set is $\{x|0 < x < 5\}$.

11. $x \le 3$ and $x \le 5$

$x \le 3$

$x \leq 5$

$x \leq 3$ and $x \leq 5$

The solution set is $\{x | x \leq 3\}$.

13. $x > 2$ or $x > 4$

$x > 2$

$x > 4$

$x > 2$ or $x > 4$

The solution set is $\{x | x > 2\}$.

15. $x \leq -1$ or $x \geq 1$

$x \leq -1$

$x \geq 1$

$x \leq -1$ or $x \geq 1$

The solution set is $\{x | x \leq -1$ or $x \geq 1\}$.

17. Let y represent the score on the fourth exam.

$$\frac{88 + 95 + 90 + y}{4} \geq 92$$

$$\frac{273 + y}{4} \geq 92$$

$$4\left(\frac{273 + y}{4}\right) \geq 4(92)$$

$$273 + y \geq 368$$

$$y \geq 95$$

Felix must score 95 or higher on the fourth exam.

19. Let w represent the number of wins for the remainder of the season. Seventy percent of 160 games is the same as 0.70(160).

$$\begin{pmatrix} \text{Number} \\ \text{of wins} \end{pmatrix} \geq \begin{pmatrix} 70\% \text{ of} \\ \text{total wins} \end{pmatrix}$$

$$w + 32 \geq 0.70(160)$$

$$w + 32 \geq 112$$

$$w \geq 80$$

They need to win at least 80 games.

CHAPTER 3

Section 3.1

1. $\dfrac{x}{18} = \dfrac{5}{6}$

   $6x = 90$

   $x = 15$

   The solution set is $\{15\}$.

3. $\dfrac{x-2}{3} = \dfrac{x-4}{5}$

   $5(x - 2) = 3(x - 4)$

   $5x - 10 = 3x - 12$

   $2x - 10 = -12$

   $2x = -2$

   $x = -1$

   The solution set is $\{-1\}$.

5. $\dfrac{8}{a-1} = \dfrac{3}{a-2}$

   $8(a - 2) = 3(a - 1)$

   $8a - 16 = 3a - 3$

   $5a - 16 = -3$

   $5a = 13$

   $a = \dfrac{13}{5}$

   The solution set is $\left\{\dfrac{13}{5}\right\}$.

7. $\dfrac{x}{4} - 1 = \dfrac{x}{2}$

   $4\left(\dfrac{x}{4} - 1\right) = 4\left(\dfrac{x}{2}\right)$

   $4\left(\dfrac{x}{4}\right) - 4(1) = 2(x)$

   $x - 4 = 2x$

   $-4 = x$

   The solution set is $\{-4\}$.

9. One inch represents 12 feet. Let x represent the distance in feet between the two driveways.

$$\dfrac{1}{3\frac{1}{4}} = \dfrac{12}{x}$$

$$x(1) = \left(3\dfrac{1}{4}\right)(12)$$

$$x = \left(\frac{13}{4}\right)\left(\frac{12}{1}\right)$$
$$x = 39$$

The distance between the two driveways is 39 feet.

11. Let a represent the amount of money in one account. Then $1500 - a$ represents the amount of money in the second account.

$$\frac{a}{1500 - a} = \frac{2}{3}$$
$$3a = 2(1500 - a)$$
$$3a = 3000 - 2a$$
$$5a = 3000$$
$$a = 600$$

If $a = 600$, then $1500 - a = 900$. Therefore, one account has $600 and the other has $900.

13. $\frac{n}{100} = \frac{13}{25}$
$$25n = 1300$$
$$n = 52$$
$$\frac{13}{25} = 52\%$$

15. $\frac{n}{100} = \frac{7}{12}$
$$12n = 700$$
$$n = \frac{700}{12} = \frac{175}{3} = 58\frac{1}{3}$$
$$\frac{7}{12} = 58\frac{1}{3}\%$$

17. What is 12% of 58?
$$n = (12\%)(58)$$
$$n = (0.12)(58)$$
$$n = 6.96$$

Twelve percent of 58 is 6.96.

19. Eight percent of what number is 56? Let n represent the number to be found.
$$(8\%)(n) = 56$$
$$0.08n = 56$$
$$8n = 5600$$
$$n = 700$$

Eight percent of 700 is 56.

21. Thirty-two is what percent of 80? Let r represent the percent to be found.
$$32 = r(80)$$
$$\frac{32}{80} = r$$
$$\frac{2}{5} = r$$
$$\frac{2}{5} = \frac{40}{100} = 0.40$$
$$r = 40\%$$

Thirty-two is 40% of 80.

23. Sixty is what percent of 40? Let r represent the percent to be found.
$$60 = r(40)$$
$$\frac{60}{40} = r$$
$$\frac{3}{2} = r$$
$$\frac{3}{2} = \frac{150}{100} = 1.50$$
$$r = 150\%$$

Sixty is 150% of 40.

Section 3.2

1. $0.4x = 36$
$$4x = 360$$
$$x = 90$$

The solution set is {90}.

3. $3x + 0.2x = 128$
$$10(3x + 0.2x) = 10(128)$$
$$30x + 2x = 1280$$
$$32x = 1280$$
$$x = 40$$

The solution set is {40}.

5. $0.04a + 0.07a = 16.5$
$$100(0.04a + 0.07a) = 100(16.5)$$
$$4a + 7a = 1650$$
$$11a = 1650$$
$$a = 150$$

The solution set is {150}.

7. $0.08y = -0.15(y + 200) + 99$
$$8y = -15(y + 200) + 9900$$
$$8y = -15y - 3000 + 9900$$
$$8y = -15y + 6900$$
$$23y = 6900$$
$$y = 300$$

The solution set is {300}.

9. Let x represent the original price of the computer.
$$\begin{pmatrix}\text{Original}\\ \text{selling price}\end{pmatrix} - (\text{Discount}) = \begin{pmatrix}\text{Discount}\\ \text{sale price}\end{pmatrix}$$
$$(100\%)(x) - (20\%)(x) = 640$$
$$1x - 0.2x = 640$$
$$10x - 2x = 6400$$
$$8x = 6400$$
$$\frac{8x}{8} = \frac{6400}{8}$$
$$x = 800$$
The original price of the computer was $800.

11. Let x represent the discount sale price. Because the sunglasses are 30% off, we must pay 70% of the original price.
$$x = (0.70)(140)$$
$$x = 98$$
The sale price of the sunglasses is $98.

13. Let s represent the selling price of the wheels.
$$\begin{pmatrix}\text{Selling}\\ \text{price}\end{pmatrix} = (\text{Cost}) + \begin{pmatrix}\text{Profit}\\ (\%\text{ of cost})\end{pmatrix}$$
$$s = 450 + (80\%)(450)$$
$$s = 450 + (0.80)(450)$$
$$s = 450 + 360$$
$$s = 810$$
The selling price of the wheels is $810.

15. Let n represent the selling price of the photo.
$$\begin{pmatrix}\text{Selling}\\ \text{price}\end{pmatrix} = (\text{Cost}) + \begin{pmatrix}\text{Profit}\\ (\%\text{ of selling price})\end{pmatrix}$$
$$n = 80 + (0.60)(n)$$
$$n = 80 + 0.60n$$
$$10n = 800 + 6n$$
$$4n = 800$$
$$n = 200$$
The selling price of the photo must be $200.

17. $\quad i = Prt$
$$1620 = 10{,}000r(3)$$
$$1620 = 30{,}000r$$
$$\frac{1620}{30{,}000} = r$$
$$\frac{54}{1000} = r$$
$$0.054 = r$$
The annual interest rate is 5.4%.

19. $\quad i = Prt$
$$715 = P(0.065)(2)$$
$$715 = P(0.13)$$
$$\frac{715}{0.13} = r$$
$$\frac{71{,}500}{13} = r$$
$$5500 = r$$
The principal must be $5500.

21. $i = Prt$
$$i = 14{,}000(0.06)\left(\frac{1}{12}\right)$$
$$i = 840\left(\frac{1}{12}\right)$$
$$i = 70$$
The monthly interest will be $70.

Section 3.3

1. $\quad d = rt \quad$ for $d = 150, r = 60$
$$150 = 60t$$
$$2.5 = t$$

3. $\quad A = \frac{1}{2}h(B_1 + 18) \qquad A = 150, h = 10$
$$150 = \frac{1}{2}(10)(B_1 + 18)$$
$$150 = 5(B_1 + 18)$$
$$150 = 5B_1 + 90$$
$$60 = 5B_1$$
$$12 = B_1$$

5. Use $A = \frac{1}{2}bh$.
$$A = \frac{1}{2}bh \qquad A = 52, b = 16$$
$$52 = \frac{1}{2}(16)(h)$$
$$52 = 8(h)$$
$$6.5 = h$$
The height of the triangle is 6.5 feet.

7. $A = \frac{1}{2}h(B_1 + B_2) \qquad h = 10, B_1 = 12, B_2 = 16$
$$A = \frac{1}{2}(10)(12 + 16)$$
$$A = 5(28)$$
$$A = 140$$
The area of the deck is 140 square feet.

9. Use $S = 2\pi r^2 + 2\pi rh$.
$$S = 2\pi r^2 + 2\pi rh \qquad h = 24, r = 15$$
$$S = 2\pi(15)^2 + 2\pi(15)(24)$$
$$S = 2\pi(225) + 2\pi(360)$$

$S = 500\pi + 720\pi = 1170\pi$

The total surface area is 1170π square feet.

11. $\begin{pmatrix}\text{Area of}\\\text{deck}\end{pmatrix} = \begin{pmatrix}\text{Area of deck}\\\text{and pool}\end{pmatrix} - \begin{pmatrix}\text{Area of}\\\text{pool}\end{pmatrix}$

$\begin{aligned}\text{Area of deck}\\\text{and pool}\end{aligned} = lw$
$= (30 + 4 + 4)(20 + 4 + 4)$
$= (38)(28)$
$= 1064$

$\begin{aligned}\text{Area of}\\\text{pool}\end{aligned} = lw$
$= (30)(20)$
$= 600$

$\begin{pmatrix}\text{Area of}\\\text{deck}\end{pmatrix} = \begin{pmatrix}\text{Area of deck}\\\text{and pool}\end{pmatrix} - \begin{pmatrix}\text{Area of}\\\text{pool}\end{pmatrix}$
$= 1064 - 600$
$= 464$

The area of the deck is 464 square feet.

13. Solve $i = Prt$ for P.

$i = Prt$

$\dfrac{i}{rt} = \dfrac{Prt}{rt}$ Divide by rt

$\dfrac{i}{rt} = P$

15. Solve $A = \dfrac{1}{2}bh$ for b.

$A = \dfrac{1}{2}bh$

$2A = 2\left(\dfrac{1}{2}bh\right)$ Multiply by 2

$2A = bh$

$2A = \dfrac{bh}{h}$ Divide by h

$\dfrac{2A}{h} = b$

17. Solve $S = 2\pi r^2 + 2\pi rh$ for h.

$S = 2\pi r^2 + 2\pi rh$

$S - 2\pi r^2 = 2\pi rh$ Subtract $2\pi r^2$

$\dfrac{S - 2\pi r^2}{2\pi r} = \dfrac{2\pi rh}{2\pi r}$ Divide by $2\pi r$

$\dfrac{S - 2\pi r^2}{2\pi r} = h$

19. Solve $2x + y = -3$ for x.

$2x + y = -3$

$2x = -3 - y$ Subtract y

$x = \dfrac{-3 - y}{2}$ Divide by 2

21. Solve $3x - 8y = 5$ for y.

$3x - 8y = 5$

$-8y = 5 - 3x$ Subtract $3x$

$y = \dfrac{5 - 3x}{-8}$ Divide by -8

$y = \dfrac{5 - 3x}{-8} \cdot \dfrac{-1}{-1} = \dfrac{-5 + 3x}{8}$ Multiply by $\dfrac{-1}{-1}$

$y = \dfrac{3x - 5}{8}$

23. Solve $y = ax + b$ for a.

$y = ax + b$

$y - b = ax$ Subtract b

$\dfrac{y - b}{x} = \dfrac{ax}{x}$ Divide by x

$\dfrac{y - b}{x} = a$

Section 3.4

1. (a) $\dfrac{2}{5}b + b + 4 = 39$

$\dfrac{2}{5}b + \dfrac{5}{5}b + 4 = 39$

$\dfrac{7}{5}b + 4 = 39$

$\dfrac{7}{5}b = 35$

$\dfrac{5}{7}\left(\dfrac{7}{5}b\right) = \dfrac{5}{7}(35)$

$b = 25$

The solution set is {25}.

(b) $n + (n + 1) + (n + 2) = 153$

$n + n + 1 + n + 2 = 153$

$3n + 3 = 153$

$3n = 150$

$n = 50$

The solution set is {50}.

(c) $\dfrac{1}{3}c + \dfrac{2}{3}(c + 4) = 15$

$3\left[\dfrac{1}{3}c + \dfrac{2}{3}(c + 4)\right] = 3(15)$

$3\left(\dfrac{1}{3}c\right) + 3\left[\dfrac{2}{3}(c + 4)\right] = 45$

$c + 2(c + 4) = 45$

$c + 2c + 8 = 45$

$3c + 8 = 45$

$$3c = 37$$
$$c = \frac{37}{3}$$

The solution set is $\left\{\frac{37}{3}\right\}$.

3. Let t represent the time needed to earn $525.
Use $i = Prt$.
$$525 = 3000(0.10)t$$
$$525 = 300t$$
$$\frac{525}{300} = t$$
$$\frac{7}{4} = t$$
$$1.75 = t$$

It will take 1.75 years to earn $525.

5. Let w represent the width and l the length of the computer screen. The length is twice the width, so $l = 2w$.
Use $P = 2l + 2w$.
$$48 = 2(2w) + 2w$$
$$48 = 4w + 2w$$
$$48 = 6w$$
$$8 = w$$

If $w = 8$, then $l = 2w = 16$. The width is 8 inches, and the length is 16 inches.

7. Let x represent the time each walked. The total distance is 3.5 miles.

Rate · Time = Distance

	Rate	Time	Distance
Justin	5	x	$5x$
Hope	2	x	$2x$

$\left(\begin{array}{c}\text{Justin's}\\\text{distance}\end{array}\right) + \left(\begin{array}{c}\text{Hope's}\\\text{distance}\end{array}\right) = 3.5$

$$5x + 2x = 3.5$$
$$7x = 3.5$$
$$70x = 35$$
$$x = \frac{35}{70} = \frac{1}{2}$$

It will take Justin and Hope $\frac{1}{2}$ hour to meet.

9. Let t represent the time the student traveled and let $t - \frac{1}{4}$ represent the time his mom traveled. They both traveled the same distance.

Rate · Time = Distance

	Rate	Time	Distance
Student	60	t	$60t$
Mom	70	$t - \frac{1}{4}$	$70\left(t - \frac{1}{4}\right)$

$\left(\begin{array}{c}\text{Student's}\\\text{distance}\end{array}\right) = \left(\begin{array}{c}\text{Mom's}\\\text{distance}\end{array}\right)$

$$60t = 70\left(t - \frac{1}{4}\right)$$
$$60t = 70t - \frac{70}{4}$$
$$4(60t) = 4\left(70t - \frac{70}{4}\right)$$
$$240t = 280t - 70$$
$$-40t = -70$$
$$t = \frac{-70}{-40} = \frac{7}{4}$$

If $t = \frac{7}{4}$, then $t - \frac{1}{4} = \frac{6}{4} = \frac{3}{2}$.

His mom drove for 1.5 hours.

Section 3.5

1. Let x represent the amount of pure orange juice added.

	Amount of solution	Percent	Amount of pure orange juice
Original solution	36	0.10	(0.10)(36)
Pure orange juice	x	1.00	$1.00x$
Final solution	$36 + x$	0.20	$0.20(36 + x)$

$\left(\begin{array}{c}\text{Pure orange juice}\\\text{in original solution}\end{array}\right) + \left(\begin{array}{c}\text{Pure orange juice}\\\text{added}\end{array}\right)$
$= \left(\begin{array}{c}\text{Pure orange juice}\\\text{in final solution}\end{array}\right)$

$$(0.10)(36) + 1.00x = 0.20(36 + x)$$
$$36 + 10x = 2(36 + x)$$
$$36 + 10x = 72 + 2x$$
$$36 + 8x = 72$$
$$8x = 36$$
$$\frac{8x}{8} = \frac{36}{8}$$
$$x = \frac{36}{8} = \frac{9}{2}$$

The amount of pure orange juice that must be added is 4.5 quarts.

3. Let x represent the amount of 10% solution and let $6000 - x$ represent the amount of 20% solution.

	Amount of solution	Percent	Amount of pure ethanol
10% solution	x	0.10	$(0.10)(x)$
20% solution	$6000 - x$	0.20	$0.20(6000 - x)$
16% solution	6000	0.16	$0.16(6000)$

$$\begin{pmatrix}\text{Pure ethanol}\\\text{in 10\% solution}\end{pmatrix} + \begin{pmatrix}\text{Pure ethanol}\\\text{in 20\% solution}\end{pmatrix}$$
$$= \begin{pmatrix}\text{Pure ethanol}\\\text{in 16\% solution}\end{pmatrix}$$

$$(0.10)(x) + 0.20(6000 - x) = 0.16(6000)$$
$$10x + 20(6000 - x) = 16(6000)$$
$$10x + 120{,}000 - 20x = 96{,}000$$
$$-10x + 120{,}000 = 96{,}000$$
$$-10x = -24{,}000$$
$$\frac{-10x}{-10} = \frac{-24{,}000}{-10}$$
$$x = 2400$$

If $x = 2400$, then $6000 - x = 3600$. There should be 2400 gallons of 10% solution and 3600 gallons of 20% solution.

5. Let y represent the amount invested at 7% and let $8000 - y$ represent the amount invested at 5%.

	Amount invested	Percent	Interest earned
7% investment	y	0.07	$(0.07)(y)$
5% investment	$8000 - y$	0.05	$0.05(8000 - y)$
Total investment	8000		524

$$\begin{pmatrix}\text{Interest earned}\\\text{at 7\%}\end{pmatrix} + \begin{pmatrix}\text{Interest earned}\\\text{at 5\%}\end{pmatrix}$$
$$= \begin{pmatrix}\text{Total interest}\\\text{earned}\end{pmatrix}$$

$$(0.07)(y) + 0.05(8000 - y) = 524$$
$$7y + 5(8000 - y) = 52{,}400$$
$$7y + 40{,}000 - 5y = 52{,}400$$
$$2y + 40{,}000 = 52{,}400$$
$$2y = 12{,}400$$
$$\frac{2y}{2} = \frac{12{,}400}{2}$$
$$y = 6200$$

If $y = 6200$, then $8000 - y = 1800$. The amount invested at 7% is $6200, and the amount invested at 5% is $1800.

7. Let x represent the amount invested at 6% and let $x - 2000$ represent the amount invested at 9%.

	Amount invested	Percent	Interest earned
6% investment	x	0.06	$(0.06)(x)$
9% investment	$x - 2000$	0.09	$0.09(x - 2000)$
			870

$$\begin{pmatrix}\text{Interest earned}\\\text{at 6\%}\end{pmatrix} + \begin{pmatrix}\text{Interest earned}\\\text{at 9\%}\end{pmatrix}$$
$$= \begin{pmatrix}\text{Total interest}\\\text{earned}\end{pmatrix}$$

$$(0.06)(x) + 0.09(x - 2000) = 870$$
$$6x + 9(x - 2000) = 87{,}000$$
$$6x + 9x - 18{,}000 = 87{,}000$$
$$15x - 18{,}000 = 87{,}000$$
$$15x = 105{,}000$$
$$\frac{15x}{15} = \frac{105{,}000}{15}$$
$$x = 7000$$

If $x = 7000$, then $x - 2000 = 5000$. Mrs. Stewart invested $7000 at 6% and $5000 at 9%.

9. Let y represent Kyle's age now. We represent the unknown quantities as follows:

y: Kyle's present age
$y + 3$: Jennifer's present age
$y + 1$: Kyle's age in one year
$y + 3 + 1 = y + 4$: Jennifer's age in one year

$$\begin{pmatrix}\text{Kyle's age}\\\text{in one year}\end{pmatrix} = \frac{2}{3} \cdot \begin{pmatrix}\text{Jennifer's age}\\\text{in one year}\end{pmatrix}$$

$$y + 1 = \frac{2}{3}(y + 4)$$
$$3(y + 1) = 3\left[\frac{2}{3}(y + 4)\right]$$
$$3y + 3 = 2(y + 4)$$
$$3y + 3 = 2y + 8$$
$$y + 3 = 8$$
$$y = 5$$

If $y = 5$, then $y + 3 = 8$. Kyle's present age is 5, and Jennifer's present age is 8.

CHAPTER 4

Section 4.1

1. $(8x^2 - 2x + 6) + (3x^2 + 5x - 10)$
   $= (8x^2 + 3x^2) + (-2x + 5x) + (6 - 10)$
   $= 11x^2 + 3x - 4$

3. $(6x + 4) + (2x - 3) + (5x - 8)$
   $= (6x + 2x + 5x) + (4 - 3 - 8)$
   $= 13x - 7$

5. $(2x^2 - 3x + 4) + (5x^3 + 7x + 2) + (-2x - 6)$
   $= (5x^3) + (2x^2) + (-3x + 7x - 2x)$
   $\quad + (4 + 2 - 6)$
   $= 5x^3 + 2x^2 + 2x$

7. $(9x^2 - 8x - 2) - (3x^2 - 2x + 1)$
   $= (9x^2 - 8x - 2) + (-3x^2 + 2x - 1)$
   $= (9x^2 - 3x^2) + (-8x + 2x) + (-2 - 1)$
   $= 6x^2 - 6x - 3$

9. $(6a + 4) - (2a^2 - 3a - 1)$
   $= (6a + 4) + (-2a^2 + 3a + 1)$
   $= (-2a^2) + (6a + 3a) + (4 + 1)$
   $= -2a^2 + 9a + 5$

11. To subtract, add the opposite.
    $\begin{array}{r} 11x^2 + 8x - 2 \\ 4x^2 - x + 3 \end{array} \Rightarrow \begin{array}{r} 11x^2 + 8x - 2 \\ -4x^2 + x - 3 \\ \hline 7x^2 + 9x - 5 \end{array}$

13. To subtract, add the opposite.
    $\begin{array}{r} 5y^3 - 2y^2 \quad\quad + 8 \\ 8y^3 \quad\quad + 6y - 4 \end{array} \Rightarrow \begin{array}{r} 5y^3 - 2y^2 \quad\quad + 8 \\ -8y^3 \quad\quad - 6y + 4 \\ \hline -3y^3 - 2y^2 - 6y + 12 \end{array}$

15. $(6x + 1) + (3x - 2) - (5x - 8)$
    $= 1(6x + 1) + 1(3x - 2) - 1(5x - 8)$
    $= 6x + 1 + 3x - 2 - 5x + 8$
    $= (6x + 3x - 5x) + (1 - 2 + 8)$
    $= 4x + 7$

17. $(x^2 + 3x - 4) - (4x^2 + 5x - 6) + (2x + 7)$
    $= x^2 + 3x - 4 - 4x^2 - 5x + 6 + 2x + 7$
    $= (x^2 - 4x^2) + (3x - 5x + 2x) + (-4 + 6 + 7)$
    $= -3x^2 + 9$

19. $10y - [7y + (y - 4)]$
    $= 10y - [7y + y - 4]$
    $= 10y - [8y - 4]$
    $= 10y - 8y + 4$
    $= 2y + 4$

21. $12 - \{4y - [6 + (3y - 2)] + 10\}$
    $= 12 - \{4y - [6 + 3y - 2] + 10\}$
    $= 12 - \{4y - [4 + 3y] + 10\}$
    $= 12 - (4y - 4 - 3y + 10)$
    $= 12 - (y + 6)$
    $= 12 - y - 6$
    $= -y + 6$

23. Area of circle: πr^2
    Area of rectangle: $10r$
    Sum of areas $= \pi r^2 + 10r$

Section 4.2

1. (a) $y^3 \cdot y^5 = y^{3+5} = y^8$
   (b) $a^2 \cdot a = a^{2+1} = a^3$

3. (a) $(-4x^2)(6x^3) = -4 \cdot 6 \cdot x^2 \cdot x^3$
       $= -24x^5$
   (b) $\left(\dfrac{2}{3}a^5b\right)\left(\dfrac{6}{7}a^4b^2\right) = \dfrac{2}{3} \cdot \dfrac{6}{7} \cdot a^5 \cdot a^4 \cdot b \cdot b^2$
       $= \dfrac{4}{7}a^9 b^3$

5. (a) $(y^3)^2 = y^{3 \cdot 2} = y^6$
   (b) $(m^2)^4 = m^{2 \cdot 4} = m^8$

7. (a) $(2x^3)^4 = (2)^4(x^3)^4 = 16x^{12}$
   (b) $(-3a^3b^5)^2 = (-3)^2(a^3)^2(b^5)^2 = 9a^6 b^{10}$

9. (a) $(5a^3)^2(-2a^4)^3 = (5)^2(a^3)^2(-2)^3(a^4)^3$
       $= 25(a^6)(-8)(a^{12})$
       $= -200a^{18}$
   (b) $(3m^4n^2)^3(-mn)^2 = (3)^3(m^4)^3(n^2)^3(-1)^2(m)^2(n)^2$
       $= 27(m)^{12}(n^6)(1)(m^2)(n^2)$
       $= 27m^{14}n^8$

11. (a) $(4a^3)(a^2 - 3ab - 5b^2)$
        $= (4a^3)(a^2) - (4a^3)(3ab) - (4a^3)(5b^2)$
        $= 4a^5 - 12a^4b - 20a^3b^2$
    (b) $(-2y)(5x^2y + 6xy + y^3)$
        $= (-2y)(5x^2y) + (-2y)(6xy) + (-2y)(y^3)$
        $= -10x^2y^2 - 12xy^2 - 2y^4$

13. Area of floor plan:
    Area of left side: $(2x)(2x) = 4x^2$
    Area of middle region: $(4x)(x) = 4x^2$
    Area of right side: $(2x)(2x) = 4x^2$
    Total area $= 4x^2 + 4x^2 + 4x^2 = 12x^2$.

Section 4.3

1. $(x + 5)(y + 2) = x(y + 2) + 5(y + 2)$
   $= x(y) + x(2) + 5(y) + 5(2)$
   $= xy + 2x + 5y + 10$

3. $(x - 4)(y + z - 3) = x(y + z - 3) - 4(y + z - 3)$
   $= x(y) + x(z) + x(-3) - 4(y)$
   $\quad - 4(z) - 4(-3)$
   $= xy + xz - 3x - 4y - 4z + 12$

5. $(x + 6)(x + 5) = x(x + 5) + 6(x + 5)$
   $= x^2 + 5x + 6x + 30$
   $= x^2 + 11x + 30$

7. $(x + 2)(x - 8) = x(x - 8) + 2(x - 8)$
   $= x^2 - 8x + 2x - 16$
   $= x^2 - 6x - 16$

9. $(x + 3)(x^2 + 2x + 6)$
   $= x(x^2 + 2x + 6) + 3(x^2 + 2x + 6)$
   $= x^3 + 2x^2 + 6x + 3x^2 + 6x + 18$
   $= x^3 + 2x^2 + 3x^2 + 6x + 6x + 18$
   $= x^3 + 5x^2 + 12x + 18$

11. $(x - 2y)(5x^2 + 3xy + 4y^2)$
    $= x(5x^2 + 3xy + 4y^2) - 2y(5x^2 + 3xy + 4y^2)$
    $= 5x^3 + 3x^2y + 4xy^2 - 10x^2y - 6xy^2 - 8y^3$
    $= 5x^3 + 3x^2y - 10x^2y + 4xy^2 - 6xy^2 - 8y^3$
    $= 5x^3 - 7x^2y - 2xy^2 - 8y^3$

13. $(x + 1)(x + 6) = x^2 + 7x + 6$

15. $(x + 3)(x - 7) = x^2 - 4x - 21$

17. $(4x + 1)(2x - 3) = 8x^2 - 10x - 3$

19. $(x + 5)^3 = (x + 5)(x + 5)(x + 5)$
    $= (x + 5)[(x + 5)(x + 5)]$
    $= (x + 5)(x^2 + 10x + 25)$
    $= x(x^2 + 10x + 25) + 5(x^2 + 10x + 25)$
    $= x^3 + 10x^2 + 25x + 5x^2 + 50x + 125$
    $= x^3 + 10x^2 + 5x^2 + 25x + 50x + 125$
    $= x^3 + 15x^2 + 75x + 125$

21. $(x + 8)^2 = (x)^2 + 2(x)(8) + (8)^2$
    $= x^2 + 16x + 64$

23. $(x - 1) = (x)^2 - 2(x)(1) + (1)^2$
    $= x^2 - 2x + 1$

25. $(3x + 5)(3x - 5) = (3x)^2 - (5)^2$
    $= 9x^2 - 25$

27. Let y represent the length of each side of the square corner piece to be removed.

    Then $10 - 2y$ represents the length of the cardboard box after the corner pieces have been removed.

    And $6 - 2y$ represents the width of the cardboard box after the corner pieces have been removed.

    Volume $= lwh$
    $= (10 - 2y)(6 - 2y)(y)$
    $= (60 - 20y - 12y + 4y^2)(y)$
    $= (60 - 32y + 4y^2)(y)$
    $= 60y - 32y^2 + 4y^3$ cubic inches

    Surface area of original piece of cardboard
    $\quad -$ Area of four cut-out pieces
    $= (10)(6) - (4)(y)(y)$
    $= 60 - 4y^2$ square inches

Section 4.4

1. (a) $\dfrac{-32b^8}{4b^2} = -8b^{8-2} = -8b^6$

   (b) $\dfrac{6a^4b^7}{-6ab^2} = -1a^{4-1}b^{7-2} = -a^3b^5$

3. (a) $\dfrac{21m^4n^5 + 14m^2n^3}{-7m^2n}$
   $= \dfrac{21m^4n^5}{-7m^2n} + \dfrac{14m^2n^3}{-7m^2n}$
   $= -3m^2n^4 - 2m^0n^2$
   $= -3m^2n^4 - 2n^2$

   (b) $\dfrac{12a^5 - 18a^4 - 36a^3}{6a^3}$
   $= \dfrac{12a^5}{6a^3} - \dfrac{18a^4}{6a^3} - \dfrac{36a^3}{6a^3}$
   $= 2a^2 - 3a - 6a^0$
   $= 2a^2 - 3a - 6$

Section 4.5

1.
```
            x + 7
      ┌─────────────
x + 4 │ x² + 11x − 28
        x² + 4x
        ─────────
              7x + 28
              7x + 28
              ───────
                    0
```
$(x^2 + 11x + 28) \div (x + 4) = x + 7$

3.
```
            3x + 1
      ┌─────────────
x − 2 │ 3x² − 5x − 2
        3x² − 6x
        ─────────
              x − 2
              x − 2
              ─────
                  0
```
$(3x^2 - 5x - 2) \div (x - 2) = 3x + 1$

5.
$$\begin{array}{r} 2x+5 \\ 4x-3 \overline{\smash{\big)}8x^2+14x-15} \\ \underline{8x^2-6x} \\ 20x-15 \\ \underline{20x-15} \\ 0 \end{array}$$

$(8x^2+14x-15) \div (4x-3) = 2x+5$

or $\dfrac{8x^2+14x-15}{4x-3} = 2x+5$

7.
$$\begin{array}{r} 4x-6 \\ x+3 \overline{\smash{\big)}4x^2+6x-4} \\ \underline{4x^2+12x} \\ -6x-4 \\ \underline{-6x-18} \\ 14 \end{array}$$

$\dfrac{4x^2+6x-4}{x+3} = 4x-6+\dfrac{14}{x+3}$

9.
$$\begin{array}{r} a^2-3a+9 \\ a+3 \overline{\smash{\big)}a^3+0a^2+0a+27} \\ \underline{a^3+3a^2} \\ -3a^2+0a \\ \underline{-3a^2-9a} \\ 9a+27 \\ \underline{9a+27} \\ 0 \end{array}$$

$\dfrac{a^3+27}{a+3} = a^2-3a+9$

11.
$$\begin{array}{r} a+4 \\ a^2+a \overline{\smash{\big)}a^3+5a^2-2a+3} \\ \underline{a^3+a^2} \\ 4a^2-2a \\ \underline{4a^2+4a} \\ -6a+3 \end{array}$$

$\dfrac{a^3+5a^2-2a+3}{a^2+a} = a+4+\dfrac{-6a+3}{a^2+a}$

Section 4.6

1. (a) $\dfrac{m^4}{m^7} = m^{4-7} = m^{-3} = \dfrac{1}{m^3}$

(b) $\dfrac{y^{-2}}{y^{-6}} = y^{-2-(-6)} = y^{-2+6} = y^4$

(c) $\dfrac{z^{-3}}{z^2} = z^{-3-2} = z^{-5} = \dfrac{1}{z^5}$

(d) $\dfrac{k^5}{k^5} = k^{5-5} = k^0 = 1$

3. (a) $3^5 \cdot 3^{-8} = 3^{5-8} = 3^{-3} = \dfrac{1}{3^3} = \dfrac{1}{27}$

(b) $10^{-2} \cdot 10^{-1} = 10^{-2+(-1)} = 10^{-3} = \dfrac{1}{10^3} = \dfrac{1}{1000}$

(c) $\dfrac{10^{-1}}{10^{-4}} = 10^{-1-(-4)} = 10^{-1+4} = 10^3 = 1000$

(d) $(2^{-5})^{-1} = 2^{(-5)(-1)} = 2^5 = 32$

5. (a) $z^{-1} \cdot z^{-4} = z^{-1+(-4)} = z^{-5} = \dfrac{1}{z^5}$

(b) $(a^5 b^{-1})^{-2} = (a^5)^{-2}(b^{-1})^{-2} = a^{-10}b^2 = \dfrac{b^2}{a^{10}}$

(c) $\left(\dfrac{m^2}{n^{-4}}\right)^{-1} = \dfrac{(m^2)^{-1}}{(n^{-4})^{-1}} = \dfrac{m^{-2}}{n^4} = \dfrac{1}{m^2 n^4}$

(d) $(2a^{-5}b^2)(-5a^3 b) = (2)(-5)(a^{-5+3})(b^{2+1})$
$= -10a^{-2}b^3$
$= \dfrac{-10b^3}{a^2}$

7. (a) $0.00000683 = (6.83)(10^{-6})$

(b) $5{,}600{,}000 = (5.6)(10^6)$

(c) $9.54 = (9.54)(10^0)$

9. (a) $(8.2)(10^7) = 82{,}000{,}000$

(b) $(3.46)(10^{-4}) = 0.000346$

11. $(50{,}000)(0.0000013) = (5)(10^4)(1.3)(10^{-6})$
$= (5)(1.3)(10^4)(10^{-6})$
$= (6.5)(10^{4-6})$
$= (6.5)(10^{-2})$
$= 0.065$

13. $\dfrac{8400}{0.042} = \dfrac{(8.4)(10^3)}{(4.2)(10^{-2})}$
$= \dfrac{(8.4)}{(4.2)} \cdot \dfrac{(10^3)}{(10^{-2})}$
$= (2)(10^{3-(-2)})$
$= (2)(10^5)$
$= 200{,}000$

15. $\dfrac{(900)(0.0006)}{(30{,}000)(0.02)} = \dfrac{(9)(10^2)(6)(10^{-4})}{(3)(10^4)(2)(10^{-2})}$
$= \dfrac{(9)(6)(10^2)(10^{-4})}{(3)(2)(10^4)(10^{-2})}$
$= \dfrac{(54)}{(6)} \cdot \dfrac{(10^{2-4})}{(10^{4-2})}$
$= (9) \cdot \dfrac{(10^{-2})}{(10^2)}$
$= (9)(10^{-2-2})$
$= (9)(10^{-4})$
$= 0.0009$

Section 5.1

1. $14a^2 = 2 \cdot 7 \cdot a \cdot a$
   $7a^5 = 7 \cdot a \cdot a \cdot a \cdot a \cdot a$
   The greatest common factor is $7 \cdot a \cdot a = 7a^2$.

3. $18m^2n^4 = 2 \cdot 3 \cdot 3 \cdot m \cdot m \cdot n \cdot n \cdot n \cdot n$
   $4m^3n^5 = 2 \cdot 2 \cdot m \cdot m \cdot m \cdot n \cdot n \cdot n \cdot n \cdot n$
   $10m^4n^3 = 2 \cdot 5 \cdot m \cdot m \cdot m \cdot m \cdot n \cdot n \cdot n$
   The greatest common factor is $2 \cdot m \cdot m \cdot n \cdot n \cdot n = 2m^2n^3$.

5. $15a^2 + 21a^6 = 3a^2(5) + 3a^2(7a^4)$
   $= 3a^2(5 + 7a^4)$

7. $8m^3n^2 + 2m^6n = 2m^3n(4n) + 2m^3n(m^3)$
   $= 2m^3n(4n + m^3)$

9. $48y^8 - 16y^6 + 24y^4 = 8y^4(6y^4) - 8y^4(2y^2) + 8y^4(3)$
   $= 8y^4(6y^4 - 2y^2 + 3)$

11. $8b^3 + 8b^2 = 8b^2(b) + 8b^2(1)$
    $= 8b^2(b + 1)$

13. (a) $ab + 5a + 3b + 15 = a(b + 5) + 3(b + 5)$
    $= (b + 5)(a + 3)$

    (b) $xy - 2x + 4y - 8 = x(y - 2) + 4(y - 2)$
    $= (y - 2)(x + 4)$

15. $y^2 - 7y = 0$
    $y(y - 7) = 0$
    $y = 0$ or $y - 7 = 0$
    $y = 0$ or $y = 7$
    The solution set is $\{0, 7\}$.

17. $a^2 = 15a$
    $a^2 - 15a = 0$
    $a(a - 15) = 0$
    $a = 0$ or $a - 15 = 0$
    $a = 0$ or $a = 15$
    The solution set is $\{0, 15\}$.

19. $5y^2 + 2y = 0$
    $y(5y + 2) = 0$
    $y = 0$ or $5y + 2 = 0$
    $y = 0$ or $5y = -2$
    $y = 0$ or $y = \dfrac{-2}{5}$
    The solution set is $\left\{-\dfrac{2}{5}, 0\right\}$.

21. $m(m - 2) + 5(m - 2) = 0$
    $(m - 2)(m + 5) = 0$
    $m - 2 = 0$ or $m + 5 = 0$
    $m = 2$ or $m = -5$
    The solution set is $\{-5, 2\}$.

23. Let s represent the length of the side of the square. Use these formulas:
    $A = s^2$ Area of a square
    $P = 4s$ Perimeter of a square
    $\left(\begin{array}{c}\text{Area of}\\ \text{square}\end{array}\right) = 3 \cdot \left(\begin{array}{c}\text{Perimeter}\\ \text{of square}\end{array}\right)$
    $s^2 = 3(4s)$
    $s^2 = 12s$
    $s^2 - 12s = 0$
    $s(s - 12) = 0$
    $s = 0$ or $s - 12 = 0$
    $s = 0$ or $s = 12$
    Discard $s = 0$, because length cannot be 0.
    The length is 12.

Section 5.2

1. $3y^2 - 12 = 3(y^2 - 4)$
   $= 3(y - 2)(y + 2)$

3. $32x^3 - 2x = 2x(16x^2 - 1)$
   $= 2x(4x - 1)(4x + 1)$

5. $a^4 - 81 = (a^2 + 9)(a^2 - 9)$
   $= (a^2 + 9)(a + 3)(a - 3)$

7. $x^2 = 64$
   $x^2 - 64 = 0$
   $(x + 8)(x - 8) = 0$
   $x + 8 = 0$ or $x - 8 = 0$
   $x = -8$ or $x = 8$
   The solution set is $\{-8, 8\}$.

9. $4x^2 = 9$
   $4x^2 - 9 = 0$
   $(2x + 3)(2x - 3) = 0$
   $2x + 3 = 0$ or $2x - 3 = 0$
   $2x = -3$ or $2x = 3$
   $x = -\dfrac{3}{2}$ or $x = \dfrac{3}{2}$
   The solution set is $\left\{-\dfrac{3}{2}, \dfrac{3}{2}\right\}$.

11.
$$3x^2 = 75$$
$$\frac{3x^2}{3} = \frac{75}{3}$$
$$x^2 = 25$$
$$x^2 - 25 = 0$$
$$(x + 5)(x - 5) = 0$$
$$x + 5 = 0 \quad \text{or} \quad x - 5 = 0$$
$$x = -5 \quad \text{or} \quad x = 5$$
The solution set is $\{-5, 5\}$.

13.
$$x^3 - 49x = 0$$
$$x(x^2 - 49) = 0$$
$$x(x + 7)(x - 7) = 0$$
$$x = 0 \quad \text{or} \quad x + 7 = 0 \quad \text{or} \quad x - 7 = 0$$
$$x = 0 \quad \text{or} \quad x = -7 \quad \text{or} \quad x = 7$$
The solution set is $\{-7, 0, 7\}$.

15. Let y represent the length of the side of one square. Then $3y$ represents the length of the side of the other square. The sum of the areas of the squares is 250 square feet.

$$\begin{pmatrix}\text{Area of}\\\text{first square}\end{pmatrix} + \begin{pmatrix}\text{Area of}\\\text{second square}\end{pmatrix} = 250$$

$$y^2 + (3y)^2 = 250$$
$$y^2 + 9y^2 = 250$$
$$10y^2 = 250$$
$$\frac{10y^2}{10} = \frac{250}{10}$$
$$y^2 = 25$$
$$y^2 - 25 = 0$$
$$(y + 5)(y - 5) = 0$$
$$y + 5 = 0 \quad \text{or} \quad y - 5 = 0$$
$$y = -5 \quad \text{or} \quad y = 5$$

Discard $y = -5$, because length cannot be negative. The length of the side of the first square is 5 feet, and the length of the side of the second square is $3(5) = 15$ feet.

Section 5.3

1. $y^2 + 9y + 20$

We need two numbers with a sum of 9 and a product of 20.

Product	Sum
$1(20) = 20$	$1 + 20 = 21$
$2(10) = 20$	$2 + 10 = 12$
$4(5) = 20$	$4 + 5 = 9$

The two numbers are 4 and 5.
$y^2 + 9y + 20 = (y + 4)(y + 5)$

3. $m^2 - 8m + 15$

We need two numbers with a sum of -8 and a product of 15.

Product	Sum
$1(15) = 15$	$1 + 15 = 16$
$3(5) = 15$	$3 + 5 = 8$
$-3(-5) = 15$	$-3 + (-5) = -8$

The two numbers are -3 and -5.
$m^2 - 8m + 15 = (m - 3)(m - 5)$

5. $y^2 + 5y - 24$

We need two numbers with a sum of 5 and a product of -24.

Product	Sum
$1(-24) = -24$	$1 + (-24) = -23$
$2(-12) = -24$	$2 + (-12) = -10$
$3(-8) = -24$	$3 + (-8) = -5$
$(-3)(8) = -24$	$-3 + 8 = 5$

The two numbers are -3 and 8.
$y^2 + 5y - 24 = (y - 3)(y + 8)$

7. $y^2 - 4y - 12$

We need two numbers with a sum of -4 and a product of -12.

Product	Sum
$1(-12) = -12$	$1 + (-12) = -11$
$2(-6) = -12$	$2 + (-6) = -4$

The two numbers are -6 and 2.
$y^2 - 4y - 12 = (y - 6)(y + 2)$

9. $y^2 - 7y + 6$

We need two numbers with a sum of -7 and a product of 6.

Product	Sum
$1(6) = 6$	$1 + 6 = 7$
$(-1)(-6) = 6$	$-1 + (-6) = -7$

The two numbers are -6 and -1.
$y^2 - 7y + 6 = (y - 6)(y - 1)$

11. $a^2 + a - 12$

We need two numbers with a sum of 1 and a product of -12.

Product	Sum
$1(-12) = -12$	$1 + (-12) = -11$
$2(-6) = -12$	$2 + (-6) = -4$
$3(-4) = -12$	$3 + (-4) = -1$
$(-3)(4) = -12$	$-3 + 4 = 1$

The two numbers are -3 and 4.
$a^2 + a - 12 = (a - 3)(a + 4)$

13. $y^2 + 7y + 18$

   We need two numbers with a sum of 7 and a product of 18.

Product	Sum
$1(18) = 18$	$1 + 18 = 19$
$2(9) = 18$	$2 + 9 = 11$
$3(6) = 18$	$3 + 6 = 9$
$-1(-18) = 18$	$-1 + (-18) = -19$
$-2(-9) = 18$	$-2 + (-9) = -11$
$(-3)(-6) = 18$	$-3 + (-6) = -9$

   No two numbers will satisfy these conditions.
   $y^2 + 7y + 18$ is not factorable.

15. $y^2 - 14y + 24 = 0$
    $(y - 12)(y - 2) = 0$
    $y - 12 = 0$ or $y - 2 = 0$
    $y = 12$ or $y = 2$
    The solution set is $\{2, 12\}$.

17. $a^2 + 3a - 28 = 0$
    $(a + 7)(a - 4) = 0$
    $a + 7 = 0$ or $a - 4 = 0$
    $a = -7$ or $a = 4$
    The solution set is $\{-7, 4\}$.

19. $x^2 - 18x = 40$
    $x^2 - 18x - 40 = 0$
    $(x - 20)(x + 2) = 0$
    $x - 20 = 0$ or $x + 2 = 0$
    $x = 20$ or $x = -2$
    The solution set is $\{-2, 20\}$.

21. Let n represent the first consecutive integer. Then $n + 1$ represents the second consecutive integer. The product of the integers is 110.
    $n(n + 1) = 110$
    $n^2 + n = 110$
    $n^2 + n - 110 = 0$
    $(n + 11)(n - 10) = 0$
    $n + 11 = 0$ or $n - 10 = 0$
    $n = -11$ or $n = 10$
    If $n = -11$, then $n + 1 = -10$.
    If $n = 10$, then $n + 1 = 11$.
    The consecutive integers are -11 and -10 or 10 and 11.

23. Let z represent the width of the rectangular plot. Then $z + 2$ represents the length. To solve, use $A = lw$.
    $80 = (z + 2)z$
    $80 = z^2 + 2z$
    $0 = z^2 + 2z - 80$
    $0 = (z - 8)(z + 10)$
    $z + 10 = 0$ or $z - 8 = 0$
    $z = -10$ or $z = 8$
    Discard $z = -10$, because width cannot be negative. The width is 8 yards, and the length is $z + 2 = 10$ yards.

25. Let x represent the length of one leg of a right triangle and let $x + 7$ represent the other leg. To solve, use the Pythagorean theorem.
    $a^2 + b^2 = c^2$
    $(x)^2 + (x + 7)^2 = (13)^2$
    $x^2 + x^2 + 14x + 49 = 169$
    $2x^2 + 14x + 49 = 169$
    $2x^2 + 14x - 120 = 0$
    $\frac{1}{2}(2x^2 + 14x - 120) = \frac{1}{2}(0)$
    $x^2 + 7x - 60 = 0$
    $(x + 12)(x - 5) = 0$
    $x + 12 = 0$ or $x - 5 = 0$
    $x = -12$ or $x = 5$
    Discard $x = -12$, because length cannot be negative. If $x = 5$, then $x + 7 = 12$. The legs of the right triangle are 5 inches and 12 inches.

Section 5.4

1. $3y^2 + 16y + 5$

   The binomial will have the form $(3y + \underline{})(y + \underline{})$. There are only two factors of 5, so there are only two possibilities.
   $(3y + 1)(y + 5)$ or $(3y + 5)(y + 1)$
   By checking we find that
   $3y^2 + 16y + 5 = (3y + 1)(y + 5)$

3. $4a^2 - 8a + 3$

   We find that $4a^2$ can be written as $4a \cdot a$ or $2a \cdot 2a$. Because the last term is positive and the middle term is negative, we have two possible forms:
   $(4a - \underline{})(a - \underline{})$ or $(2a - \underline{})(2a - \underline{})$
   The factors of 3 are 3 and 1, so we have the following possibilities:
   $(4a - 1)(a - 3)$ $(2a - 1)(2a - 3)$
   $(4a - 3)(a - 1)$
   By checking we find that
   $4a^2 - 8a + 3 = (2a - 1)(2a - 3)$

5. $6y^2 + 27y + 30$ has a common factor of 3. So $6y^2 + 27y + 30 = 3(2y^2 + 9y + 10)$.
   Now factor $2y^2 + 9y + 10$.

We find that $2y^2$ can be written as $2y \cdot y$. The binomial will have the form $(2y + \_\_)(y + \_\_)$ because the last term is positive and the middle term is positive.

The possible factors of 10 are $1 \cdot 10$ and $2 \cdot 5$, which will give the following possibilities:

$(2y + 1)(y + 10)$ $(2y + 2)(y + 5)$
$(2y + 10)(y + 1)$ $(2y + 5)(y + 2)$

By checking we find that
$2y^2 + 9y + 10 = (2y + 5)(y + 2)$
Therefore, $6y^2 + 27y + 30 = 3(2y + 5)(y + 2)$.

7. $2y^2 + 7y + 6 = 0$
$(2y + 3)(y + 2) = 0$
$2y + 3 = 0$ or $y + 2 = 0$
$2y = -3$ or $y = -2$
$\dfrac{2y}{2} = \dfrac{-3}{2}$ or $y = -2$
$y = -\dfrac{3}{2}$ or $y = -2$

The solution set is $\left\{-2, -\dfrac{3}{2}\right\}$.

9. $2a^2 - 5a - 12 = 0$
$(2a + 3)(a - 4) = 0$
$2a + 3 = 0$ or $a - 4 = 0$
$2a = -3$ or $a = 4$
$\dfrac{2a}{2} = \dfrac{-3}{2}$ or $a = 4$
$a = -\dfrac{3}{2}$ or $a = 4$

The solution set is $\left\{-\dfrac{3}{2}, 4\right\}$.

Section 5.5

1. **(a)** $a^2 + 10a + 25 = (a)^2 + 2(a)(5) + (5)^2$
$= (a + 5)(a + 5)$
$= (a + 5)^2$

   (b) $36x^2 - 6x + 1 = (6x)^2 - 2(6x)(1) + (1)^2$
$= (6x - 1)(6x - 1)$
$= (6x - 1)^2$

   (c) $49m^2 + 56mn + 16n^2$
$= (7m)^2 + 2(7m)(4n) + (4n)^2$
$= (7m + 4n)(7m + 4n)$
$= (7m + 4n)^2$

3. $3x^2 - 6x - 72 = 3(x^2 - 2x - 24)$
$= 3(x - 6)(x + 4)$

5. $4x^2 + 100 = 4(x^2 + 25)$

7. $9x^2 - 30x + 25 = (3x)^2 - 2(3x)(5) + (5)^2$
$= (3x - 5)(3x - 5)$
$= (3x - 5)^2$

9. $x^2 - 8x - 24$

   We need two numbers with a sum of -8 and a product of -24.

Product	Sum
$1(-24) = -24$	$1 + (-24) = -23$
$2(-12) = -24$	$2 + (-12) = -10$
$3(-8) = -24$	$3 + (-8) = -5$
$(4)(-6) = -24$	$4 + (-6) = -2$

   There are no numbers that satisfy those conditions. Therefore, $x^2 - 8x - 24$ is not factorable.

11. $3x^2 - 10x - 8$

    We find that $3x^2$ can be written as $3x \cdot x$. Because the last term is negative and the middle term is negative, the two binomials will have different signs.

    The possible factors of 8 are $1 \cdot 8$ and $2 \cdot 4$, so we must test for both possibilities.

    Testing for $1 \cdot 8$:
    $(3x + 1)(x - 8)$ $(3x - 1)(x + 8)$
    $(3x + 8)(x - 1)$ $(3x - 8)(x + 1)$
    None of these gives $-10x$ as the middle term.

    Testing for $2 \cdot 4$:
    $(3x + 2)(x - 4)$ $(3x - 2)(x + 4)$
    $(3x + 4)(x - 2)$ $(3x - 4)(x + 2)$

    By checking we find that
    $3x^2 - 10x - 8 = (3x + 2)(x - 4)$

13. $4x^2 - 100 = 4(x^2 - 25)$
$= 4(x + 5)(x - 5)$

15. $y^2 = 4y$
$y^2 - 4y = 0$
$y(y - 4) = 0$
$y = 0$ or $y - 4 = 0$
$y = 0$ or $y = 4$
The solution set is $\{0, 4\}$.

17. $z^3 - 25z = 0$
$z(z^2 - 25) = 0$
$z(z + 5)(z - 5) = 0$
$z = 0$ or $z + 5 = 0$ or $z - 5 = 0$
$z = 0$ or $z = -5$ or $z = 5$
The solution set is $\{-5, 0, 5\}$.

19. $3y^2 + 2y - 8 = 0$
$(3y - 4)(y + 2) = 0$

$3y - 4 = 0$ or $y + 2 = 0$
$3y = 4$ or $y = -2$
$y = \dfrac{4}{3}$ or $y = -2$

The solution set is $\left\{-2, \dfrac{4}{3}\right\}$.

21. $25y^2 + 20y + 4 = 0$
$(5y + 2)(5y + 2) = 0$
$5y + 2 = 0$ or $5y + 2 = 0$
$5y = -2$ or $5y = -2$
$y = \dfrac{-2}{5}$ or $y = \dfrac{-2}{5}$

The solution set is $\left\{-\dfrac{2}{5}\right\}$.

23. $(x + 1)(x + 2) = 20$
$x^2 + 3x + 2 = 20$
$x^2 + 3x - 18 = 0$
$(x + 6)(x - 3) = 0$
$x + 6 = 0$ or $x - 3 = 0$
$x = -6$ or $x = 3$

The solution set is $\{-6, 3\}$.

25. $3y^2 - 12y - 36 = 0$
$\dfrac{1}{3}(3y^2 - 12y - 36) = \dfrac{1}{3}(0)$
$y^2 - 4y - 12 = 0$
$(y - 6)(y + 2) = 0$
$y - 6 = 0$ or $y + 2 = 0$
$y = 6$ or $y = -2$

The solution set is $\{-2, 6\}$.

27. Let x represent one number. Then $2x + 1$ represents the other number. Their product is 36.

$x(2x + 1) = 36$
$2x^2 + x = 36$
$2x^2 + x - 36 = 0$
$(2x + 9)(x - 4) = 0$
$2x + 9 = 0$ or $x - 4 = 0$
$2x = -9$ or $x = 4$
$x = \dfrac{-9}{2}$ or $x = 4$

If $x = -\dfrac{9}{2}$, then $2x + 1 = 2\left(-\dfrac{9}{2}\right) + 1 = -9 + 1$
$= -8$.

If $x = 4$, then $2x + 1 = 2(4) + 1 = 8 + 1 = 9$.

Therefore, the numbers are $-\dfrac{9}{2}$ and -8 or 4 and 9.

29. Let a represent the altitude and let $a + 4$ represent the side. To solve, use $A = \dfrac{1}{2}bh$.

$30 = \dfrac{1}{2}(a)(a + 4)$
$60 = (a)(a + 4)$
$60 = a^2 + 4a$
$0 = a^2 + 4a - 60$
$0 = (a - 6)(a + 10)$
$a - 6 = 0$ or $a + 10 = 0$
$a = 6$ or $a = -10$

Discard $a = -10$, because length cannot be negative. If $a = 6$, then $a + 4 = 10$. The altitude is 6 inches, and the side is 10 inches.

CHAPTER 6

Section 6.1

1. $\dfrac{2a^2 + 8a}{a^2 - 16} = \dfrac{2a(a+4)}{(a+4)(a - 4)} = \dfrac{2a}{a - 4}$

3. $\dfrac{x - 3}{x^2 - 5x + 6} = \dfrac{1(x-3)}{(x-3)(x - 2)} = \dfrac{1}{x - 2}$

5. $\dfrac{a^2 + a - 20}{2a^2 + 11a + 5} = \dfrac{(a+5)(a - 4)}{(2a + 1)(a+5)} = \dfrac{a - 4}{2a + 1}$

7. $\dfrac{xy^2 + x^2y}{x^2 + xy} = \dfrac{xy(y+x)}{x(x+y)} = \dfrac{y}{1} = y$

9. $\dfrac{5a^2 - 8a - 4}{3a^3b - 12ab} = \dfrac{(5a + 2)(a - 2)}{3ab(a^2 - 4)}$
$= \dfrac{(5a + 2)(a-2)}{3ab(a + 2)(a-2)} = \dfrac{5a + 2}{3ab(a + 2)}$

11. $\dfrac{-5x - 10}{6x + 12} = \dfrac{-5(x+2)}{6(x+2)} = -\dfrac{5}{6}$

13. $\dfrac{20 - 5y}{y - 4} = \dfrac{5(4 - y)}{y - 4} = \dfrac{5(-1)(y-4)}{1(y-4)}$
$= \dfrac{-5}{1} = -5$

15. $\dfrac{12 + a - a^2}{a^2 - 2a - 8} = \dfrac{(4 - a)(3 + a)}{(a - 4)(a + 2)}$
$= \dfrac{-1(a-4)(3 + a)}{(a-4)(a + 2)}$
$= \dfrac{-1(3 + a)}{(a + 2)} = -\dfrac{a + 3}{a + 2}$

Section 6.2

1. $\dfrac{a}{a^2 - 5a + 4} \cdot \dfrac{a-4}{b} = \dfrac{a}{(a-1)(a-4)} \cdot \dfrac{(a-4)}{b}$

$= \dfrac{a}{b(a-1)}$

3. $\dfrac{y^2 - 4y - 21}{2} \cdot \dfrac{y}{y^2 + 3y}$

$= \dfrac{(y-7)(y+3)}{2} \cdot \dfrac{\overset{1}{y}}{y(y+3)} = \dfrac{y-7}{2}$

5. $\dfrac{2x-6}{3x+12} \cdot \dfrac{x^2 + 5x + 4}{x^2 - 2x - 3}$

$= \dfrac{2(x-3)}{3(x+4)} \cdot \dfrac{(x+1)(x+4)}{(x-3)(x+1)}$

$= \dfrac{2(x-3)(x+1)(x+4)}{3(x+4)(x-3)(x+1)} = \dfrac{2}{3}$

7. $\dfrac{y^2 - 25}{2y^2 - 7y - 15} \cdot \dfrac{2y^2 + y}{y + 5}$

$= \dfrac{(y-5)(y+5)}{(2y+3)(y-5)} \cdot \dfrac{y(2y+1)}{y+5}$

$= \dfrac{y(y-5)(y+5)(2y+1)}{(2y+3)(y-5)(y+5)}$

$= \dfrac{y(2y+1)}{2y+3}$

9. $\dfrac{5a + 15}{20b} \div \dfrac{a^2 - 9}{ab + 4b} = \dfrac{5a+15}{20b} \cdot \dfrac{ab+4b}{a^2-9}$

$= \dfrac{5(a+3)}{20b} \cdot \dfrac{b(a+4)}{(a+3)(a-3)}$

$= \dfrac{\overset{1}{5} \cdot b(a+3)(a+4)}{\underset{4}{20} \cdot b(a+3)(a-3)}$

$= \dfrac{a+4}{4(a-3)}$

11. $\dfrac{x^2 + 4x - 12}{2x + 3} \div \dfrac{1}{4x + 6} = \dfrac{x^2 + 4x - 12}{2x + 3} \cdot \dfrac{4x + 6}{1}$

$= \dfrac{(x+6)(x-2)}{(2x+3)} \cdot \dfrac{2(2x+3)}{1}$

$= \dfrac{2(x+6)(x-2)(2x+3)}{(2x+3)}$

$= 2(x+6)(x-2)$

13. $\dfrac{3x^2 - 5x + 2}{3x^2 + x - 2} \div (x - 1) = \dfrac{3x^2 - 5x + 2}{3x^2 + x - 2} \cdot \dfrac{1}{x - 1}$

$= \dfrac{(3x-2)(x-1)}{(3x-2)(x+1)} \cdot \dfrac{1}{x-1}$

$= \dfrac{1(3x-2)(x-1)}{(3x-2)(x+1)(x-1)}$

$= \dfrac{1}{x+1}$

Section 6.3

1. (a) $\dfrac{2x + 4}{3} - \dfrac{x + 1}{3} = \dfrac{2x + 4 - (x+1)}{3}$

$= \dfrac{2x + 4 - x - 1}{3}$

$= \dfrac{x + 3}{3}$

(b) $\dfrac{5(y+6)}{3y} + \dfrac{2(4y-1)}{3y} = \dfrac{5(y+6) + 2(4y-1)}{3y}$

$= \dfrac{5y + 30 + 8y - 2}{3y}$

$= \dfrac{13y + 28}{3y}$

3. (a) $\dfrac{8x + 4}{9} - \dfrac{5x + 4}{9} = \dfrac{8x + 4 - (5x + 4)}{9}$

$= \dfrac{8x + 4 - 5x - 4}{9}$

$= \dfrac{3x}{9} = \dfrac{x}{3}$

(b) $\dfrac{-3x + 8}{x^2 + 5x + 6} + \dfrac{4x - 6}{x^2 + 5x + 6}$

$= \dfrac{-3x + 8 + 4x - 6}{x^2 + 5x + 6}$

$= \dfrac{1(x+2)}{(x+3)(x+2)}$

$= \dfrac{1}{x+3}$

5. $\dfrac{4}{7} + \dfrac{1}{3} = \dfrac{4}{7}\left(\dfrac{3}{3}\right) + \dfrac{1}{3}\left(\dfrac{7}{7}\right)$

$= \dfrac{12}{21} + \dfrac{7}{21} = \dfrac{12 + 7}{21} = \dfrac{19}{21}$

7. LCD $= 2 \cdot 2 \cdot 3 \cdot 5 = 60$.

$\dfrac{5}{12} + \dfrac{7}{30} = \dfrac{5}{12}\left(\dfrac{5}{5}\right) + \dfrac{7}{30}\left(\dfrac{2}{2}\right)$

$= \dfrac{25}{60} + \dfrac{14}{60} = \dfrac{25 + 14}{60} = \dfrac{39}{60}$

$= \dfrac{3 \cdot 13}{3 \cdot 20} = \dfrac{13}{20}$

9. LCD = $3 \cdot 5 = 15$.
$$\frac{2x+7}{5} + \frac{x-4}{3} = \frac{2x+7}{5}\left(\frac{3}{3}\right) + \frac{x-4}{3}\left(\frac{5}{5}\right)$$
$$= \frac{3(2x+7)}{15} + \frac{5(x-4)}{15}$$
$$= \frac{3(2x+7) + 5(x-4)}{15}$$
$$= \frac{6x + 21 + 5x - 20}{15} = \frac{11x+1}{15}$$

11. LCD = $2 \cdot 3 = 6$.
$$\frac{5x+3}{6} - \frac{x+1}{2} = \frac{5x+3}{6} - \left(\frac{x+1}{2}\right)\left(\frac{3}{3}\right)$$
$$= \frac{5x+3}{6} - \frac{3(x+1)}{6}$$
$$= \frac{5x+3 - 3(x+1)}{6}$$
$$= \frac{5x+3-3x-3}{6}$$
$$= \frac{2x}{6} = \frac{2 \cdot x}{2 \cdot 3} = \frac{x}{3}$$

13. LCD = $6x$.
$$\frac{5}{2x} + \frac{1}{3x} = \frac{5}{2x}\left(\frac{3}{3}\right) + \left(\frac{1}{3x}\right)\left(\frac{2}{2}\right)$$
$$= \frac{15}{6x} + \frac{2}{6x} = \frac{15+2}{6x} = \frac{17}{6x}$$

15. LCD = $40a$.
$$\frac{3}{8a} - \frac{7}{10a} = \frac{3}{8a}\left(\frac{5}{5}\right) - \left(\frac{7}{10a}\right)\left(\frac{4}{4}\right)$$
$$= \frac{15}{40a} - \frac{28}{40a} = \frac{15-28}{40a} = \frac{-13}{40a} = -\frac{13}{40a}$$

17. LCD = $n(n+1)$.
$$\frac{4}{n+1} + \frac{14}{n} = \frac{4}{n+1}\left(\frac{n}{n}\right) + \frac{14}{n}\left(\frac{n+1}{n+1}\right)$$
$$= \frac{4n}{n(n+1)} + \frac{14(n+1)}{n(n+1)}$$
$$= \frac{4n + 14n + 14}{n(n+1)}$$
$$= \frac{18n + 14}{n(n+1)}$$

19. LCD = $(a+5)(a+4)$.
$$\frac{2}{a+5} - \frac{3}{a+4} = \frac{2}{a+5}\left(\frac{a+4}{a+4}\right) - \frac{3}{a+4}\left(\frac{a+5}{a+5}\right)$$
$$= \frac{2(a+4)}{(a+5)(a+4)} - \frac{3(a+5)}{(a+4)(a+5)}$$
$$= \frac{2(a+4) - 3(a+5)}{(a+5)(a+4)}$$
$$= \frac{2a + 8 - 3a - 15}{(a+5)(a+4)} = \frac{-a-7}{(a+5)(a+4)}$$

Section 6.4

1. $\frac{4}{y} + \frac{2}{y^2 + 3y} = \frac{4}{y} + \frac{2}{y(y+3)}$
LCD = $y(y+3)$.
$$\frac{4}{y} + \frac{2}{y(y+3)} = \frac{4}{y}\left(\frac{y+3}{y+3}\right) + \frac{2}{y(y+3)}$$
$$= \frac{4(y+3)}{y(y+3)} + \frac{2}{y(y+3)} = \frac{4y + 12 + 2}{y(y+3)}$$
$$= \frac{4y + 14}{y(y+3)}$$

3. $\frac{7}{a+2} - \frac{8}{a^2+5a+6} = \frac{7}{a+2} - \frac{8}{(a+2)(a+3)}$
LCD = $(a+2)(a+3)$.
$$\frac{7}{a+2} - \frac{8}{(a+2)(a+3)}$$
$$= \left(\frac{7}{a+2}\right)\left(\frac{a+3}{a+3}\right) - \frac{8}{(a+2)(a+3)}$$
$$= \frac{7(a+3)}{(a+2)(a+3)} - \frac{8}{(a+2)(a+3)}$$
$$= \frac{7(a+3) - 8}{(a+2)(a+3)} = \frac{7a + 21 - 8}{(a+2)(a+3)}$$
$$= \frac{7a + 13}{(a+2)(a+3)}$$

5. $\frac{3}{x^2-4} + \frac{5}{x^2+6x+8}$
$$= \frac{3}{(x+2)(x-2)} + \frac{5}{(x+2)(x+4)}$$
LCD = $(x+2)(x-2)(x+4)$.
$$\frac{3}{(x+2)(x-2)} + \frac{5}{(x+2)(x+4)}$$
$$= \frac{3}{(x+2)(x-2)}\left(\frac{x+4}{x+4}\right)$$
$$+ \frac{5}{(x+2)(x+4)}\left(\frac{x-2}{x-2}\right)$$
$$= \frac{3(x+4)}{(x+2)(x-2)(x+4)}$$
$$+ \frac{5(x-2)}{(x+2)(x+4)(x-2)}$$
$$= \frac{3(x+4) + 5(x-2)}{(x+2)(x-2)(x+4)}$$
$$= \frac{3x + 12 + 5x - 10}{(x+2)(x-2)(x+4)}$$
$$= \frac{8x + 2}{(x+2)(x-2)(x+4)}$$

7. $\frac{4y}{y^2 - 2y - 3} + \frac{6}{y-3} - \frac{5}{y+1}$
$$= \frac{4y}{(y-3)(y+1)} + \frac{6}{y-3} - \frac{5}{y+1}$$

$$\text{LCD} = (y-3)(y+1).$$

$$\frac{4y}{(y-3)(y+1)} + \frac{6}{y-3} - \frac{5}{y+1}$$

$$= \frac{4y}{(y-3)(y+1)} + \frac{6}{y-3}\left(\frac{y+1}{y+1}\right)$$
$$- \frac{5}{y+1}\left(\frac{y-3}{y-3}\right)$$

$$= \frac{4y}{(y-3)(y+1)} + \frac{6(y+1)}{(y-3)(y+1)}$$
$$- \frac{5(y-3)}{(y-3)(y+1)}$$

$$= \frac{4y + 6(y+1) - 5(y-3)}{(y-3)(y+1)}$$

$$= \frac{4y + 6y + 6 - 5y + 15}{(y-3)(y+1)}$$

$$= \frac{5y + 21}{(y-3)(y+1)}$$

9. $\dfrac{\frac{3}{8}}{\frac{7}{16}} = \dfrac{3}{8} \div \dfrac{7}{16} = \dfrac{3}{8} \cdot \dfrac{16}{7} = \dfrac{3 \cdot \overset{2}{\cancel{16}}}{\underset{1}{\cancel{8}} \cdot 7} = \dfrac{3 \cdot 2}{1 \cdot 7} = \dfrac{6}{7}$

11. $\dfrac{\frac{2}{a}}{\frac{5}{b}} = \dfrac{2}{a} \div \dfrac{5}{b} = \dfrac{2}{a} \cdot \dfrac{b}{5} = \dfrac{2 \cdot b}{a \cdot 5} = \dfrac{2b}{5a}$

13. $\dfrac{\frac{4}{5} - \frac{1}{2}}{\frac{3}{10} + \frac{1}{5}}$ LCM $= 2 \cdot 5 = 10$

$$\dfrac{\frac{4}{5} - \frac{1}{2}}{\frac{3}{10} + \frac{1}{5}} = \dfrac{10}{10} \cdot \dfrac{\left(\frac{4}{5} - \frac{1}{2}\right)}{\left(\frac{3}{10} + \frac{1}{5}\right)} = \dfrac{10\left(\frac{4}{5} - \frac{1}{2}\right)}{10\left(\frac{3}{10} + \frac{1}{5}\right)}$$

$$= \dfrac{10\left(\frac{4}{5}\right) - 10\left(\frac{1}{2}\right)}{10\left(\frac{3}{10}\right) + 10\left(\frac{1}{5}\right)} = \dfrac{8 - 5}{3 + 2} = \dfrac{3}{5}$$

15. $\dfrac{\frac{3}{a^2} + \frac{2}{b}}{\frac{1}{a} - \frac{5}{b}}$ LCM $= a^2 b$

$$\dfrac{\frac{3}{a^2} + \frac{2}{b}}{\frac{1}{a} - \frac{5}{b}} = \dfrac{a^2 b}{a^2 b} \cdot \dfrac{\left(\frac{3}{a^2} + \frac{2}{b}\right)}{\left(\frac{1}{a} - \frac{5}{b}\right)} = \dfrac{a^2 b\left(\frac{3}{a^2} + \frac{2}{b}\right)}{a^2 b\left(\frac{1}{a} - \frac{5}{b}\right)}$$

$$= \dfrac{a^2 b\left(\frac{3}{a^2}\right) + a^2 b\left(\frac{2}{b}\right)}{a^2 b\left(\frac{1}{a}\right) - a^2 b\left(\frac{5}{b}\right)} = \dfrac{\frac{3a^2 b}{a^2} + \frac{2a^2 b}{b}}{\frac{1a^2 b}{a} - \frac{5a^2 b}{b}}$$

$$= \dfrac{3b + 2a^2}{ab - 5a^2} \quad \text{or} \quad \dfrac{3b + 2a^2}{a(b - 5a)}$$

17. $\dfrac{6 + \frac{2}{m}}{\frac{3}{n} - 7}$ LCM $= mn$

$$\dfrac{6 + \frac{2}{m}}{\frac{3}{n} - 7} = \dfrac{mn}{mn} \cdot \dfrac{\left(6 + \frac{2}{m}\right)}{\left(\frac{3}{n} - 7\right)} = \dfrac{mn\left(6 + \frac{2}{m}\right)}{mn\left(\frac{3}{n} - 7\right)}$$

$$= \dfrac{6mn + \frac{2mn}{m}}{\frac{3mn}{n} - 7mn} = \dfrac{6mn + 2n}{3m - 7mn} \quad \text{or} \quad \dfrac{2n(3m + 1)}{m(3 - 7n)}$$

19. **(a)** To express Selena's rate in miles per hour, divide the number of miles by the number of hours. Therefore, Selena's jogging rate is expressed as $\dfrac{x}{y}$.

(b) If m pounds of apples cost n dollars, then the price per pound can be expressed by dividing the price by the number of pounds. Therefore, the price per pound for apples is $\dfrac{n}{m}$.

(c) If Taylor washes b automobiles in a hours, to find his rate for washing cars, divide the number of cars by the number of hours. Therefore, Taylor's rate for washing cars is $\dfrac{b}{a}$.

Section 6.5

1. $\dfrac{x-3}{2} + \dfrac{2x-7}{3} = \dfrac{5}{6}$

The LCD is 6.

$$6\left(\dfrac{x-3}{2} + \dfrac{2x-7}{3}\right) = 6\left(\dfrac{5}{6}\right)$$

$$6\left(\dfrac{x-3}{2}\right) + 6\left(\dfrac{2x-7}{3}\right) = 5$$

$$3(x-3) + 2(2x-7) = 5$$

$$3x - 9 + 4x - 14 = 5$$

$$7x - 23 = 5$$

$$7x = 28$$

$$x = 4$$

The solution set is $\{4\}$.

3. $\dfrac{5}{x} + \dfrac{1}{4} = \dfrac{7}{x}$

The LCD is $4x$; $x \neq 0$.

$$4x\left(\dfrac{5}{x} + \dfrac{1}{4}\right) = 4x\left(\dfrac{7}{x}\right)$$

$$4x\left(\dfrac{5}{x}\right) + 4x\left(\dfrac{1}{4}\right) = \dfrac{28x}{x}$$

$$\dfrac{20x}{x} + \dfrac{4x}{4} = 28$$

$$20 + x = 28$$

$$x = 8$$

The solution set is $\{8\}$.

5. $\dfrac{4}{x-2} = \dfrac{10}{x+4}$

The LCD is $(x-2)(x+4)$; $x \neq -4$, $x \neq 2$.

$$(x-2)(x+4)\left(\dfrac{4}{x-2}\right) = (x-2)(x+4)\left(\dfrac{10}{x+4}\right)$$

$$4(x+4) = 10(x-2)$$

$$4x + 16 = 10x - 20$$

$$-6x + 16 = -20$$

$$-6x = -36$$

$$x = 6$$

The solution set is $\{6\}$.

7. $\dfrac{5}{x-5} + 3 = \dfrac{x}{x-5}$

The LCD is $x-5$; $x \neq 5$.

$$(x-5)\left(\dfrac{5}{x-5} + 3\right) = (x-5)\left(\dfrac{x}{x-5}\right)$$

$$(x-5)\left(\dfrac{5}{x-5}\right) + (x-5)(3) = \dfrac{x(x-5)}{x-5}$$

$$\dfrac{5(x-5)}{x-5} + 3(x-5) = x$$

$$5 + 3x - 15 = x$$

$$3x - 10 = x$$

$$-10 = -2x$$

$$5 = x$$

But $x \neq 5$, so there is no solution. The solution set is $\varnothing$.

9. $\dfrac{70-x}{x} + 2 = \dfrac{84}{x}$

The LCD is x; $x \neq 0$.

$$x\left(\dfrac{70-x}{x} + 2\right) = x\left(\dfrac{84}{x}\right)$$

$$x\left(\dfrac{70-x}{x}\right) + x(2) = 84$$

$$70 - x + 2x = 84$$

$$70 + x = 84$$

$$x = 14$$

The solution set is $\{14\}$.

11. Let x represent the smaller number. Then $x+4$ represents the larger number.

$$\dfrac{x}{x+4} = \dfrac{2}{3}$$

The LCD is $3(x+4)$; $x \neq -4$.

$$3(x+4)\left(\dfrac{x}{x+4}\right) = 3(x+4)\left(\dfrac{2}{3}\right)$$

$$\dfrac{3x(x+4)}{x+4} = \dfrac{3(x+4) \cdot 2}{3}$$

$$3x = 2(x+4)$$

$$3x = 2x + 8$$

$$x = 8$$

If $x = 8$, then $x + 4 = 12$. The smaller number is 8, and the larger number is 12.

13. If one angle has measure $30°$, then the sum of the other two angles of the triangle is $180 - 30 = 150°$. Let y represent the measure of the second angle and let $150 - y$ represent the measure of the third angle.

$$\dfrac{y}{150-y} = \dfrac{2}{3}$$

The LCD is $3(150-y)$; $y \neq 150$.

$$3(150-y)\left(\dfrac{y}{150-y}\right) = 3(150-y)\left(\dfrac{2}{3}\right)$$

$$\dfrac{3y(150-y)}{150-y} = \dfrac{3(150-y) \cdot 2}{3}$$

$$3y = 2(150-y)$$

$$3y = 300 - 2y$$

$$5y = 300$$

$$y = 60$$

If $y = 60$, then $150 - y = 90$. The other two angles are $60°$ and $90°$.

15. Let r represent the rate Aldrin rides his bike and let $r + 2$ represent the rate Javier rides his bike. They both ride the same amount of time.

Rate $\cdot$ Time = Distance

$$\text{Time} = \dfrac{\text{Distance}}{\text{Rate}}$$

	Rate	Time	Distance
Javier	$r + 2$	$\dfrac{24}{r+2}$	24
Aldrin	r	$\dfrac{18}{r}$	18

$$\begin{pmatrix}\text{Aldrin's}\\\text{time}\end{pmatrix} = \begin{pmatrix}\text{Javier's}\\\text{time}\end{pmatrix}$$

$$\frac{18}{r} = \frac{24}{r+2}$$

The LCD is $r(r + 2)$; $r \neq 0$, $r \neq -2$.

$$r(r+2)\left(\frac{18}{r}\right) = r(r+2)\left(\frac{24}{r+2}\right)$$

$$\frac{18r(r+2)}{r} = \frac{24r(r+2)}{r+2}$$

$$18(r+2) = 24r$$
$$18r + 36 = 24r$$
$$36 = 6r$$
$$6 = r$$

If $r = 6$, then $r + 2 = 8$. Aldrin's rate is 6 mph, and Javier's rate is 8 mph.

Section 6.6

1. $\dfrac{7}{6x+2} - \dfrac{3}{3x+1} = \dfrac{1}{20}$

$$\frac{7}{2(3x+1)} - \frac{3}{3x+1} = \frac{1}{20}$$

The LCD is $20(3x + 1)$; $x \neq -\dfrac{1}{3}$.

$$20(3x+1)\left[\frac{7}{2(3x+1)} - \frac{3}{3x+1}\right] = 20(3x+1)\left(\frac{1}{20}\right)$$

$$20(3x+1)\left[\frac{7}{2(3x+1)}\right]$$
$$- 20(3x+1)\left(\frac{3}{3x+1}\right) = \frac{20(3x+1)}{20}$$

$$\frac{140(3x+1)}{2(3x+1)} - \frac{60(3x+1)}{3x+1} = 3x+1$$

$$70 - 60 = 3x + 1$$
$$10 = 3x + 1$$
$$9 = 3x$$
$$3 = x$$

The solution set is $\{3\}$.

3. $\dfrac{x}{x+2} + \dfrac{x}{x^2-4} = 1$

$$\frac{x}{x+2} + \frac{x}{(x+2)(x-2)} = 1$$

The LCD is $(x + 2)(x - 2)$; $x \neq 2$, $x \neq -2$.

$$(x+2)(x-2)\left[\frac{x}{x+2} + \frac{x}{(x+2)(x-2)}\right] = 1(x+2)(x-2)$$

$$(x+2)(x-2)\left(\frac{x}{x+2}\right)$$
$$+ (x+2)(x-2)\left[\frac{x}{(x+2)(x-2)}\right] = (x+2)(x-2)$$

$$x(x-2) + x = x^2 - 4$$
$$x^2 - 2x + x = x^2 - 4$$
$$x^2 - x = x^2 - 4$$
$$-x = -4$$
$$x = 4$$

The solution set is $\{4\}$.

5. $y + \dfrac{1}{y} = \dfrac{5}{2}$

The LCD is $2y$; $y \neq 0$.

$$2y\left(y + \frac{1}{y}\right) = 2y\left(\frac{5}{2}\right)$$

$$2y(y) + 2y\left(\frac{1}{y}\right) = \frac{2y \cdot 5}{2}$$

$$2y^2 + 2 = 5y$$
$$2y^2 - 5y + 2 = 0$$
$$(2y - 1)(y - 2) = 0$$
$$2y - 1 = 0 \quad \text{or} \quad y - 2 = 0$$
$$2y = 1 \quad \text{or} \quad y = 2$$
$$y = \frac{1}{2} \quad \text{or} \quad y = 2$$

The solution set is $\left\{\dfrac{1}{2}, 2\right\}$.

7. Let x represent the number. Then $\dfrac{1}{x}$ represents its reciprocal.

$$x + \frac{1}{x} = \frac{50}{7}$$

The LCD is $7x$; $x \neq 0$.

$$7x\left(x + \frac{1}{x}\right) = 7x\left(\frac{50}{7}\right)$$

$$7x(x) + 7x\left(\frac{1}{x}\right) = \frac{7x \cdot 50}{7}$$

$$7x^2 + 7 = 50x$$
$$7x^2 - 50x + 7 = 0$$
$$(7x - 1)(x - 7) = 0$$
$$7x - 1 = 0 \quad \text{or} \quad x - 7 = 0$$
$$7x = 1 \quad \text{or} \quad x = 7$$
$$x = \frac{1}{7} \quad \text{or} \quad x = 7$$

If $x = \frac{1}{7}$, then $\frac{1}{x} = 1 \div \frac{1}{7} = 1 \cdot \frac{7}{1} = 7$.

If $x = 7$, then $\frac{1}{x} = \frac{1}{7}$.

If the number is $\frac{1}{7}$, then its reciprocal is 7. If the number is 7, then its reciprocal is $\frac{1}{7}$.

9. Let r represent the rate Kim jogs and let $r + 4$ represent the rate Ryan rides his bicycle.

   Rate · Time = Distance

   Time = $\frac{\text{Distance}}{\text{Rate}}$

	Rate	Time	Distance
Kim	r	$\frac{16}{r}$	16
Ryan	$r + 4$	$\frac{36}{r+4}$	36

   $\left(\begin{array}{c}\text{Kim's}\\\text{time}\end{array}\right) = \left(\begin{array}{c}\text{Ryan's}\\\text{time}\end{array}\right) - 1$

   $\frac{16}{r} = \frac{36}{r+4} - 1$

   The LCD is $r(r+4)$; $r \neq 0, r \neq -4$.

   $r(r+4)\left(\frac{16}{r}\right) = r(r+4)\left(\frac{36}{r+4} - 1\right)$

   $16(r+4) = r(r+4)\left(\frac{36}{r+4}\right) - 1r(r+4)$

   $16r + 64 = 36r - r(r+4)$

   $16r + 64 = 36r - r^2 - 4r$

   $16r + 64 = 32r - r^2$

   $-16r + 64 = -r^2$

   $r^2 - 16r + 64 = 0$

   $(r-8)(r-8) = 0$

   $r - 8 = 0$ or $r - 8 = 0$

   $r = 8$ or $r = 8$

   If $r = 8$, then $r + 4 = 12$. Kim's rate is 8 mph, and Ryan's rate is 12 mph.

11. Let m represent the number of minutes Derek spends making sandwiches. Then $m - 10$ represents the number of minutes Marlene spends making sandwiches.

	Rate	Time	Quantity = Rate × Time
Derek	3	m	$3m$
Marlene	5	$m - 10$	$5(m-10)$

    Total sandwiches = 110

    $3m + 5(m - 10) = 110$

    $3m + 5m - 50 = 110$

    $8m - 50 = 110$

    $8m = 160$

    $m = 20$

    If $m = 20$, then $m - 10 = 10$. Therefore, Marlene worked 10 minutes making sandwiches.

13. Let y represent the number of minutes Dean and Tasha worked together. Dean's rate is $\frac{1}{60}$ of the job per minute, and Tasha's rate is $\frac{1}{90}$ of the job per minute.

	Minutes	Rate/min
Dean	60	$\frac{1}{60}$
Tasha	90	$\frac{1}{90}$
Together	y	$\frac{1}{y}$

    The sum of individual rates must equal the rate working together.

    $\frac{1}{60} + \frac{1}{90} = \frac{1}{y}$

    The LCD is $180y$; $y \neq 0$.

    $180y\left(\frac{1}{60} + \frac{1}{90}\right) = 180y\left(\frac{1}{y}\right)$

    $180y\left(\frac{1}{60}\right) + 180y\left(\frac{1}{90}\right) = \frac{180y}{y}$

    $3y + 2y = 180$

    $5y = 180$

    $y = 36$

    It should take them 36 minutes working together.

15. Let c represent the number of minutes it takes Julio to clean the office. Then $3c$ represents the number of minutes it takes Martin to clean the office. Julio's rate is $\frac{1}{c}$ of the job per minute, and Martin's rate is $\frac{1}{3c}$ of the job per minute.

	Minutes	Rate/min
Julio	c	$\frac{1}{c}$
Martin	$3c$	$\frac{1}{3c}$
Together	30	$\frac{1}{30}$

The sum of individual rates must equal the rate working together.

$$\frac{1}{c} + \frac{1}{3c} = \frac{1}{30}$$

The LCD is $30c$; $c \neq 0$.

$$30c\left(\frac{1}{c} + \frac{1}{3c}\right) = 30c\left(\frac{1}{30}\right)$$

$$30c\left(\frac{1}{c}\right) + 30c\left(\frac{1}{3c}\right) = \frac{30c}{30}$$

$$30 + 10 = c$$

$$40 = c$$

If $c = 40$, then $3c = 120$. Julio can clean the office in 40 minutes, and Martin can clean the office in 120 minutes.

17. Let h represent the number of hours it takes Carlos to build the deck by himself. Carlos's working rate is $\frac{1}{h}$. Because David can build the deck in 16 hours, his working rate is $\frac{1}{16}$. However, David actually worked 12 hours (7 alone, 5 with Carlos).

$$\begin{pmatrix}\text{Working}\\\text{rate}\end{pmatrix} \cdot \begin{pmatrix}\text{Time}\\\text{working}\end{pmatrix} = \begin{pmatrix}\text{Fractional part}\\\text{of job done}\end{pmatrix}$$

	Time to do entire job	Working rate	Time working	Fractional part of the job done
David	16 hours	$\frac{1}{16}$	12	$\frac{12}{16}$
Carlos	h hours	$\frac{1}{h}$	5	$\frac{5}{h}$

Adding fractional parts = 1 whole job.

$$\begin{pmatrix}\text{Fractional part}\\\text{David does}\end{pmatrix} + \begin{pmatrix}\text{Fractional part}\\\text{Carlos does}\end{pmatrix} = 1$$

$$\frac{12}{16} + \frac{5}{h} = 1$$

The LCD is $16h$; $h \neq 0$.

$$16h\left(\frac{12}{16} + \frac{5}{h}\right) = 16h(1)$$

$$16h\left(\frac{12}{16}\right) + 16h\left(\frac{5}{h}\right) = 16h$$

$$12h + 16(5) = 16h$$

$$80 = 4h$$

$$20 = h$$

Carlos can build the deck by himself in 20 hours.

CHAPTER 7

Section 7.1

1. The ordered pairs are

Tampa $(-2, 3)$ Naples $(-1, -4)$
Orlando $(1, 3)$ Tallahassee $(3, -1)$
Miami $(0, -2)$

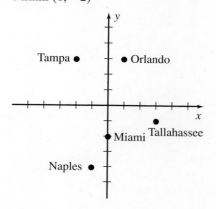

3. Graph $y = 3x - 2$.

x	y	(x, y)
0	-2	$(0, -2)$
1	1	$(1, 1)$
2	4	$(2, 4)$
3	7	$(3, 7)$
-1	-5	$(-1, -5)$

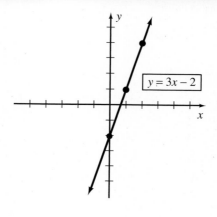

5. Graph $y = \frac{1}{3}x - 1$.

Select values of x that are divisible by 3.

x	y	(x, y)
0	-1	$(0, -1)$
3	0	$(3, 0)$
6	1	$(6, 1)$
-3	-2	$(-3, -2)$

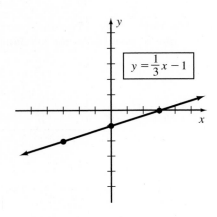

7. Solve $2x - 3y = 9$ for y.

$2x - 3y = 9$

$-3y = -2x + 9$

$\dfrac{-3y}{-3} = \dfrac{-2x}{-3} + \dfrac{9}{-3}$

$y = \dfrac{2}{3}x - 3$

9. Solve $3x + 5y = 7$ for x.

$3x + 5y = 7$

$3x = -5y + 7$

$\dfrac{3x}{3} = \dfrac{-5y}{3} + \dfrac{7}{3}$

$x = -\dfrac{5}{3}y + \dfrac{7}{3}$

11. Graph $4x - 2y = -8$.

$4x - 2y = -8$

$-2y = -4x - 8$

$\dfrac{-2y}{-2} = \dfrac{-4x}{-2} + \dfrac{-8}{-2}$

$y = 2x + 4$

x	y
0	4
1	6
2	8
-1	2
-2	0

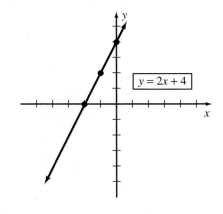

Section 7.2

1. Graph $2x + 3y = 12$.

To find the y intercept, set $x = 0$.
If $x = 0$, then

$2(0) + 3y = 12$

$3y = 12$

$y = 4$

To find the x intercept, set $y = 0$.
If $y = 0$, then

$2x + 3(0) = 12$

$2x = 12$

$x = 6$

To find a check point, let $x = 3$.
If $x = 3$, then

$2(3) + 3y = 12$

$6 + 3y = 12$

$3y = 6$

$y = 2$

Three solutions are $(0, 4)$, $(6, 0)$, and $(3, 2)$.

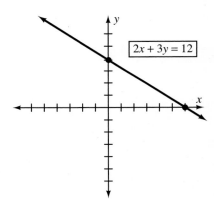

3. Graph $2x - y = 6$.

To find the y intercept, set $x = 0$.
If $x = 0$, then

$2(0) - y = 6$

$-y = 6$

$y = -6$

To find the x intercept, set $y = 0$.
If $y = 0$, then

$2x - (0) = 6$

$2x = 6$

$x = 3$

To find a check point, let $x = 2$.
If $x = 2$, then

$2(2) - y = 6$

$4 - y = 6$

$-y = 2$
$y = -2$

Three solutions are $(0, -6)$, $(3, 0)$, and $(2, -2)$.

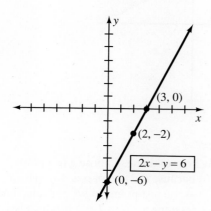

5. Graph $x - 2y = 5$.

To find the y intercept, set $x = 0$.
If $x = 0$, then
$(0) - 2y = 5$
$-2y = 5$
$y = -\dfrac{5}{2}$

To find the x intercept, set $y = 0$.
If $y = 0$, then
$x - 2(0) = 5$
$x = 5$

To find a check point, let $x = 3$.
If $x = 3$, then
$(3) - 2y = 5$
$-2y = 2$
$y = -1$

Three solutions are $\left(0, -\dfrac{5}{2}\right)$, $(5, 0)$, and $(3, -1)$.

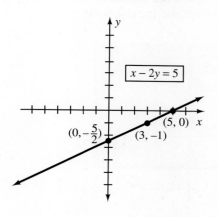

7. Graph $y = -3x$.

Because the graph goes through $(0, 0)$, we need to find more points.

x	y
0	0
2	-6
-2	6

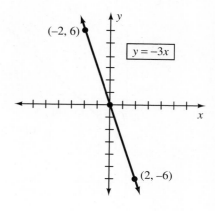

9. Graph $x = -2$.

For any value of y, $x = -2$

x	y
-2	0
-2	2
-2	-4

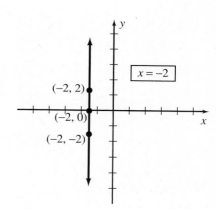

11. Graph $y = 4$.

For any value of x; $y = 4$.

x	y
0	4
2	4
-3	4

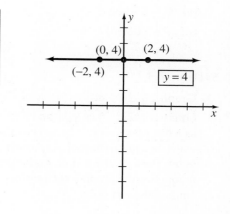

13. (a) $C = m + 5$.

m	1	3	5	10
C	6	8	10	15

(b) $C = m + 5$.

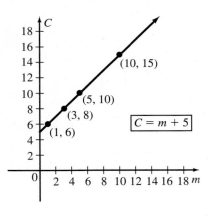

(c) $C \approx 7$ when $m = 2.5$.

(d) $C = 2.5 + 5 = 7.5$.

15. (a) $D = 1.5e$.

e	2	3	5	10
D	3	4.5	7.5	15.00

(b) $D = 1.5e$.

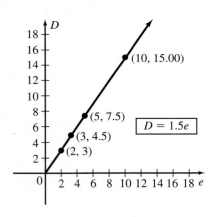

(c) $D \approx 9$ when $e = 6$.

(d) $D = (1.5)(6) = \$9$.

Section 7.3

1. Graph $x + 2y > 6$.

Graph $x + 2y = 6$ as a dashed line. Choose $(0, 0)$ as a test point, so $x + 2y > 6$ becomes $(0) + 2(0) > 6$, which is equivalent to $0 > 6$, which is a false statement. Because the test point does not satisfy the inequality, the graph is the half-plane that does *not* contain the test point.

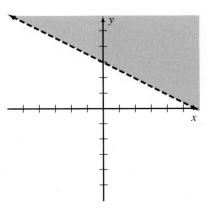

3. Graph $y \geq -3x$.

Graph $y = -3x$ as a solid line. Choose a test point other than $(0, 0)$ because the line passes through $(0, 0)$. Use $(2, 2)$ as the test point, so $y \geq -3x$ becomes $2 \geq -3(2)$, which is equivalent to $2 \geq -6$, which is a true statement. Because the test point does satisfy the inequality, the graph is the line along with the half-plane that does contain the test point.

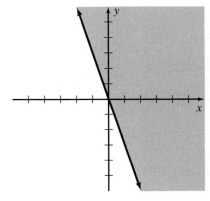

5. Graph $y \geq -2$

Graph $y = -2$ as a solid line. Choose $(0, 0)$ as the test point, so $y \geq -2$ becomes $0 \geq -2$, which is a true statement. Because the test point does satisfy the inequality, the graph is the line along with the half-plane that does contain the test point.

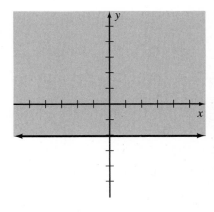

7. Graph $x \geq 3$.

Graph $x = 3$ as a solid line. Choose $(0, 0)$ as the test point, so $x \geq 3$ becomes $0 \geq 3$, which is a false statement. Because the test point does not satisfy the inequality, the graph is the line along with the half-plane that does not contain the test point.

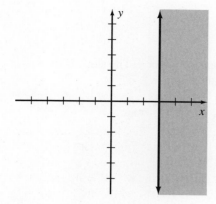

Section 7.4

1. **(a)** Let $(1, 2)$ be P_1 and let $(3, 8)$ be P_2.
$$m = \frac{y_2 - y_1}{x_2 - x_1} = \frac{8 - 2}{3 - 1} = \frac{6}{2} = \frac{3}{1} = 3$$

(b) Let $(5, 3)$ be P_1 and let $(2, 4)$ be P_2.
$$m = \frac{y_2 - y_1}{x_2 - x_1} = \frac{4 - 3}{2 - 5} = \frac{1}{-3} = -\frac{1}{3}$$

(c) Let $(-4, 2)$ be P_1 and let $(1, 2)$ be P_2.
$$m = \frac{y_2 - y_1}{x_2 - x_1} = \frac{2 - 2}{1 - (-4)} = \frac{0}{5} = 0$$

3. Find the intercepts of $2x - y = 4$.

To find the y intercept:
If $x = 0$, then
$2(0) - y = 4$
$-y = 4$
$y = -4$

To find the x intercept:
If $y = 0$, then
$2x - (0) = 4$
$2x = 4$
$x = 2$

Use $(0, -4)$ and $(2, 0)$ to find the slope.
$$m = \frac{y_2 - y_1}{x_2 - x_1} = \frac{0 - (-4)}{2 - (0)} = \frac{4}{2} = 2$$

5. Graph the line with $m = \frac{2}{5}$ that passes through $(-2, -1)$. Plot $(-2, -1)$. Locate another point by moving 2 units up and 5 units right to obtain $(3, 1)$.

We can locate another point using $m = \frac{2}{5} = \frac{-2}{-5}$.

From $(-2, -1)$ move 2 units down and 5 units left to obtain $(-7, -3)$.

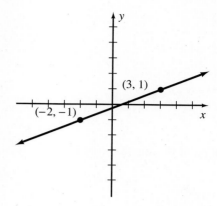

7. Graph the line with $m = -\frac{3}{2}$ that passes through $(-3, 4)$.
$$m = -\frac{3}{2} = \frac{-3}{2}$$

Plot $(-3, 4)$. Locate another point by moving 3 units down and 2 units right to obtain $(-1, 1)$. Locate a third point by again moving 3 units down and 2 units right to obtain $(1, -2)$.

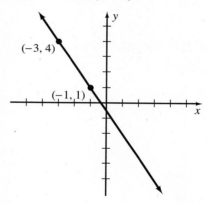

9. A 2% grade means a slope of $\frac{2}{100}$. Therefore, let x represent the unknown vertical distance.
$$\frac{x}{1250} = \frac{2}{100}$$
$$100x = 2(1250)$$
$$100x = 2500$$
$$x = 25$$

The vertical distance is 25 feet.

Section 7.5

1. Find the equation of a line that has a slope of $\frac{2}{3}$ and contains the point $(4, 2)$. Let (x, y) represent any point on the line other than $(4, 2)$. The slope of the line connecting $(4, 2)$ and (x, y) is $\frac{2}{3}$.

$$\frac{y-2}{x-4} = \frac{2}{3}$$
$$2(x-4) = 3(y-2)$$
$$2x - 8 = 3y - 6$$
$$2x - 8 - 3y = -6$$
$$2x - 3y = 2$$

3. Find the equation of the line that contains $(-2, 3)$ and $(1, 5)$.
$$m = \frac{y_2 - y_1}{x_2 - x_1} = \frac{5 - 3}{1 - (-2)} = \frac{2}{3}$$

Use a slope of $\frac{2}{3}$ and either ordered pair to write the equation.
$$\frac{y-5}{x-1} = \frac{2}{3}$$
$$2(x-1) = 3(y-5)$$
$$2x - 2 = 3y - 15$$
$$2x - 2 - 3y = -15$$
$$2x - 3y = -13$$

5. A y intercept of -4 means the line passes through $(0, -4)$. The slope is $\frac{1}{3}$.
$$\frac{y - (-4)}{x - 0} = \frac{1}{3}$$
$$\frac{y+4}{x} = \frac{1}{3}$$
$$3(y+4) = x$$
$$3y + 12 = x$$
$$12 = x - 3y$$

7. Use the point-slope form to write the equation of the line that contains the points (x, y) and $(-1, 3)$.
$$y_2 - y_1 = m(x_2 - x_1)$$
$$y - 3 = \frac{5}{2}[x - (-1)]$$
$$y - 3 = \frac{5}{2}(x + 1)$$
$$2(y - 3) = 2\left[\frac{5}{2}(x+1)\right]$$
$$2y - 6 = 5(x+1)$$
$$2y - 6 = 5x + 5$$
$$-6 = 5x - 2y + 5$$
$$-11 = 5x - 2y$$

9. Use the slope-intercept form to write the equation of the line with a slope of $-\frac{2}{3}$ and a y intercept of 5.
$$m = -\frac{2}{3} \quad b = 5$$
$$y = mx + b$$
$$y = -\frac{2}{3}x + 5$$

11. To find the slope, solve for y. The slope is the numerical coefficient of x.
$$4x - 2y = 3$$
$$-2y = -4x + 3$$
$$\frac{-2y}{-2} = \frac{-4x}{-2} + \frac{3}{-2}$$
$$y = 2x - \frac{3}{2}$$

The slope is 2.

13. Graph $y = \frac{3}{5}x - 2$.

The y intercept is -2, which gives the point $(0, -2)$. Use $m = \frac{3}{5}$ to move from $(0, -2)$ three units up and five units right to obtain $(5, 1)$.

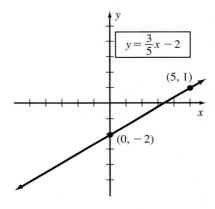

15. (a) To find the slope and y intercept, solve for y.
$$3x + 2y = 6$$
$$2y = -3x + 6$$
$$\frac{2y}{2} = \frac{-3x}{2} + \frac{6}{2}$$
$$y = -\frac{3}{2}x + 3$$
$$m = -\frac{3}{2} \quad b = 3$$

From (0, 3) move 3 units down and 2 units right to obtain (2, 0).

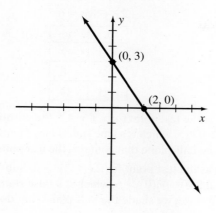

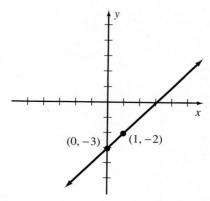

(b) To find the slope and y intercept, solve for y.

$-y = -x + 3$

$\dfrac{-y}{-1} = \dfrac{-x}{-1} + \dfrac{3}{-1}$

$y = x - 3$

$m = \dfrac{1}{1} \quad b = -3$

From (0, −3) move 1 unit up and 1 unit right to obtain (1, −2).

(c) To find the slope and y intercept, solve for y.

$y = -4$

$m = 0$; y intercept is -4.

Because the line has slope of 0, it is horizontal. From (0, −4) graph the horizontal line.

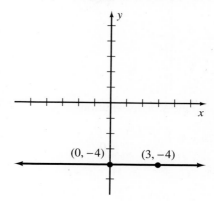

CHAPTER 8

Section 8.1

1.

Substitute 3 for x and -1 for y.

$2x + y = 5$ becomes $2(3) + (-1) = 5$

$5x - 2y = 17$ becomes $5(3) - 2(-1) = 17$

Because the ordered pair does satisfy both equations, $(3, -1)$ is a solution of the system.

3. Solve the system $\begin{pmatrix} x + y = 2 \\ x + 2y = 6 \end{pmatrix}$.

$x + y = 2$ $x + 2y = 6$

x	y
0	2
2	0
−1	3

x	y
0	3
6	0
−6	6

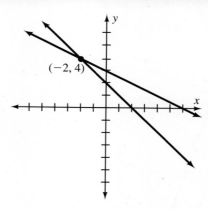

The solution set is $\{(-2, 4)\}$.

5. Solve the system $\begin{pmatrix} 2x + 4y = 8 \\ x + 2y = 4 \end{pmatrix}$.

$2x + 4y = 8$ $\qquad$ $x + 2y = 4$

x	y
0	2
4	0
2	1

x	y
0	2
4	0
-2	3

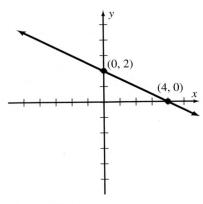

There are infinitely many solutions.

7. Solve the system $\begin{pmatrix} 3x + y = 6 \\ 3x + y = 0 \end{pmatrix}$.

$3x + y = 6$ $\qquad$ $3x + y = 0$

x	y
0	6
2	0
1	3

x	y
0	0
1	-3
-1	3

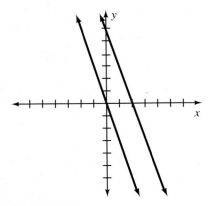

The solution set is ∅.

9. Solve the system $\begin{pmatrix} x + y < 5 \\ 2x - 3y > 6 \end{pmatrix}$.

Graph the following lines as broken lines.

$x + y = 5$ $\qquad$ $2x - 3y = 6$

x	y
0	5
5	0
2	3

x	y
0	-2
3	0
-3	-4

Use $(0, 0)$ as the test point for $x + y < 5$. Substituting, we find $0 + 0 < 5$, which is a true statement. Therefore, we shade the half-plane that contains the test point.

Use $(0, 0)$ as the test point for $2x - 3y > 6$. Substituting, we find $2(0) - 3(0) > 6$, which is a false statement. Therefore, we shade the half-plane that does not contain the test point.

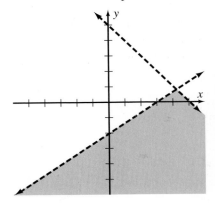

Section 8.2

1. Solve the system $\begin{pmatrix} 2x + y = 6 \\ x - y = 6 \end{pmatrix}$.

Add the equations.

$$\begin{aligned} 2x + y &= 6 \\ x - y &= 6 \\ \hline 3x &= 12 \\ x &= 4 \end{aligned}$$

Substitute 4 for x in the first equation.

$2(4) + y = 6$
$8 + y = 6$
$y = -2$

The solution set is $\{(4, -2)\}$.

3. Solve the system $\begin{pmatrix} x - 2y = 7 \\ 3x + 2y = 13 \end{pmatrix}$.

Add the equations.

$$\begin{aligned} x - 2y &= 7 \\ 3x + 2y &= 13 \\ \hline 4x &= 20 \\ x &= 5 \end{aligned}$$

Substitute 5 for x in the first equation.
$$(5) - 2y = 7$$
$$-2y = 2$$
$$y = -1$$
The solution set is $\{(5, -1)\}$.

5. Solve the system $\begin{pmatrix} y = x + 7 \\ x + 2y = 5 \end{pmatrix}$.

   In the first equation, subtract x from both sides. Leave the second equation alone. Add the equations.
   $$-x + y = 7$$
   $$\underline{x + 2y = 5}$$
   $$3y = 12$$
   $$y = 4$$
   Substitute 4 for y in the first equation.
   $$4 = x + 7$$
   $$-3 = x$$
   The solution set is $\{(-3, 4)\}$.

7. Solve the system $\begin{pmatrix} 3x + 2y = -3 \\ 4x - y = 18 \end{pmatrix}$.

   Leave the first equation alone. Multiply the second equation by 2. Then add the equations.
   $$3x + 2y = -3$$
   $$\underline{8x - 2y = 36}$$
   $$11x = 33$$
   $$x = 3$$
   Substitute 3 for x in the first original equation.
   $$3(3) + 2y = -3$$
   $$9 + 2y = -3$$
   $$2y = -12$$
   $$y = -6$$
   The solution set is $\{(3, -6)\}$.

9. Solve the system $\begin{pmatrix} 3x + 2y = 4 \\ 4x - 5y = 13 \end{pmatrix}$.

   Multiply the first equation by 5. Multiply the second equation by 2. Then add the equations.
   $$15x + 10y = 20$$
   $$\underline{8x - 10y = 26}$$
   $$23x = 46$$
   $$x = 2$$
   Substitute 2 for x in the first original equation.
   $$3(2) + 2y = 4$$
   $$6 + 2y = 4$$
   $$2y = -2$$
   $$y = -1$$
   The solution set is $\{(2, -1)\}$.

11. Solve the system $\begin{pmatrix} 5x - 2y = 13 \\ 2x + 3y = 10 \end{pmatrix}$.

    Multiply the first equation by 3. Multiply the second equation by 2. Then add the equations.
    $$15x - 6y = 39$$
    $$\underline{4x + 6y = 20}$$
    $$19x = 59$$
    $$x = \frac{59}{19}$$
    Substitute $\frac{59}{19}$ for x in the second original equation.
    $$2\left(\frac{59}{19}\right) + 3y = 10$$
    $$\frac{118}{19} + 3y = 10$$
    $$3y = 10 - \frac{118}{19}$$
    $$3y = \frac{190}{19} - \frac{118}{19}$$
    $$3y = \frac{72}{19}$$
    $$\frac{1}{3}(3y) = \left(\frac{72}{19}\right)\left(\frac{1}{3}\right)$$
    $$y = \frac{24}{19}$$
    The solution set is $\left\{\left(\frac{59}{19}, \frac{24}{19}\right)\right\}$.

13. Let x represent the smaller number and let y represent the larger number.

    The sum of two numbers is 31.
    $$x + y = 31$$
    Twice the smaller number is subtracted from the larger number, with a result of 1.
    $$y - 2x = 1$$
    Solve the system $\begin{pmatrix} x + y = 31 \\ -2x + y = 1 \end{pmatrix}$.

    Multiply the first equation by -1. Leave the second equation alone.
    $$-x - y = -31$$
    $$\underline{-2x + y = 1}$$
    $$-3x = -30$$
    $$x = 10$$
    Substitute 10 for x in the first original equation to find y.
    $$10 + y = 31$$
    $$y = 21$$
    The smaller number is 10, and the larger number is 21.

15. Let x represent the cost of one pizza and let y represent the cost of one basket of wings.

The cost of two pizzas and three baskets is $29.

$2x + 3y = 29$

The cost of five pizzas and two baskets is $45.

$5x + 2y = 45$

Multiply the first equation by -2. Multiply the second equation by 3. Add the equations.

$$-4x - 6y = -58$$
$$\underline{15x + 6y = 135}$$
$$11x = 77$$
$$x = 7$$

Substitute 7 for x in the first original equation to find y.

$2(7) + 3y = 29$
$14 + 3y = 29$
$3y = 15$
$y = 5$

The cost of one pizza is $7, and the cost of one basket of wings is $5.

Section 8.3

1. Solve the system $\begin{pmatrix} x = y + 11 \\ x + 2y = -10 \end{pmatrix}$.

Substitute $y + 11$ for x in the second equation.

$(y + 11) + 2y = -10$
$3y + 11 = -10$
$3y = -21$
$y = -7$

Substitute -7 for y in the first original equation to find x.

$x = y + 11$
$x = -7 + 11$
$x = 4$

The solution set is $\{(4, -7)\}$.

3. Solve the system $\begin{pmatrix} 4x + 3y = -15 \\ y = 2x + 5 \end{pmatrix}$.

Substitute $2x + 5$ for y in the first original equation to find x.

$4x + 3(2x + 5) = -15$
$4x + 6x + 15 = -15$
$10x + 15 = -15$
$10x = -30$
$x = -3$

Substitute -3 for x in the second original equation to find y.

$y = 2x + 5$
$y = 2(-3) + 5$
$y = -6 + 5 = -1$

The solution set is $\{(-3, -1)\}$.

5. Solve the system $\begin{pmatrix} x + 4y = 22 \\ y = \frac{1}{2}x + 4 \end{pmatrix}$.

Substitute $\frac{1}{2}x + 4$ for y in the first original equation to find x.

$x + 4\left(\frac{1}{2}x + 4\right) = 22$
$x + 2x + 16 = 22$
$3x + 16 = 22$
$3x = 6$
$x = 2$

Substitute 2 for x in the second original equation to find y.

$y = \frac{1}{2}x + 4$
$y = \frac{1}{2}(2) + 4$
$y = 1 + 4 = 5$

The solution set is $\{(2, 5)\}$.

7. Solve the system $\begin{pmatrix} 2x + 3y = 4 \\ x + y = -1 \end{pmatrix}$.

Solve the second equation for x.

$x + y = -1$
$x = -y - 1$

Substitute $-y - 1$ for x in the first original equation to find y.

$2(-y - 1) + 3y = 4$
$-2y - 2 + 3y = 4$
$y - 2 = 4$
$y = 6$

Substitute 6 for y in the second original equation to find x.

$x + y = -1$
$x + 6 = -1$
$x = -1 - 6 = -7$

The solution set is $\{(-7, 6)\}$.

9. Solve the system $\begin{pmatrix} 2x + 3y = 8 \\ 3x - 2y = -14 \end{pmatrix}$.

Solve the second equation for y.

$3x - 2y = -14$
$-2y = -3x - 14$

$$\frac{-2y}{-2} = \frac{-3x}{-2} - \frac{14}{-2}$$

$$y = \frac{3x}{2} + 7$$

Substitute $\frac{3x}{2} + 7$ for y in the first original equation to find x.

$$2x + 3y = 8$$
$$2x + 3\left(\frac{3x}{2} + 7\right) = 8$$
$$2x + \frac{9x}{2} + 21 = 8$$
$$4x + 9x + 42 = 16$$
$$13x + 42 = 16$$
$$13x = -26$$
$$x = -2$$

Substitute -2 for x in the first original equation to find y.

$$2x + 3y = 8$$
$$2(-2) + 3y = 8$$
$$-4 + 3y = 8$$
$$3y = 12$$
$$y = 4$$

The solution set is $\{(-2, 4)\}$.

11. Solve the system $\begin{pmatrix} x = 2y - 6 \\ 4x - 3y = 16 \end{pmatrix}$.

Use substitution to solve.

Substitute $2y - 6$ for x in the second original equation to find y.

$$4(2y - 6) - 3y = 16$$
$$8y - 24 - 3y = 16$$
$$5y - 24 = 16$$
$$5y = 40$$
$$y = 8$$

Substitute 8 for y in the first original equation to find x.

$$x = 2y - 6$$
$$x = 2(8) - 6$$
$$x = 16 - 6 = 10$$

The solution set is $\{(10, 8)\}$.

13. Solve the system $\begin{pmatrix} 9x + 5y = -15 \\ 3x - 2y = -27 \end{pmatrix}$.

Use elimination-by-addition to solve.

Multiply the first equation by 2. Multiply the second equation by 5. Then add the equations.

$$18x + 10y = -30$$
$$15x - 10y = -135$$
$$\overline{33x = -165}$$
$$x = -5$$

Substitute -5 for x in the first original equation.

$$9x + 5y = -15$$
$$9(-5) + 5y = -15$$
$$-45 + 5y = -15$$
$$5y = 30$$
$$y = 6$$

The solution set is $\{(-5, 6)\}$.

15. Solve the system $\begin{pmatrix} y = 3x + 2 \\ 3x - y = 4 \end{pmatrix}$.

Use substitution to solve.

Substitute $3x + 2$ for y in the second original equation to find x.

$$3x - (3x + 2) = 4$$
$$3x - 3x - 2 = 4$$
$$-2 = 4$$

The statement is false. Therefore, the solution set is $\emptyset$.

17. Solve the system $\begin{pmatrix} x - 2y = 4 \\ 3x - 6y = 12 \end{pmatrix}$.

Multiply the first equation by -3. Leave the second equation alone. Then add the equations.

$$-3x + 6y = -12$$
$$3x - 6y = 12$$
$$\overline{0 = 0}$$

This is a true statement. Therefore, there are infinitely many solutions.

19. Let x represent the amount invested at 5% and let y represent the amount invested at 7%.

Simoni invested $7000.

$$x + y = 7000$$

The total interest earned was $440.

$$0.05x + 0.07y = 440$$

Multiply the second equation by 100. Multiply the first equation by -5. Then add the equations.

$$-5x - 5y = -35{,}000$$
$$5x + 7y = 44{,}000$$
$$\overline{2y = 9000}$$
$$y = 4500$$

Substitute 4500 for y in the first original equation.

$$x + y = 7000$$
$$x + 4500 = 7000$$
$$x = 2500$$

Simoni invested $2500 at 5% and $4500 at 7%.

21. Let x represent the number of liters of 6% saline solution and let y represent the number of liters of 12% saline solution.

The total number of liters is 30.

$x + y = 30$

The total amount of saline is 10% of 30.

$0.06x + 0.12y = 0.10(30)$

Multiply the first equation by -6. Multiply the second equation by 100. Then add the equations.

$$\begin{aligned} -6x - 6y &= -180 \\ 6x + 12y &= 300 \\ \hline 6y &= 120 \\ y &= 20 \end{aligned}$$

Substitute 20 for y in the first original equation.

$x + y = 30$

$x + 20 = 30$

$x = 10$

There are 10 liters of 6% saline solution and 20 liters of 12% saline solution.

23. Let w represent the width of the rectangle and let l represent the length of the rectangle.

The length is 6 more than twice the width.

$l = 6 + 2w$

The perimeter is 168 inches.

Use $P = 2l + 2w$.

$168 = 2l + 2w$

Use substitution to solve.

Substitute $6 + 2w$ for l in the second original equation.

$168 = 2(6 + 2w) + 2w$

$168 = 12 + 4w + 2w$

$168 = 12 + 6w$

$156 = 6w$

$26 = w$

Substitute 26 for w in the first original equation.

$l = 6 + 2w$

$l = 6 + 2(26)$

$l = 6 + 52 = 58$

The length is 58 inches, and the width is 26 inches.

Section 8.4

1. Solve the system $\begin{pmatrix} 3x - y + z = 15 \\ y + 2z = 6 \\ 3z = 12 \end{pmatrix}$.

Solve the third equation.

$3z = 12$

$z = 4$

Substitute 4 for z in the second original equation.

$y + 2z = 6$

$y + 2(4) = 6$

$y + 8 = 6$

$y = -2$

Substitute 4 for z and -2 for y in the first original equation.

$3x - y + z = 15$

$3x - (-2) + 4 = 15$

$3x + 6 = 15$

$3x = 9$

$x = 3$

The solution set is $\{(3, -2, 4)\}$.

3. Solve the system $\begin{pmatrix} 3x + 4y - z = 14 \\ y + 4z = 17 \\ 2y + 3z = 19 \end{pmatrix}$.

Multiply the second original equation by -2. Leave the third original equation alone. Then add the equations.

$$\begin{aligned} -2y - 8z &= -34 \\ 2y + 3z &= 19 \\ \hline -5z &= -15 \\ z &= 3 \end{aligned}$$

Substitute 3 for z in the second original equation.

$y + 4z = 17$

$y + 4(3) = 17$

$y + 12 = 17$

$y = 5$

Substitute 3 for z and 5 for y in the first original equation.

$3x + 4y - z = 14$

$3x + 4(5) - 3 = 14$

$3x + 20 - 3 = 14$

$3x + 17 = 14$

$3x = -3$

$x = -1$

The solution set is $\{(-1, 5, 3)\}$.

5. Solve the system $\begin{pmatrix} x + 3y - 2z = -1 \\ 2x - y + 3z = 5 \\ 3x + 2y - z = 8 \end{pmatrix}$ (1) (2) (3)

Multiply equation (3) by 3. Leave equation (2) alone. Then add the equations.

$$\begin{aligned} 9x + 6y - 3z &= 24 \\ 2x - y + 3z &= 5 \\ \hline 11x + 5y &= 29 \quad (4) \end{aligned}$$

Multiply equation (3) by −2. Leave equation (1) alone. Then add the equations.

$$-6x - 4y + 2z = -16$$
$$\underline{x + 3y - 2z = -1}$$
$$-5x - y = -17 \quad (5)$$

Solve the system $\begin{pmatrix} 11x + 5y = 29 \\ -5x - y = -17 \end{pmatrix}$.

$$11x + 5y = 29 \quad (4)$$
$$-5x - y = -17 \quad (5)$$

Multiply equation (5) by 5. Leave equation (4) alone. Then add the equations.

$$11x + 5y = 29$$
$$\underline{-25x - 5y = -85}$$
$$-14x = -56$$
$$x = 4$$

Substitute 4 for x in equation (4) to find y.

$$11x + 5y = 29$$
$$11(4) + 5y = 29$$
$$44 + 5y = 29$$
$$5y = -15$$
$$y = -3$$

Substitute 4 for x and −3 for y in equation (1) to find z.

$$x + 3y - 2z = -1$$
$$(4) + 3(-3) - 2z = -1$$
$$4 - 9 - 2z = -1$$
$$-5 - 2z = -1$$
$$-2z = 4$$
$$z = -2$$

The solution set is $\{(4, -3, -2)\}$.

7. Solve the system $\begin{pmatrix} 2x + 3y + z = 9 \\ 3x - 2y - 2z = -16 \\ 5x + 4y + 3z = 9 \end{pmatrix}$ $\begin{matrix} (1) \\ (2) \\ (3) \end{matrix}$

Multiply equation (1) by 2. Leave equation (2) alone. Then add the equations.

$$4x + 6y + 2z = 18$$
$$\underline{3x - 2y - 2z = -16}$$
$$7x + 4y = 2 \quad (4)$$

Multiply equation (1) by −3. Leave equation (3) alone. Then add the equations.

$$-6x - 9y - 3z = -27$$
$$\underline{5x + 4y + 3z = 9}$$
$$-x - 5y = -18 \quad (5)$$

Solve the system $\begin{pmatrix} 7x + 4y = 2 \\ -x - 5y = -18 \end{pmatrix}$.

$$7x + 4y = 2 \quad (4)$$
$$-x - 5y = -18 \quad (5)$$

Multiply equation (5) by 7. Leave equation (4) alone. Then add the equations.

$$7x + 4y = 2$$
$$\underline{-7x - 35y = -126}$$
$$-31y = -124$$
$$y = 4$$

Substitute 4 for y in equation (4) to find x.

$$7x + 4y = 2$$
$$7x + 4(4) = 2$$
$$7x + 16 = 2$$
$$7x = -14$$
$$x = -2$$

Substitute −2 for x and 4 for y in equation (1) to find z.

$$2x + 3y + z = 9$$
$$2(-2) + 3(4) + z = 9$$
$$-4 + 12 + z = 9$$
$$8 + z = 9$$
$$z = 1$$

The solution set is $\{(-2, 4, 1)\}$.

9. Solve the system $\begin{pmatrix} x + 2y - z = 4 \\ 3x - 4y + 2z = 5 \\ 2x + 4y - 2z = 10 \end{pmatrix}$ $\begin{matrix} (1) \\ (2) \\ (3) \end{matrix}$

Multiply equation (1) by −2. Leave equation (3) alone. Then add the equations.

$$-2x - 4y + 2z = -8$$
$$\underline{2x + 4y - 2z = 10}$$
$$0 = 2$$

This is a false statement, which means the system is inconsistent. The solution set is $\varnothing$.

11. Solve the system $\begin{pmatrix} 2x + y + z = 8 \\ x + y - z = 5 \\ x - y + 5z = 1 \end{pmatrix}$ $\begin{matrix} (1) \\ (2) \\ (3) \end{matrix}$

Add equation (1) to equation (3).

$$2x + y + z = 8$$
$$\underline{x - y + 5z = 1}$$
$$3x + 6z = 9 \quad (4)$$

Add equation (2) to equation (3).

$$x + y - z = 5$$
$$\underline{x - y + 5z = 1}$$
$$2x + 4z = 6 \quad (5)$$

Solve the system $\begin{pmatrix} 3x + 6z = 9 \\ 2x + 4z = 6 \end{pmatrix}$.

$$3x + 6z = 9 \quad (4)$$
$$2x + 4z = 6 \quad (5)$$

Multiply equation (4) by $\frac{1}{3}$. Multiply equation (5) by $-\frac{1}{2}$. Then add the equations.

$$x + 2z = 3$$
$$-x - 2z = -3$$
$$\overline{0 = 0}$$

This is a true statement. Therefore, the system has infinitely many solutions.

13. Let a represent the measure of $\angle A$, let b represent the measure of $\angle B$, and let c represent the measure of $\angle C$.

The measure of $\angle A$ is 15 less than the measure of $\angle B$.

$a = b - 15$

The sum of the measures of $\angle A$ and $\angle B$ is 10 more than the measure of $\angle C$.

$a + b = 10 + c$

The sum of the measures of the angles of a triangle is 180°.

$a + b + c = 180$

Solve the system $\begin{pmatrix} a - b = -15 & (1) \\ a + b - c = 10 & (2) \\ a + b + c = 180 & (3) \end{pmatrix}$.

Add equation (2) to equation (3).

$$a + b - c = 10$$
$$a + b + c = 180$$
$$\overline{2a + 2b = 190 \quad (4)}$$

Multiply equation (1) by 2. Add to equation (4).

$$2a - 2b = -30$$
$$2a + 2b = 190$$
$$\overline{4a = 160}$$
$$a = 40 \quad (5)$$

Substitute 40 for a in equation (1).

$$a - b = -15$$
$$40 - b = -15$$
$$-b = -55$$
$$b = 55$$

Substitute 40 for a and 55 for b in equation (2).

$$a + b - c = 10$$
$$40 + 55 - c = 10$$
$$95 - c = 10$$
$$-c = -85$$
$$c = 85$$

The measure of $\angle A$ is 40°, the measure of $\angle B$ is 55°, and the measure of $\angle C$ is 85°.

CHAPTER 9

Section 9.1

1. (a) $\sqrt{289} = 17$ because $(17)^2 = 289$

   (b) $\sqrt{\frac{25}{81}} = \frac{5}{9}$ because $\left(\frac{5}{9}\right)^2 = \frac{25}{81}$

   (c) $\sqrt{0.04} = 0.2$ because $(0.2)^2 = 0.04$

3. (a) $\sqrt[3]{64} = 4$ because $(4)^3 = 64$

   (b) $\sqrt[3]{\frac{27}{1000}} = \frac{3}{10}$ because $\left(\frac{3}{10}\right)^3 = \frac{27}{1000}$

   (c) $\sqrt[3]{0.008} = 0.2$ because $(0.2)^3 = 0.008$

5. (a) $\sqrt{425} \approx 20.616$

   (b) $-\sqrt{138} \approx -11.747$

   (c) $\sqrt[3]{-450} \approx -7.663$

   (d) $\sqrt[3]{212} \approx 5.963$

7. $6\sqrt{5} + 12\sqrt{3} - 2\sqrt{5} - 4\sqrt{3}$
   $= 6\sqrt{5} - 2\sqrt{5} + 12\sqrt{3} - 4\sqrt{3}$
   $= (6 - 2)\sqrt{5} + (12 - 4)\sqrt{3}$
   $= 4\sqrt{5} + 8\sqrt{3}$

9. $6\sqrt{2} + 4\sqrt{7} + 5\sqrt{2} - 2\sqrt{7} - 3\sqrt{2}$
   $= 6\sqrt{2} + 5\sqrt{2} - 3\sqrt{2} + 4\sqrt{7} - 2\sqrt{7}$
   $= (6 + 5 - 3)\sqrt{2} + (4 - 2)\sqrt{7}$
   $= 8\sqrt{2} + 2\sqrt{7}$
   $\approx 8(1.414) + 2(2.646)$
   $\approx 11.312 + 5.292$
   ≈ 16.604

   To the nearest tenth, 16.6.

11. Substitute 3.2 for L in the formula. Use 3.14 as an approximation for π.

    $T = 2\pi\sqrt{\frac{L}{32}}$

    $T = 2(3.14)\sqrt{\frac{3.2}{32}} = 6.28\sqrt{0.10} = 1.9859$

    The period to the nearest tenth is 2.0 seconds.

13. Substitute 0.35 for f and 200 for D in the formula $S = \sqrt{30Df}$

    $= \sqrt{30(200)(0.35)}$
    $= \sqrt{2100} \approx 45.8275$

    To the nearest mile per hour, the car was traveling at approximately 46 mph.

Section 9.2

1. (a) $\sqrt{24} = \sqrt{4 \cdot 6} = \sqrt{4}\sqrt{6} = 2\sqrt{6}$
   (b) $\sqrt{63} = \sqrt{9 \cdot 7} = \sqrt{9}\sqrt{7} = 3\sqrt{7}$
   (c) $\sqrt{32} = \sqrt{16 \cdot 2} = \sqrt{16}\sqrt{2} = 4\sqrt{2}$

3. (a) $\sqrt{20} = \sqrt{4 \cdot 5} = \sqrt{4}\sqrt{5} = 2\sqrt{5}$
   (b) $\sqrt{72} = \sqrt{36 \cdot 2} = \sqrt{36}\sqrt{2} = 6\sqrt{2}$
   (c) $6\sqrt{50} = 6\sqrt{25 \cdot 2} = 6\sqrt{25}\sqrt{2} = 6(5)\sqrt{2}$
   $= 30\sqrt{2}$

5. (a) $\sqrt[3]{16} = \sqrt[3]{8 \cdot 2} = \sqrt[3]{8}\sqrt[3]{2} = 2\sqrt[3]{2}$
   (b) $\sqrt[3]{40} = \sqrt[3]{8 \cdot 5} = \sqrt[3]{8}\sqrt[3]{5} = 2\sqrt[3]{5}$
   (c) $\sqrt[3]{81} = \sqrt[3]{27 \cdot 3} = \sqrt[3]{27}\sqrt[3]{3} = 3\sqrt[3]{3}$

7. (a) $\sqrt{12x^2 y} = \sqrt{4x^2 \cdot 3y} = \sqrt{4x^2}\sqrt{3y} = 2x\sqrt{3y}$
   (b) $\sqrt{75a^3 b^2} = \sqrt{25a^2 b^2 \cdot 3a} = \sqrt{25a^2 b^2}\sqrt{3a}$
   $= 5ab\sqrt{3a}$
   (c) $\sqrt[3]{16x^4 y^3} = \sqrt[3]{8x^3 y^3 \cdot 2x} = \sqrt[3]{8x^3 y^3}\sqrt[3]{2x}$
   $= 2xy\sqrt[3]{2x}$

9. $2\sqrt{45} + 4\sqrt{5} = 2\sqrt{9}\sqrt{5} + 4\sqrt{5}$
   $= 2(3)\sqrt{5} + 4\sqrt{5}$
   $= 6\sqrt{5} + 4\sqrt{5}$
   $= (6+4)\sqrt{5}$
   $= 10\sqrt{5}$

11. $\sqrt{18} + 3\sqrt{50} - \sqrt{72}$
    $= \sqrt{9}\sqrt{2} + 3\sqrt{25}\sqrt{2} - \sqrt{36}\sqrt{2}$
    $= 3\sqrt{2} + 3(5)\sqrt{2} - 6\sqrt{2}$
    $= 3\sqrt{2} + 15\sqrt{2} - 6\sqrt{2}$
    $= (3 + 15 - 6)\sqrt{2}$
    $= 12\sqrt{2}$

13. $\frac{1}{5}\sqrt{27} + \frac{3}{4}\sqrt{3} = \frac{1}{5}\sqrt{9 \cdot 3} + \frac{3}{4}\sqrt{3}$
    $= \frac{1}{5}\sqrt{9}\sqrt{3} + \frac{3}{4}\sqrt{3}$
    $= \frac{1}{5}(3)\sqrt{3} + \frac{3}{4}\sqrt{3}$
    $= \frac{3}{5}\sqrt{3} + \frac{3}{4}\sqrt{3}$
    $= \left(\frac{3}{5} + \frac{3}{4}\right)\sqrt{3}$
    $= \left(\frac{12}{20} + \frac{15}{20}\right)\sqrt{3}$
    $= \frac{27}{20}\sqrt{3}$

15. $7\sqrt[3]{24} + 4\sqrt[3]{81} = 7\sqrt[3]{8}\sqrt[3]{3} + 4\sqrt[3]{27}\sqrt[3]{3}$
    $= 7(2)\sqrt[3]{3} + 4(3)\sqrt[3]{3}$
    $= 14\sqrt[3]{3} + 12\sqrt[3]{3}$
    $= 26\sqrt[3]{3}$

Section 9.3

1. $\sqrt{\frac{7}{9}} = \frac{\sqrt{7}}{\sqrt{9}} = \frac{\sqrt{7}}{3}$

3. $\sqrt[3]{\frac{81}{8}} = \frac{\sqrt[3]{81}}{\sqrt[3]{8}} = \frac{\sqrt[3]{27}\sqrt[3]{3}}{2} = \frac{3\sqrt[3]{3}}{2}$

5. $\sqrt{\frac{18}{81}} = \frac{\sqrt{18}}{\sqrt{81}} = \frac{\sqrt{9 \cdot 2}}{9} = \frac{\sqrt{9}\sqrt{2}}{9} = \frac{3\sqrt{2}}{9} = \frac{\sqrt{2}}{3}$

7. $\sqrt{\frac{3}{5}} = \frac{\sqrt{3}}{\sqrt{5}} \cdot \frac{\sqrt{5}}{\sqrt{5}} = \frac{\sqrt{15}}{\sqrt{25}} = \frac{\sqrt{15}}{5}$

9. $\sqrt{\frac{5}{12}} = \frac{\sqrt{5}}{\sqrt{12}} = \frac{\sqrt{5}}{\sqrt{4}\sqrt{3}} = \frac{\sqrt{5}}{2\sqrt{3}} \cdot \frac{\sqrt{3}}{\sqrt{3}}$
   $= \frac{\sqrt{15}}{2\sqrt{9}} = \frac{\sqrt{15}}{2(3)} = \frac{\sqrt{15}}{6}$

11. (a) $\frac{4}{\sqrt{a}} = \frac{4}{\sqrt{a}} \cdot \frac{\sqrt{a}}{\sqrt{a}} = \frac{4\sqrt{a}}{\sqrt{a^2}} = \frac{4\sqrt{a}}{a}$
    (b) $\sqrt{\frac{5m}{2n}} = \frac{\sqrt{5m}}{\sqrt{2n}} \cdot \frac{\sqrt{2n}}{\sqrt{2n}} = \frac{\sqrt{10mn}}{\sqrt{4n^2}} = \frac{\sqrt{10mn}}{2n}$
    (c) $\frac{5\sqrt{3}}{\sqrt{10}} = \frac{5\sqrt{3}}{\sqrt{10}} \cdot \frac{\sqrt{10}}{\sqrt{10}} = \frac{5\sqrt{30}}{\sqrt{100}} = \frac{5\sqrt{30}}{10} = \frac{\sqrt{30}}{2}$
    (d) $\sqrt{\frac{9a^2}{16b}} = \frac{\sqrt{9a^2}}{\sqrt{16b}} = \frac{3a}{\sqrt{16b}} = \frac{3a}{4\sqrt{b}} \cdot \frac{\sqrt{b}}{\sqrt{b}}$
    $= \frac{3a\sqrt{b}}{4\sqrt{b^2}} = \frac{3a\sqrt{b}}{4b}$

13. $7\sqrt{3} + \frac{2}{\sqrt{3}} = 7\sqrt{3} + \frac{2}{\sqrt{3}} \cdot \frac{\sqrt{3}}{\sqrt{3}} = 7\sqrt{3} + \frac{2\sqrt{3}}{\sqrt{9}}$
    $= 7\sqrt{3} + \frac{2\sqrt{3}}{3} = \left(7 + \frac{2}{3}\right)\sqrt{3}$
    $= \frac{23}{3}\sqrt{3} = \frac{23\sqrt{3}}{3}$

15. $\sqrt{\frac{2}{5}} + \sqrt{90} = \frac{\sqrt{2}}{\sqrt{5}} + \sqrt{9 \cdot 10}$
    $= \frac{\sqrt{2}}{\sqrt{5}} \cdot \frac{\sqrt{5}}{\sqrt{5}} + \sqrt{9}\sqrt{10}$
    $= \frac{\sqrt{10}}{\sqrt{25}} + 3\sqrt{10} = \frac{\sqrt{10}}{5} + 3\sqrt{10} = \left(\frac{1}{5} + 3\right)\sqrt{10}$
    $= \frac{16\sqrt{10}}{5}$

Section 9.4

1. (a) $\sqrt{2}\sqrt{14} = \sqrt{2 \cdot 14} = \sqrt{28} = \sqrt{4}\sqrt{7} = 2\sqrt{7}$
   (b) $(5\sqrt{3})(2\sqrt{7}) = 10\sqrt{21}$
   (c) $\sqrt[3]{10}\sqrt[3]{4} = \sqrt[3]{40} = \sqrt[3]{8 \cdot 5} = \sqrt[3]{8}\sqrt[3]{5} = 2\sqrt[3]{5}$

(d) $(2\sqrt[3]{5})(3\sqrt[3]{50}) = 6\sqrt[3]{250} = 6\sqrt[3]{125}\sqrt[3]{2}$
$= 6(5)\sqrt[3]{2} = 30\sqrt[3]{2}$

3. (a) $\sqrt{3}(\sqrt{7} - \sqrt{5}) = \sqrt{21} - \sqrt{15}$

(b) $\sqrt{2}(\sqrt{18} + 5) = \sqrt{36} + 5\sqrt{2}$
$= 6 + 5\sqrt{2}$

(c) $\sqrt{a}(\sqrt{a} - \sqrt{b}) = \sqrt{a^2} - \sqrt{ab}$
$= a - \sqrt{ab}$

(d) $\sqrt[3]{3}(\sqrt[3]{3} + \sqrt[3]{18}) = \sqrt[3]{9} + \sqrt[3]{54}$
$= \sqrt[3]{9} + \sqrt[3]{27 \cdot 2}$
$= \sqrt[3]{9} + \sqrt[3]{27}\sqrt[3]{2}$
$= \sqrt[3]{9} + 3\sqrt[3]{2}$

5. (a) $(\sqrt{6} - \sqrt{2})(\sqrt{3} + \sqrt{7})$
$= \sqrt{18} + \sqrt{42} - \sqrt{6} - \sqrt{14}$
$= \sqrt{9}\sqrt{2} + \sqrt{42} - \sqrt{6} - \sqrt{14}$
$= 3\sqrt{2} + \sqrt{42} - \sqrt{6} - \sqrt{14}$

(b) $(\sqrt{5} + 4)(\sqrt{5} + 2) = \sqrt{25} + 2\sqrt{5} + 4\sqrt{5} + 8$
$= 5 + 6\sqrt{5} + 8$
$= 13 + 6\sqrt{5}$

7. (a) $(\sqrt{5} - 6)(\sqrt{5} + 6) = \sqrt{25} + 6\sqrt{5} - 6\sqrt{5} - 36$
$= 5 - 36$
$= -31$

(b) $(4 - \sqrt{3})(4 + \sqrt{3}) = 16 + 4\sqrt{3} - 4\sqrt{3} - \sqrt{9}$
$= 16 - 3$
$= 13$

(c) $(\sqrt{6} + \sqrt{2})(\sqrt{6} - \sqrt{2})$
$= \sqrt{36} - \sqrt{12} + \sqrt{12} - \sqrt{4}$
$= 6 - 2$
$= 4$

9. $\dfrac{3}{\sqrt{7} + \sqrt{2}} = \dfrac{3}{(\sqrt{7} + \sqrt{2})} \cdot \dfrac{(\sqrt{7} - \sqrt{2})}{(\sqrt{7} - \sqrt{2})}$
$= \dfrac{3(\sqrt{7} - \sqrt{2})}{(\sqrt{7} + \sqrt{2})(\sqrt{7} - \sqrt{2})}$
$= \dfrac{3(\sqrt{7} - \sqrt{2})}{\sqrt{49} - \sqrt{14} + \sqrt{14} - \sqrt{4}}$
$= \dfrac{3(\sqrt{7} - \sqrt{2})}{7 - 2}$
$= \dfrac{3(\sqrt{7} - \sqrt{2})}{5}$

11. $\dfrac{\sqrt{2}}{\sqrt{10} + 3} = \dfrac{\sqrt{2}}{\sqrt{10} + 3} \cdot \dfrac{(\sqrt{10} - 3)}{(\sqrt{10} - 3)}$
$= \dfrac{\sqrt{2}(\sqrt{10} - 3)}{(\sqrt{10} + 3)(\sqrt{10} - 3)}$
$= \dfrac{\sqrt{20} - 3\sqrt{2}}{\sqrt{100} - 3\sqrt{10} + 3\sqrt{10} - 9}$
$= \dfrac{\sqrt{4}\sqrt{5} - 3\sqrt{2}}{10 - 9}$
$= \dfrac{2\sqrt{5} - 3\sqrt{2}}{1} = 2\sqrt{5} - 3\sqrt{2}$

13. $\dfrac{\sqrt{a} + 3}{\sqrt{a} - 4} = \dfrac{\sqrt{a} + 3}{\sqrt{a} - 4} \cdot \dfrac{(\sqrt{a} + 4)}{(\sqrt{a} + 4)}$
$= \dfrac{(\sqrt{a} + 3)(\sqrt{a} + 4)}{(\sqrt{a} - 4)(\sqrt{a} + 4)}$
$= \dfrac{\sqrt{a^2} + 4\sqrt{a} + 3\sqrt{a} + 12}{\sqrt{a^2} - 4\sqrt{a} + 4\sqrt{a} - 16}$
$= \dfrac{a + 7\sqrt{a} + 12}{a - 16}$

15. $\dfrac{5 - \sqrt{6}}{\sqrt{6} - 2} = \dfrac{5 - \sqrt{6}}{\sqrt{6} - 2} \cdot \dfrac{(\sqrt{6} + 2)}{(\sqrt{6} + 2)}$
$= \dfrac{(5 - \sqrt{6})(\sqrt{6} + 2)}{(\sqrt{6} - 2)(\sqrt{6} + 2)}$
$= \dfrac{5\sqrt{6} + 10 - \sqrt{36} - 2\sqrt{6}}{\sqrt{36} - 2\sqrt{6} + 2\sqrt{6} - 4}$
$= \dfrac{3\sqrt{6} + 10 - 6}{6 - 4}$
$= \dfrac{3\sqrt{6} + 4}{2}$

Section 9.5

1. $\sqrt{y} = 9$
$(\sqrt{y})^2 = (9)^2$
$y = 81$
Because $\sqrt{81} = 9$, the solution set is $\{81\}$.

3. $\sqrt{a} = -12$
$(\sqrt{a})^2 = (-12)^2$
$a = 144$
Because $\sqrt{144} \neq -12$, 144 is not the solution, so the solution set is $\varnothing$.

5. $\sqrt{3y + 1} = 4$
$(\sqrt{3y + 1})^2 = (4)^2$
$3y + 1 = 16$
$3y = 15$
$y = 5$

Check:
$\sqrt{3y + 1} = 4$
$\sqrt{3(5) + 1} \stackrel{?}{=} 4$
$\sqrt{15 + 1} \stackrel{?}{=} 4$
$\sqrt{16} \stackrel{?}{=} 4$
$4 = 4$

The solution set is $\{5\}$.

7. $\sqrt{2y + 8} = -10$
$(\sqrt{2y + 8})^2 = (-10)^2$
$2y + 8 = 100$
$2y = 92$
$y = 46$

Check:
$\sqrt{2y + 8} = -10$
$\sqrt{2(46) + 8} \stackrel{?}{=} -10$
$\sqrt{92 + 8} \stackrel{?}{=} -10$
$\sqrt{100} \stackrel{?}{=} -10$
$10 \neq -10$

Because $y = 46$ does not check, there is no real number solution. So the solution set is $\varnothing$.

9. $2\sqrt{3x - 6} = 24$
$\sqrt{3x - 6} = 12$
$(\sqrt{3x - 6})^2 = (12)^2$
$3x - 6 = 144$
$3x = 150$
$x = 50$

Check:
$2\sqrt{3x - 6} = 24$
$2\sqrt{3(50) - 6} \stackrel{?}{=} 24$
$2\sqrt{150 - 6} \stackrel{?}{=} 24$
$2\sqrt{144} \stackrel{?}{=} 24$
$2(12) \stackrel{?}{=} 24$
$24 = 24$

The solution set is {50}.

11. $\sqrt{2y + 5} = \sqrt{4y - 7}$
$(\sqrt{2y + 5})^2 = (\sqrt{4y - 7})^2$
$2y + 5 = 4y - 7$
$-2y + 5 = -7$
$-2y = -12$
$y = 6$

Check:
$\sqrt{2y + 5} = \sqrt{4y - 7}$
$\sqrt{2(6) + 5} \stackrel{?}{=} \sqrt{4(6) - 7}$
$\sqrt{12 + 5} \stackrel{?}{=} \sqrt{24 - 7}$
$\sqrt{17} = \sqrt{17}$

The solution set is {6}.

13. $\sqrt{2y - 8} = y - 4$
$(\sqrt{2y - 8})^2 = (y - 4)^2$
$2y - 8 = y^2 - 8y + 16$

$0 = y^2 - 10y + 24$
$0 = (y - 6)(y - 4)$
$y - 6 = 0$ or $y - 4 = 0$
$y = 6$ or $y = 4$

Check $y = 6$.
$\sqrt{2y - 8} = y - 4$
$\sqrt{2(6) - 8} \stackrel{?}{=} (6) - 4$
$\sqrt{12 - 8} \stackrel{?}{=} 2$
$\sqrt{4} \stackrel{?}{=} 2$
$2 = 2$

Check $y = 4$.
$\sqrt{2y - 8} = y - 4$
$\sqrt{2(4) - 8} \stackrel{?}{=} (4) - 4$
$\sqrt{8 - 8} \stackrel{?}{=} 0$
$\sqrt{0} \stackrel{?}{=} 0$
$0 = 0$

The solution set is {4, 6}.

15. $2 + \sqrt{y} = y$
$\sqrt{y} = y - 2$
$(\sqrt{y})^2 = (y - 2)^2$
$y = y^2 - 4y + 4$
$0 = y^2 - 5y + 4$
$0 = (y - 1)(y - 4)$
$y - 1 = 0$ or $y - 4 = 0$
$y = 1$ or $y = 4$

Check $y = 1$.
$2 + \sqrt{y} = y$
$2 + \sqrt{(1)} \stackrel{?}{=} (1)$
$2 + 1 \stackrel{?}{=} 1$
$3 \neq 1$

Check $y = 4$.
$2 + \sqrt{(4)} = (4)$
$2 + 2 \stackrel{?}{=} 4$
$4 = 4$

The solution set is {4}.

17. Use $D = \dfrac{S^2}{30f}$.

$D = \dfrac{(55)^2}{30(0.40)}$

$D = \dfrac{(55)(55)}{12}$

$D \approx 252$

Section 10.1

1. $y^2 + 7y = 0$
$y(y + 7) = 0$
$y = 0$ or $y + 7 = 0$
$y = 0$ or $y = -7$
The solution set is $\{-7, 0\}$.

3. $x^2 - 4x - 21 = 0$
$(x - 7)(x + 3) = 0$
$x - 7 = 0$ or $x + 3 = 0$
$x = 7$ or $x = -3$
The solution set is $\{-3, 7\}$.

5. $a^2 - 10x + 25 = 0$
$(a - 5)(a - 5) = 0$
$a - 5 = 0$ or $a - 5 = 0$
$a = 5$ or $a = 5$
The solution set is $\{5\}$.

7. $n^2 = 144$
$n^2 - 144 = 0$
$(n - 12)(n + 12) = 0$
$n - 12 = 0$ or $n + 12 = 0$
$n = 12$ or $n = -12$
The solution set is $\{-12, 12\}$.

9. $y^2 = 64$
$y = \pm\sqrt{64}$
$y = \pm 8$
The solution set is $\{-8, 8\}$.

11. $n^2 = 12$
$n = \pm\sqrt{12}$
$n = \pm\sqrt{12} = \pm\sqrt{4}\sqrt{3}$
$y = \pm 2\sqrt{3}$
The solution set is $\{-2\sqrt{3}, 2\sqrt{3}\}$.

13. $3y^2 = 8$
$y^2 = \dfrac{8}{3}$
$y = \pm\sqrt{\dfrac{8}{3}} = \pm\dfrac{\sqrt{8}}{\sqrt{3}} \cdot \dfrac{\sqrt{3}}{\sqrt{3}} = \pm\dfrac{\sqrt{24}}{\sqrt{9}} = \pm\dfrac{\sqrt{4}\sqrt{6}}{3}$
$y = \pm\dfrac{2\sqrt{6}}{3}$
The solution set is $\left\{-\dfrac{2\sqrt{6}}{3}, \dfrac{2\sqrt{6}}{3}\right\}$.

15. $(a + 5)^2 = 9$
$a + 5 = \pm\sqrt{9}$
$a + 5 = \pm 3$
$a + 5 = 3$ or $a + 5 = -3$
$a = -2$ or $a = -8$
The solution set is $\{-8, -2\}$.

17. $(y - 7)^2 = 20$
$y - 7 = \pm\sqrt{20}$
$y - 7 = \pm\sqrt{4}\sqrt{5}$
$y - 7 = \pm 2\sqrt{5}$
$y - 7 = 2\sqrt{5}$ or $y - 7 = -2\sqrt{5}$
$y = 7 + 2\sqrt{5}$ or $y = 7 - 2\sqrt{5}$
The solution set is $\{7 - 2\sqrt{5}, 7 + 2\sqrt{5}\}$.

19. $3(5n + 1)^2 + 4 = 58$
$3(5n + 1)^2 = 54$
$(5n + 1)^2 = 18$
$5n + 1 = \pm\sqrt{18}$
$5n + 1 = \pm\sqrt{9}\sqrt{2}$
$5n + 1 = \pm 3\sqrt{2}$
$5n + 1 = 3\sqrt{2}$ or $5n + 1 = -3\sqrt{2}$
$5n = -1 + 3\sqrt{2}$ or $5n = -1 - 3\sqrt{2}$
$n = \dfrac{-1 + 3\sqrt{2}}{5}$ or $n = \dfrac{-1 - 3\sqrt{2}}{5}$
The solution set is $\left\{\dfrac{-1 - 3\sqrt{2}}{5}, \dfrac{-1 + 3\sqrt{2}}{5}\right\}$.

21. $a = 5$, $b = 12$, c

Use the Pythagorean theorem to solve.
$c^2 = a^2 + b^2$
$c^2 = (5)^2 + (12)^2$
$c^2 = 25 + 144$
$c^2 = 169$
$c = \pm\sqrt{169}$
$c = \pm 13$

Discard $c = -13$, because length cannot be negative. The length of c is 13 inches.

23. Let x represent the length.

30 feet, x, 12 feet

Use the Pythagorean theorem to solve.

$a^2 + b^2 = c^2$

$x^2 + (12)^2 = (30)^2$

$x^2 + 144 = 900$

$x^2 = 756$

$x = \pm\sqrt{756}$

$x \approx \pm 27.5$

Discard $x = -27.5$, because length cannot be negative. The length of x is 27.5 feet.

25. Let a represent the length of each leg of the isosceles triangle.

Use the Pythagorean theorem to solve.

$a^2 + b^2 = c^2$

$a^2 + a^2 = (12)^2$

$2a^2 = 144$

$a^2 = 72$

$a = \pm\sqrt{72}$

$c = \pm\sqrt{36}\sqrt{2}$

$c = \pm 6\sqrt{2}$

Discard $-6\sqrt{2}$, because length cannot be negative. The length is $6\sqrt{2}$ feet.

27. In a 30–60 right triangle, the side opposite the 30° angle is equal in measure to one-half the length of the hypotenuse.

The ladder is 16 feet in length, and the distance from the base of the ladder to the corner of the building is 8 feet.

To solve for x, use the Pythagorean theorem.

$a^2 + b^2 = c^2$

$(8)^2 + x^2 = (16)^2$

$64 + x^2 = 256$

$x^2 = 192$

$x = \pm\sqrt{192}$

$x \approx \pm 13.9$

Discard $x = -13.9$, because length cannot be negative. The top of the ladder reaches 13.9 feet.

Section 10.2

1. $y^2 + 10y - 6 = 0$

$y^2 + 10y = 6$

$y^2 + 10y + 25 = 6 + 25$

$(y + 5)^2 = 31$

$y + 5 = \pm\sqrt{31}$

$y = -5 \pm \sqrt{31}$

The solution set is $\{-5 - \sqrt{31}, -5 + \sqrt{31}\}$.

3. $y^2 - 6y + 2 = 0$

$y^2 - 6y = -2$

$y^2 - 6y + 9 = -2 + 9$

$(y - 3)^2 = 7$

$y - 3 = \pm\sqrt{7}$

$y = 3 \pm \sqrt{7}$

The solution set is $\{3 - \sqrt{7}, 3 + \sqrt{7}\}$.

5. $a^2 + 5a - 3 = 0$

$a^2 + 5a = 3$

$a^2 + 5a + \dfrac{25}{4} = 3 + \dfrac{25}{4}$

$\left(a + \dfrac{5}{2}\right)^2 = \dfrac{37}{4}$

$a + \dfrac{5}{2} = \pm\sqrt{\dfrac{37}{4}}$

$a + \dfrac{5}{2} = \pm\dfrac{\sqrt{37}}{2}$

$a = -\dfrac{5}{2} \pm \dfrac{\sqrt{37}}{2}$

$a = \dfrac{-5 \pm \sqrt{37}}{2}$

The solution set is $\left\{\dfrac{-5 - \sqrt{37}}{2}, \dfrac{-5 + \sqrt{37}}{2}\right\}$.

7. $3y^2 + 6y - 2 = 0$

Multiply both sides by $\dfrac{1}{3}$.

$y^2 + 2y - \dfrac{2}{3} = 0$

$y^2 + 2y = \dfrac{2}{3}$

$y^2 + 2y + 1 = \dfrac{2}{3} + 1$

$(y + 1)^2 = \dfrac{5}{3}$

$y + 1 = \pm\sqrt{\dfrac{5}{3}}$

$y + 1 = \pm\dfrac{\sqrt{5}}{\sqrt{3}} \cdot \dfrac{\sqrt{3}}{\sqrt{3}} = \dfrac{\sqrt{15}}{\sqrt{9}}$

$$y + 1 = \pm\frac{\sqrt{15}}{3}$$

$$y = -1 \pm \frac{\sqrt{15}}{3} = -\frac{3}{3} \pm \frac{\sqrt{15}}{3}$$

$$y = \frac{-3 \pm \sqrt{15}}{3}$$

The solution set is $\left\{\frac{-3 - \sqrt{15}}{3}, \frac{-3 + \sqrt{15}}{3}\right\}$.

9. $n^2 - n - 12 = 0$

$n^2 + n = 12$

$n^2 + n + \frac{1}{4} = 12 + \frac{1}{4}$

$\left(n + \frac{1}{2}\right)^2 = \frac{49}{4}$

$n + \frac{1}{2} = \pm\sqrt{\frac{49}{4}}$

$n + \frac{1}{2} = \pm\frac{7}{2}$

$n = -\frac{1}{2} \pm \frac{7}{2}$

$n = -\frac{1}{2} - \frac{7}{2}$ or $n = -\frac{1}{2} + \frac{7}{2}$

$n = -\frac{8}{2}$ or $n = \frac{6}{2}$

$n = -4$ or $n = 3$

$n^2 + n - 12 = 0$

$(n + 4)(n - 3) = 0$

$n + 4 = 0$ or $n - 3 = 0$

$n = -4$ or $n = 3$

The solution set is $\{-4, 3\}$.

11. $y^2 + 6y + 22 = 0$

$y^2 - 6y = -22$

$y^2 - 6y + 9 = -22 + 9$

$(y - 3)^2 = -13$

Any value of y will give a nonnegative value for $(y - 3)^2$. Therefore, there are no real number solutions.

Section 10.3

1. $x^2 + 3x - 18 = 0$

Use the quadratic formula.

$a = 1$, $b = 3$, $c = -18$

$x = \dfrac{-b \pm \sqrt{b^2 - 4ac}}{2a}$

$x = \dfrac{-(3) \pm \sqrt{(3)^2 - 4(1)(-18)}}{2(1)}$

$x = \dfrac{-3 \pm \sqrt{9 + 72}}{2}$

$x = \dfrac{-3 \pm \sqrt{81}}{2}$

$x = \dfrac{-3 \pm 9}{2}$

$x = \dfrac{-3 - 9}{2}$ or $x = \dfrac{-3 + 9}{2}$

$x = -\dfrac{12}{2}$ or $x = \dfrac{6}{2}$

$x = -6$ or $x = 3$

The solution set is $\{-6, 3\}$.

3. $x^2 + 5x = 3$

$x^2 + 5x - 3 = 0$

Use the quadratic formula.

$a = 1$, $b = 5$, $c = -3$

$x = \dfrac{-b \pm \sqrt{b^2 - 4ac}}{2a}$

$x = \dfrac{-(5) \pm \sqrt{(5)^2 - 4(1)(-3)}}{2(1)}$

$x = \dfrac{-5 \pm \sqrt{25 + 12}}{2}$

$x = \dfrac{-5 \pm \sqrt{37}}{2}$

The solution set is $\left\{\dfrac{-5 - \sqrt{37}}{2}, \dfrac{-5 + \sqrt{37}}{2}\right\}$.

5. $6y^2 + 5y - 4 = 0$

Use the quadratic formula.

$a = 6$, $b = 5$, $c = -4$

$y = \dfrac{-b \pm \sqrt{b^2 - 4ac}}{2a}$

$y = \dfrac{-(5) \pm \sqrt{(5)^2 - 4(6)(-4)}}{2(6)}$

$y = \dfrac{-5 \pm \sqrt{25 + 96}}{12}$

$y = \dfrac{-5 \pm \sqrt{121}}{12}$

$y = \dfrac{-5 \pm 11}{12}$

$y = \dfrac{-5 - 11}{12}$ or $y = \dfrac{-5 + 11}{12}$

$y = -\dfrac{16}{12}$ or $y = \dfrac{6}{12}$

$y = -\dfrac{4}{3}$ or $y = \dfrac{1}{2}$

The solution set is $\left\{-\dfrac{4}{3}, \dfrac{1}{2}\right\}$.

7. $x^2 + 4x - 6 = 0$

Use the quadratic formula.

$a = 1, b = 4, c = -6$

$x = \dfrac{-b \pm \sqrt{b^2 - 4ac}}{2a}$

$x = \dfrac{-(4) \pm \sqrt{(4)^2 - 4(1)(-6)}}{2(1)}$

$x = \dfrac{-4 \pm \sqrt{16 + 24}}{2}$

$x = \dfrac{-4 \pm \sqrt{40}}{2}$

$x = \dfrac{-4 \pm \sqrt{4}\sqrt{10}}{2}$

$x = \dfrac{-4 \pm 2\sqrt{10}}{2} = \dfrac{2(-2 \pm \sqrt{10})}{2}$

$x = -2 \pm \sqrt{10}$

The solution set is $\{-2 - \sqrt{10}, -2 + \sqrt{10}\}$.

9. $y^2 + 3y + 11 = 0$

Use the quadratic formula.

$a = 1, b = 3, c = 11$

$y = \dfrac{-b \pm \sqrt{b^2 - 4ac}}{2a}$

$y = \dfrac{-(3) \pm \sqrt{(3)^2 - 4(1)(11)}}{2(1)}$

$y = \dfrac{-3 \pm \sqrt{9 - 44}}{2}$

$y = \dfrac{-3 \pm \sqrt{-35}}{2}$

Because $\sqrt{-35}$ is not a real number, there are no real number solutions.

Section 10.4

1. $3x^2 - 3x - 6 = 0$

Multiply both sides by $\dfrac{1}{3}$.

$x^2 - x - 2 = 0$

$(x - 2)(x + 1) = 0$

$x - 2 = 0$ or $x + 1 = 0$

$x = 2$ or $x = -1$

The solution set is $\{-1, 2\}$.

3. $(5x - 1)^2 = 49$

$5x - 1 = \pm\sqrt{49}$

$5x - 1 = \pm 7$

$5x - 1 = 7$ or $5x - 1 = -7$

$5x = 8$ or $5x = -6$

$x = \dfrac{8}{5}$ or $x = -\dfrac{6}{5}$

The solution set is $\left\{-\dfrac{6}{5}, \dfrac{8}{5}\right\}$.

5. $y - \dfrac{2}{y} = -3$

Multiply both sides by y.

$y^2 - 2 = -3y$

$y^2 + 3y - 2 = 0$

Use the quadratic formula.

$a = 1, b = 3, c = -2$

$y = \dfrac{-b \pm \sqrt{b^2 - 4ac}}{2a}$

$y = \dfrac{-(3) \pm \sqrt{(3)^2 - 4(1)(-2)}}{2(1)}$

$y = \dfrac{-3 \pm \sqrt{9 + 8}}{2}$

$y = \dfrac{-3 \pm \sqrt{17}}{2}$

The solution set is $\left\{\dfrac{-3 - \sqrt{17}}{2}, \dfrac{-3 + \sqrt{17}}{2}\right\}$.

7. $x^2 = 10x$

$x^2 - 10x = 0$

$x(x - 10) = 0$

$x = 0$ or $x - 10 = 0$

$x = 0$ or $x = 10$

The solution set is $\{0, 10\}$.

9. $t^2 + 30t + 216 = 0$

Use the quadratic formula.

$a = 1, b = 30, c = 216$

$t = \dfrac{-b \pm \sqrt{b^2 - 4ac}}{2a}$

$t = \dfrac{-(30) \pm \sqrt{(30)^2 - 4(1)(216)}}{2(1)}$

$t = \dfrac{-(30) \pm \sqrt{900 - 864}}{2}$

$t = \dfrac{-30 \pm \sqrt{36}}{2}$

$t = \dfrac{-30 \pm 6}{2}$

$$t = \frac{-30 - 6}{2} \quad \text{or} \quad t = \frac{-30 + 6}{2}$$

$$t = \frac{-36}{2} \quad \text{or} \quad t = \frac{-24}{2}$$

$$t = -18 \quad \text{or} \quad t = -12$$

The solution set is $\{-18, -12\}$.

11. $y^2 + 10y = 15$

$y^2 + 10y - 15 = 0$

Use the quadratic formula.

$a = 1, b = 10, c = -15$

$$y = \frac{-b \pm \sqrt{b^2 - 4ac}}{2a}$$

$$y = \frac{-(10) \pm \sqrt{(10)^2 - 4(1)(-15)}}{2(1)}$$

$$y = \frac{-(10) \pm \sqrt{100 + 60}}{2}$$

$$y = \frac{-(10) \pm \sqrt{160}}{2} = \frac{-10 \pm \sqrt{16}\sqrt{10}}{2}$$

$$y = \frac{-(10) \pm 4\sqrt{10}}{2} = \frac{2(-5 \pm 2\sqrt{10})}{2}$$

$$y = -5 \pm 2\sqrt{10}$$

The solution set is $\{-5 - 2\sqrt{10}, -5 + 2\sqrt{10}\}$.

Section 10.5

1. Let w represent the width of the rectangle and let l represent the length.

The length is 3 inches more than the width.

$l = 3 + w$

The area is 88 square inches.

$88 = lw$

Solve the system $\begin{pmatrix} l = 3 + w \\ 88 = lw \end{pmatrix}$.

Use substitution to solve.

$88 = (3 + w)w$

$88 = 3w + w^2$

$0 = w^2 + 3w - 88$

$0 = (w - 8)(w + 11)$

$w - 8 = 0 \quad \text{or} \quad w + 11 = 0$

$w = 8 \quad \text{or} \quad w = -11$

Discard $w = -11$, because length cannot be negative. If $w = 8$, then $l = 3 + 8 = 11$.

The length is 11 inches, and the width is 8 inches.

3. Let n represent the first whole number and let $n + 1$ represent the second consecutive whole number. Their product is 342.

$n(n + 1) = 342$

$n^2 + n = 342$

$n^2 + n - 342 = 0$

$(n - 18)(n + 19) = 0$

$n - 18 = 0 \quad \text{or} \quad n + 19 = 0$

$n = 18 \quad \text{or} \quad n = -19$.

Discard $n = -19$, because it is not a whole number.

The two consecutive whole numbers are 18 and 19.

5. Let w represent the width of the garden and let l represent the length.

The perimeter is 84 feet.

$84 = 2l + 2w$

$\frac{1}{2}(84) = \frac{1}{2}(2l + 2w)$

$42 = l + w$

The area is 360 square feet.

$360 = lw$

Solve the system $\begin{pmatrix} 42 = l + w \\ 360 = lw \end{pmatrix}$.

Solve $42 = l + w$ for l.

$42 - w = l$.

Substitute $42 - w$ for l in the second equation.

$360 = (42 - w)w$

$360 = 42w - w^2$

$0 = -w^2 + 42w - 360$

Multiply both sides by -1

$0 = w^2 - 42w + 360$

$0 = (w - 12)(w - 30)$

$w - 12 = 0 \quad \text{or} \quad w - 30 = 0$

$w = 12 \quad \text{or} \quad w = 30$

If $w = 12$, then $l = 42 - w = 30$.

If $w = 30$, then $l = 42 - w = 12$.

The dimensions of the garden are 30 feet by 12 feet.

7. Let x represent the first number and let y represent the second number.

Their sum is 10.

$x + y = 10$

Their product is 6.

$xy = 6$

Solve the system $\begin{pmatrix} x + y = 10 \\ xy = 6 \end{pmatrix}$.

Solve the first equation for y.

$y = 10 - x$

Substitute $10 - x$ for y in the second equation.

$x(10 - x) = 6$

$10x - x^2 = 6$

$-x^2 + 10x - 6 = 0$

Multiply both sides by -1.

$x^2 - 10x + 6 = 0$

Use the quadratic formula.

$a = 1, \ b = -10, \ c = 6$

$x = \dfrac{-b \pm \sqrt{b^2 - 4ac}}{2a}$

$x = \dfrac{-(-10) \pm \sqrt{(-10)^2 - 4(1)(6)}}{2(1)}$

$x = \dfrac{10 \pm \sqrt{100 - 24}}{2}$

$x = \dfrac{10 \pm \sqrt{76}}{2} = \dfrac{10 \pm \sqrt{4}\sqrt{19}}{2}$

$x = \dfrac{10 \pm 2\sqrt{19}}{2} = \dfrac{2(5 \pm \sqrt{19})}{2}$

$x = 5 \pm \sqrt{19}$

The two numbers are $5 - \sqrt{19}$ and $5 + \sqrt{19}$.

9. Let r represent the rate for Mark's car and let t represent the time Mark drove.

Use Rate × Time = Distance.

$rt = 80$

Let $r + 10$ represent the rate of John's car and let $t - 1$ represent the time John drove.

Use Rate × Time = Distance.

$(r + 10)(t - 1) = 90$

Solve the system $\begin{pmatrix} rt = 80 \\ (r + 10)(t - 1) = 90 \end{pmatrix}$.

Solve the first equation for r.

$r = \dfrac{80}{t}$

Substitute $\dfrac{80}{t}$ for r in the second equation and solve.

$(r + 10)(t - 1) = 90$

$\left(\dfrac{80}{t} + 10\right)(t - 1) = 90$

$80 - \dfrac{80}{t} + 10t - 10 = 90$

$70 - \dfrac{80}{t} + 10t = 90$

$-20 - \dfrac{80}{t} + 10t = 0$

Multiply both sides by t.

$-20t - 80 + 10t^2 = 0$

Multiply both sides by $\dfrac{1}{10}$.

$-2t - 8 + t^2 = 0$

$t^2 - 2t - 8 = 0$

$(t - 4)(t + 2) = 0$

$t - 4 = 0 \quad \text{or} \quad t + 2 = 0$

$\quad t = 4 \quad \text{or} \quad \quad t = -2$

Discard $t = -2$, because time is nonnegative.

If $t = 4$, then $r = \dfrac{80}{t} = \dfrac{80}{4} = 20$.

If $r = 20$, then $r + 10 = 20 + 10 = 30$.

The rate for Mark's car is 20 mph, and the rate for John's car is 30 mph.

Answers to All Chapter Review, Chapter Practice Test, and Cumulative Review Problems

CHAPTER 1

Chapter 1 Review Problem Set (page 53)

1. $\{0, 1, 2, 3, 4\}$ **2.** $\{1, 3, 5, 7, 9\}$ **3.** $\{2, 3, 5, 7\}$
4. $\{2, 4, 6, \ldots\}$ **5.** $\neq$ **6.** $\neq$ **7.** $=$ **8.** $=$
9. Real, rational, integer, and negative
10. Real, rational, integer, and zero
11. Real, rational, noninteger, and negative
12. Real, rational, noninteger, and positive
13. Real, irrational, and positive
14. Real, irrational, and negative **15.** -3 **16.** -25
17. -13 **18.** -12 **19.** -22 **20.** 6 **21.** -25 **22.** -39
23. 24 **24.** -10 **25.** 5 **26.** -15 **27.** -1 **28.** 2
29. 45 **30.** -11 **31.** -56 **32.** 12 **33.** 34 **34.** -45
35. -156 **36.** 252 **37.** 6 **38.** -13 **39.** 48 **40.** -7
41. 175°F **42.** 20,602 feet **43.** $2(6) - 4 + 3(8) - 1 = 31$
44. $3444 **45.** Commutative property of addition
46. Distributive property **47.** Prime **48.** Composite
49. Composite **50.** Composite **51.** Composite
52. Composite **53.** $2^3 \cdot 3$ **54.** $3^2 \cdot 7$ **55.** $2^3 \cdot 3 \cdot 5$
56. $3^3 \cdot 5$ **57.** $2^2 \cdot 3 \cdot 7$ **58.** 2^6 **59.** 18 **60.** 12
61. 180 **62.** 60 **63.** $-\dfrac{3}{10}$ **64.** $\dfrac{5}{16}$ **65.** $\dfrac{6}{7}$ **66.** $-\dfrac{3}{4}$
67. 18 **68.** $-\dfrac{12}{5}$ **69.** $\dfrac{16}{15}$ **70.** $-\dfrac{2}{3}$ **71.** $-\dfrac{20}{9}$ **72.** $-\dfrac{7}{5}$
73. $-\dfrac{1}{4}$ **74.** $-\dfrac{12}{7}$ **75.** $\dfrac{7}{8}$ **76.** $\dfrac{2}{3}$ **77.** $\dfrac{7}{12}$ **78.** $-\dfrac{1}{5}$
79. $\dfrac{13}{36}$ **80.** $-\dfrac{23}{72}$ **81.** $-\dfrac{17}{18}$ **82.** $\dfrac{5}{24}$ **83.** 77.52
84. 31.4 **85.** 4.08 **86.** 26.05 **87.** 0.776 **88.** 2.287
89. 29.52 **90.** 11.556 **91.** 0.1521 **92.** 0.009
93. -0.1008 **94.** -0.04 **95.** 14.1 **96.** 1.582 **97.** 3.16
98. 5.32 **99.** 100 **100.** 720 **101.** 64 **102.** 81 **103.** 25
104. 0.16 **105.** -4 **106.** -14 **107.** 28 **108.** -25
109. $\dfrac{29}{12}$ **110.** $\dfrac{1}{2}$ **111.** 25 **112.** 33 **113.** 4 **114.** 16
115. $-5x + 4y$ **116.** $-3xy - y$ **117.** $2x + 7$
118. $-7x - 18$ **119.** $-3x + 12$ **120.** $-2a + 4$
121. $\dfrac{1}{8}x - \dfrac{7}{10}y$ **122.** $\dfrac{1}{2}x + \dfrac{1}{6}y$ **123.** $2.2a - 0.6b$
124. $-0.58ab + 0.36bc$ **125.** 22 **126.** -16 **127.** 37
128. -39 **129.** 2 **130.** 1 **131.** $-\dfrac{1}{72}$ **132.** 38
133. $168 **134.** $\dfrac{5}{8}$ yard **135.** 3.377 inches **136.** 297 euros

Chapter 1 Practice Test (page 56)

1. -11 **2.** 54 **3.** -26 **4.** 25 **5.** -20 **6.** -6°F
7. Commutative property of addition **8.** $-12x + 6y$
9. $-13x - 21$ **10.** $2a^2 - 11b^2$ **11.** $5x^2 + 7x - 5$ **12.** $\dfrac{7}{9}$
13. Prime **14.** $2^3 \cdot 3^2 \cdot 5$ **15.** 12 **16.** 72
17. $\dfrac{2}{9}$ **18.** $\dfrac{2}{5}$ **19.** $-\dfrac{5}{24}$ **20.** -2.58 **21.** 2.85 **22.** 37
23. -4 **24.** 14 teaspoons **25.** $200.31

CHAPTER 2

Chapter 2 Review Problem Set (page 99)

1. $\{-3\}$ **2.** $\{1\}$ **3.** $\left\{-\dfrac{3}{4}\right\}$ **4.** $\{9\}$ **5.** $\{-4\}$ **6.** $\{-12\}$
7. $\left\{\dfrac{40}{3}\right\}$ **8.** $\left\{\dfrac{9}{4}\right\}$ **9.** $\left\{-\dfrac{15}{8}\right\}$ **10.** $\{-32\}$ **11.** $\{21\}$
12. $\left\{\dfrac{-11}{4}\right\}$ **13.** $\left\{\dfrac{5}{21}\right\}$ **14.** $\{-7\}$ **15.** $\left\{\dfrac{2}{41}\right\}$
16. $\left\{\dfrac{19}{7}\right\}$ **17.** $\left\{\dfrac{1}{2}\right\}$ **18.** $\{-60\}$ **19.** $\{10\}$ **20.** $\left\{-\dfrac{8}{5}\right\}$
21. $\varnothing$ **22.** {All reals} **23.** {All reals} **24.** $\varnothing$
25. {All reals} **26.** $\varnothing$
27. $\{x \mid x > 1\}$
28. $\{x \mid x \leq -2\}$
29. $\{x \mid x \leq 0\}$
30. $\{x \mid x > -3\}$

31. $\{x|x > 4\}$ 32. $\{x|x > -4\}$ 33. $\{x|x \geq 13\}$
34. $\left\{x|x \geq \dfrac{11}{2}\right\}$ 35. $\{y|y < 24\}$ 36. $\{x|x > 10\}$
37. $\{x|x > 35\}$ 38. $\left\{x|x < \dfrac{26}{5}\right\}$ 39. $\left\{n|n < \dfrac{2}{11}\right\}$
40. $\{n|n > 33\}$ 41. $\left\{x|x > \dfrac{9}{2}\right\}$ 42. $\left\{x|x < -\dfrac{43}{3}\right\}$
43. $\{n|n < 2\}$ 44. $\left\{n|n > \dfrac{5}{11}\right\}$ 45. $\{n|n \leq 120\}$
46. $\left\{n|n \leq -\dfrac{180}{13}\right\}$

47. ⟵──○────○──⟶ −3   2
48. ⟵──●────●──⟶ −2   5
49. ⟵────●──⟶ 3
50. ⟵────●────⟶ 1
51. ⟵──○────○──⟶ −1   4
52. ⟵────────⟶ All reals
53. ⟵──○────⟶ −3
54. ⟵────●──⟶ 4

55. $n - 5$ 56. $5 - n$ 57. $10(x - 2)$ 58. $10x - 2$
59. $x - 3$ 60. $\dfrac{d}{r}$ 61. $x^2 + 9$ 62. $(x + 9)^2$ 63. $x^3 + y^3$
64. $xy - 4$ 65. 24 66. 7 67. 33 68. 8
69. 16 and 24 70. 18 71. 8 nickels and 22 dimes
72. 8 nickels, 25 dimes, and 50 quarters 73. 52°
74. 700 miles 75. 89 or better 76. 88 or better

Chapter 2 Practice Test (page 101)

1. $\{2\}$ 2. $\{3\}$ 3. $\{-9\}$ 4. $\{-5\}$ 5. $\{-53\}$ 6. $\{-18\}$
7. $\left\{-\dfrac{5}{2}\right\}$ 8. $\left\{\dfrac{35}{18}\right\}$ 9. $\{12\}$ 10. $\left\{\dfrac{11}{5}\right\}$ 11. $\{22\}$
12. $\left\{\dfrac{31}{2}\right\}$ 13. $\{x|x < 5\}$ 14. $\{x|x \leq 1\}$
15. $\{x|x \geq -9\}$ 16. $\{x|x < 0\}$ 17. $\left\{x|x > -\dfrac{23}{2}\right\}$
18. $\{n|n \geq 12\}$
19. ⟵──●────●──⟶ −2   4
20. ⟵──○────○──⟶ 1   3
21. $72.00 per hour
22. 15 meters, 25 meters, and 30 meters 23. 96 or better
24. 17 nickels, 33 dimes, and 53 quarters 25. 60°, 30°, 90°

Chapters 1–2 Cumulative Review Problem Set (page 103)

1. 3 2. -128 3. -2.4 4. $-\dfrac{5}{12}$ 5. 20 6. $-\dfrac{19}{90}$
7. 0.09 8. 12 9. 36 10. $-\dfrac{1}{24}x - \dfrac{9}{28}y$ 11. $3x + 40$
12. $22x - 16$ 13. $16x - 1$ 14. $-5x - 22$ 15. $\{16\}$
16. $\{35\}$ 17. $\left\{-\dfrac{17}{11}\right\}$ 18. $\varnothing$ 19. $\{x|x \leq 16\}$
20. $\{x|x < -1\}$ 21. $2^2 \cdot 3 \cdot 5^2$
22. ⟵──●────●──⟶ −1   3
23. 9 on Friday and 33 on Saturday
24. 67 or fewer people 25. 11 and 13

CHAPTER 3

Chapter 3 Review Problem Set (page 136)

1. 20 2. 21 3. 6 4. 8 5. $\dfrac{23}{5}$ 6. $\dfrac{1}{2}$ 7. 20 8. 12
9. -34 10. $\dfrac{1}{3}$ 11. 14 feet by 17 feet 12. 130 miles
13. 300 miles 14. $4000 15. 25 pounds
16. $9\dfrac{3}{5}$ feet and $14\dfrac{2}{5}$ feet 17. $300 and $450
18. 2750 females and 1650 males 19. 65% 20. 32%
21. 87.5% 22. 12.5% 23. 175% 24. 18.75%
25. $66\dfrac{2}{3}\%$ 26. $233\dfrac{1}{3}\%$ 27. 60% 28. 16% 29. 40
30. 560 31. 600 32. 8.4 33. $\{5\}$ 34. $\{3.5\}$ 35. $\{5\}$
36. $\{800\}$ 37. $\{16\}$ 38. $\{73\}$ 39. $52 40. $60
41. 40% 42. 35% 43. $8 44. $36.25 45. $110
46. $1.60 a pound 47. $2720 48. 4 years 49. $3000
50. 8.2% 51. $w = 6$ 52. $C = 25$
53. $r = 0.0625$ or $r = 6.25\%$ 54. $t = 2.5$
55. 77 square inches 56. 6 centimeters 57. 354 yards
58. 15 feet 59. $w = \dfrac{V}{lh}$ 60. $h = \dfrac{V}{\pi r^2}$ 61. $b = \dfrac{2A}{h}$
62. $B = \dfrac{3V}{h}$ 63. $y = \dfrac{-2x - 12}{5}$ 64. $y = 3x - 2$
65. $x = \dfrac{2y + 5}{6}$ 66. $x = \dfrac{-4y + 7}{3}$
67. 6 meters by 17 meters
68. 15 centimeters by 40 centimeters
69. 29 yards by 10 yards 70. 34° and 99° 71. $1\dfrac{1}{2}$ hours
72. 58 miles per hour and 65 miles per hour 73. 5 hours
74. $3\dfrac{1}{2}$ hours 75. 20 liters 76. 18 gallons 77. 26%
78. 4 liters 79. $2\dfrac{2}{9}$ cups 80. 25 milliliters
81. $675 at 3% and $1425 at 5%
82. $3000 at 5% and $4500 at 6%
83. $1000 at 8% and $3000 at 9% 84. $1000
85. Shane is 20 years old and Caleb is 12 years old
86. Melinda is 22 years old and her mother is 42 years old

87. Kaitlin is 12 years old and Nikki is 4 years old
88. Jessie is 18 years old and Annilee is 12 years old

Chapter 3 Practice Test (page 139)

1. $\{-22\}$ 2. $\left\{-\dfrac{17}{18}\right\}$ 3. $\{-77\}$ 4. $\left\{\dfrac{4}{3}\right\}$ 5. $\{14\}$
6. $\left\{\dfrac{12}{5}\right\}$ 7. $\{100\}$ 8. $\{70\}$ 9. $\{250\}$ 10. $\left\{\dfrac{11}{2}\right\}$
11. $C = \dfrac{5F - 160}{9}$ 12. $x = \dfrac{y + 8}{2}$ 13. $y = \dfrac{9x + 47}{4}$
14. 64π square centimeters 15. 576 square inches
16. 14 yards 17. 125% 18. 70 19. $189 20. $52
21. 40% 22. 875 women 23. 10 hours
24. 4 centiliters 25. 11.1 years

Chapters 1–3 Cumulative Review Problem Set (page 141)

1. $-16x$ 2. $4a - 6$ 3. $12x + 27$ 4. $-5x + 1$
5. $9n - 8$ 6. $14n - 5$ 7. $\dfrac{1}{4}x$ 8. $-\dfrac{1}{10}n$ 9. $-0.1x$
10. $0.7x + 0.2$ 11. -65 12. -51 13. 20 14. 32
15. $\dfrac{7}{8}$ 16. $-\dfrac{5}{6}$ 17. -0.28 18. $-\dfrac{1}{4}$ 19. 5 20. $-\dfrac{1}{4}$
21. 81 22. -64 23. 0.064 24. $-\dfrac{1}{32}$ 25. $\dfrac{25}{36}$
26. $-\dfrac{1}{512}$ 27. $\{-4\}$ 28. $\{-2\}$ 29. $\varnothing$ 30. $\{-8\}$
31. $\left\{\dfrac{25}{2}\right\}$ 32. $\left\{-\dfrac{4}{7}\right\}$ 33. $\left\{\dfrac{34}{3}\right\}$ 34. $\{200\}$
35. $\varnothing$ 36. $\{11\}$ 37. $\left\{\dfrac{3}{2}\right\}$ 38. $\{0\}$
39. $\{x | x > 7\}$ 40. $\{x | x > -6\}$ 41. $\left\{n \bigm| n \geq \dfrac{7}{5}\right\}$
42. $\{x | x \geq 21\}$ 43. $\{t | t < 100\}$ 44. $\{x | x < -1\}$
45. $\{n | n \geq 18\}$ 46. $\left\{x \bigm| x < \dfrac{5}{3}\right\}$
47. $15,000 48. 45° and 135°
49. 8 nickels and 17 dimes 50. 130 or higher
51. 12 feet and 18 feet 52. $40 per pair
53. 45 miles per hour and 50 miles per hour 54. 5 liters

CHAPTER 4

Chapter 4 Review Problem Set (page 175)

1. C 2. F 3. E 4. D 5. B 6. A
7. $8x^2 - 13x + 2$ 8. $3y^2 + 11y - 9$ 9. $3x^2 + 2x - 9$
10. $-8x^2 + 18$ 11. $11x + 8$ 12. $-9x^2 + 8x - 20$
13. $2y^2 - 54y + 18$ 14. $-13a - 30$ 15. $-27a - 7$
16. $n - 2$ 17. $-5n^2 - 2n$ 18. $17n^2 - 14n - 16$
19. $35x^6$ 20. $-54x^8$ 21. $24x^3y^5$ 22. $-6a^4b^9$
23. $8a^6b^9$ 24. $9x^2y^4$ 25. $16x^2y^6$ 26. $125a^6b^3$
27. $35x^2 + 15x$ 28. $-24x^3 + 3x^2$
29. $3x^3 - 5x^2 - 3x + 2$ 30. $2x^3 + 17x^2 + 26x - 24$
31. $x^3 - 3x^2 + 8x - 12$ 32. $2x^3 + 7x^2 + 10x - 7$
33. $x^4 + x^3 + 2x^2 - 7x - 5$
34. $n^4 - 5n^3 - 11n^2 - 30n - 4$ 35. $x^2 + 17x + 72$
36. $x^2 + 10x + 24$ 37. $x^2 - 8x + 12$
38. $x^2 - 11x + 30$ 39. $x^2 - 5x - 14$ 40. $x^2 + 6x - 40$
41. $2x^2 + 11x + 5$ 42. $3x^2 + 14x + 8$
43. $5x^2 + 14x - 3$ 44. $4x^2 - 17x - 15$
45. $14x^2 - x - 3$ 46. $20a^2 - 3a - 56$
47. $30n^2 + 19n - 5$ 48. $12n^2 + 13n - 4$
49. $a^3 + 15a^2 + 75a + 125$ 50. $a^3 - 18a^2 + 108a - 216$
51. $y^2 + 16y + 64$ 52. $m^2 - 8m + 16$
53. $9a^2 - 30a + 25$ 54. $4x^2 + 28x + 49$ 55. $4n^2 - 1$
56. $16n^2 - 25$ 57. $600 - 4x^2$ square feet
58. $4x^3 - 60x^2 + 216x$ cubic inches 59. $-12x^3y^3$
60. $7a^3b^4$ 61. $-3x^2y - 9x^4$ 62. $10a^4b^9 - 13a^3b^7$
63. $14x^2 - 10x - 8$ 64. $4a^2 - 5a - 1$
65. $x + 4, R = -21$ 66. $7x - 6$ 67. $2x^2 + x + 4, R = 4$
68. $3x + 5, R = -4$ 69. 13 70. 25 71. $\dfrac{1}{16}$ 72. 1
73. -1 74. 9 75. $\dfrac{16}{9}$ 76. $\dfrac{1}{4}$ 77. -8 78. $\dfrac{11}{18}$ 79. $\dfrac{5}{4}$
80. $\dfrac{1}{25}$ 81. $\dfrac{1}{x^3}$ 82. $12x^3$ 83. x^2 84. $\dfrac{1}{x^2}$ 85. $-\dfrac{9y^3}{x^2}$
86. $\dfrac{4}{x^2y^2}$ 87. $8a^6$ 88. $\dfrac{4}{n}$ 89. $\dfrac{x^2}{y}$ 90. $\dfrac{b^6}{a^4}$ 91. $\dfrac{1}{2x}$
92. $\dfrac{1}{9n^4}$ 93. $\dfrac{n^3}{8}$ 94. $-12b$ 95. 610 96. 56,000
97. 0.08 98. 0.00092 99. $(9)(10^3)$ 100. $(4.7)(10)$
101. $(4.7)(10^{-2})$ 102. $(2.1)(10^{-4})$ 103. 0.48 104. 4.2
105. 2000 106. 0.00000002

Chapter 4 Practice Test (page 177)

1. $-2x^2 - 2x + 5$ 2. $-3x^2 - 6x + 20$ 3. $-13x + 2$
4. $-28x^3y^5$ 5. $12x^5y^5$ 6. $x^2 - 7x - 18$
7. $n^2 + 7n - 98$ 8. $40a^2 + 59a + 21$
9. $9x^2 - 42xy + 49y^2$ 10. $2x^3 + 2x^2 - 19x - 21$
11. $81x^2 - 25y^2$ 12. $15x^2 - 68x + 77$ 13. $8x^2y^4$
14. $-7x + 9y$ 15. $x^2 + 4x - 5$ 16. $4x^2 - x + 6$
17. $\dfrac{27}{8}$ 18. $1\dfrac{5}{16}$ 19. 16 20. $-\dfrac{24}{x^2}$ 21. $\dfrac{x^3}{4}$ 22. $\dfrac{x^6}{y^{10}}$
23. $(2.7)(10^{-4})$ 24. 9,200,000 25. 0.006

Chapters 1–4 Cumulative Review Problem Set (page 178)

1. 130 2. -1 3. 27 4. -16 5. 81 6. -32 7. $\dfrac{3}{2}$
8. 16 9. 36 10. $1\dfrac{3}{4}$ 11. 0 12. $\dfrac{13}{40}$ 13. $-\dfrac{2}{13}$ 14. 5

15. -1 **16.** -33 **17.** $-15x^3y^7$ **18.** $12ab^7$
19. $-8x^6y^{15}$ **20.** $-6x^2y + 15xy^2$ **21.** $15x^2 - 11x + 2$
22. $21x^2 + 25x - 4$ **23.** $-2x^2 - 7x - 6$ **24.** $49 - 4y^2$
25. $3x^3 - 7x^2 - 2x + 8$ **26.** $2x^3 - 3x^2 - 13x + 20$
27. $4n^2 + 12n + 9$ **28.** $1 - 6n + 12n^2 - 8n^3$
29. $2x^4 + x^3 - 4x^2 + 42x - 36$ **30.** $-4x^2y^2$ **31.** $14ab^2$
32. $7y - 8x^2 - 9x^3y^3$ **33.** $2x^2 - 4x - 7$
34. $x^2 + 6x + 4$ **35.** $-\dfrac{6}{x}$ **36.** $\dfrac{2}{x}$ **37.** $\dfrac{xy^2}{3}$ **38.** $\dfrac{z^2}{x^2y^4}$
39. 0.12 **40.** 0.0000000018 **41.** 200 **42.** $\{11\}$
43. $\{-1\}$ **44.** $\{48\}$ **45.** $\left\{-\dfrac{3}{7}\right\}$ **46.** $\{9\}$ **47.** $\{13\}$
48. $\left\{\dfrac{9}{14}\right\}$ **49.** $\{500\}$ **50.** $\{x | x \leq -1\}$ **51.** $\{x | x > 0\}$
52. $\left\{x \Big| x < \dfrac{4}{5}\right\}$ **53.** $\{x | x < -2\}$ **54.** $\left\{x \Big| x \geq \dfrac{12}{7}\right\}$
55. $\{x | x \geq 300\}$ **56.** 3 **57.** 40
58. 8 dimes and 10 quarters
59. \$700 at 8% and \$800 at 9% **60.** 3 gallons
61. $3\dfrac{1}{2}$ hours
62. The length is 15 meters and the width is 7 meters.

CHAPTER 5

Chapter 5 Review Problem Set (page 207)

1. $2x^2y$ **2.** $3xy^2$ **3.** $4ab^2$ **4.** $12m^3n^2$ **5.** $3x^2(5x^2 + 7)$
6. $4a^3(6 - 5a^3)$ **7.** $5x^2y(2x^2 - 10xy + y^2)$
8. $4m^2n(3 + 5mn - 6m^2n^2)$ **9.** $(a + 3)(b - 4)$
10. $(2x + 5)(y + 6)$ **11.** $(x^2 + y^2)(x + y)$
12. $(2m + 3n)(4m + 1)$ **13.** $(5x + 4y)(5x - 4y)$
14. $2(3x + 5)(3x - 5)$ **15.** $(x^2 + 4)(x + 2)(x - 2)$
16. $x(x + 7)(x - 7)$ **17.** $(x - 3)(x - 8)$
18. $(x + 12)(x + 1)$ **19.** $(x - 8)(x + 3)$
20. $(x + 10)(x - 2)$ **21.** $(2x + 3)(x + 1)$
22. $(5x + 2)(x + 3)$ **23.** $(4x - 3)(x + 2)$
24. $(3x + 2)(x - 4)$ **25.** $(3x + 2)^2$ **26.** $(2x - 5)^2$
27. $(4x - 1)^2$ **28.** $(x + 6y)^2$ **29.** $(x - 2)(x - 7)$
30. $3x(x + 7)$ **31.** $(3x + 2)(3x - 2)$ **32.** $(2x - 1)(2x + 5)$
33. $(5x - 6)^2$ **34.** $n(n + 5)(n + 8)$ **35.** $(y + 12)(y - 1)$
36. $3xy(y + 2x)$ **37.** $(x + 1)(x - 1)(x^2 + 1)$
38. $(6n + 5)(3n - 1)$ **39.** Not factorable
40. $(4x - 7)(x + 1)$ **41.** $3(n + 6)(n - 5)$
42. $x(x + y)(x - y)$ **43.** $(2x - y)(x + 2y)$
44. $2(n - 4)(2n + 5)$ **45.** $(x + y)(5 + a)$
46. $(7t - 4)(3t + 1)$ **47.** $2x(x + 1)(x - 1)$
48. $3x(x + 6)(x - 6)$ **49.** $(4x + 5)^2$ **50.** $(y - 3)(x - 2)$
51. $(5x + y)(3x - 2y)$ **52.** $n^2(2n - 1)(3n - 1)$
53. $\{-6, 2\}$ **54.** $\{0, 11\}$ **55.** $\left\{-4, \dfrac{5}{2}\right\}$ **56.** $\left\{-\dfrac{8}{3}, \dfrac{1}{3}\right\}$
57. $\{-2, 2\}$ **58.** $\left\{-\dfrac{5}{4}\right\}$ **59.** $\{-1, 0, 1\}$ **60.** $\left\{-\dfrac{9}{4}, -\dfrac{2}{7}\right\}$
61. $\{-7, 4\}$ **62.** $\{-5, 5\}$ **63.** $\left\{-6, \dfrac{3}{5}\right\}$ **64.** $\left\{-\dfrac{7}{2}, 1\right\}$
65. $\{-2, 0, 2\}$ **66.** $\{8, 12\}$ **67.** $\left\{-5, \dfrac{3}{4}\right\}$ **68.** $\{-2, 3\}$
69. $\left\{-\dfrac{7}{3}, \dfrac{5}{2}\right\}$ **70.** $\{-9, 6\}$ **71.** $\left\{-5, \dfrac{3}{2}\right\}$ **72.** $\left\{\dfrac{4}{3}, \dfrac{5}{2}\right\}$
73. $-\dfrac{8}{3}$ and $-\dfrac{19}{3}$ or 4 and 7
74. The length is 8 centimeters and the width is 2 centimeters.
75. A 2-by-2-inch square and a 10-by-10-inch square
76. 8 by 15 by 17 **77.** $-\dfrac{13}{6}$ and -12 or 2 and 13
78. 7, 9, and 11 **79.** 4 shelves
80. A 5-by-5-yard square and a 5-by-40-yard rectangle
81. -18 and -17 or 17 and 18 **82.** 6 units
83. 2 meters and 7 meters **84.** 9 and 11

Chapter 5 Practice Test (page 209)

1. $(x + 5)(x - 2)$ **2.** $(x + 3)(x - 8)$ **3.** $2x(x + 1)(x - 1)$
4. $(x + 9)(x + 12)$ **5.** $3(2n + 1)(3n + 2)$
6. $(x + y)(a + 2b)$ **7.** $(4x - 3)(x + 5)$ **8.** $6(x^2 + 4)$
9. $2x(5x - 6)(3x - 4)$ **10.** $(7 - 2x)(4 + 3x)$ **11.** $\{-3, 3\}$
12. $\{-6, 1\}$ **13.** $\{0, 8\}$ **14.** $\left\{-\dfrac{5}{2}, \dfrac{2}{3}\right\}$ **15.** $\{-6, 2\}$
16. $\{-12, -4, 0\}$ **17.** $\{5, 9\}$ **18.** $\left\{-12, \dfrac{1}{3}\right\}$ **19.** $\left\{-4, \dfrac{2}{3}\right\}$
20. $\{-5, 0, 5\}$ **21.** $\left\{\dfrac{7}{5}\right\}$ **22.** 14 inches
23. 12 centimeters **24.** 16 chairs per row
25. $-6, -5, -4$ or $4, 5, 6$

CHAPTER 6

Chapter 6 Review Problem Set (page 245)

1. $\dfrac{7x^2}{9y^2}$ **2.** $\dfrac{x}{x + 3}$ **3.** $\dfrac{3n + 5}{n + 2}$ **4.** $\dfrac{4a + 3}{5a - 2}$ **5.** $-(x + 4)$
6. $5 - y$ **7.** -6 **8.** $-\dfrac{x + 5}{x + 1}$ **9.** $\dfrac{3x}{8}$ **10.** $x(x - 3)$
11. $\dfrac{3(x + 3)}{x + 1}$ **12.** $\dfrac{x(x + y)}{3}$ **13.** $\dfrac{n - 7}{n^2}$ **14.** $\dfrac{2a + 1}{a + 6}$
15. $\dfrac{7b^3}{3}$ **16.** $\dfrac{8xy^3}{y + 2}$ **17.** $2n + 3$ **18.** $\dfrac{5}{7}$ **19.** $\dfrac{22x - 19}{20}$

20. $\dfrac{43x - 3}{12x^2}$ 21. $\dfrac{10n - 7}{n(n - 1)}$ 22. $\dfrac{-a + 8}{(a - 4)(a - 2)}$
23. $\dfrac{5x + 9}{4x(x - 3)}$ 24. $\dfrac{5x - 4}{(x + 5)(x - 5)(x + 2)}$
25. $\dfrac{6x - 37}{(x - 7)(x + 3)}$ 26. $\dfrac{12x - 23}{(x - 7)(x + 4)}$ 27. $\dfrac{3y^2 - 4x}{4xy + 5y^2}$
28. $\dfrac{2y - xy}{3xy + 5x}$ 29. $\dfrac{6 + 3n}{5n + 2m}$ 30. $\dfrac{x^2 + 6x}{x + 12}$ 31. $\dfrac{x}{y}$
32. $\dfrac{d}{g}$ 33. $\dfrac{40}{m}$ 34. $\dfrac{p}{k}$ 35. $\left\{\dfrac{20}{17}\right\}$ 36. $\left\{-\dfrac{2}{9}\right\}$ 37. $\{6\}$
38. $\left\{-\dfrac{61}{60}\right\}$ 39. $\{9\}$ 40. $\left\{\dfrac{28}{3}\right\}$ 41. $\left\{\dfrac{3}{4}\right\}$ 42. $\{1\}$
43. $\{-7\}$ 44. $\left\{\dfrac{1}{7}\right\}$ 45. $\left\{\dfrac{1}{2}, 2\right\}$ 46. $\{-5, 10\}$
47. $\left\{-\dfrac{1}{5}\right\}$ 48. $\{-1\}$ 49. Becky $2\dfrac{2}{3}$ hours, Nancy 8 hours
50. 1 or 2 51. $\dfrac{36}{72}$
52. Todd's rate is 15 miles per hour and Lanette's rate is 22 miles per hour. 53. 8 miles per hour 54. 60 minutes

Chapter 6 Practice Test (page 247)

1. $\dfrac{8x^2y}{9}$ 2. $\dfrac{x}{x - 6}$ 3. $\dfrac{2n + 1}{3n + 4}$ 4. $\dfrac{2x - 3}{x - 5}$ 5. $2x^2y^2$
6. $\dfrac{x - 2}{x + 3}$ 7. $\dfrac{(x + 4)^2}{x(x + 7)}$ 8. $\dfrac{6x + 5}{24}$ 9. $\dfrac{9n - 4}{30}$
10. $\dfrac{41 - 15x}{18x}$ 11. $\dfrac{2n - 6}{n(n - 1)}$ 12. $\dfrac{5x - 18}{4x(x + 6)}$
13. $\dfrac{5x - 11}{(x - 4)(x + 8)}$ 14. $\dfrac{-13x + 43}{(2x - 5)(3x + 4)(x - 6)}$
15. $\{-5\}$ 16. $\left\{-\dfrac{19}{16}\right\}$ 17. $\left\{\dfrac{4}{3}, 3\right\}$ 18. $\{-6, 8\}$
19. $\left\{-\dfrac{1}{5}, 2\right\}$ 20. $\{-23\}$ 21. $\{2\}$ 22. $\left\{-\dfrac{3}{2}\right\}$
23. $\dfrac{2}{3}$ or 3 24. 14 miles per hour 25. 12 minutes

Chapters 1–6 Cumulative Review Problem Set (page 249)

1. $\dfrac{5}{2}$ 2. 6 3. $\dfrac{17}{12}$ 4. 0.6 5. 20 6. 0 7. 2 8. -3
9. $\dfrac{1}{27}$ 10. $\dfrac{3}{2}$ 11. 1 12. $\dfrac{12}{7}$ 13. $-\dfrac{1}{16}$ 14. $\dfrac{9}{4}$
15. $\dfrac{4}{25}$ 16. $-\dfrac{1}{27}$ 17. $\dfrac{19}{10x}$ 18. $\dfrac{2y}{3x}$ 19. $\dfrac{7x - 2}{(x - 6)(x + 4)}$
20. $\dfrac{-x + 12}{x^2(x - 4)}$ 21. $\dfrac{x - 7}{3y}$ 22. $\dfrac{-3x - 4}{(x - 4)(x + 3)}$
23. $-35x^5y^5$ 24. $81a^2b^6$ 25. $-15n^4 - 18n^3 + 6n^2$
26. $15x^2 + 17x - 4$ 27. $4x^2 + 20x + 25$
28. $2x^3 + x^2 - 7x - 2$ 29. $x^4 + x^3 - 6x^2 + x + 3$
30. $-6x^2 + 11x + 7$ 31. $3xy - 6x^3y^3$ 32. $7x + 4$
33. $3x(x^2 + 5x + 9)$ 34. $(x + 10)(x - 10)$
35. $(5x - 2)(x - 4)$ 36. $(4x + 7)(2x - 9)$
37. $(n + 16)(n + 9)$ 38. $(x + y)(n - 2)$
39. $3x(x + 1)(x - 1)$ 40. $2x(x - 9)(x + 6)$
41. $(6x - 5)^2$ 42. $(3x + y)(x - 2y)$ 43. $\left\{\dfrac{16}{3}\right\}$
44. $\{-11, 0\}$ 45. $\left\{\dfrac{1}{14}\right\}$ 46. $\{15\}$ 47. $\{-1, 1\}$
48. $\{-6, 1\}$ 49. $\{2\}$ 50. $\left\{\dfrac{11}{12}\right\}$ 51. $\{1, 2\}$ 52. $\left\{-\dfrac{1}{18}\right\}$
53. $\left\{-\dfrac{7}{2}, \dfrac{1}{3}\right\}$ 54. $\left\{\dfrac{1}{2}, 8\right\}$ 55. $\{-9, 2\}$ 56. $\left\{\dfrac{16}{5}\right\}$
57. $\left\{\dfrac{1}{2}\right\}$ 58. 6 inches, 8 inches, 10 inches 59. 75
60. 100 milliliters 61. 6 feet by 8 feet 62. 2.5 hours
63. 7.5 centimeters 64. 27 gallons 65. 92 or better
66. $\{1, 2, 3\}$ 67. All real numbers greater than $\dfrac{7}{2}$

CHAPTER 7

Chapter 7 Review Problem Set (page 287)

1. x intercept is -6, y intercept is -2
2. x intercept is 4, y intercept is -2
3. x intercept is 5, y intercept is -4
4. x intercept is $-\dfrac{1}{2}$, y intercept is 1
5. x intercept is 2, y intercept is -6
6. x intercept is 4, y intercept is 8

7.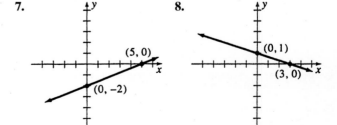
8.

9.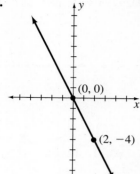
10.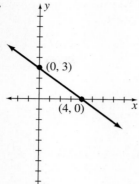
11.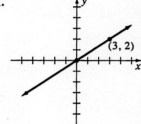
12.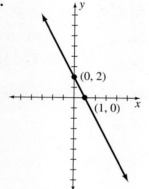
13.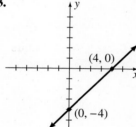
14.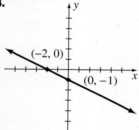
15.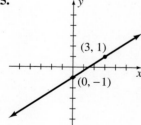
16.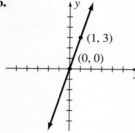
17.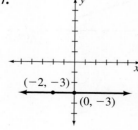
18.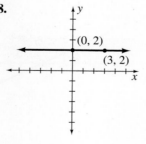
19.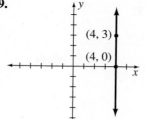
20.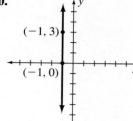
21.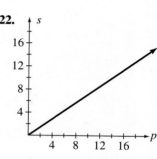
22.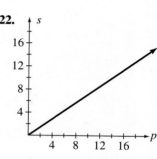
23.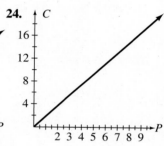
24.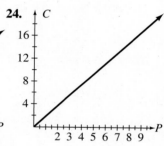
25.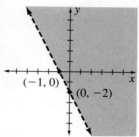
26.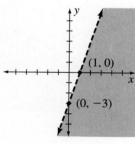
27.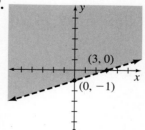
28.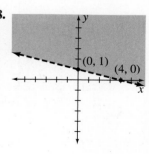

29. 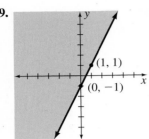 **30.**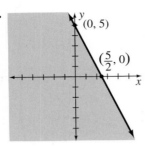

31. 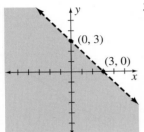 **32.**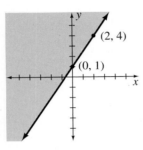

33. 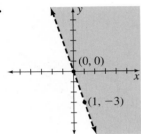 **34.**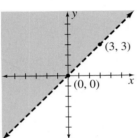

35. 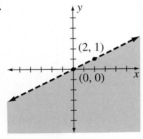 **36.**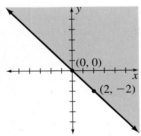

37. 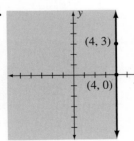 **38.**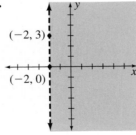

39. 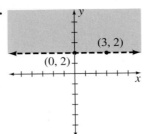 **40.**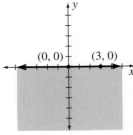

41. -3 **42.** 3 **43.** $-\dfrac{1}{3}$ **44.** $\dfrac{3}{2}$ **45.** 0 **46.** -1

47. Undefined **48.** $-\dfrac{5}{2}$ **49.** 50 feet

50. 42.9 centimeters **51.** $\dfrac{16}{75}$; No **52.** 7500 feet

53. $3x - 4y = 11$ **54.** $2x + y = -7$ **55.** $y = -4x + 2$

56. $y = -\dfrac{3}{5}x$ **57.** $7x + 2y = 12$ **58.** $3x - 2y = -9$

59. $y = -\dfrac{1}{4}x$ **60.** $y = 0x + 5$

61. $m = \dfrac{2}{5}, b = -2$

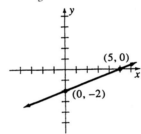

62. $m = -\dfrac{1}{3}, b = 1$

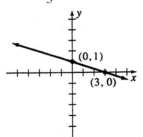

63. $m = -\dfrac{1}{2}, b = 1$

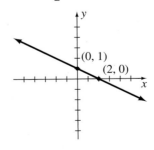

64. $m = -3, b = -2$

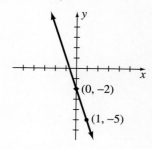

65. $m = 2, b = -4$

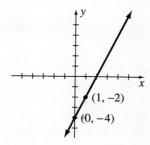

66. $m = \frac{3}{4}, b = -3$

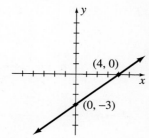

Chapter 7 Practice Test (page 289)

1. $m = -\frac{5}{3}, b = 5$

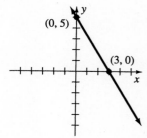

2. $m = 2, b = -4$

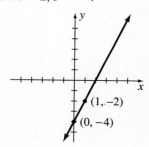

3. $m = -\frac{1}{2}, b = -2$

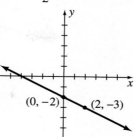

4. $m = -3, b = 0$

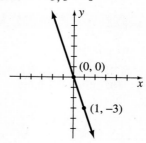

5. $-\frac{3}{2}$ **6.** 0 **7.** 8 **8.** 5 **9.** $(7, 6), (11, 7)$ or others
10. $(3, -2), (4, -5)$ or others **11. (a)** A **(b)** C **(c)** B
(d) D **12.** 4.6% **13.** -2 **14.** -4

15. 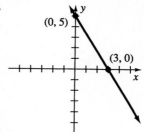 **16.**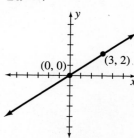

17. 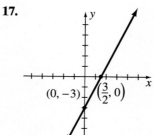 **18.**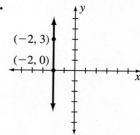

19. 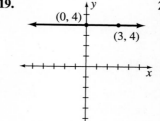 **20.**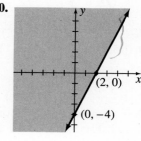

21. 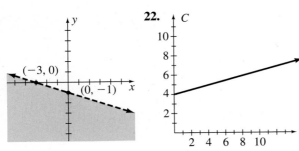 **22.**   **23.** $3x + 5y = 20$ **24.** $4x - 9y = 34$ **25.** $3x - 2y = 0$

CHAPTER 8

Chapter 8 Review Problem Set (page 321)

1. Yes **2.** No **3.** No **4.** Yes **5.** $\{(-1, 4)\}$ **6.** $\{(2, 4)\}$
7. $\varnothing$ **8.** $\varnothing$ **9.** Infinitely many **10.** Infinitely many

11. 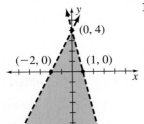 **12.**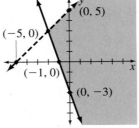

13. 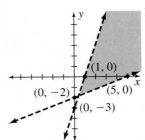 **14.**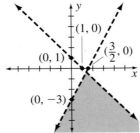

15. 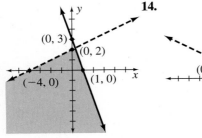 **16.**

17. $\{(7, 13)\}$ **18.** $\{(10, 25)\}$ **19.** $\{(-6, -2)\}$
20. $\{(5, -4)\}$ **21.** $\{(-3, 1)\}$ **22.** $\{(0, -2)\}$
23. $\{(16, -5)\}$ **24.** $\left\{\left(\dfrac{5}{16}, -\dfrac{17}{16}\right)\right\}$ **25.** $\{(-9, 6)\}$ **26.** $\varnothing$
27. $\{(7, -1)\}$ **28.** $\{(3, 8)\}$ **29.** Elimination-by-addition
30. Elimination-by-addition **31.** Substitution
32. Substitution **33.** $\{(0, 2, 5)\}$ **34.** $\{(2, 1, 3)\}$
35. $\{(1, 1, -4)\}$ **36.** $\{(-4, 0, 1)\}$ **37.** 38 and 75
38. $250 at 6% and $300 at 8%
39. 18 nickels and 25 dimes
40. Length of 19 inches and width of 6 inches
41. 32° and 58° **42.** 50° and 130°
43. $3.25 for a cheeseburger and $2.50 for a milkshake
44. $1.59 for orange juice and $0.99 for water

Chapter 8 Practice Test (page 323)

1. No **2.** Yes **3.** Yes **4.** $\{(2, 3)\}$ **5.** $\{(0, -3)\}$ **6.** $\varnothing$
7. Infinitely many

8. 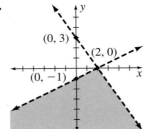 **9.**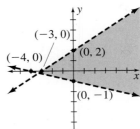

10.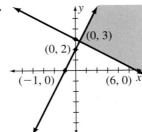

11. $\{(4, 6)\}$ **12.** $\{(3, 4)\}$ **13.** $\{(-2, 1)\}$ **14.** $\{(-2, -6)\}$
15. $\left\{\left(\dfrac{1}{2}, -3\right)\right\}$ **16.** Infinitely many **17.** $\varnothing$
18. $\{(4, 3, -2)\}$ **19.** $\{(1, 3, -5)\}$ **20.** $\{(-2, 0, 4)\}$
21. 25 and 71 **22.** Paper $3.49, notebook $2.29
23. 13 inches **24.** 3 liters
25. $2500 at 5% and $3000 at 6%

CHAPTER 9

Chapter 9 Review Problem Set (page 353)

1. 8   2. −7   3. 40   4. $\frac{9}{5}$ 5. $-\frac{2}{3}$ 6. $\frac{7}{6}$ 7. 3   8. −5
9. 4   10. $\frac{2}{5}$ 11. 12.57   12. 35.36   13. −29.33
14. 15.81   15. 8.49   16. −9.24   17. $2\sqrt{5}$ 18. $4\sqrt{2}$
19. $10\sqrt{2}$ 20. $4\sqrt{5}$ 21. $2\sqrt[3]{3}$ 22. $5\sqrt[3]{2}$ 23. $2x\sqrt{3}$
24. $5x\sqrt{2y}$ 25. $6ab\sqrt{5a}$ 26. $4a^2\sqrt{3}$ 27. $2y\sqrt[3]{x^2}$
28. $-3x\sqrt[3]{x}$ 29. $18\sqrt{2}$ 30. $12\sqrt{3}$ 31. $-7\sqrt{2x}$
32. $2\sqrt{5y}$ 33. $-x\sqrt[3]{x}$ 34. $-9\sqrt[3]{x^2}$ 35. 2   36. $\sqrt[3]{5}$
37. $\frac{x\sqrt{2x}}{3}$ 38. $\frac{3\sqrt{2a}}{2b}$ 39. $\frac{6\sqrt{7}}{7}$ 40. $\frac{3\sqrt{10}}{5}$ 41. $\frac{\sqrt{14}}{4}$
42. $\frac{\sqrt{3}}{3}$ 43. $\frac{2\sqrt{x}}{x}$ 44. $\frac{\sqrt{10xy}}{5y}$ 45. 2   46. $\frac{5\sqrt{6}}{6}$
47. $\frac{-\sqrt{6}}{3}$ 48. $\frac{2\sqrt{2}}{3}$ 49. $\frac{3\sqrt{xy}}{4y^2}$ 50. $\frac{-2\sqrt{x}}{5}$
51. $\frac{-17\sqrt{6}}{3}$ 52. $\frac{16\sqrt{10}}{5}$ 53. $\frac{52\sqrt{5}}{5}$ 54. $\frac{31\sqrt{15}}{5}$
55. $6\sqrt{2}$ 56. $18\sqrt{2}$ 57. −40   58. $10\sqrt[3]{28}$
59. $\sqrt[3]{6} + 2$ 60. $6\sqrt{10} - 12\sqrt{15}$
61. $3 + \sqrt{21} + \sqrt{15} + \sqrt{35}$ 62. $-24 - 7\sqrt{6}$
63. $4 + 5\sqrt{42}$ 64. $-18 - \sqrt{5}$ 65. $\frac{5(\sqrt{7} + \sqrt{5})}{2}$
66. $3\sqrt{2} + 2\sqrt{3}$ 67. $\frac{3\sqrt{2} + \sqrt{6}}{6}$ 68. $\frac{3\sqrt{42} - 4\sqrt{15}}{23}$
69. {6}   70. {85}   71. {−28}   72. {−32}   73. ∅
74. ∅   75. {4}   76. $\left\{\frac{11}{2}\right\}$ 77. {−5, −4}   78. {10}
79. {3}   80. {1}   81. 2.5 seconds, 4.7 seconds, 5.6 seconds
82. $b = \frac{c^2}{22a}$

13. $\frac{5y}{2}$ 14. $\frac{1}{x - 4}$ 15. $\frac{-8x - 41}{(x + 6)(x - 3)}$ 16. $60x^4y^4$
17. $\frac{-8}{x^2}$ 18. $\frac{-3a^3}{b}$ 19. $\frac{1}{3n^4}$ 20. $27x^2 + 30x - 8$
21. $-5x^2 - 12x - 7$ 22. $6x^3 - x^2 - 13x - 4$
23. $3x^3y^6 - 4y^3$ 24. $2x^2 - 2x - 3$ 25. $y - x$
26. 4.67 gallons   27. 25%   28. 108   29. 130 feet
30. 314 square inches   31. (a) $(8.5)(10^4)$ (b) $(9)(10^{-4})$
(c) $(1.04)(10^{-6})$ (d) $(5.3)(10^7)$ 32. $2x(2x + 5)(3x - 4)$
33. $3(2x + 3)(2x - 3)$ 34. $(y + 3)(x - 2)$
35. $(5 - x)(6 + 5x)$ 36. $4(x + 1)(x - 1)(x^2 + 1)$
37. $(7x - 2)(3x + 4)$ 38. $8\sqrt{7}$ 39. $-3\sqrt{5}$ 40. $\frac{6\sqrt{5}}{5}$
41. $\frac{5\sqrt{6}}{18}$ 42. $6y^2\sqrt{2xy}$ 43. $-\frac{2\sqrt{ab}}{5}$ 44. 48
45. $216 - 36\sqrt{6}$ 46. 11   47. $4\sqrt{3} - 4\sqrt{2}$
48. $-\frac{2(3\sqrt{5} + \sqrt{6})}{13}$ 49. $\sqrt{2}$ 50. $\frac{37\sqrt{5}}{12}$

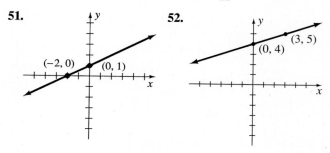

51. 52.
53. 54.

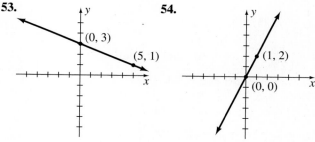

Chapter 9 Practice Test (page 355)

1. $-\frac{8}{7}$ 2. 0.05   3. 2.8   4. −5.7   5. 2.1   6. $3\sqrt{5}$
7. $-12\sqrt[3]{2}$ 8. $\frac{\sqrt{2}}{3}$ 9. $\frac{5\sqrt{2}}{2}$ 10. $\frac{\sqrt{6}}{3}$ 11. $\frac{\sqrt{10}}{4}$
12. $-5xy\sqrt[3]{2x}$ 13. $\frac{\sqrt{15xy}}{5y}$ 14. $3xy\sqrt{3x}$ 15. $4\sqrt{6}$
16. $24\sqrt[3]{10}$ 17. $12\sqrt{2} - 12\sqrt{3}$ 18. $1 - 5\sqrt{15}$
19. $\frac{3\sqrt{2} - \sqrt{3}}{5}$ 20. $4\sqrt{6}$ 21. {5}   22. ∅
23. {3}   24. {−2, 1}   25. 2.1 seconds

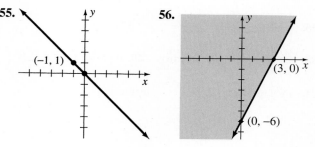

55. 56.

Chapters 1–9 Cumulative Review Problem Set (page 357)

1. −64   2. 64   3. 144   4. −8   5. $\frac{2}{3}$ 6. $\frac{13}{9}$ 7. −9
8. −49   9. −29   10. 49   11. $-\frac{15}{4x}$ 12. $\frac{-x + 17}{(x - 2)(x + 3)}$

57. 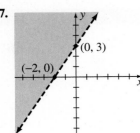 **58.**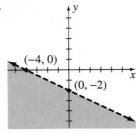

59. -2 **60.** $\dfrac{4}{7}$ **61.** $2x - 3y = 8$ **62.** $4x + 3y = -13$

63. $x + 4y = -12$ **64.** $\dfrac{3}{2}$ **65.** $\{(1, -2)\}$ **66.** $\{(-2, 4)\}$

67. $\{(-6, 12)\}$ **68.** $\left\{\left(\dfrac{1}{2}, 3\right)\right\}$ **69.** $\{17\}$ **70.** $\left\{-\dfrac{23}{5}\right\}$

71. $\{5\}$ **72.** $\left\{-\dfrac{19}{10}\right\}$ **73.** $\left\{\dfrac{47}{7}\right\}$ **74.** $\{500\}$ **75.** $\{53\}$

76. $\left\{-\dfrac{2}{3}\right\}$ **77.** $\{-6, 2\}$ **78.** $\{-2, 2\}$ **79.** $\{29\}$ **80.** $\{27\}$

81. $\{n \mid n \geq -5\}$ **82.** $\left\{n \mid n > \dfrac{1}{4}\right\}$ **83.** $\left\{x \mid x > -\dfrac{2}{5}\right\}$

84. $\{n \mid n > 6\}$ **85.** $\left\{x \mid x < \dfrac{5}{16}\right\}$ **86.** $\left\{x \mid x \geq -\dfrac{16}{3}\right\}$

87. 65° and 115° **88.** 13 and 37 **89.** 7, 9

90. 7 nickels, 15 dimes, and 25 quarters **91.** $3780

92. $48 **93.** $4\dfrac{1}{2}$ hours **94.** 12.5 milliliters

95. 91 or better **96.** More than 11 **97.** 12 minutes

CHAPTER 10

Chapter 10 Review Problem Set (page 385)

1. $\{-6, 4\}$ **2.** $\{5, 8\}$ **3.** $\{-7, 2\}$ **4.** $\{-4, 0\}$

5. $\{-3, 13\}$ **6.** $\{-12, 4\}$ **7.** $\{-4, 4\}$ **8.** $\left\{\dfrac{-1 \pm 2\sqrt{10}}{3}\right\}$

9. $\{-7 \pm \sqrt{69}\}$ **10.** $\{-4 \pm 2\sqrt{10}\}$ **11.** $\{1 \pm 2\sqrt{2}\}$

12. $\{5 \pm \sqrt{29}\}$ **13.** $\left\{\dfrac{-3 \pm \sqrt{29}}{2}\right\}$ **14.** $\left\{\dfrac{-7 \pm \sqrt{73}}{2}\right\}$

15. $\left\{\dfrac{1 \pm \sqrt{13}}{3}\right\}$ **16.** $\left\{\dfrac{-2 \pm \sqrt{10}}{2}\right\}$ **17.** $\{-6, -1\}$

18. $\{-4 - \sqrt{13}, -4 + \sqrt{13}\}$ **19.** $\left\{\dfrac{2}{7}, \dfrac{1}{3}\right\}$ **20.** $\{0, 17\}$

21. $\{-4, 1\}$ **22.** $\{11, 15\}$ **23.** $\left\{\dfrac{-7 - \sqrt{61}}{6}, \dfrac{-7 + \sqrt{61}}{6}\right\}$

24. $\left\{\dfrac{1}{2}\right\}$ **25.** No real number solutions **26.** $\{-5, -1\}$

27. $\{2 - \sqrt{2}, 2 + \sqrt{2}\}$ **28.** $\{-2 - 3\sqrt{2}, -2 + 3\sqrt{2}\}$

29. $\{-3\sqrt{5}, 3\sqrt{5}\}$ **30.** $\{-3, 9\}$ **31.** $\{0, 1\}$

32. $\{2 - \sqrt{13}, 2 + \sqrt{13}\}$ **33.** $\{20, 24\}$

34. $\{2 - 2\sqrt{2}, 2 + 2\sqrt{2}\}$ **35.** $\left\{-\dfrac{8}{5}, 1\right\}$

36. $\left\{\dfrac{-1 - \sqrt{73}}{12}, \dfrac{-1 + \sqrt{73}}{12}\right\}$ **37.** $\left\{\dfrac{1}{2}, 4\right\}$

38. $\left\{-\dfrac{4}{3}, -1\right\}$ **39.** 30 yards **40.** 4 yards and 6 yards

41. $6\sqrt{2}$ inches **42.** $\dfrac{16\sqrt{3}}{3}$ centimeters

43. 9 inches by 12 inches **44.** 18 and 19 **45.** 7, 9, and 11

46. $\sqrt{5}$ meters and $3\sqrt{5}$ meters **47.** 10 meters

48. Jay's rate was 45 mph and Jean's rate was 48 mph; or Jay's rate was $7\dfrac{1}{2}$ mph and Jean's rate was $10\dfrac{1}{2}$ mph

Chapter 10 Practice Test (page 386)

1. $2\sqrt{13}$ inches **2.** 13 meters **3.** 7 inches

4. $4\sqrt{3}$ centimeters **5.** $\left\{-3, \dfrac{5}{3}\right\}$ **6.** $\{-4, 4\}$

7. $\left\{\dfrac{1}{2}, \dfrac{3}{4}\right\}$ **8.** $\left\{\dfrac{3 - \sqrt{29}}{2}, \dfrac{3 + \sqrt{29}}{2}\right\}$

9. $\{-1 - \sqrt{10}, -1 + \sqrt{10}\}$

10. No real number solutions **11.** $\{-12, 2\}$

12. $\left\{\dfrac{3 - \sqrt{41}}{4}, \dfrac{3 + \sqrt{41}}{4}\right\}$ **13.** $\left\{\dfrac{1 - \sqrt{57}}{2}, \dfrac{1 + \sqrt{57}}{2}\right\}$

14. $\left\{\dfrac{1}{5}, 2\right\}$ **15.** $\{13, 15\}$ **16.** $\left\{\dfrac{3}{4}, 4\right\}$ **17.** $\left\{0, \dfrac{1}{6}\right\}$

18. $\left\{-1, \dfrac{3}{7}\right\}$ **19.** $\left\{\dfrac{1 - 3\sqrt{3}}{4}, \dfrac{1 + 3\sqrt{3}}{4}\right\}$

20. No real number solutions **21.** 15 seats per row

22. 14 miles per hour **23.** 15 and 17 **24.** 9 feet

25. Length 18 inches, width 10 inches

Index

A

Abscissa, 253
Absolute value of a number, 11
Addition:
   of algebraic fractions, 219
   of decimals, 47
   of fractions, 38
   of integers, 10
   of polynomials, 145
   of radicals, 329
   of rational expressions, 219
   of rational numbers, 38
Addition method of solving systems, 297
Addition properties, 60, 89
Additive inverse, 10
Additive inverse property, 21, 46
Age problems, 134
Algebraic equation, 59
Algebraic expressions, 2, 24, 25, 50, 60
Algebraic fraction, 211
Algebraic inequalities, 87
Analytic geometry, 251
Approximation of square roots, 329
Area, 148
Associative properties, 20, 21, 46
Axes of a coordinate system, 252

B

Base of a power, 5
Binominal, 144, 163

C

Cartesian coordinate system, 253
Circle, 119
Circumference, 119
Common binomial factor, 182
Commutative properties, 20, 46
Complementary angles, 78
Completing the square, 367, 376
Complex fractions, 226
Composite number, 28
Compound inequality, 93
Cone, right circular, 119
Conjugates, 345
Consecutive number problems, 76, 194
Consistent system of equations, 294
Contraction, 83

Coordinate geometry, 251
Coordinates of a point, 46, 253
Cross products, 105
Cube of a number, 327
Cube root, 327
Cylinder, right circular, 119

D

Decimal notation, 172
Decimals, 44, 47, 50, 112
Declaring the variable, 72
Degree, 144
Denominator, 38, 39, 340
Dependent equations, 294
Descartes, René, 251
Difference of two squares, 186
Discounts, 113
Distributive property, 22, 47, 80, 94, 155
Divisibility, 27, 29
Division:
   of algebraic fractions, 216
   of decimals, 47
   of integers, 15
   of monomials, 161
   of polynomials, 162
   of rational expressions, 216
   of rational numbers, 36
Division properties, 61, 90

E

Elements of a set, 3
Elimination-by-addition method, 297
Empty set, 59
English system of measure, 67
Equations:
   dependent, 294
   equivalent, 59
   first-degree of one variable, 59
   linear, 259
   quadratic, 360
   radical, 348
   rational, 231
   root of, 59
   second-degree in one variable, 360
   solution of, 59, 69, 75, 76, 80, 81, 83
Equivalent equation, 59
Equivalent factors, 38

Equivalent fractions, 38
Equivalent inequalities, 88
Evaluating algebraic expressions, 6, 25, 50
Exponents:
    definition of, 5
    integers as, 168
    negative, 168
    positive integers as, 150
    properties of, 151, 152, 168, 169, 170
    zero as an, 168
Extraneous solutions, 349

F

Factor polynomials, 179
Factoring:
    common factor, 180
    completely, 182, 187
    composite number, 28
    difference of squares, 186
    by grouping, 182
    quadratic equations, 372
    trinomials, 190
Factors, 28
First-degree equation of one variable, 59
FOIL, 157
Formulas, 117
Fractional equation, 80
Fractional forms, 81
Fractions:
    addition of, 38
    algebraic, 211
    complex, 226
    division of, 37
    equivalent, 38
    fundamental principle of, 35, 211
    inequalities, 94
    least common denominator of, 39
    lowest terms of, 35
    multiplication of, 214
    reduced form of, 33
    subtraction of, 38
Fundamental principle of fractions, 35, 211

G

Geometric formula/problems, 118, 127, 159
Graphing system on linear equations, 293
Graphs:
    of an equation, 254
    of inequalities, 89, 267
    of a solution set, 87
Greatest common factor, 29, 180
Greatest common monomial factor, 180, 202

H

Half-plane, 267
Hypotenuse, 195, 364

I

Identity, 83
Identity properties, 21, 46
Inconsistent system of equations, 294
Index of a radical, 328
Inequalities:
    addition-subtraction property of, 89
    compound, 95
    equivalent, 88
    graphs of, 89, 267
    open sentences as, 87
    properties of, 89, 90
    solutions of, 87
    statements, 87, 97
    systems of, 295
Integers, 10, 17, 20
Intercepts, 259
Interest, simple, 115, 127
Interest problems, 133
Intersection of sets, 95
Irrational numbers, 45
Isosceles right triangle, 365

L

Least common denominator, 39
Least common multiple, 31
Legs, 195
Like terms, 24
Linear equations, 259, 283, 293
Linear inequalities, 267
Literal factors, 24
Lowest terms, 35

M

Members of the set, 3
Metric system of measure, 67
Mixture problems, 131
Monomial, 144
Motion problems, 128
Multiplication:
    of algebraic fractions, 214
    of decimals, 48
    of fractions, 34
    of integers, 14
    of monomials, 149, 153
    of polynomials, 152, 155
    of radicals, 343

of rational expressions, 214
of rational numbers, 34, 36
Multiplication properties, 21, 47, 61, 90
Multiplicative inverse property, 47
Multiplicative property of negative one, 22, 47

N

Negative, 168
Negative slope, 274
Nonrepeating decimals, 44
Null set, 3, 59
Number line, 9, 252
Numbers:
   absolute value, 11
   composite, 28
   integers, 10
   irrational, 45
   negative, 9
   positive, 10
   prime, 28
   rational, 33
   real, 45
   whole, 3
Numerical coefficient, 24
Numerical expressions, 1, 2, 6, 172
Numerical statement, 59

O

Open sentence, 87
Operations, order of, 6
Opposite of an integer, 11
Ordered pairs, 253, 292
Ordered triples, 312
Ordinate, 253
Origin, 252

P

Parallelogram, 118
Percent, 108
Perfect squares, 201, 327, 368
Perimeter of geometric figures, 148
Point-slope form, 282
Polynomials:
   addition of, 145, 146
   area or perimeter, 148
   classifying by size and degree, 144
   completely factored form of, 187
   definition of, 144
   degree of, 144
   division of, 162
   equations for, 143
   factor, 179

   multiplication of, 155
   subtraction of, 145, 146
Positive slope, 274
Power of a base, 5
Prime factored form, 28
Prime number, 28
Principal cube root, 328
Principal square root, 327
Prism, 119
Problem-solving suggestions, 76, 126, 380
Profit, 114
Properties, 11
Properties of equality, 60
Properties of exponents, 149, 151, 159, 161
Properties of roots of quotients, 338
Proportion, 105, 107
Pyramid, 119
Pythagorean theorem, 195, 364

Q

Quadrant, 252
Quadratic equations, 360, 376
Quadratic formula, 372, 376
Quotients, roots of, 338

R

Radical sign, 326
Radicals:
   addition of, 329
   definition of, 326
   equation, 348
   formulas for, 352
   multiplication of, 343
   roots of quotients, 332
   simplest form of, 332
   subtraction of, 329
Radicand, 326
Ratio, 105, 275
Rational approximation, 329
Rational equation, 231
Rational expressions:
   addition of, 219
   definition of, 211
   division of, 216
   multiplication of, 214
   simplifying, 211
   subtraction of, 219
Rational number, 33, 43, 44, 52
Rationalizing the denominator, 340
Real number, 45
Real number line, 46
Reciprocal, 35, 47, 216
Rectangle, 118

Rectangular coordinate system, 253
Rectangular prism, 119
Reflexive property of equality, 60
Repeating decimals, 44
Right triangle, 195, 365
Root of a number, 326, 327
Root of an equation, 59
Roots, 349
Roots of quotients, 338

S

Scientific notation, 171
Second-degree equations in one variable, 360
Set, 2
Set-builder notation, 87
Shortcut pattern, 156
Similar terms, 24
Simplest radical form, 339
Simplifying:
   expressions, 6, 22, 24, 211
   fractions, 211, 229
   radicals, 332, 338
Slope, 272, 274, 282
Slope-intercept form, 283
Solution of an equation, 59
Solution of an inequality, 87
Solution set, 59, 87
Special product patterns, 158
Specific variable, 121, 122
Sphere, 119
Square of number, 326
Square root, 326
Square-root property, 362
Standard form of quadratic equation, 360
Statement of inequalities, 87, 97
Substitution method of solving systems, 304
Substitution property of equality, 60
Subtraction:
   of algebraic fractions, 219
   of decimals, 48
   of fractions, 38
   of integers, 13
   of polynomials, 145
   of radicals, 329
   of rational expressions, 219
   of rational numbers, 37

Subtraction properties, 60, 89
Suggestions for solving word problems, 76, 126, 380
Supplementary angles, 78
Symmetric property of equality, 61
Systems of linear equations, 293
Systems of linear inequalities, 295
Systems of three equations, 312

T

Term, 24
Terminating decimals, 44
Transitive property of equality, 60
Translating English phrases, 65
Trapezoid, 118
Triangle, 118
Trinomial, 145

U

Union of sets, 95

V

Value, absolute, 11
Variable, 1, 121

W

Whole numbers, 3, 28

X

x intercept, 259

Y

y intercept, 259

Z

Zero:
   addition property of, 12
   as an exponent, 168
   multiplication property of, 21, 47
   slope, 274